P9-DVG-269

A Dictionary of

Geography

Susan Mayhew is a geography teacher and the author of *Masterstudies Geography*. Her current research interests are in feminist and historical geography and she is dedicated to the demystification of her favourite subject.

 SEE WEB LINKS

For recommended web links for this title, visit www. oxfordreference.com/page/geog when you see this sign.

VISIT ONLINE
OXFORD
QUICK
REFERENCE

The most authoritative and up-to-date reference books for both students and the general reader.

Many of these titles are also available online at www.oxfordreference.com

A Dictionary of
Geography

FIFTH EDITION

SUSAN MAYHEW

OXFORD
UNIVERSITY PRESS

UNIVERSITY PRESS

Great Clarendon Street, Oxford, OX2 6DP,
United Kingdom

Oxford University Press is a department of the University of Oxford.
It furthers the University's objective of excellence in research, scholarship,
and education by publishing worldwide. Oxford is a registered trade mark of
Oxford University Press in the UK and in certain other countries

First published 1992 as *The Concise Oxford Dictionary of Geography*
First edition (with Anne Penny) 1992
Second edition 1997
Third edition 2004
Fourth edition 2009
Fifth edition 2015

Impression: 1

Published in the United States of America by Oxford University Press
198 Madison Avenue, New York, NY 10016, United States of America

British Library Cataloguing in Publication Data

Data available

Library of Congress Control Number: 2014947483

ISBN 978-0-19-968085-6

Printed in Great Britain by
Clays Ltd, St Ives plc

Links to third party websites are provided by Oxford in good faith and
for information only. Oxford disclaims any responsibility for the materials
contained in any third party website referenced in this work.

How to use this book

This dictionary provides coverage in one volume of the terms used in both human and physical geography. There are over 3,100 definitions across the following fields: cartography, surveying, remote sensing, statistics, meteorology, climatology, biogeography, ecology, simple geology, soils, geomorphology, population, migration, settlement, agriculture, industry, transport, development, and diffusion.

Headwords are printed in bold type and appear in alphabetical order. However, some entries contain further definitions. These have been included under the headword to avoid unnecessary repetition and to indicate some of the wider applications of the headword. Many terms have been marked with an asterisk. This points the reader to cross-references. In a very few cases the cross-reference as indicated does not have the exact wording as in the entry, but is close enough to make further reference possible.

References to geographical authors include surname, initials, and date. The use of these details in a card or electronic library catalogue will yield the book referred to.

Most journals are easily identifiable by their abbreviations, less simple abbreviations include:

AAAG   *Annals of the Association of American Geographers*
ESPL   *Earth Surface Processes and Landforms*
TIBG   *Transactions of the Institute of British Geographers*

To find recommended web links for this title, visit www.oxfordreference.com/page/geog when you see this sign.

 SEE WEB LINKS

Acknowledgement

I would like to thank my editor Jamie Crowther for his patience and understanding and Abigail Humphries Robertson, my production editor, for her moral support and humour during difficult times.

aa *See* BLOCK LAVA.

abiotic Not living, non-biological, usually describing factors in an ecosystem: atmospheric gases, humidity, salinity, soil mineral particles, water, and so on. An **abiotic environment** is one without any life. M. Gray (2003) argues that abiotic diversity is valuable, but threatened by human agency, which can destroy, damage, or pollute physical systems.

ablation Loss of snow and ice from a glacier by *sublimation, melting, and evaporation; and from the *calving of icebergs, and avalanches. In temperate and subpolar regions melting is the major form of ablation; in the Antarctic, it is calving. The rate of loss varies with air temperature, relative humidity, wind speed, insolation, aspect, and the nature of the surface. In snowfields, ablation includes snow removed by the wind, and is affected by aspect, depth of snow, and the nature of the underlying surface.

The **ablation sub-system**, where annual ablation exceeds annual accumulation, lies between the *firn line and the glacier snout. Where ablation occurs at the edges of glaciers, debris accumulates to form **ablation moraine/ablation till**. An **ablation valley** is a subsidiary valley formed beneath the crest of a lateral moraine and the valley side. Iturrizaga (2001) *Geogr. J.* 54 notes that 'true ablation valleys' can result from differences in *insolation, and may act in the development of lateroglacial moraine valleys.

aborigine A member of an indigenous people existing in a land before invasion or colonization from outside. For Canadian aboriginal peoples, see *Atlas of Urban Original Peoples*; for Australian, the Aboriginal Environments Research Centre, U. Adelaide; for Bolivian, Denevan (1966) *Nat. Tech. Info. Service.*

abrasion Also known as corrasion, this is the grinding away of bedrock by fragments of rock which may be incorporated in ice (*glacial abrasion), water (marine abrasion, *fluvial abrasion), or wind (*aeolian abrasion). In fluvial environments, the main agent of abrasion is the *bed load. The mass of solid material removed varies with the size, density, and velocity of the particles, and the density of the vector carrying these particles. See Sklar and Dietrich (2004) *Water Resour. Res.* 40 on a model for bedrock abrasion.

(((⊕))) SEE WEB LINKS

• For Sklar on the efficiency of particle abrasion in rivers.

absolute humidity The density of the water vapour present in a mixture of air and water vapour. Cold air cannot contain as much water vapour as warm air, so has a lower absolute humidity than warm air. Low absolute humidity results in more evaporation than in air with high absolute humidity. *See also* RELATIVE HUMIDITY.

absolute plate motion The movement of a crustal plate in relation to a fixed point, such as a hot spot, or latitude and longitude.

((()) SEE WEB LINKS
- Has an absolute plate motion calculator.
- Shows an absolute plate motion calculator.

abstraction The selection and conceptualization of a phenomenon, or some aspect of it; a way of viewing a real world object, usually a simplification— such as a *model. **Data abstraction** captures the essential information needed to describe a spatial phenomenon and is fundamental in *geographic information science.

See Smaalen in M. J. Kraak and M. Molenaar, eds (1996) for a hierarchical rule for geographic information abstraction. McCormack (2012) *PHG* 36, 6, 715, argues that 'abstraction can be affirmed as a necessary element of understandings of lived worlds in the making'.

abundance The total number of individuals of a certain species present in an area, generally estimated by sampling methods (such as capture–recapture), and which may vary according to competition, predation, and resources.

An **index of abundance** is a relative measure of the size of a population or sub-unit of the population, such as a year class. The abundance, for example, of fish may be measured as number/weight of fish caught per standard unit of fishing effort, in a given area, and for a given time span, and then compared to earlier data. See Goel et al. (2012) *CIM-SiM*, for models of species abundance in *biogeography.

abyssal, abysso- Of or belonging to the ocean depths, especially below 2000 m. The **abyssopelagic zone** is that part of deep lakes, oceans, or seas characterized by specific forms of plankton and nekton which inhabit open water, and the **abyssobenthic zone** is the bottom of a deep lake, ocean, or sea.

An **abyssal plain** is the deep sea-floor, formed of abyssal deposits, with a gradient of less than 1 in 10 000. The Canada Abyssal Plain, lying between Canada and the Alpha ridge, is the largest of the Arctic sub-basins, with an average depth of 3658 m. **Abyssal hills**, which interrupt an abyssal plain, are 50–250 m high.

accelerator A factor which increases the momentum of a boom or slump in an economy, so that a small change in demand, for example, may lead to a greater industrial growth (or decline).

((()) SEE WEB LINKS
- Has an interesting map of growth acceleration in Britain.

access Connection or entry to a service. A. Sen (1999) measures **access to advantage** by one's access to basic needs: satisfying goods—like food; freedoms— as in a labour market; and capabilities. S. W. Allard (2004), in an examination of access to social services in three American cities, finds that poor populations in urban centres generally have greater spatial access to social services than poor populations in the suburbs.

accessibility The ease of approach to one location from other locations: in terms of the distance travelled, the cost of travel, or the time taken. The improvement of infrastructures produces an increase in accessibility that positively affects the competitiveness of the economic system and favours the appearance of specialization benefits and scale economies. In fact, regions with high levels of accessibility to the locations of input materials and markets tend to be more productive and competitive than peripheral and remote regions; see Gutiérrez in R. Kitchin and N. Thrift, eds (2009). Accessibility also refers to the ability to connect with information, knowledge, and people.

Topological accessibility is measured in relation to a system of nodes and paths (transportation network), assuming that accessibility is significant only to specific elements of a transport system, such as airports or ports. **Contiguous accessibility** is a measurable attribute of every location, as space is considered in a contiguous manner. See Miller and Wu (2000) *GeoInformatica* 4, 2 on *GIS software for **space-time accessibility** measures.

Physical accessibility is the spatial separation of people from the supply of goods and services; see, for example, Orcao and Diez-Cornago (2007) *Area* 39, 3 on physical access to health services in Spain, and Gage and Calixte (2006) *Pop. Studs* 60, 3 on maternal health services in Haiti. **Social accessibility** is the ability of an individual to reach a resource or location, as affected by class structures, income, age, educational background, gender, or race. See Parks (2004) *Econ. Geog.* 80, 2 on social accessibility in Los Angeles, and the UK Transport Studies Group's *Social Accessibility Mapping Project*.

accordant Complying with; thus, **accordant drainage** has evolved in conformity with the underlying geological structure: domes show a radial pattern, for example. The **law of accordant junctions** (Playfair's law) states that tributaries join a stream or river at the same elevation as that of the larger watercourse; there is no sudden 'drop' in the level of the tributary, which is therefore *graded to the level of the junction. See Niemann et al. (2001) *ESPL* 26, 12 for a quantitative evaluation of Playfair's law. **Accordant summits** are hill or mountain tops of approximately the same elevation, whose presence has been seen as confirmation of the cycle of erosion theory (Ojany (1978) *Geo. J.* 2, 5).

accreting margin *See* CONSTRUCTIVE MARGIN.

accretion 1. The growth of land by the offshore deposition of sediment, forming *spits and *tombolos. Accretion is most active in *estuaries, particularly within the tropics. See Robin et al. (2007) *U. Caen Coastal Morphodynamics Gp.* on *swash bar and hook spit formation.

2. The increase in size of a continent by the addition of **accretion terranes**.

3. The growth of a landform by the addition of deposits; *seif dunes grow by accretion. See H. S. Edgell (2006).

4. The increase in size of particles by additions to the exterior, as in the formation of *hailstones. See Zheng and List (1995) *Procs Conf. Cloud Phys.*

acculturation The adaptation to, and adoption of, a new culture. This may occur simultaneously as two cultures meet, but occurs more often as an immigrant group takes to the behaviour patterns and standards of the receiving group. A major example is acculturation among African Americans; see H. Landrine and E. Klonoff (1996), and Sclemper (2007) *J. Hist. Geog.* 33, 2 on acculturation at the regional scale.

accumulated temperature One of the key factors for crop growth, this is the length of time, from a specific date, for which mean daily temperatures have been above, or below, a stated temperature; the total time for which temperatures varied from that standard.

accumulation 1. The input of ice to a glacier. Weaver (1975) *Arct. & Alpine Res.* 7, 3 suggests that it is greatest in shaded uplands. The **accumulation zone** (**accumulation sub-system**) is between the source and the *firn line, where the input of snow, firn, and ice exceeds *ablation. For snowfall in an accumulation zone to emerge as ice at the glacier snout takes about 100 years (Tuffen et al. (2002) *Sed. Geol.* 149).

2. The reinvestment of surplus value, in the form of capital, in order to increase that capital. Accumulation is a key feature of *capitalism because, in order to remain in business, capitalists have not only to preserve the value of their capital, but to add to it. 'The accumulation of [a] firm's net worth determines the growth rate of capital and the growth rate of the economy' (Chatelain (2004) *Econ. Letts.* 85). D. Harvey (1982) argues that the obligatory accumulation of capitalism has fostered *uneven development. Somel (2004) *ERC W. Paper* 0411 argues that the terms of trade between North and South help maintain a gap in capital accumulation. See also Hart-Landsberg and Burkett (2006) *Hist. Materialism* 14, 3 on China and the dynamics of transnational accumulation.

Smith's views on accumulation and phases of capitalist development (2005, *Hist. Materialism* 13, 4) are contested by O'Brien (2007) *Hist. Materialism* 15, 1. See Holden et al. (2011) *Geografiska B* 93, 2, 141 on **accumulation by dispossession**.

acidification A soil-forming process whereby organic acids (from humus) increase hydrogen ion concentration, as in the transformation of *brown-earth soils into acid brown earth, in humid temperate forest regions. For the Model of Acidification of Groundwaters in Catchments (MAGIC), see Cosby (1995) *Hydrol. & Earth Sys. Scis* 2, 4. See Eidukeviciene et al. (2010) *Nat. Hazards Earth Syst. Sci.* 10, 1477 on effective soil acidification management.

acid rain When *fossil fuels are burned, dioxides of sulphur and nitrogen are released into the air; these dissolve in atmospheric water to form acid rain. In addition, nitrogen oxides combine with volatile organic compounds to form ground-level smog. These pollutants lead to the acidification of lakes and streams (making some of them incapable of supporting aquatic life), impair visibility, weaken forests, and degrade buildings.

Thermal power plants in India, which generally use coal with relatively high sulphur content (0.5 to 3%), are the major source of oxides of sulphur—they release about 2500 tons per year. Oxides of nitrogen are produced during high-temperature combustion. The greatest source of nitrogen oxides is road vehicles. Sing and Madhoolika Agrawal (2008) *J. Enviro. Biol.* 29, 1, 15, have a useful article, available online. In China, sulphur dioxide emissions from coal combustion are a primary contributor to acid rain. By the mid-1990s, an estimated 19% of the agricultural land in seven provinces (Jiangsu, Zhejiang, Anhui, Fujian, Hunan, Hubei, and Jiangxi) in southern China has been affected by sulphur dioxide and acid rain; the average decrease in crop yield attributable to the combined effects of SO_2 and acid rain was 4.3% (S. K. Guttikunda et al. 2004). Oxides of sulphur and nitrogen may also be 'exported'; more than half of Taiwanese acid rain is from mainland China (Govt. Info. Office, Rep. China (Taiwan)). However, the UK Environment Agency and the US Environmental Protection Agency both record sharp falls in atmospheric sulphur dioxide.

acid soil Soil with a pH under 7, such as *podzols and *brown earths. Acidity in a soil may be due to the *leaching out of cations when *precipitation exceeds *evapotranspiration. Other factors include the nature of the vegetation (and thus the *humus) and of the parent rock. Adams and Evans (1989) *Eur. J. Soil Sci.* 40, 3 propose a model of soil acidification.

action research (participatory action research) A collaborative research process whereby people with a particular issue work with academics; 'it seeks to democratize knowledge production and foster opportunities for empowerment by those involved' (Kindon, Pain, and Kesby [2008] in *International Encyclopedia of Human Geography*). An excellent example is provided by Young and Barrett (2001) *Area* 33, 2, 141, who examine the use of four 'action' methods for eliciting and acting upon information from street children in Kampala. In comparison with *activist research, action research is guided by a pragmatic, problem-solving approach, and is not necessarily underpinned by radical politics.

action space (activity space) The area in which an individual moves and makes decisions about her or his life. Wong and Shaw (2011) *J. Geog. Syst.* 13, 2, 127 use this concept to assess segregation in south-east Florida, and include statistical methods of determining activity space. A useful and powerful paper.

active layer A highly mobile layer, periodically thawing, located above the *permafrost in *tundra regions, and ranging in depth from a few centimetres to 3 m. Its thickness depends on slope angle and aspect, drainage, rock and/or soil type, depth of snow cover, and ground-moisture conditions; clearing vegetation will increase the depth and mobility of the active layer. See Zhang et al. (2006) *Nat. Snow & Ice Data Center* on active layers in the Russian Arctic, and M. Lecki (2006). The mobility of the active layer is due to the tundra vegetation, which scarcely binds it together, so that it moves on slopes as gentle as 2°. Thawing may occur daily or only in summer; on refreezing, the active layer may expand, especially if *silt-sized particles predominate.

*Periglacial processes in the active layer include *frost heaving, *frost thrusting, *ice wedging, *gelifluction,

and the formation of *patterned ground. *See* THERMOKARST.

active margin Also called Pacific margin, this is a type of destructive *plate margin characterized by *ocean trenches, *earthquakes, andesitic *volcanic chains, and young *fold mountains (Schellart and Lister (2005) *Earth & Planet. Sci. Lett.* 231).

active volcano A volcano known to have erupted in recent times, or which is likely to erupt. Examples include Mt Etna, Sicily; see Behncke et al. (2004) *Geomorph.* 63, 3–4.

activist research A method of collaborative, self-reflective problem-solving in a community context with academic geographers using their position and expertise to influence events beyond the university world. L. Hunt, E. Emerald, and G. Martin (2013) 'provide an all-inclusive road map to generating and implementing politically active grass-roots research activities'. *See* ACTION RESEARCH.

actor–network theory (ANT) Focusing primarily on the actor—human or nonhuman, individual or group, conscious or unconscious—this theory explores the interconnectedness of all things. ANT recognizes that all objects and things exhibit consciousness, and through a consciousness, interact heterogeneously in space; the location of the interaction(s), where they are performed homogeneously, is the landscape. If, as ANT promotes, all objects and things exhibit consciousness, then the closer in space they are to one another, the more essential they are to each other (Allen (2011), *Area* 43, 3, 274). The basis of every action and decision depends on the actor's subjective interpretations, which are made and mutually adjusted in interaction with others. Jóhannesson and Bærenholdt (in N. Thrift and R. Kitchin 2009) clarify that 'the ANT approach should be understood as a theory of what to study rather than an interpretive framework of the world'. See Sheehan (2011) *Area* 43, 3, 336 on ANT as a reflexive tool. For a clear demonstration of the application of ANT to a geographical topic, see Bosco in S. Aitken and G. Valentine (2006), pp. 200–201.

actualism The supposition that no biological and geological processes other than those operating today have operated in the past. However, H. G. Reading (1978) observes that 'the majority of past environments differ in some respects from modern environments. We must therefore be prepared, and have the courage, to develop non-actualistic models, unlike any that exist today'.

adaptive environmental management A management strategy that emphasizes the use of environmental feedback in planning strategies. 'The recommended means for continuing ecosystem management and use of natural resources, especially in the context of integrated natural resource management. Conceptually, adaptive management is simply learning from past management actions to improve future planning and management. However, adaptive management has proved difficult to achieve in practice' (C. Allan and G. H. Stankey, eds, 2009).

adaptive radiation The diversification of species to fill a wide range of ecological *niches. Perhaps the most famous example is that of Darwin's finches in the Galapagos Islands: see K. T. Grant and G. B. Estes (2009). Kassen et al. (2004) *Nature* 431, 7011 argue that the ecological gradient may limit the size of adaptive radiations.

adiabat A line plotted on a thermodynamic diagram, usually on a *tephigram, showing as a continuous sequence the temperature and pressure states of a parcel of air with changing height. An **adiabatic change** is a change in temperature, pressure, or volume, involving no transfer of energy to or from another material or system. In an adiabatic process, compression is accompanied by warming, and expansion by cooling. An **adiabatic temperature change** thus results from a pressure change. The speed at which the temperature of rising air falls with altitude is the **adiabatic lapse rate**. Dry, rising air expands with height. The energy needed for this expansion comes from the air itself in the form of heat.

The resulting change in temperature is expressed in the equation:

$$\frac{dT}{dz} = \frac{-g}{Cp}$$

where dT is the temperature change, g is the acceleration due to gravity, dz is the height change, and Cp is the specific heat of the air parcel. This change is the **dry adiabatic lapse rate** (DALR): 9.84 °C/1 000 m. The temperature change sustained by any parcel of dry air is calculated using Poisson's equation.

If the rising air becomes saturated to *dew point, condensation of vapour will begin. This condensation is accompanied by the release of *latent heat, which partly offsets the cooling with height, so that the rate of cooling of moist air—the **saturated adiabatic lapse rate** (SALR)—is lower than the DALR. In the lower *troposphere, the vapour content of air is high so the latent heat of condensation is high; SALRs may be as low as 5 °C/1 000 m. In the cold, dry, high troposphere, though, there is little vapour ready for condensation, so the SALR may be close to the DALR. Quantitative expressions of the SALR are therefore quite complex. See B. Haurwitz (2007).

adsorption In *soil science, the addition of ions or molecules to the electrically charged surface of a particle of clay or humus. In this way, minerals become bonded to soil particles (M. Ashman and G. Puri 2002). An **adsorption complex** is an 'organic and inorganic substance in soil that can absorb ions or molecules' (K. Peverill et al. 1999).

advanced capitalism A key feature of advanced capitalism is the possession of capital by fewer and fewer owners. M. Castells (2002) stresses the fundamental importance of collective consumption in advanced capitalism. See R. Hernandez (2002) on the mobility of workers under advanced capitalism.

advanced economy (economically advanced economy) 'A term used by the International Monetary Fund (IMF) for the top group in its hierarchy: advanced economies, countries in transition, and developing countries. It includes: Australia, Austria, Belgium, Canada, Denmark, Finland, France, Germany, Greece, Hong Kong, Iceland, Ireland, Israel, Italy, Japan, South Korea, Luxembourg, Netherlands, NZ, Norway, Portugal, Singapore, Spain, Sweden, Switzerland, Taiwan, UK, and USA' (CIA (2007) *World Factbook, appendix B*). The IMF also includes Cyprus and Slovenia (IMF (2007) *WEO Groups & Aggregates Info.*).

advection The 'horizontal transport of energy and mass by parcels of air in convective flow systems' (Sivertsen (2007) *EMS7/ECAM8 Abstracts* 4). In meteorology, the term usually means the horizontal transfer of heat. **Advective transfers** occur in air streams, ocean currents, and surface *run-off, partly redressing the *insolation imbalance

between the tropics and the poles. For **advection fog**, *see* FOG.

aeolian Of the wind: aeolian processes include erosion, transport, and deposition, and work best with sparse, or absent vegetation. Baas (2007) *Geomorph.* 91, 3–4 describes aeolian sand transport as 'a classic dissipative process with non-linear dynamics'. **Aeolian landforms** include desert *zeugen and *yardangs. See I. Livingstone, ed. (1998).

aeration zone The zone between the soil-moisture zone and the capillary zone, just above the water-table, which acts as a dispenser and buffer in the infiltration of groundwater flow (Diankov and Nitcheva (2005) *Bulgarian. Acad. Scis*). See also Sobotovich et al. (1991) *Atom. Energy* 70, 6 on the reduction of groundwater contamination since Chernobyl, 1986.

aerobic Describing any living organism which depends on atmospheric oxygen to release energy from foodstuffs during respiration. *Compare with* ANAEROBIC.

aerological diagram A chart plotting the factors which determine the movement of air. Variations of temperature, pressure, dry and saturated *adiabatic lapse rates, and *saturated mixing ratio lines are plotted against height in a *tephigram. An adiabatic chart may be used to predict the *convective condensation level.

aerosol A suspension of droplets or particles in a gas; more precisely, of particles with a maximum diameter of 1 μm (*fog and *mist are thus aerosols). In meteorology, the term is often used to describe the particles suspended within the air, such as minute fragments of sea-salt, dust, organic matter, and smoke. These enter the atmosphere by natural processes such as vulcanicity, and by human agency such as burning *fossil fuels. Aerosols have a role in the radiative forcing of climate; they can absorb and scatter both infrared and solar radiation in the atmosphere; and they can affect and change the processes that control cloud and precipitation formation through their role as condensation nuclei. See I. T. Houghton, Y. Ding, and D. I. Griggs, et al. (2001) on aerosols and climate change. NASA posts a daily global aerosol map on the web.

affect Some consider that affect describes a process of change within a person; this person is affected and can affect. However, Thrift (2004) claims that 'there is no stable definition of affect'; and Bondi (2009) *Int. Encycl. Hum. Geog.* 446, describes affect as a 'feeling, disposition, or mood that exists prior to cognition or rational thought. Often referred to as a mental state, affect is also bodily or sensory and unconscious or nonconscious in character. To be affected is to feel or to be touched or moved'. Shouse (2005) *MC J.* 8, 6 explains that 'feelings are *personal* and *biographical*, emotions are *social*, and affects are *prepersonal*'—in other words, affect is our unconscious response which precedes our conscious feelings and decisions. 'The importance of affect', Shouse explains, is that 'in many cases the message consciously received may be of less import to the receiver of that message than his or her non-conscious affective resonance with the source of the message'. See Thrift (2004) *Geografiska B* 86, 1 and Ash and Thrift (2005) *Antipode* 37. Pile (2010) *TIBG* 35, 1, 5 reviews **affectual geography** and is not impressed; read the conclusion to his paper.

afforestation Planting trees. Tree-planting can stabilize soils by increasing *interception and reducing *run-off, reduce flooding through the reduction of silting, improve soil fertility, and provide timber and firewood. See Zhang et al. (2007) *Indus. Rept.* 1/07, eWater CRC.

aftershock A smaller tremor or series of tremors occurring after an *earthquake. Aftershocks result from static stress changes and 'dynamic' stresses (transient, oscillatory stress changes), that can weaken faults. Such dynamically weakened faults may fail after the seismic waves have passed, and might even cause earthquakes that would not otherwise have occurred (Kilb et al. (2003) *Nature* 408). See also Johnson et al. (2008) *Nature* 451. See Abe and Suzuki (2012) *Acta Geophysica* 60, 3, 547 on aftershocks in modern perspectives.

age dependency The dependency of people too young or too old to be employed full-time on the contributions of those in full employment. 'The actual dependency burden is determined not just by the sizes of the various functional age groups, but also by a host of social, cultural, and institutional factors' (Yap et al. (2005) *J. Cross-Cult. Gerontol.* 20, 4).

ageing, geographies of Spatial variations in demographic ageing, and the distribution of the elderly at local, regional, and national scales, are key issues in geographies of ageing: see the United Nations Population Division *World Population and Ageing: 1950–2050* (2001) and Ferry et al. (2006) *EPRC* on variations within the EU.

Mulder (2007) *Pop. Space & Place* 13, 4 notes that when elderly people change residence they more often move closer to their children than further away. 'However, adult children changing residence move towards or away from their parents equally frequently.' See also Williams et al. (2000) *Tourism Geog.* 2, 1 on international retirement migration.

Sunley (2000) *TIBG* 25, 4 reports a north–south divide in income and benefit take-up among pensioners in the UK, with lower private pension coverage in low-income areas. 'The social status of people throughout life is reflected in the way their deaths are perceived and in the practices concerned with their dying' (Lloyd (2004) *Ageing & Soc.* 24, 2).

agency The capacity of the individual to act independently: a capacity that is meaningful, and has both intent and purpose. Clearly, the degree of agency each individual can exert varies with social class and geographical location, and these variations are a major concern for human geographers; see, for example, Plummer and Sheppard (2006) *Econ. Geog.* 6, 5, 619, or Herbert (2000), *PHG* 24, 4, 550.

The **agency-structure debate** refers to the relative merits of agency and *structure—the relative importance of social structure or human agency—in producing particular outcomes. From the early 1980s, human geographers sought to move beyond this 'structure versus agency' impasse by citing A. Giddens (1984), who provided logical arguments for taking agency and structure equally seriously.

Social structures and human agency are interpenetrated and mutually determined. 'Structures provide human agents with a limited horizon of capabilities and possibilities, but those structures only exist in everyday practices' (Herbert, 2000). Cumbers et al. (2011, *Antipode* 42, 46) deplore the tendency of scholars to consider only capital and state actors. 'There is little sense of agency for individuals and communities . . . they are usually treated as passive victims of deeper underlying processes. In this paper, our purpose is to highlight the autonomy and agency of workers, people and communities in old industrial cities.'

Agenda 21 A set of proposals, made at the 1992 UN Conference on Environment and Development, Rio de Janeiro, to promote *sustainable development. It sets out environmental strategies for managing coasts, oceans, and water, monitoring and reducing chemical waste, eradicating radioactive waste, and conserving natural vegetation and soils (through sustainable farming). Other proposals include measures to improve health care, reduce poverty, and develop fair and environmentally friendly trade policies. The aim is to promote global sustainable development at a more fundamental level than traditional aid programmes, based on common needs and interests. The requirement is for local governments to develop their own 'Local Agenda 21' in order to promote and implement sustainable development.

(⊕) SEE WEB LINKS
• Agenda 21 online.

agent An entity which can take a role; a function, service, or identity. Müller (2007) *Area* 39, 2 calls the geographer 'an active agent who moulds the production of meaning'. Bondi (2005) *Antipode* 37, 3 holds that neoliberalism envisages people as autonomous, individualized, self-directing, decision-making agents. 'Interdependencies between agents are constrained by social and spatial structure, but, over time, structure and agency are mutually constituted' (Plummer and Sheppard (2006) *J. Econ. Geog.* 6). The term '**agent-based model**' refers to 'a broad class of computational models known also as "multi-agent models". As the name suggests, the model consists of a number of different "agents" that represent real-world entities such as households, businesses, individuals, and so forth. Each agent follows a set of relatively simple rules (an algorithm) controlling its interactions with other adjacent agents and its immediate environment over time' (Green (2013) *TIBG* 38, 1, 2).

age–sex pyramid A set of two histograms set back to back on a vertical axis, depicting the numbers of the two sexes by age group. In many less developed countries the pyramid will have a very wide base; these are termed **progressive pyramids** because they suggest future population growth. **Regressive pyramids** characteristic of advanced economies, where there are fewer children and more old people, are more cylindrical; population is likely to fall.

age structure The composition of a nation by age groups. The age structure of the UK population has become older, the median age rising from 34.1 years in 1971 to 38.4 in 2003 and projected to rise to 43.3 in 2031 (*UK National Statistics*). Europe, with more than 15% of the population over 60, suffers from *age dependency, while South Asia, with 45–55% of its population under 16, is concerned to limit population growth. *See* AGE–SEX PYRAMID.

agglomerate A coarse-grained volcanic rock, made of sharp/sub-angular lava fragments, set in a fine matrix. Some are *pyroclastic; others are deposits from *mud flows or *lahars.

agglomeration 1. Within a geographical area, a concentration of economic activity brought about by, among

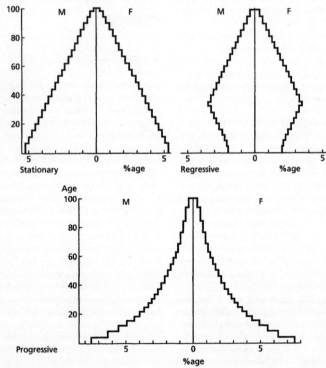

Age–sex pyramid

others, *external economies (such as a pool of skilled labour), economies of scale, *cumulative causation, local authority planning, and chance. 'The agglomeration of economic activities is a phenomenon as old as cities themselves' (E. Soja 2000). Phelps and Ozawa (2003) *PHG* 27, 5 distinguish between proto-industrial, industrial, and post-industrial agglomerations, with no common, formative processes underlying each. Pflüger and Südekum (2008) *J. Econ. Geog.* 8, 1 believe that agglomeration processes depend on specific technical properties; Roos (2005) *J. Econ. Geog.* 5, 5 claims that the influence of geography on agglomerations is small: 'at most, ∼36% of the total variation can be explained by the effects of geography'. Fan and Scott (2003) *Econ. Geog.* 79, 3 find a strong positive relationship between spatial agglomeration and productivity in China. See Feldman in G. Clark et al., eds. (2000).

2. In meteorology, the process by which cloud droplets grow by assimilating other droplets.

agglomeration economies For an industry, the benefits of locating in a densely peopled and highly industrialized situation, where the market is large, but spatially concentrated. Transport costs are thus low, and specialized industries can evolve, since local demand is high. Further benefits arise because of *functional linkages and *external economies. See Park (2002) *AI & Soc.* 16, 3 on a Swedish Science Park which has intensified knowledge generation by facilitating technical innovation, education, and research.

Roos (2005) *J. Econ. Geog.* 5, 5 claims that the effect of agglomeration economies is over five times larger than the effects of geography, while Viladecans-Marsal (2004) *J. Econ. Geog.* 4, 5 shows that, in Spain, agglomeration economies influence the location of manufacturing

activity. 'Empirical analyses often find that factor endowments are more powerful in accounting for trade and production patterns than are agglomeration economies, but that both are important' (Hanink and Cromley (2005) *AAAG* 95, 3).

aggradation The deposition of unconsolidated deposits by *aeolian, marine, or *fluvial processes, when the quantity or calibre of the load is greater than the competence of the transporting medium to carry it, or as a result of *mass movement. See Daniels (2003) *Geomorph.* 56, 3–4 on aggradational *pedogenesis on floodplains. For **aggradation surface**, see Wang and Grapes (2008) *Geomorph.* 96, 1–2.

aggregate A group of soil particles held together by electrostatic forces, *polysaccharide gums, and *cementation by carbonates and iron oxides.

aggregation A way of combining a collection of very large datasets into classes, both to make subsequent analysis simpler and to provide a more generalized account of a phenomenon. For any analyses based on areal units, the problem is one of *scale: that plotting geographical information on the wrong scale can conceal information that only becomes apparent at other scales and resolutions. Openshaw (1977) *TIBG* 2, 4, 459 offers a closely argued mathematical inquiry into this problem and suggests a solution to it; his conclusion is important to statisticians and non-statisticians alike: 'geographers and other spatial disciplines should pay far more attention to the way in which they design their zoning systems for spatial study'.

aggressivity In *karst geomorphology, the ability of water to dissolve calcium carbonate. See Stafford et al. (2008) *Int. J. Speleol.* 37, 2.

agoraphobia A disorder whereby the sufferer avoids places or situations where he or she previously had a panic attack; an agoraphobic typically avoids public places. Between 70 and 90% of people diagnosed with agoraphobia are female. Holmes (2008) *Area* 40, 3, 375 and Davidson (2005) *Area* 32, 1, 31 tackle this topic, but both articles are jargon-heavy and diffuse. P. Carter (2002) finds that '"space fear" ultimately results from the inhibition of movement', a conclusion that signals a fascinating link with *feminist geography.

agribusiness An integrated farming system, linking farm operations with upstream and downstream manufacturing and distribution: a total production system from inputs (such as seed and fertilizers) through cultivation on the farm to agro-processing and marketing (Sutton (2001) *Sing. J. Trop. Geog.* 22, 1). See Kristiansen (2002) *Sing. J. Trop. Geog.* 23, 1 on small-scale agribusiness entrepreneurs in Indonesia, and Wilson and Thompson (2003) *Rev. Ag. Econ.* 25, 1 on timing in agribusiness.

agricultural geography The study of spatial patterns in agricultural activity. 'Agriculture is an integral part of a wider food and fibre production system, encompassing input supply, farming, food processing, wholesaling and retailing' (G. Robinson 2004). Rural land-use patterns, cultural landscapes, and encroachment of urban sprawl continue to inform work in rural geography; see, for example, Hart (2003) *Geogr. Rev.* 93.

Antle and Stoorvogel in N. Heerink et al., eds (2001) discuss models of an agricultural system; see also Antle and Stoorvogel (2006) *Env. & Dev. Econ.* 11. Other agricultural geographers are concerned with the relationship of agricultural systems and economic development; see Tsakok and Gardner (2007) *Am. J. Ag. Econ.* 89, 5, but see also Evans et al. (2002) *PHG* 26, 3.

agriculture, industrialized Agriculture with a 'manufacturing' approach: units are large, technology is standardized, and the producer is responding to the processers' requirements. Green et al. (2003) *J. Environ. Policy & Plan.* 5, 145 write 'within agriculture, there is high energy use (especially in machinery) and a large demand for water, requiring substantial investment in irrigation systems. Crop production relies heavily on synthetic chemicals as fertilizers and pesticides, though chemical use has become more targeted. Since 1960, there has been a dramatic upsurge in the use of "hybrid" seeds as part of the "green revolution"; in the 1990s, in some countries and for a small number of crops, new seeds have been produced based on methods of genetic modification (GM). Animal rearing is similarly industrialized, as meat production soars, with corresponding criticisms of inadequate attention being paid to animal welfare. Farms will typically be large, with low (and reducing) labour inputs and with integrated forms of farm and crop management. Applying the classic "Fordist" model of manufacturing to the primary sector, high-input efficiency and labour productivity of agriculture is matched to the production of uniform products geared for mass markets. High-yield mono-cropping reduces pressure on marginal land and reduces the amount of land needed for agriculture'. This is an excellent article from which to begin.

agriculture, organic An approach to farming and ranching that avoids the use of herbicides, pesticides, growth hormones, and other similar synthetic inputs. An excellent summary by the Soil Association UK concludes that 'collectively, the evidence supports the hypothesis that organically grown crops are significantly different in terms of food safety, nutritional content, and nutritional value from those produced by non-organic farming' and that 'feeding trials have shown significant improvements in the growth, reproductive health, and recovery from illness of animals fed organically produced feed'. Furthermore 'a small body of observational and clinical evidence supports the hypothesis that consumption of organically produced food is beneficial to human health'. But J. Guthman (2004) argues that, at least in California, organic farming has replicated what it set out to oppose.

(⊕) SEE WEB LINKS
- A pdf of *Organic farming, food quality and human health: A review of the evidence*, by the Soil Association.

agriculture, sustainable Agriculture strategies that are profitable, environmentally sound, and supportive to communities. *Sustainable development 'meets the needs of the present without compromising the ability of future generations to meet their own needs' (1987 UN Convention on Environment and Development). Despite this confident definition, Fish et al. (2006) *TGJ* 172, 3, 183 note that 'sustainability, it is often claimed, is a nebulous and slippery resource for thought and practice, one able to marshal together a range of competing, and indeed contradictory, narratives about society–nature relations'. However, India's Centre for Sustainable Agriculture is much more confident and outlines a number of strategies.

(⊕) SEE WEB LINKS
- Website of Centre for Sustainable Agriculture, India.

agrobiodiversity A subset of *biodiversity that includes all forms of life directly relevant to agriculture.

Farm biodiversity includes rare seeds and animal breeds, fauna, weeds, pests, predators. **Wild biodiversity** includes all the native plants and animals existing on and flowing through the farm. Kothari (1999), in G. M. Pillai (ed.) *Maharashtra Council of Agricultural Education and Research*, is good on agrobiodiversity in India.

agroclimatology The study of the interactions between atmospheric variables and biological systems in agriculture, and the application of this knowledge to increase food production and improve food quality. See O. P. Bishnoi (2010).

agroecology A scientific discipline that uses ecological theory to study, design, manage, and evaluate agricultural systems that are productive but also resource conserving. It is concerned with the maintenance of a productive agriculture that sustains yields and optimizes the use of local resources while minimizing the negative environmental and socio-economic impacts of modern technologies.

(⊕) SEE WEB LINKS
- An excellent site on agroecology in the Timbaktu, Andhra Pradesh, India.

agro-food system A system that delivers food from the farm to our tables. This can include credit for the farmer, fertilizer, seeds, and animals, and the complex system from producer to supermarket: 'interdependent sets of enterprises, institutions, activities, and relationships which collectively develop and deliver material inputs to the farming sector, produce primary commodities, and subsequently handle, process, transport, market, and distribute food and other agro-based products to consumers' (World Bank, Africa Reg. Working Paper Series 44 2003). For a developing country, agro-food systems might account for 50% or more of national GDP, and Grimes, in Wharf (ed.) 2006, refers to 'the increasingly long and complicated path that food takes to get to our table'. Agriculture and Agri-Food Canada describes the Canadian agrofood system as 'modern, highly complex, integrated, internationally competitive', and is full of information. In contrast agro-food and agribusiness account for only 20% of GDP for Sub-Saharan Africa. The World Bank paper on agro-food systems in Africa is comprehensive, clear, and recommended.

(⊕) SEE WEB LINKS

• Website of Agriculture and Agri-Food Canada.

agroforestry Any agricultural system which includes planting, or encouraging, trees. Tree roots can bind soil and limit soil erosion, deep-rooted trees can tap new nutrient sources, leguminous trees can fix atmospheric nitrogen and improve soil fertility, leaf litter can add organic matter, and tree cover can moderate temperatures; trees may also provide food, fodder, firewood, and timber. See Dove (2004) *Agroforest. Sys.* 61.

aid Transferring resources from developed to less developed countries. **Bilateral aid** is from one donor to a recipient country, while **multilateral aid** comes from a group of countries. **Emergency aid** is short-term, generally given in response to disasters: 'windfalls in relief and reconstruction aid can also increase the risk of (perceived) fraud, corruption, mismanagement and dispossession by the government and its cronies, aggravating the plight of the most vulnerable' (LeBillon and Waizenegger (2007) *TIBG* 32, 3). **Structural aid** is given for long-term *development; see Seguiti (2003) *Public Budget. & Finance* 23, 2 on EU structural aid. A donor of **tied aid** can

transfer resources more cheaply as its aid is tied to its own exports (Kemp (2005) *Pacif. Econ. Rev.* 10, 3).

Bebbington (2004) *PHG* 28, 6 criticizes NGOs: 'their geographies of intervention do not reflect the geographies of poverty and livelihood in the Andes; and their strategies of intervention do not respond to the economic and spatial dynamics of poor people's livelihoods . . . they do less and less to address the deeper processes of . . . development that produce poverty and inequality of opportunity.' See Swain (2006) *TIBG* 31, 2 on aid and market principles 'designed by the government of the United States and the international financial institutions . . . and implemented by them in various ways, for the more vulnerable societies, often as stringent structural adjustment programs' (N. Chomsky 1999).

The WTO-sponsored **Aid for Trade** programme assists developing countries in expanding their trade to help growth and reduce poverty, since many developing countries face internal constraints such as a lack of productive capacity, poor infrastructure, inefficient customs management, excessive red tape, and difficulties to meet technical standards in high-value export markets. This has a negative impact on their ability to trade and on the competitiveness of their exports. The Aid for Trade initiative can include support for building new transport, energy, or telecommunications infrastructure, investments in agriculture, fisheries, and services, as well as assistance in managing any balance of payments shortfalls due to changes in the world trading environment.

(⊕) SEE WEB LINKS

• European Commission's Aid for Trade site.

AIDS (Acquired Immunodeficiency Syndrome) A condition which makes the body more susceptible to infections that a healthy person would be able to fight off.

'In order to develop AIDS an individual first has to be carrying the HIV virus. But not everyone who is HIV positive will develop AIDS and the use of antiretroviral drugs has significantly extended life expectancy for people who are HIV positive. An estimated 34 million people were estimated to be living with HIV/AIDS as of 2011, 30 million of these live in low- and middle-income countries. Sub-Saharan Africa (particularly southern and eastern Africa) has been worst affected globally by the epidemic to date. An estimated 15 million children have lost one or both parents to AIDS in Sub-Saharan Africa. This is over a quarter of the total numbers of orphaned children in Sub-Saharan Africa However, HIV prevalence is declining or has stabilized in many African countries, and the number of people acquiring new HIV infections in 2011 was 25% lower than in 2001' (R. Evans (2013), RGS).

Moran (2005) *Euras. Geog. & Econs.* 46, 7, 525 offers some explanations for the distribution of prevalence rates in Russia and considers the potential regional impacts of the epidemic.

(((⊕))) SEE WEB LINKS
• Royal Geographical Society website pages on HIV/AIDS.

aiguille French for a *pyramidal peak, sharply pointed due to *nivation and *frost wedging.

air mass An area of the *atmosphere extending for hundreds of kilometres that, horizontally, has generally uniform properties, especially of temperature and *humidity, and with similar vertical variations of temperature and pressure throughout. Air masses obtain these attributes from their areas of origin, known as source regions, and become less uniform with movement into different areas, and with the effects of *wind shear.

Air masses are classified by the source region: **continental air mass**, **c**, or **maritime air mass**, **m**; and by the latitude of the source area: the arctic, **A**, the poles, **P**, and the tropics, **T**. These two combine to distinguish most air masses, such as **cA**— **continental arctic**, **mP**— **maritime polar**, **mT**—**maritime tropical**, and so on. **Equatorial**, **E**, and **monsoon**, **M**, tags also exist. Browell et al. (2003) *J. Geophys. Res.* 108, D20 identify nine air mass types, quantifying their frequency as a function of altitude, and Chen et al. (2007) *J. Geophys. Res.* 112, D10S05 classify air mass types along coastal New England. See Peterson et al. (1981) *J. Appl. Met.* 20, 3 on turbidity and synoptic air masses.

In the case of Bangalore, India, for instance, a continental polar (cP) and continental arctic (cA) mass is predominant in the autumn and winter, which produces dry conditions, while a maritime equatorial (mE) is dominant in the spring and summer, bringing moister conditions.

air pollution An atmospheric condition in which substances (air pollutants) are present at concentrations higher than their normal ambient (clean atmosphere) levels that produce measurable adverse effects on humans, animals, vegetation, or materials (J. H. Seinfeld, 1986). For example, the conclusion of Grass and Kane (2008) *Int. J. Climatol.* 28, 1113, neatly summarizes the effects of stressful weather and elevated air pollution levels on cause-specific mortality in Santiago, Chile, 1988–96. Common pollutants include carbon dioxide, carbon monoxide, lead, nitrogen oxides, ozone, smoke, and sulphur dioxide. Givati and Rosenfeld (2004) *J. Appl. Meteor.* 43, 1038 suggest that air pollution in orographic clouds slows down cloud-drop coalescence and hence delays the conversion of cloud water into precipitation. This effect explains the pattern of greatest loss of precipitation at the midlevel of the upwind slopes, smaller losses at the crest, and enhancement at the downslope side of the hills—quite the opposite of the orographic process of precipitation. See Gulliver and Briggs (2005) *Env. Res.* 97 on modelling traffic management plans, emission technologies, fuel pricing strategies, and transport policies.

⊕ SEE WEB LINKS

• Description of types of air pollution.
• EU emission standards.

air pressure *See* ATMOSPHERIC PRESSURE.

air–sea interaction *See* OCEAN–ATMOSPHERE INTERACTION.

Aitken nuclei *See* NUCLEI.

alas In a *periglacial landscape, a large, steep-sided, flat-bottomed depression which can be several kilometres across. Lakes often occur in alases, which may then develop into *pingos.

albedo That proportion of solar radiation reflected from a surface, such as clouds (low from thin stratus cloud; up to 80% from thick strato-cumulus), or bare rock. Lighter, whiter bodies have higher albedos than darker, blacker bodies. Earth's total albedo is about 35%. The **ice-albedo feedback** mechanism occurs when a reduction in greenhouse gases in the atmosphere, principally carbon dioxide and methane, make the global climate colder, creating larger areas of ice and snow. Ice and snow reflect more solar radiation than does bare ground or liquid water, which creates a 'positive feedback'. If the Earth ever became half-covered by ice or snow, the feedback would become

self-sustaining and glacial ice would rapidly spread to the equator

Zhang et al. (2012) find that, over the last 300 years, surface albedo decreased due to conversions from grassland to cropland in the Northeast China Plain and increased due to conversions from forests to cropland in the surrounding mountains.

Land surface albedo is an important variable in General Circulation Models; see Roesch (2002) *J. Geophys. Res.* 107, 4221. See also Tasum et al. (2008) *J. Hydrol. Eng.* 13, 2, on the retrieval of land surface albedo from satellite observations.

Surface albedo (percentage of incoming short-wave radiation which is reflected)	
Fresh, dry snow	80–95
Sea ice	30–40
Dry light sandy soils	35–45
Meadows	15–25
Dry steppe	20–30
Coniferous forest	10–15
Deciduous forest	15–20

alcohol, geographies of Alcohol is at once a social problem, a leisure activity, a pleasure, an accelerator of violence, and central to identity formation (Jayne et al. (2008) *PHG* 32, 2). In geographical studies of alcohol, drunkenness features heavily. It's good, therefore, to read that Eldridge and Roberts (2008) *Area* 40, 3 challenge the urban savage/social drinker binary, 'which wrongly presumes consumers cannot be both'. In Europe, a broad division was made between a northern dry area, where beer is the leading beverage, consumed at weekends and outside mealtimes; and a southern wet area, where wine is the main beverage, usually drunk at meals. Subsequently, European alcohol choices have started to converge (Allamani et al. (2000) *Substance Abuse* 21, 4).

P. Chatterton and R. Hollands (2003) see alcohol-related establishments as the driving force in recent city-centre regeneration; Roberts (2006) *Cities* 23 reports that cities at night now attract vast numbers of people. See also D. Hobbs et al. (2003). However, Roberts et al. (2006) *Urb. Studs* 43, 7 conclude that British free market attitudes to extended licensing hours will undermine the government's professed aspirations for an 'urban renaissance' of cultural inclusion and animation. Smith (2003) *Tijdschrift* 94, 4 makes the expected link between alcohol and deprivation in the city—where alcohol use is more visible. Working-class cities are deemed to be places for excessive partying and wild stag or hen nights (Chatterton and Hollands (2001) *U. Newcastle on Tyne*), and a drinking culture remains firmly tied to the North-East industrial heritage (Nayak (2003) *Env. & Plan. D* 21, 1). Shaw, (2000) *Australian Geogr. Studs* 38, 3 observes that when aboriginality was conflated with alcoholism, it tended to be exoticized and romanticized in its traditional 'bush' setting, but demonized in its urban setting. Kokorev (1997, cited in Round (2006) *Geografiska B* 88, 1) finds that Magadan Oblast, in Russia's far north-east, has the country's highest alcohol consumption rates; Magadan is the location of a notorious gulag. Henry (2006) *Paper IGU Conf.* argues that differences between a city Central Business District in Hobart, Tasmania, where alcohol is banned, and the Wine and Food Festival, where it is not, result from the 'different ontological claims made about their spaces'.

S. Winlow and S. Hall (2006) discuss alcohol's central connection to interpersonal violence and disorder; see also Silverman and Della-Giustina (2001) *Urban Studies* 38. Tatlow et al. (2000)

J. Community Health 25 demonstrate a clear link between the density of shops selling alcohol and alcohol-related hospital admissions in San Diego county. Alavaikko and Osterberg (2000) *Addiction*, 95 (suppl. 4) consider the way the increasingly globalized alcohol supply and international supply-side initiatives impact on strategies to reduce alcohol-related harm, and how international trade treaties affect alcohol regulation and policy.

alcove A semicircular, steep-sided cavity on a rock outcrop, caused by water erosion, especially spring-sapping. See Reneau (2000) *Geomorph.* 32, 1 for photographs.

Aleutian current A cold ocean current of the North Pacific. The **Aleutian trench** lies at the boundary of the North American and Pacific plates; try Choi (1990) *Pac. Geol.* 5. The **Aleutian low** is a semi-permanent 'centre of action' of persistently lower pressure (less than 1013 mb) and higher humidity and relatively higher temperatures for that latitude. In conjunction with its changing relationship with the North Pacific High in the eastern part of the Pacific, it is the major feature of the general circulation controlling the weather over parts of North America.

alfisol A soil order of the *US soil classification. Alfisols are young, acid soils, with a clay-rich B horizon, commonly occurring beneath deciduous forests. *See* BROWN EARTH.

algae A diverse group of simple plants that contain chlorophyll and can photosynthesize. An **algal bloom** is a dense spread of algae caused by changes in water chemistry and/or temperature; Shukla et al. (2008) *Appl. Maths & Comput.* 196, 2 find a positive correlation between increasing nutrients in water and algal density—leading to *eutrophication. Moran et al. (2001) *CSIRO Land & Water Tech. Rept.* 20/01 question the proposed link between phosphorus entering channel systems and algal blooms.

alienation The estrangement or separation of individuals from one another, said to include isolation, powerlessness, and meaninglessness. K. Marx (1844, 1975) saw this as a result of separating the workers from the benefits of industrialization as enjoyed by the owners of the means of production. L. Wirth (1938) saw alienation as a product of urban life. Others cite political causes; see *Human Rights Watch*, 20 December 2000, on Israeli Arab students' feelings of alienation from their education; McEwan (2003) *Geoforum* 34, 4 on cultures of alienation amongst black women in South Africa, and Davies and Niemann (2002) *New Pol. Sci.* 24, 4 on alienation—especially as it relates to globalization—in everyday life. A major contemporary theme is the alienation of some Muslims from 'Western' lifestyles; 'young British Muslims are increasingly found to be in the precarious position of having to choose their loyalties, being impacted by radical Islamism on the one hand, and . . . British multicultural citizenship on the other' (Abbas (2007) *J. Intercult. Stud.* 28). See also Polat (2004) *Midwest Polit. Sci. Assoc.* on Muslim alienation and modernity.

alkaline soil Any soil, such as a *rendzina, which has a pH above 7. Alkalinity usually reflects a high concentration of carbonates, notably of sodium and calcium.

Allerød A north-west European *interstadial, from *c.*12000 to 10800 years BP; in Britain not usually distinct from the *Bølling. Pollen records point to a cool temperate flora, with birch. See Willard et al. (2007) *Pal. Pal. & Pal.* 251.

allochthonous Of rock formed of materials brought from outside, for example from sediments transported by ice; *till is an example. *Compare with* AUTOCHTHONOUS.

allogenic Caused externally. Thus, an **allogenic stream** is one—for example, the Nile—fed from outside the local area, where precipitation and run-off generate the flow. **Allogenic succession** is an alteration in a plant *succession brought about by some external factor.

allometry The study of relative rates of change of two variables of a system. This term is specifically used to designate such changing proportions correlated with changes in absolute magnitude of a system or of specific parts of interest; that is, scale-related distortion. In the study of landforms, or the processes acting upon them, the power function $Y = a X^b$ is useful for correlating changes in variables. Allometry is very clearly explained by Church and Mark (1980) *PPG* 4, 342.

See Bull (2003) *GSA Bull.* on allometric change of landforms, Hood (2002) *Restoration Ecol.* 10, 2 on landscape allometry and the restoration of tidal channels, and Evans and McClean (1995) *Zeitschrift. Suppl.-Band*, 101 on *cirque scale and allometry.

allopatric Describes species or populations of the same species that occur in habitats that are geographically separated and do not overlap. **Allopatry** is the occurrence of species or populations of a species in geographically separate, non-overlapping habitats. Marko (1998) *Evolution* 52, 3, 757 explains how this happened for two recently diverged Pacific snails.

alluvial Referring to *fluvial contexts, processes, and products. The form of an **alluvial channel** (cut in alluvium) reflects the *load and discharge of the river rather than the *bedrock. Dade (2000) *Geomorph.* 35, 1 relates sediment grain size and transport mode to the channel pattern of a river. An **alluvial fan** is a triangular landform; its shape results from the river swinging back and forth during *aggradation while the apex is fixed at the point where the river emerges from the uplands. Arzani (2012) *ESPL* 37, 7, 726 finds that alluvial fans in Iran developed their geometry mainly in response to different weathering intensities of their catchment bedrock lithologies. See also Pope et al. (2008) *Geomorph.* 94, 1–2. An **alluvial cone**, similar in origin, tends to have a steeper slope.

Alluvial filling is *sedimentation from fluvial channels. During low water, **alluvial flood plains** are mostly dry, but when most of the *flood plain is swamped, coarse material is transported short distances, temporarily deposited, and picked up during the next flood. Major controls on alluvial flood plains include sediment, flow regime, and differing base levels (O'Connor et al. (2003) *Geomorph.* 51, 1–3). **Alluvium** is a general term for deposits laid down by present-day rivers, characterized by sorting (coarser alluvium is found in the upper course of rivers; finer in the lower courses), and *stratification (coarse material is overlain by finer). **Alluvial landforms** include *deltas, river beds, and flood plains.

alluviation The accumulation of sediment. See DeLong et al. (2007) *Geomorph.* 90, 1–2. See Blum and Törnqvist (2000) *Sedimentol.* 47, Suppl. on change in sediment yield and discharge, and alluviation during relative sea-level fall. **Cut and fill alluviation** results in a shifting channel eroding old surfaces and creating new surfaces (Hauer et al. (2007) *J. Am. Water Resources Ass.* 43, 1).

Alonso model William Alonso's 1964 explanation of urban land use and land values is grounded on the concept of *bid rents. If the level of goods and services is constant, land prices should decrease with distance from the centre. The poor will live near the city centre, where high housing densities more than offset high land costs; the better-off will live at lower densities at the edge of the city (on more, but cheaper, land). Each household represents a balance between land, goods, and *accessibility to the workplace. However, commuting costs rise with distance from the centre, reducing disposable income, and higher-income groups may prefer the accessibility to the *CBD offered by the inner city to the space, quiet, and cheaper land of the suburbs, thus causing *gentrification. Hua (2001) *Int. Reg. Sci. Rev.* 24, 3 reviews and re-presents Alonso's model.

Alps A chain of *fold mountains running from eastern France and northern Italy through Switzerland to Austria. An **alpine glacier** is a valley glacier, and **alpine topography** describes features of upland glaciation: *cirques, *glacial troughs, *hanging valleys, and *truncated spurs. The **alpine orogeny** occurred in the *tertiary, creating fold mountains as far apart as the Andes, Japan, and East Indian Ocean *island arcs. See Rosenbaum (2002) *Tectonophys.* 359 on the movements of Africa and Europe during the Alpine orogeny.

alternative economic geographies These are forms of economic activity pursued as alternatives to those that currently dominate peoples' lives, notably *capitalism, and usually have goals of greater environmental sustainability and social justice. See D. Fuller, A. Jonas, and R. Lee, eds (2011) and Lee (2000) *Env. & Plan. A* 32, 6, 991.

alternative economies More sustainable economies that do not define success by growth, that challenge unnecessary consumption, invest in localized systems of exchange, and minimize environmental damage. It may be that

a

the processes of production, exchange, labour/compensation, finance, and consumption are intentionally different from mainstream capitalist economic activity (Healy, in R. Kitchin and N. Thrift 2009), or an alternative representation of economy as a social space; *see* ALTERNATIVE ECONOMIC GEOGRAPHIES.

alternative food networks Associations of food producers, distributors, and retailers committed to quality, *fair trade, and ecology, to which Jarosz (2008) *J. Rural Studies* 24, 231 adds 'the spatial proximity between farmers and consumers, the existence of retail venues such as farmers markets, community-supported agriculture (CSA), and a commitment to sustainable food production and consumption'. See Morris and Kirwan (2011) *J. Rural Studies* 27, 322 on ecological embeddedness and alternative food networks.

alto- Prefix referring to *clouds between 3000 and 6000 m high.

ambient Surrounding, of the surroundings. **Ambient temperature** is the temperature surrounding a given point. The **ambient air standard** is a quality standard; see J. Stedman et al. (2005), DEFRA, on ambient air quality assessment in the UK.

ambivalence In *post-colonial theory, 'the mixture of charm and revulsion in the relationship between the colonizer and the colonized' (H. Bhabha 2004). It is inherent in all colonial relationships; colonizers don't care to be copied exactly, and ambivalence can lead to *hybridization in the colonizers ('going native'). Jorgenson and Tylecote (2007) *Landsc. Res.* 32, 4 use the concept in exploring urban responses to 'areas in which the spontaneous growth of vegetation through natural succession suggests that nature is in control'.

amenity Pleasantness; those aspects of an area such as housing, space, and recreational and leisure activities which make it attractive to live in. **Amenity migration** is the purchase of first or second homes in rural areas prized for their aesthetic and/or recreational values (McCarthy (2008) *PHG* 32, 1). It has been aided by rising incomes, the easing of limits on foreign property ownership in many countries, developments in transport and communications technologies, and increasing exposure to images of prized rural landscapes. 'Amenity migration and much associated tourism is driven at least in part by temporarily deterritorialized and significantly globalized visions of the rural' (P. Cloke et al. 2006).

amphidromic point The point around which *tides oscillate; while there are tides along the coasts of East Anglia and the Netherlands, there is a point in the sea between the two where there is no change in the height of the water. High water rotates around the amphidromic point; anticlockwise in the Northern Hemisphere, clockwise in the Southern.

ana- Rising. An **anabatic wind** is an upslope wind driven by heating—usually daytime insolation—at the slope surface under fair-weather conditions (AMS (1959) *Glossary of Meteorology*; which see). These winds tend to be less frequent, and weaker, than their corresponding, descending, *katabatic winds.

anabranching channel A *distributary channel which leaves the main channel, sometimes running parallel to it for several kilometres, and then rejoins it; a channel 'separated by vegetated semi-permanent alluvial islands, excised from an existing floodplain, or formed by within-channel or deltaic accretion' (Nanson and Knighton (1998) *ESPL*

21, 3). Jansen and Nanson (2004) *Water Resources Res.* 40, W04503 show that anabranching can sustain a stable river pattern where a single channel could not transport sediment. Anabranching channels differ from *anastomosing channels in that they are undivided.

anaerobic Describing any organism or process which can or must exist without free oxygen from the air, such as the anaerobic bacteria which are responsible for the process of *gleying.

ana-front A front at which the warm air is ascending. An ana-front is normally the initial state of a cold front, developing into a kata-front as the depression becomes more occluded. See Sansom (1951) *Q. J. Royal Met. Soc.* 77.

anastomosis The division of a river into a stable multi-channel system with levées, backswamps, and large, stable islands. An **anastomosing channel**, unlike an *anabranching channel, has distributaries of its own. Valley incision starts at outflow points; up-valley incision begins where the flow rejoins the main channel (Schumm et al. (1996) *GSA Bull.* 108). In time, the channel segments join, making a new anabranch of the anastomosing-channel system; see Kleinhans (2012) *ESPL* 37, 12, 1337. Many anastomosed systems cannot be categorized; one common factor seems to be low specific stream powers, thus explaining the river's inability to maintain channel capacity via sediment transport and erosion; see Makaske (1998) *Earth Sci. Rev.* 53.

anchor tenant A pioneer tenant in a new development. A state may promote the 'anchor' with preferential treatments in terms of land prices, tax exemptions, infrastructure support, and administrative services, in order to persuade the downstream and upstream firms along its supply chain to follow suit (Lee and Saxenian (2008) *J. Econ. Geog.* 8, 2). See Rice, in R. Kitchin and N. Thrift (2009) on this.

angle of repose The angle at which granular material comes to rest, varying inversely with fragment size in perfectly sorted materials, but directly in those imperfectly sorted; inversely with fragment density; directly with fragment angularity, roughness, and compaction; inversely with height of fall of material on free cones; and directly with moisture content up to saturation point, but inversely thereafter (van Burkalow (1945) *GSA Bull.* 56, 6).

angular momentum The momentum of a body taking a curved path; for a rotating planet, angular momentum is defined as the moment of the linear momentum of a particle about a point, thus:

$$M = \overrightarrow{r} \, xm \, \overrightarrow{V}$$

where M is the angular momentum about a point 0, $\overrightarrow{r}$ the position vector from 0 to the particle, m the mass of the particle, and $\overrightarrow{V}$ the velocity. As Rossby (1940) *Q. J. Royal Met. Soc.* 66, 68 explains, the Earth's angular momentum can easily be visualized by taking a marble on the end of a string and rotating it on a smooth table. If the string is shortened the rate increases; if lengthened it slows. The speed is inversely proportional to radius. Thus, part of the atmospheric envelope, rotating with the Earth at the equator, moves polewards, it will tend to speed up; see *conservation of angular momentum.

animal geographies Animals occupy human social life and space in many ways, yet, it is argued, humans have turned them into commodities (J. Wolch and J. Emel 1998), or have degraded them by using them as pets (J. Berger 1980). Major themes in animal geographies include: the relationships and interactions between animals and humans; the spaces and places occupied by animals in human culture; the environmental, social, and cultural implications of the representation and use of animals; the meanings and identities ascribed to animals and animal issues, and questions of animal welfare and rights. For a summary of animal geographies, see Emel et al. (2002) *Soc. & Animals* 10, 4.

Geographers using *actor–network theory argue that the divisions between people and animals are subject to change and negotiation (S. Whatmore and L. Thorne 1998; Hobson (2007) *Pol. Geog.* 26, 3) while ecofeminists see the 'human chauvinism' towards animals as part of a hierarchy topped by European-American males (G. Gaard 1993). Wolch and Emel (op. cit.) argue that Western representations of animals such as wolves underpin the racism and sexism which label 'others' as savage. (Interestingly, in Russian representations, the wolf is the benevolent 'caretaker' of the forest, cleaning up the carrion).

annular drainage *See* DRAINAGE PATTERNS.

annular modes Two hemisphere scale patterns of climate variability: a northern annular mode (NAM) and a *southern annular mode (SAM). Both annular modes explain more of the week-to-week, month-to-month, and year-to-year variance in the extratropical atmospheric flow than any other climate phenomenon. The annular modes describe variability in the atmospheric flow that is not associated with the seasonal cycle. These modes are characterized by north-south shifts in atmospheric pressure mass between the polar regions and the middle latitudes (Thompson and Woodworth, *Annular Modes* website).

 SEE WEB LINKS

- Thompson and Woodworth's excellent Annular Modes website.

anomie The lack of traditional social patterns within a group; a lack of norms leading to conflict and confusion. In a study of second-generation British-Barbadians moving to Barbados, Potter and Phillips (2006) *AAAG* 96, 3 note that they are often thought by the locals to be mad. L. Wirth (1938) sees anomie as the result of urbanism; Entriken (2002) *Geografiska B* 84, 1 thinks anomie and multiculturalism are 'two related, yet often competing, visions of modern society' (but gives no more supporting evidence than Wirth).

Antarctic Denoting regions south of the Antarctic circle, 66° 32′ S. Within the **Antarctic circle**, the sun does not rise on 21 June (winter solstice in the Southern Hemisphere) or set on 22 December (summer solstice in that hemisphere). **Antarctic meteorology** is characterized by severe winters with double temperature minima, i.e. two separate occasions of minimum temperature, due to the absence of *insolation for several winter months, and to the frequent exchange of air with that of lower latitudes. *Blizzards are common. Temperatures rise in late summer as the long waves of the *westerlies bring incursions of warmer air. Nevertheless, precipitation is still almost always in the form of snow, since maximum temperatures, occurring at the summer (December) solstice, rarely exceed 0°C. See J. King and J. Turner (1997). Antarctica seems to be part of a single tectonic **Antarctic plate**, but there is clear palaeomagnetic evidence that in the past there has been a large relative rotation between different parts of the continent (Gordon (2000) *Nature* 404). Whittaker et al. (2007) *Science* 318, 83 detail major Australian and Antarctic plate reorganizations.

antecedent Prior to, before, as in **antecedent drainage** patterns: the Wind River, Wyoming, for example, probably pre-dates the dome it now cuts through, the river's erosion keeping pace with the uplift of the dome. **Antecedent moisture** is the amount of moisture already present in the soil before a specified rainstorm. An **antecedent boundary** is one drawn before an area is populated, as opposed to a superimposed boundary, fixed by colonial powers.

anthropocene era The current epoch of Earth's history when humankind has joined with the other environmental forces in shaping the planet (Crutzen, (2002) *Nature* 415: 23; Crutzen and Stoermer (2000) *Global Change Newsletter* 41: 17). When did such an epoch commence? Crutzen places great weight on atmospheric composition and assumes the origin of Anthropocene to be synchronized with the appearance of the steam engine in the late-eighteenth century, which prompted a significant upturn in the burning of fossil fuels, as shown by the composition of the air trapped in the polar ice sheets and ice caps, corresponding to a sharp increase in carbon dioxide and methane concentrations in the atmosphere. However, Certini and Scalenghe (2011) *Holocene* 21, 1269, propose that the Anthropocene be defined as the last 2000 years of the late Holocene and characterized on the basis of anthropogenic soils.

During this era humankind has: increased its population tenfold; exhausted 40% of the known oil reserves; contributed to a 30% increase in atmospheric CO_2; transformed nearly 50% of the land surface (with significant consequences for biodiversity, nutrient cycling, soil structure, soil biology, and climate); dramatically altered coastal and marine habitats (50% of mangroves have been removed and wetlands have shrunk by one-half); and increased

extinction rates (Sanderson et al. (2002) *Bioscience* 52, 10). The entire issue of the *Holocene*, 2011, 21, is devoted to early Anthropocene levels of CO_2 and CH_4.

The **early anthropogenic hypothesis** suggests that increased human agricultural production altered the natural interglacial trends in CO_2 and CH_4 during the latter part of the Holocene (Ruddiman, 2003, *Climatic Change* 6, 261).

anthropogenic Brought about by human agency.

anti-capitalism The clue is in the words, but essentially this is *anti-globalization, a global political project defined by notions of autonomy, ecology, democracy, self-organization, and direct action. See CAPITALISM.

(((⊕))) SEE WEB LINKS
• Anti-capitalism website containing key text 'Notes from Nowhere; We Are Everywhere'.

anticline *See* FOLD.

anticlinorium *See* FOLD.

anticyclone A region of relatively high *atmospheric pressure, frequently thousands of kilometres across, often formed as a response to *convergence in the upper *atmosphere, and also known as a high. As air near ground level flows into an anticyclone, its absolute *vorticity decreases, causing *divergence, and the descent of air (Tan and Curry (1993) *Monthly Weather Rev.* 121). **Anticyclonic circulation** is clockwise in the Northern Hemisphere, and anticlockwise in the Southern Hemisphere.

Cold anticyclones (continental highs) form as continental interiors, such as Siberia, lose heat in winter through *terrestrial radiation, and cool the air above. They are marked by subsidence, which inhibits cloud formation and maximizes radiative cooling, making them self-sustaining. Keegan (1958) *J. Meteorol.* 15.

Subtropical anticyclones are warm, and form due to subsidence below convergence resulting from the westerly sub-polar *jet stream at the northern limit of the *Hadley cell. The semi-permanent subtropical anticyclone over the North Atlantic (the *Azores high*) strongly influences weather and climate of much of North America, western Europe, and north-west Africa (Davis et al. (1997) *J. Climatol.* 10, 4). Subtropical anticyclones bring stable atmospheric conditions, and fine, hot, dry weather. In *mid-latitudes, anticyclones often locate beneath the leading edge of ridges in the upper-air *westerlies, where they may be associated with *blocking weather patterns; see Wiedmann et al. (2002) *J. Climate* 15, 23.

anticyclonic gloom In winter, in north-west Europe, fog, or poor visibility caused by broad, persistent sheets of strato-cumulus cloud at the base of an *inversion trapping polluted air below. The inversion results from the arrival of a cold *anticyclone.

antidune A ripple on the bed of a *fluvial channel, travelling upstream. Antidunes are sediment bedforms, formed beneath *standing waves by fast, shallow water flows. A critical *Froude number of 0·84 may be taken as the discriminator between dunes and anti-dunes (Carling and Shvidchenko (2002) *Sedimentol.* 49, 6).

anti-globalization A colloquial term for the stance of those who, internation-ally, coordinate, organize, and mobilize opposition to *neoliberalism and the political power of *transnational corpor-ations. Most anti-globalization activists use other titles: 'the Global Justice Movement'; the 'Movement of Move-ments' (a critical alliance of disparate

movements espousing anti-capitalist politics, anti-globalization, anti-war, and radical ecological issues); and the 'alter-globalist movement' (promoting humanist values and 'economic justice'). A major concern is that past North–South trade agreements have been unfair to the South.

J. Agnew et al., eds (2003) think the claim that anti-globalization protests are global is a bold one: 'in the absence of a locatable centre of authority, anti-globalization protesters mirror the "non-place" of power by choosing targets that symbolize global imperial sovereignty'. De Filippis (*On-Line Conf. Commun. Organ. & Dev.*) thinks *local* organizers should fight neoliberal globalization, while M. Neocleous (2003) argues that 'a non-territorial democratization of power . . . a geography of insurrection as opposed to the geography of order' is needed to fight globalization. Wissen-burg (2004) *Org. & Env.* 17, 4 calls for a 'radical articulation of environmental interests' by anti-globalization activists.

anti-humanism Broadly, anti-humanism encompasses a range of different perspectives that all share a critique of humanism and an effort to displace the human subject as the centre of philosophical and social inquiry. Within academic geography, it has 'pri-marily taken the form of (1) a travel from the self-containing subject towards sub-jectivities as relational effects of ar-rangements or assemblages, and (2) an emphasis on materiality or more broadly the "non-human" or "more-than-human"' (Simonsen (2013) *PHG* 37,1, 10).

anti-natalist Limiting population growth. The People's Republic of China has pursued anti-natalist policies, nota-bly the 'one-child' strategy (although J. Dreze and A. Sen (1989) claim that economic security for China's rural population hastened the fall in birth rates). Barber (2003) *Cross-Cultural Res.* 37 believes that countries with large populations are more likely to pursue anti-natalist policies in general, and to discourage single parenthood in particular, to justify controlling population. However, Coe (2004) *Reproductive Health Matters* 12, 24 argues that pro- and anti-natalism are 'the greatest obstacles to making real and long-lasting improvements to sexual and reproductive health and rights'.

antitrades Westerly winds in the upper atmosphere above, and in contrast to, the easterly *trade winds at ground level.

apartheid The system of racial segregation first promulgated by the largely Afrikaner National Party of South Africa in 1948. **Petty apartheid** meant the separation of facilities such as lavatories, transport, parks, and theatres into white and non-white. On a much larger scale was the allocation of 12% of the land area into independent republics ('homelands, or bantustans') for the African population, which comprised 69% of the population when the policy began, in 1954. These 'homelands' were to be governed and developed separately from white South Africa, while allowing African workers strictly limited rights to live in the white areas, as and when their labour was required. J. Robinson (1996) suggests that the organization of urban space into racially segregated living areas was central to the persistence of the racial state. 'Without a gathering of the racially defined African population into spatially contained areas and the evolution of specific methods of administration and governance in these areas, the implementation of various racial policies would have been held hostage to the racial and physical "chaos" of the early twentieth-century city.'

With the election of South Africa's first democratic government in 1994, the last vestiges of apartheid were officially removed, but the policy will leave a mark on the South African landscape and its society for many years to come; see Benjaminsen et al. (2006) *AAAG* 96, 3 and Ramutsindela (2007) *Tijdschrift* 98, 4. 'What investigations into, say, apartheid in South Africa . . . reveal is the way in which these racist geographies were embedded in a wider field of utter indifference' (Jacobs (2000) *TIBG* 25, 4).

aphelion The point furthest from the sun of any body orbiting the sun; the Earth is at aphelion 152 000 000 km from the sun in July. *See* MILANKOVITCH CYCLES.

aphotic zone In any watery environment, the deeper zone, which is not penetrated by light.

applied climatology This describes, defines, interprets, and explains the relationships between climate conditions and countless weather-sensitive activities. For example, in business, applied climatology embraces interactions with marketing, sales, customer services, research support, and the delivery arm for climatological data and information. See the journal *Theoretical and Applied Climatology;* see also McKendry (2002) *PPG* 26, 3, 462 for a review, and Souch and Grimmond (2006) *PPG* 30, 2, 270, on applied climatology: urban climate.

applied geography A type of problem-solving that uses geographical thinking. Four main areas of study are: the description and auditing of contemporary environmental conditions; the evaluation of the value of particular environments for specified future uses; the identification and analysis of environmental impacts, mainly of human action, actual and proposed; and the prediction and design of environmental works.

Applied geography is, according to *Applied Geography*: 'research which utilizes geographic approaches (human, physical, nature-society and GIScience) to resolve human problems that have a spatial dimension. These problems may be related to the assessment, management and allocation of the world's physical and/or human resources. The underlying rationale . . . is that only through a clear understanding of the relevant societal, physical, and coupled natural-humans systems can we resolve such problems'.

(⊕) SEE WEB LINKS

• Website of the Applied Geography Specialty Group of the American Association of Geographers.

applied geomorphology 'As geomorphology is the study of the surface of the Earth, its form, evolution and the processes that sculpture it, so applied geomorphology can be defined as the application of this knowledge in the resolution of engineering, planning and environmental/resource management problems' (Jones (1983) *Quart. J. Engineering Geol. & Hydrogeol.* 16, 3). 'The social relevance of geomorphology' (anon., U. Victoria, Canada). See P. Fookes et al., eds (2005).

applied meteorology The use of climate information in decision-making, impact assessments, seasonal climate forecast applications and verification, climate risk and vulnerability, development of climate monitoring tools, urban and local climates, and climate as it relates to the environment and society. It includes weather modification, satellite meteorology, radar meteorology, boundary layer processes, air pollution meteorology, agricultural and forest meteorology, and applied meteorological numerical models.

appropriate technology A technology which evolves or is developed in response to a particular set of needs, and in accordance with prevailing circumstances (Intermediate Techn. Dev. Group); 'a development strategy that would be appropriate to the material, technical, and organizational resources of nations poor but proud, and that at the same time would enhance local and national autonomy' (Brown (1979) *Dev. Econs.* 17, 1).

aquiclude A rock, such as London clay, which does not allow the passage of water through it; an impermeable rock. Such a rock will act as a boundary to an *aquifer.

aquifer A rock, such as chalk, which will hold water and let it through. Water runs into aquifers where the rock is exposed to the surface or lies below the *water-table. See Water Science and Technology Board (2008), *Prospects for Managed Underground Storage of Recoverable Water*. A **confined aquifer** is sandwiched between two impermeable rocks, and the water-table marks the upper limit of an **unconfined aquifer**. See A. Klimchouck, ed. (2000).

arch Along a coast, an arch is made when caves on either side of a headland meet. Arches are temporary; roof falls cut off the seaward end of the arch, which is then left as a *stack.

Arctic Denoting regions within the Arctic Circle, i.e. north of 66° 32′ N (often taken as 66½° N). Within these regions the sun does not set on 21 June (the summer solstice in the Northern Hemisphere) nor rise on 22 December (the winter solstice in the Northern Hemisphere). In climatology the Arctic is defined in terms of the treeless zone of *tundra and of the regions of *permafrost in the Northern Hemisphere. **Arctic air masses** are exceedingly cold, with the Arctic Ocean as their source region.

During winter, they are very dry, and little different from a continental *polar (cP) air mass. In summer, the air is more *maritime. A **maritime Arctic (mA)** air mass is generally similar to maritime polar air. Initially it has continental characteristics, but when it crosses warmer water, it is warmed from below, and in consequence becomes unstable.

Arctic meteorology The Arctic regions experience an annual cycle of winter 'night' and summer 'day'. Most weather results from the intensely cold ground air which is chilled by contact with land, which loses heat from strong *terrestrial radiation—winter clouds are scarce. Only infrequently do depressions penetrate the *inversions so formed. Winter temperatures are close to –40 °C. Snowfall is slight, but winds cause frequent blizzards and drifting. In spring, days are longer and sunny, but temperatures remain low because of the high *albedo of the snow surface. In summer, some depressions bring thicker cloud and light rain, and snow- and ice-melt in June and July keep air temperatures low.

Since the early 1970s, sea-level pressure over the Arctic has decreased; the stratospheric polar vortex has become colder and has been persisting longer into spring; and water mass characteristics of the Arctic Ocean have changed. Ostermeier and Wallace (2003) *J. Climate* 16, 2 link these changes to a shift in the *North Atlantic Oscillation.

Arctic sea smoke A form of *steam fog. As very cold air passes over a warmer sea, it is rapidly heated. Convection lifts moisture up, which quickly recondenses to form fog.

areal scouring Landscapes of areal glacial *scouring have been described as comprising irregular depressions with intervening bosses scraped by ice and labelled 'knock and lochan' topography,

but Johansson et al. (2001) *Geografiska A* 83 suggest that an etched bedrock surface is a prerequisite for this type of landscape to develop.

Area of Outstanding Natural Beauty (AONB) In the UK, an area in which development is very carefully considered, so that the beauty of the landscape is not diminished. The Countryside and Rights of Way Act 2000 empowers local authorities to establish a Conservation Board for their own AONB, in order to conserve it, enhance it, and increase public understanding and enjoyment of its qualities. See McKenna et al. (2005) *Area* 37, 3.

area studies Interdisciplinary programmes specializing in research and teaching on specific regions of the world. Gibson-Graham (2004) *Env. & Plan. A* 2004, 36, 405 provides a handy table:

Thematic studies	Area studies
general	specific
universal	particular
abstract	concrete
nomothetic	ideographic
theoretical	empirical
deductive	inductive
social scientific	humanistic
quantitative	qualitative
hypothesis testing	case study
economic	cultural
contemporary	historical
the West	the rest
developed	underdeveloped
modern	premodern

areic Without surface drainage, that is, without streams or rivers. Areas of permeable rocks, such as limestone, often lack surface drainage.

arête A steep knife-edge ridge between *corries or *glacial troughs in a glacially eroded, mountainous region. Arêtes are

Areas of Outstanding Natural Beauty

possibly formed by the backward extension of corries into a mountain mass, or
moulded by *nivation and *frost
wedging.

argillaceous Clay-like in composition
and texture, referring to rocks containing

*clay minerals and clay-sized particles,
for example, shale.

arid geomorphology The investigation and explanation of landforms and
geomorphological processes in drylands.
Tooth (2009) *PPG* 33, 2 identifies three

emerging research themes that highlight the growing links between arid geomorphology and other disciplines: the role of fire in arid geomorphological systems; arid fluvial sedimentary systems, characterized by investigations that commonly focus on processes, landforms, and sedimentary products at larger spatial scales and longer timescales; and arid geomorphology on Mars.

aridisol A soil order of the *US soil classification, found in arid environments. It is a *desert soil, predominantly composed of minerals, and often high in accumulations of water-soluble salts.

aridity Dryness; aridity is the most common criterion for the definition and classification of drylands, occurring when *evapotranspiration exceeds precipitation (Adamo and Crews-Meyer (2006) *Appl. Geog.* 26, 1). Thomas, in J. Holden (ed.) (2012) considers the annual overall net negative moisture balance to be the defining characteristic of deserts and drylands.

An **aridity index** is a numerical indicator of the degree of dryness of the climate at a given location, expressed as the ratio of annual potential evaporation to precipitation, and the UNEP index of aridity (1992 *World Atlas of Desertification*) is defined as:

$$AI_U = \frac{P}{PET}$$

where *PET* is the potential evapotranspiration and *P* is the average annual precipitation. *PET* and *P* must be expressed in millimetres.

Moriyama et al. (2010, *Int. Archs Photogrammetry, Remote Sensing & Spatial Information Science*, 38, part 8) propose a satellite-derived aridity index, derived from the day and night surface temperatures and the maximum solar flux (See Arora (2002) *J. Hydrol.* 265).

armouring The coarse surface material which protects the finer material below (Gibbins (2007) *Freshwater Biol.* 52, 12). The armouring of a river bed is caused by selective entrainment of smaller sediments, and reduces further entrainment of smaller sediments. An **armour layer** is generally well sorted, and one or two grains in thickness. Migratory during floods, it will re-form when normal flow resumes (Vericat et al. (2006) *Geomorph* 76, 1–2).

arroyo A *gully, rectangular in cross-section, cut into debris-choked valleys, usually with a sandy floor, made of fine-grained cohesive sediments in arid environments. It results from the entrenchment of stream channels when run-off increases. *Badlands are associated with arroyos. Perroy et al. (2012) *AAAG* 102, 6, 1229 untangle the relative importance of climatic, tectonic, and anthropogenic drivers as triggers of arroyo formation in California.

art, geographies of Geographers of art are concerned with where art—architecture, painting, and sculpture—is or was located, and how material and indigenous culture shaped the human environment and the nature of human civilization. 'The ready enrolment of geographic discourses within art world critiques, and how geographers' own expanding fields of practice have come to intersect with those of art' (Hawkins (2013) *PHG*. 37, 1, 52). This may include: paying close attention to landscape; the layers of social and cultural meaning and power that could be discerned via visual interpretation and contextual investigation of symbolic imagery (D. Cosgrove and S. Daniels 1988); geographical understandings of art and other definitions of environment, often through a focus on monumental forms that have centred interpretations of art in urban spaces and the role of art in urban regeneration (Hall and Robertson (2001) *Landscape*

Res. 26, 1, 5) and place promotion; and the variety of ways in which people's experiences, knowledge, and exploration of the environments which surround them may be understood through embodied–sensuous experience. See Cant and Harris (2006) *Soc. & Cult. Geog.* 7, 6, 857.

artesian basin A *syncline of permeable rocks, outcropping at the crest. Water from rain or streams seeps into this *aquifer; the rock becomes saturated, and the water is under pressure. If a hole is sunk to tap the water, an **artesian well** forms—water will initially flow upwards without pumping. Try B. Radke et al. (2000) on the Great Australian Artesian Basin.

artificial neural network (ANN) A computing methodology which simulates the human brain as it processes spatial data problems. Thurston has written an easily accessible summary, noting that 'in almost every instance where *GIS is being used, [artificial intelligence] applications could potentially be developed for the purpose of enhancing decision-making capabilities'.

((●)) SEE WEB LINKS

• Website that explains artificial neural networks in plain English.

ash *See* PYROCLAST.

ash cone *See* SCORIA.

ash fall The precipitation of volcanic ash (tephra) during and after a volcanic eruption; see Kimura et al. (2005) *Island Arc* 14, 4. Ash may sterilize soils, bury crops, and cause roofs to collapse under its weight (Cronin and Neil (2000) *Bull. Volcanol.* 62, 3).

ash flow Watenabe et al. (1999) *J. Volc. & Geotherm. Res.* 89, 1 suggest that the ash flows from Mt Unzen, 1991, were

formed by the fracturing of the cooling lava blocks, and were thereafter entrained into the plumes at the tops of the moving flows. *See* PYROCLAST.

aspect The direction in which a valley side or slope faces. In deeply cut east-west oriented valleys, the slopes facing the equator receive more sun and are more attractive to settlement than the shaded sides of the valley.

Aspect may be an important factor in the formation of landforms, since slopes facing away from the equator may be 6 °C colder than their opposites. Beaty (1961) *J. Geol.* 51 estimates gradational processes to be two to three times as active on northward-facing slopes in the Northern Hemisphere; in the Jotunheim of Norway, 70% of corries lie on the north side of the massif (Evans (1969) *Geografiska A* 59). Dikau in J. Raper, ed. (1989) uses aspect as a criterion in the description of landforms, as do Dymond et al. (1995) *ESPL* 20, 2.

assarting In medieval Britain, the expansion of agriculture into the forested areas (Beaver in S. Wooldridge and S. H. Beaver 1950).

assemblage A word used in so many different ways that Anderson and McFarlane (2011) *Area* 43, 2, 124 reflect that 'there is no single "correct" way to deploy the term, nor does any one theoretical tradition or style hold an exclusive right to it'. The *Oxford English Dictionary* has 'the contingent, non-necessary coming together of various phenomena, actors, institutions, or processes to form a temporarily stable bundle of relationships and capacities'. This seems to mean a chance or accidental assembly or collection of things; 'of all sorts of social and material bits and pieces' (D. Gregory et al.).

assimilation Also known as *accul-turation, or integration, this is the inte-gration of an immigrant, outsider, or subordinate group into the dominant, host community.

The concept of assimilation implies that the minority group eventually take on the values of the host, or *charter group; Akinwumi (2006) *Geografiska B* 88, 2 asserts that the USA requires im-migrants to put aside previous national identities and heritages; 'for immigrants to the United States, "correct" con-sumption became a symbol of assimila-tion' (Domosh (2004) *TIBG* 29, 4). 'In today's debates about immigrant assim-ilation, the degree to which immigrants remain spatially concentrated is treated as a measure of their assimilation, or willingness to assimilate' (Ellis (2006) *Tijdschrift* 97, 1). Sometimes the indigen-ous population is required to assimilate with a non-indigenous group; Lester (2006) *Geogr. Res.* 44, 3 argues that Aus-tralian aborigines were to retain their lands as sites upon which missionaries could prepare them for assimilation as black, Christian Britons.

Wei Li (2006) challenges traditional notions of the urban-to-suburban tran-sition in assimilation, revealing emerg-ing ethnic spatial patterns and friction within the host community. R. Smith (2005) argues for transnationalism as a crucial component of assimilation; 'as-similation and pluralism are not mutu-ally exclusive' (Burnley (2005) *Geogr. Res.* 43, 4).

associated number (König num-ber) The associated number of a *node is the number of *edges from that node to the furthest node from it. A low num-ber indicates a high degree of *connec-tivity; for an example of its use, see Wood (1975) *Geografiska B* 57, 109–19.

association (plant association) A *plant community unit; a floral assemblage with a characteristic dom-inant and persistent species.

asthenosphere That zone of the Earth's *mantle which lies beneath the relatively rigid *lithosphere, between 50 and 300 km below the surface, and representing a mechanical boundary between more rigid regions above and below. It is approximately commensu-rate with that zone of the mantle which transmits seismic waves at low velocity. The asthenosphere represents the loca-tion in the mantle where the melting point is most closely approached, and the transition between a rigid litho-sphere and a viscous asthenosphere is gradational (Stüwe (1999) *Surveys Geo-phys.* 20). The asthenosphere is com-posed of hot, semi-molten, and therefore deformable, rock, and may be plate- or density-driven, or both (Hansen et al. (2006) *Geology* 34, 10).

asylum migration The international movement of *refugees and persons who, while having suffered generalized repression, violence, and poverty, do not qualify as refugees under the strict re-quirements of the 1976 UN protocol. 'Asylum seeking is embedded in com-plex ways in the linkages and structures of transnational entrepreneurism, infor-mal economic activity and the shadow labour markets of the host economy' (B. Jordan and F. Düvell 2002).

Economic hardship and economic discrimination against ethnic minorities lead to higher inflows of asylum seekers; political oppression, human rights abuse, violent conflict, and state failure are also important determinants, as are migration networks and geographical proximity (Neumayer (2005) *Int. Studs Qly.* 49, 3). Attempts by the EU to reduce access to asylum systems or to curtail the rights of asylum seekers have generated an increase in illegal migration (Boswell (2003) *Int. Affairs* 79, 3). 'If Western

European countries want to tackle the root causes of asylum migration, then they need to undertake policy measures that promote economic development, democracy, respect for human rights and peaceful conflict resolution in countries of origin' (Neumayer, op. cit.).

asymmetry Lacking symmetry. An **asymmetrical valley**, common in present or past *periglacial environments, has one side steeper than the other, since *aspect can be critical in determining where frost forms; see Meikeljohn (1992) S. African Geogr. J. 74.

A-tents Raised rock slabs or plates of varying thickness, mostly resulting from the buckling of laminae, plates, and slabs. These forms are generally cracked by tectonic activity, although decreases in lithostatic pressure through unloading, or, more rarely, from surficial expansion of plates caused by intense heat. Smaller plates and slabs may be changed in part through freeze-thaw. A-tents are also known as pop-ups, but Twidale and Bourne (2009) PPG 33, 2, restrict this term to those forms developed instantaneously during earthquakes.

Atlantic Period In north-west Europe, the period from around 7500 to 5000 years BP of wet, oceanic climate, when temperatures were warmer than at present. See Uścinowicz et al. (2007) GSA Special Paper 426 on sea-level oscillations in the Atlantic period.

Atlantic-type coast A coastline where the trend of ridges and valleys runs transverse to the coast. If the coastal lowlands are inundated by the sea, a *ria or *fiord coastline may result. The coast of south-west Eire is an example. Compare with PACIFIC COAST.

atmosphere A gaseous envelope gravitationally bound to a celestial body; in the case of the Earth, with an average composition, by volume, of 79% nitrogen, 20% oxygen, 0.03% carbon dioxide, and traces of rare gases. This surprisingly uniform composition is achieved by *convection in the *turbosphere and by *diffusion above it, especially above 100 km, where diffusion is rapid in the thin atmosphere, and stirring is weak. Also present are atmospheric moisture, ammonia, ozone, and salts and solid particles. The atmosphere is commonly divided into the *troposphere, the *stratosphere, and the *ionosphere. Since the troposphere contains the majority of the atmospheric mass, and virtually all of the atmospheric water vapour, most weather events occur within it. The **atmospheric energy balance** is the balance between incoming short-wave energy from the Sun and outgoing long-wave energy from the Earth. Outgoing energy loss is greater than incoming energy gain; this is equivalent to a cooling of 1°C or 2°C per day. However, this heat loss is mostly compensated for by the latent heat released by the condensation of atmospheric water vapour into *precipitation.

Typical annual global values for the energy balance of the atmosphere are:

- net radiative loss = -99.4 W m^{-2}
- sensible heat gain = 20.4 W m^{-2}
- latent heat gain = 79.0 W m^{-2}

Annual global precipitation is therefore linked closely to the atmospheric energy balance and can change only if there are associated changes in the global energy balance (see Lockwood in J. Holden 2012).

atmospheric boundary layer The lowest part of the atmosphere; as such, it is shaped through contact with the planetary surface. J. C. Kaimal and J. J. Finnigan (1994) define it as the 'lowest 1–2 km of the atmosphere, the region most directly influenced by the exchange of momentum, heat, and water vapour at the Earth's surface'. However, R. B. Stull (1988) notes that this layer 'responds to surface forcings with a timescale of about an hour or less'. *Acta Geographica* 5, 2012 is devoted to the atmospheric boundary layer.

atmospheric cells Air may move, with a circular motion, northwards or southwards in a vertical cell; see Trump et al. (1982) *Boundary-Layer Met.* 24 for a model of a closed mesoscale atmospheric cell. Atmospheric cells are major components in the transfer of heat and momentum in the atmosphere from the equator to the poles. Horizontal cells (winds) also fulfil this role; see Abitol

and Ben-Yosef (1994) *J. Opt. Soc. Am. A* 11, 7.

atmospheric correction The retrieval of surface characteristics from remotely sensed imagery by removing the atmospheric effects. See Lane et al. (1999) *J. Appl. Met.* 38, 12.

atmospheric energetics The relations between internal, potential, and kinetic energy in the atmosphere. The bulk of the energy is heat, latent heat, and potential energy, but some of this can be converted into kinetic wind energy. Ackerman (*AIP Conf. Proc.* 247, pp. 1–23) reviews basic concepts.

Atmospheric cells

atmospheric forcing See FORCING.

atmospheric heat engine The system of energy which drives and controls the nature of the pressure, winds, and climatic belts of the Earth's surface, with *solar radiation as the heat source, and *terrestrial radiation as the sink. Mechanical energy is expended by the 'engine' in the form of atmospheric processes. Haine and Marshall (1998) *J. Phys. Oceanog.* 28, 4 characterize the poleward heat transport as a diffusion by baroclinic eddies. Barry et al. (2002) *Nature* 415 regard the baroclinic zone as a heat engine which generates eddy kinetic energy by transporting heat from a warmer to a colder region.

atmospheric pressure The pressure exerted by the *atmosphere as a result of gravitational attraction exerted on the column of air lying above a particular point. Atmospheric pressure, measured in millibars, decreases logarithmically with height.

atoll A *coral reef, ring- or horseshoe-shaped, enclosing a tropical *lagoon. Most of the world's atolls are found in the Indian and Pacific oceans, and are said to be sensitive to fluctuations in relative sea level—although Woodruffe et al. (1999) *Marine Geol.* 160, 1, in a study of West Island, Cocos, dispute this assumption. Single colony, semicircular corals (**microatolls**) grow to a level constrained by exposure at low spring tides (Spencer and Viles (2002) *Geomorph.* 48, 1–3).

Atterberg limits Limits used to classify soils. The **liquid limit** is the minimum moisture content at which the soil can flow under its own weight; the **plastic limit** is the minimum moisture content at which the soil can be rolled into a 3 mm diameter thread without breaking; the **shrinkage limit** is the moisture content at which further water loss will not cause further shrinkage. See Fener et al. (2005) *Cana. Geotech. J.* on the Atterberg limits of cohesive soils.

attrition The wearing away or fragmentation of particles of debris by contact with other such particles, as with river pebbles.

aulacogen A long-lived, sediment-filled graben oriented at a high angle to either a neighbouring modern ocean or a neighbouring orogenic belt; an intra-continental graben (Umeji (1988) *Int. J. Earth Scis* 77, 2). An aulacogen can be seen as the failed arm of a triple-limbed spreading centre (Jansen (1975) *Trans. Geol. Soc. S. Afr.* 78).

aureole See METAMORPHIC AUREOLE.

autecology The ecology of particular species and individual organisms, particularly the relationships between species and their environments and within species, and especially the way organisms act within communities. See Sundberg et al. (2006) *J. Biogeog.* 33, 8.

autochthonous Referring to features and processes occurring within, rather than outside, an environment. An **autochthonous rock** has been formed *in situ*; coal is an example.

autogenic succession A succession which begins as a result of disturbance by an external physical factor, but whose sequence in community composition results primarily from interactions among the species. Sometimes the early species actually enhance the success of the later ones, perhaps by increasing the nutrient status or water-holding capacity of the soil, as in a psammosere (C. Townsend et al. 2003).

automobility The study of the economic, technological, social, and political consequences of the pervasive and accelerating culture of the car. The analysis of automobility extends beyond congestion, commuting, suburbanization, and environmental impacts to consider also the role of driving in cultural identity formation, social standing, and status; it shows that the automobile is more than just a technological object, it is in a very real sense 'a whole way of life' (A. Latham et al. 2008).

autonomous spaces (autonomous geographies) Self-managed spaces; those spaces where there is a desire to constitute 'non-capitalist, collective forms of politics, identity, and citizenship, which are created through a combination of resistance and creation, and the questioning and challenging of dominant laws and social norms' (in the words of Autonomous Geographies).

autotrophe An organism which uses light energy to synthesize sugars and proteins from inorganic substances. Green plants are by far the most common autotrophes.

avalanche A rapidly descending mass, usually of snow, down a mountainside. **Powder avalanches** consist of a moving amorphous mass of snow. **Slab avalanches** occur when a large block of snow moves down a slope and can cut a swathe through the soil and sometimes erode the bedrock if the snow is wet. See A. V. Briuchanov (1967).

Avalanches of other substances are forms of *mass movement, and are distinguished by the type of material involved. **Debris avalanches** are typically triangular in shape, and are associated with morphological discontinuities, such as scarps and road cuts; see Guadagno et al. (2005) *Geomorph.* 66, 1–4 on man-made cuts and debris slides. **Rock avalanches** occur when a jointed rock loses internal cohesion, until some sections are held together only by the friction between blocks. If this frictional force is lessened through water seepage, or if lateral support is removed, failure will occur. See Mitchell et al. (2007) *Landslides* 4, 3.

avalanche wind The rush of air formed in front of an *avalanche. Its most destructive form, the **avalanche blast**, occurs when an avalanche stops abruptly.

avulsion The abandonment of all or part of a channel belt in favour of a new course (Stouthaven and Berensen (2007) *Sed. Geol.* 198). Makaske (2001) *Earth-Science Revs* 53 argues that bed aggradation reduces channel carrying capacity, increases overbank flooding and sedimentation, and eventually results in avulsion. See Jerolmack and Paola (2007) *Geomorph.* 91, 3–4 for a cellular model of river avulsion.

azonal Referring to a soil without *soil horizons, such as a young soil developing on a bare rock surface. Alluvium and *sand dunes are examples of azonal soils.

B

back-arc basin A zone of tectonic enlargement, and substantial, and strongly asymmetric, sedimentation. A back-arc basin forms when an *island arc is split along the line of its magmatic axis, forming a remnant arc, which then glides away. Well-developed back-arc basins widen by **back-arc spreading**, which is thought to resemble normal oceanic crust formation, but in conjunction with a *subducting plate. See Ohara (2006) *Island Arc* 15, 1 for a general view of the mantle process beneath a back-arc basin spreading ridge.

backreef A sea area landward of a reef, usually including a lagoon. Dahlgren and Marr (2004) *Bull. Marine Sci.* 75, 2 argue that backreefs are of critical ecological value to a coral reef ecosystem. See K. Sullivan-Sealey et al. (2002).

backshore The part of a beach between the ordinary limit of high tides and the point reached by the very highest tides. On cliffed coastlines, the backshore is the section of cliff foot and *shore platform affected only by storm tides; on low, shingle coasts it can take the form of a *berm. Udo and Yamawaki (2007) *Geomorph.* 60, 1–2 find that waves have larger effects than winds on backshore sedimentation.

backswamp On a *flood plain, a marshy area where floodwater may be confined between the river *levées and higher ground. See Takehiko et al. (2006) *Geog. Res.* 44, 1 on mangroves and backswamps, and Patrick and Ellery (2007) *Afr. J. Ecol.* 45, 3.

backwall The curved, steep head wall of a landslip scar or *tafone (Matsukura and Tanaka (2000) *Geografiska A* 82, 1), of *nivation hollows (Thorn and Hall (2002) *PPG* 26, 4), or of *cirques (Hughes et al. (2007) *Geomorph.* 88, 3–4).

backwash The return flow of water downslope to the sea after the *swash has moved upshore. Steep waves and long waves are linked with stronger backwash, and flatter beach gradients; see the great C. King (1959), and Hardisty et al. (1984) *Marine Geol.* 61, 1. When the path taken by the backwash differs from the path of the swash, *longshore drift occurs. See Bird et al. (2007) *Geogr. J.* 173, 2, on backwash erosion.

badlands Arid lands, generally bare of vegetation, dissected to form ravines and sharp-crested hills developed best in weakly consolidated sediments. See B. Rorke and A. Yair (1989). This dissection is aided by the lack of vegetation, high *run-off, and heavy sedimentation, which increases *abrasion. Non-fluvial processes, such as mass wasting, *piping, and tunnel erosion, are also important.

Badlands are useless for agriculture because of their intensely dissected landscape (Gallart et al. in L. J. Bull and M. J. Kirkby, eds 2002).

bajada (bahada) A series of *alluvial fans, typically laid down by *ephemeral streams, which has coalesced along the foot of the mountains to form a gently sloping plain of unconsolidated sediments. The *weathering of the mountain front may also supply debris. Alluvium on the lower part of a *pediment may also make a bajada. See Bull's classic paper (1977), *PPG* 1; Leeder and Mack (2001) *J. Geol. Soc.* 158 for a clear review, plus a great title; and Wainwright et al. (2002) *J. Arid Envs.* 51.

balance of payments The ratio between the payments made by one country to other nations, and the revenue it receives from them. If receipts exceed outgoings, the balance is positive. Bristow (2005) *J. Econ. Geog.* 5, 3 thinks that modern socio-economic systems have not only to achieve a sustainable balance of payments, but social objectives, such as income redistribution, and basic health care. Ramos (2006) *ersa* 06 concludes that regional differences in balances of payments are not significant.

balance of trade The relationship between the value of a country's imports and exports, generally expressed in monetary terms. The balance of trade is sometimes known as the visibles (actual goods) account. Together with the *invisibles (account and capital transfers), it makes up the balance of payments.

bank erosion The erosion of material from the side of a river channel. Hooke (1979) *J. Hydrology* 42 identifies corrasion and slumping as the major processes, and these appear to be associated with river flow levels and antecedent precipitation conditions, respectively. Rates of erosion vary with speed of flow (rates are highest on the outer bank of meander bends or where bars in the channel have diverted the *thalweg), bank vegetation and moisture content,

and bank composition (Parker et al. (2008) *Geomorph.* 101, 4).

bankfull discharge The *discharge of a river which is just contained within the banks. This is the state of maximum velocity in the channel, and of maximum *competence (V. P. Singh, ed. 1987). Timár (2003) *Quat. Sci. Revs* 22, 10 shows that sinuosity is a function of the channel slope and bankfull discharge; Wu et al. (2008) *Geomorph.* 100, 3–4 show that bankfull channel dimensions result from the accumulative effect of several consecutive years' discharge and sediment load conditions.

bank retreat The advance or retreat of a channel bank line (Rüther and Olsen (2007) *Geomorph.* 89, 3–4). See Amiri-Tokaldany et al. (2003) *J. Am. Water Resources Ass.* 39, 4.

bank storage Water held in the river bank which may contribute to stream flow, but, in arid conditions, may cause a decrease in discharge as water percolates from the river to the banks; see Burt et al. (2002) *J. Hydrol.* 262, 1. Chen et al. (2006) *J. Hydrol.* 327, 3–4 present a technical paper on bank storage and baseflow separation.

banner cloud 'Clouds resembling a banner, or flag, which form on the leeward side of ridges or peaked mountains' (Schween (2007) *Geophys. Res. Abstr.* 9).

bar 1. On a gently sloping coastline, a submarine accumulation of marine sediment, which may be exposed at low tide. Most bars form where steep, *destructive waves break; such bars can be called **break-point bars**, with crests generally running parallel to the coast. **Bay bars** extend across an *estuary or a bay: 'it seems likely that [Chesil Beach, UK] began as a bay bar . . . when the sea level was lower' (I. West 2007). Some bay bars entirely enclose the inlet, and a *lagoon may then form on the landward side; see Roberts and Slater (2007) *Holocene* 17. **Offshore bars**, located further out to sea, are thought to result from the breaking of larger waves, which erode the sea bed and throw up material ahead of them to form ridges (Hoefel and Elgar (2003) *Science* 29).

2. In a *glacial trough, such as the Nant Ffrancon of North Wales, a transverse rocky barrier (Embleton (1961) *TIBG*, 29).

3. Within a river, a deposit of alluvium, dropped where velocity and turbulence are low, and which may form temporary islands. **Alternating bars** develop as patches of alluvium, often regularly spaced, along opposite sides of a straight channel. **Braid bars**, roughly diamond shaped, are generally aligned along the channel course; Best et al. (2003) *J. Sed. Res.* 73, 4 provide 'a *unique* [authors' italics] insight into the process-product relationship of braid-bar'. **Point bars** form on the inner curves of a meandering river where *discharge is low; see Pyrce and Ashmore (2005) *Sedimentol.* 52, 4.

barchans *See* SAND DUNE.

baroclinic A term applied to sections of the *atmosphere where trends in pressure (pressure surfaces) are at an angle to trends in temperature. (The precise definition refers to the intersection of isobars and isopycnals—levels, or surfaces, of equal density—but, for most purposes, isotherms can replace isopycnals. The number of intersecting isobars and isopycnals is a measure of **baroclinicity** (Nakamura (1992) *J. Atmos. Sci.* 17). Moving polewards, temperatures fall very rapidly, while pressure remains constant in that direction, but falls with height (Garreaud (1999) *Monthly Weather Rev.* 127). At some point, the slope for isotherms intersects with the

slope for isobars, so that the two intersect in a **baroclinic zone**; most often in *mid-latitudes in winter. Barry et al. (2002) *Nature* 415 see the baroclinic zone as a transport mechanism.

In the mid-latitude baroclinic zones spontaneous generation of weather systems such as depressions and thunderstorms is common. (Sanders and Hoffman (2002) *Weather & Forecast.* 17, 4). These are **baroclinic disturbances**, characteristically, on *synoptic charts, with strong meridional pressure gradients in the constant-pressure surfaces and vertical *wind shear. When the temperature gradient along the *meridians is very steep, *atmospheric cells break down into cyclonic and anticyclonic *eddies. This failure is known as **baroclinic instability** (Lesiur et al. (2000) *J. Turbulence* 1, 1), and is characterized by the ascent of warmer, and the descent of colder, air.

barometric gradient *See* PRESSURE GRADIENT.

barotropic Describing *atmospheric conditions where trends in pressure align with trends in temperature, as in the ideal *air mass; the reverse of *baroclinic.

barrage A structure across a river or estuary, built in order to restrain or use water. Brichieri-Columbi and Bradnock (2003) *Geog. J.* 169, 1 discuss the development of a barrage for the Ganges–Brahmaputra–Meghna waters: which 'would cause negligible population displacement, and making maximum use of existing river channels would minimize the environmental impacts associated with large canal or dam construction'.

barrier In *geomorphology, a general term for any offshore depositional form, usually running parallel to the coast or across an *estuary, and above water at normal tides. A **barrier coast** is a series of barrier islands, barrier spits, and *barrier beaches. **Transgressive barriers** are those that migrate landwards under the influence of rising sea level and/or a negative sediment budget. **Regressive barriers**, or strandplains, are large mounds of sediment that have developed along a shoreline with a surface exhibiting well-defined parallel or semi-parallel sand ridges separated by shallow swales. Landwards sediments are deposited on top of more seaward ones. Regressive barriers are linked to a positive sediment budget and/or falling sea level.

barrier beach A long, narrow beach, running parallel to the coastline, and unsubmerged by high tides. See Chadwick et al. (2005) *Procs Inst. Civil Engineer. Maritime Engineer.* 158, 4 on the barrier beach at Slapton Ley.

🌐 **SEE WEB LINKS**
• West (2008) shows a barrier beach.

barrier island A long, narrow, offshore island, usually having beaches and *dunes on the seaward side, often with *lagoons on the landward side, such as the islands enclosing Pamlico Sound, off the coast of North Carolina. The formation of barrier islands has been explained as the drowning of beach ridges (Stutz and Pilky (2002) *J. Coastal Res.* SI 36) or the *progradation of *spits (van Mauren (2005) *Marine Geol.* 224). Most barrier islands and spits are in a state of landwards retreat in response to a combination of *eustatic sea-level rise and local land subsidence. See Lentz and Hapke (2011) *Geomorph.* 126 on the longevity and success of short-term anthropogenic modifications to a coastal barrier system.

barrier reef A *coral reef, stretching along a line parallel with the coastline but separated from it by a wide, deep *lagoon, the most famous of which is the

Great Barrier Reef off north-west Australia (see D. Hopley et al. 2007).

barrow A communal burial mound built from the Stone Age until Saxon times. Long barrows, up to 100 m long and 20 m wide, were the earlier form, while round barrows were introduced during the Bronze Age; see P. Ashbee (1970 and 1960, respectively).

basal complex (basement complex) The ancient *igneous and metamorphic rocks which lie beneath Precambrian rocks and constitute the *shield area of the Earth's continents. See Gutiérrez et al. (2006) *GSA Bull.* 118 on a basal complex in the Canary Islands.

basal contact pressure In a glacier, the contact pressure between a *clast in basal transport and the underlying bed: the greater the pressure, the more abrasion will occur. Boulton (in D. R. Coates, ed. 1974) assumes that the pressure between a particle in contact with the glacier bed is related to the *effective normal pressure, while Hallet (1981 *Annals. Glaciol* 2, 3) argues that clasts are completely surrounded by the ice and can be considered to be essentially floating within it. In this case the contact pressure between a clast and the glacier bed is a function of the rate at which ice flows towards the bed; see M. Bennett and N. F. Glasser (2009) p. 97.

basal ice Many glaciers display basal layers of debris-rich ice. Lawson et al. (1998) *J. Glaciol.* 44 argue that supercooling is the principal mechanism responsible for creating debris-laden basal ice at the Matanuska Glacier. Hall and Glasser (2003) *Boreas* 32, 1 model basal ice temperatures.

basal slipping (basal sliding) The advance of a glacier by comparatively rapid creep close to its bed, occurring where the ice at the base is at its *pressure melting point. The major types are enhanced basal creep, where ice, deformed by stresses created by jagged bedrock, bends round obstructions; regelation (pressure melting); and slippage over a layer of water at the bed. Basal sliding, occurring in sporadic movements, accounts for the majority of flow in warm glaciers, especially where the gradient is steep, the ice thin, or where *meltwater is present. Glasser and Bennett (2004) *PPG* 28, 1 believe that patterns of basal sliding strongly influence patterns of glacial erosion. See also Cohen et al. (2000) *J. Glaciol.* 46, 155, and Cooke et al. (2006) *PPG* 30, 5.

basal thermal regime The general temperature pattern of ice at the base of a glacier. Glaciers are commonly divided into three types according to their temperature: cold-based, warm-based, and polythermal. Typically, the thicker, higher-level ice in the accumulation area is warm-based and the snout, lateral margins, and surface layer of the glacier are below the pressure-melting point.

base flow The usual, reliable, background level of a river, maintained by seepage from groundwater and by *throughflow, which means that the river can maintain the base flow during dry periods. The **base flow index** is the contribution of base flow to a stream. Total flow to stream systems consists of surface run-off, interflow, and base flow; streamflow partitioning methods are used for finding the relative contributions of each (try Morgan (1972) *TIBG* 56). With prolonged drought, base flow itself will diminish, the rate of flow falling in a depletion curve (Gebert et al. (2007) *J. Am. Water Resources Ass.* 43, 1).

base level The theoretically lowest level to which the course of a river can cut down. This level may be sea level, the junction between a tributary and the main river, or the level of a waterfall or lake, but streams rarely erode as far as base level. Base level may alter due to *eustatic or *isostatic change, and be termed **positive changes of base level** if the land sinks relative to the sea, or **negative changes of base level** if it rises. (Try Schumm (1992) *J. Geol.* 101 on river response to base level change; *see also* REJUVENATION). **Marine base level** is the lowest point at which marine erosion occurs, perhaps as low as 180 m below the surface.

Base flow

basement (crystalline basement)
Rocks below a sedimentary platform or cover, that are metamorphic or igneous in origin. Nageswara et al. (2012) *Zeitschr. Geom.* discuss the possible influence of the basement framework in shaping the Krishna-Godavari deltas in India.

base saturation The degree of saturation of the *adsorption complex of the soil; that is, its occupation by exchangeable basic cations, or cations other than aluminium and hydrogen, expressed as a percentage of the total cation-exchange capacity. Halvorson (1980) *EM2894* explains exchangeable cations, cation exchange capacity and base saturation, and the relationship to soil fertility.

base surge A type of pyroclastic surge that forms at the base of volcano eruption columns and travels outward during some hydroclastic eruptions. As base surges move laterally away from the eruptive vent, they lose heat, sediment load, and velocity, which directly affects the resulting deposits. The sedimentary structures observed within surge deposits are also directly related to the amount of liquid water involved (Brand (2004) *Rocky Mt. & Cordill. Meeting GSA Abstract* 37–5). See also R. Fisher and H.-U. Schmincke (1984).

basin A major relief *depression, guided by structure, or formed by *erosion. **Basin and range** describes a landscape where ridges made of asymmetric *fault blocks alternate with lowland basins. In the USA, the basin and range country lies between the Sierra Nevada and the Wasatch Mountains; see W. S. Baldridge (2004).

bastide A planned, fortified strong point and centre of economic development created in the Middle Ages, mostly in France. The rectilinear street pattern contrasts strongly with the irregular, cramped layout of most medieval towns. This morphology was used by French colonizers in Canada (Louder (1979) *Cahiers de géo Québec* 23).

batholith A massive, frequently *discordant, intrusion of coarsely textured plutonic rocks, at least 100 km^2 in area and extending 20–30 km down into the layer of *magma, which may be composed of several plutons.

(⊕ SEE WEB LINKS
• The Idaho Batholith.

bathymetry Measuring water depth, mainly of seas and oceans but sometimes of deep lakes, such as the Aral Sea (Shibuo et al. (2006) *Water Resources Res.* 42). GEBCO (General Bathymetric Chart of the Oceans) consists of an international group working on the development of a range of bathymetric data sets and data products.

(⊕ SEE WEB LINKS
• Website of GEBCO.

Bayesian Referring to methods in probability and statistics, particularly methods related to statistical inference. T. Ando (2010) shows how to use various Bayesian inferences.

beach An accumulation of *sediment deposited by waves and *longshore drift along the coast. The upper limit is roughly the limit of high tides; the lower, of low tides. The beach classifications **dissipative** and **reflective** seem to have fallen out of favour, but see Sherman and Bauer (1993) *PPG* 17 if necessary. *See also* DRIFT-ALIGNED BEACH. Almeida et al. (2011 *ESPL* 36, 4, 523) show a clear relationship between offshore wave height, and the beach profile. They reveal three thresholds for morphological change: waves higher than 2.3 m are responsible for important changes to the

Beach

berm and beach face; waves higher than 3.2 m are responsible for important changes to the sub-tidal terrace and long-shore bar; and waves over 4.1 m are needed to effect morphological change further seaward, near the depth of beach closure.

beach budget (beach sediment budget)
The balance between the material deposited on, and eroded from, a beach; see Davidson-Arnott and Van Heyningen (2003) *Sedimentol.* 50, 6. During winter, waves tend to be destructive, and gradients steepen, but, with summer conditions, waves become constructive, and gradients slacken; a sequence detailed in Engels and Roberts (2005) *J. Sed. Res.* 75, 6. See Martinez and Psuty (2004) *Ecol. Studs* 171 on *fore-dunes and beach budgets.

beach-dune interaction
The interconnections between beaches and dunes. These involve the aeolian transport of sand from the beach to the erosion of the dune by storm waves. Houser (2009) *PPG* 33, 6 considers the role of nearshore processes and morphological change as a flexible and dynamic constraint on the supply and transport of sediment between beach and dune. See Servera and Gelabert (2009) *Geomorph.* 110, 3–4 on the development and setting of a beach–dune system in Mallorca.

beach morphology
Beach form. The occurrence of different types of beach morphology can be parameterized by the dimensionless fall velocity Ω, The equation is:

$$\Omega = \frac{H}{w_s T}$$

where H is the wave height, w_s is the speed at which a sediment particle falls through still water, and T is the wave period. *Reflective beaches tend to develop when $\Omega < 1.5$, *intermediate beaches are those where $\Omega = 1.5$–5.5, and *dissipative beaches form when $\Omega > 5.5$.

beach ridge
A wave-deposited ridge, made of sand and reworked underlying beach material, running parallel to a shoreline, beyond the present limit of ordinary tides. Changes in beach ridge morphology and orientation reflect environmental factors such as changes in wave climate and wind regime (Bristow (2006) *Sedimentol.* 53, 4).

bedding plane
The surface separating distinct rock *strata.

bedforms (sedimentary bedforms)
1. In hydrology, accumulations of sediment on the river bed, formed by variations in flow competence, and ranging in size from ripples in the sand, a few centimetres apart, to 'dunes' tens of metres in length.

River bed form is related to energy level; as flow energy and sediment transport competence increase, bedforms change from ripples to dunes, which alters the ratio of inertial to gravitational forces in the flow field. This form-flow feedback is a reflection of river energy reorganizing the sediment mass in the most effective configuration for that flow condition; see Nelson et al. (2011) *ESPL* 36, 14, 1928.

2. Migrating megaripples are bedforms that appear in the surf zone of sandy coasts; see Gallagher et al. (1998) *Nature* 394.

SEE WEB LINKS
- The excellent USGS bedform sedimentology site.

bed load Material moved along a river bed by traction and *saltation. Bed load usually comprises sands and pebbles, but when the water level is high and the current strong, large material boulders may be moved (Laronne (2001) *Water Resources Res.* 37, 5). Recking et al. (2012, *ESPL* 37, 774) describe bed-load transport as 'a highly fluctuating temporal phenomenon, even under constant (mean) flow conditions, as a consequence of stochasticity, bedform migration, grain sorting, *hysteresis, or sediment supply limitation', and consider how timescale can affect bed-load prediction.

bed profile See LONG PROFILE.

bed return flow Part of a sea-water circulation characterized by onshore flow in the upper part of the water column and a seaward flow near the bottom. This latter is the bed flow, sometimes termed the undertow. See Masselink and Black (1995) *Coastal Engineering* 25, I, 3–4, 165.

bedrock The unweathered rock which underlies the soil and *regolith or which may be exposed at the land surface. **Bedrock gouges and cracks** are small cracks, gouges, and indentations created

on bedrock surfaces. Both *chattermarks* and *crescentic gouges* are shallow crescentic furrows: in crescentic gouges the convexity is turned forward in the direction of ice flow, in chattermarks it is turned backwards (Glasser and Bennett (2004) *PPG* 28, 1). *Crescentic cracks* are vertical fractures of the rock without the removal of bedrock fragments, with a forward-turned concavity.

A **bedrock channel** cannot substantially widen, lower, or shift its bed without eroding bedrock (Turowski et al. (2008) *Geomorph.* 99 (1–4), 26. These channels erode through a suite of processes that are poorly understood (A. Goudie and K. J. Gregory 2011), but bedrock may be removed by plucking along pre-existing joints or fractures, by *cavitation, *abrasion, and dissolution (Whipple et al. (2000) *Geol. Soc. Amer. Bull.* 112, 490).

bed roughness (bed-floor roughness) The frictional force of a river bed (also called *channel roughness*). A rough bed of boulders, pebbles, and potholes exerts more friction than a smooth, silky channel. Lamarre and Roy (2005) *Geomorph.* 68, 1–2 examine the effects of roughness on mean flow velocity. *See* MANNING'S ROUGHNESS COEFFICIENT. The term is also applied to a glacier bed; see Rippin et al. (2004) *Ann. Glaciol.* 39.

behavioural geography This is characterized by an interest in seeking process explanations for specific spatial actions. These processes include perception, learning, forming attitudes, memorizing, recalling, and using spatial thinking and reasoning to explain human activities in different environmental settings. See Golledge, in S. Aitken and G. Valentine (2006). It sets out to 'humanize *homo economicus*, to recognize that people do not have complete information, are not always distance minimizing, are embedded in networks of social relations, and therefore may base decisions on factors other than sheer economic rationality. It specifically recognized the role of cognition and the importance of social and cultural values and constraints, plus all the institutional, economic, and physical factors that characterize the public, 'objective' environment' (Couclelis and Golledge (1983) *AAAG* 73, 3). Behavioural geography was criticized as inherently positivist, severing individuals from their social and cultural contexts. Nonetheless, behavioural approaches have been gaining ground in the field of economics; see Strauss (2008) *J. Econ. Geog.* 8, 2, who calls for a renewed behavioural economic geography, 'without adopting, wholesale, its ways of conceptualising economic man' (sic).

(((∙))) SEE WEB LINKS

• Blogspot on the Environmental Perception and Behavioral Geography Specialty Group of the American Association of Geographers website.

behaviourism The view that the actions of an individual occur as responses to stimuli. Through constant repetition, the individual learns to make the same response to a given stimulus. Mental maps have especially been associated with the behavioural approach; see Soini (2001) *Landsc. & Urb. Plan.* 57, 3-4. Norton (2003) *Area* 35, 4 thinks that cultural geography might benefit from consideration of radical behaviourism.

belonging The concept of belonging is intimately tied to place, as any understanding of community and affinity with specific landscapes (place-making) simultaneously constructs a sense of socially recognized membership (Trudeau (2006) *Cult. Geogs* 13). Belonging captures 'the desire for some sort of attachment, be it to other people, places, or modes of being, and ways in which individuals and groups are caught within wanting to belong, wanting to become'

(Probyn1996, quoted in Nelson and Hiemstra (2008) *Soc. & Cult. Geog.* 9, 3).

belonging, geographies of The geographical study of the practices, meanings, and processes of belonging: from the body, the house or home, the neighbourhood, the region, the nation, and the global. See Till (1999) *Ecumene* 6; Zilberg (2004) *Landsc. Res.* 29, 3 on diasporic geographies of belonging and being; Veronis (2006) *Env. Plan A* 38, 10 on the way immigrant, ethnic/racial, visible minority groups build identity and community; and Holston and Appadurai (in J. Holston, ed. 1999) on belonging, everyday experiences, and tussles over urban space.

Amster and Lindquist (2005) *Asia Pac. J. Anthrop.* 6, 1 consider transnational relationships in South-East Asia; Ikeotuonye (2007) *SGIR Turin Conf.* examines borders, frontiers, and the state's own image of itself; and M. Joseph (1999) illustrates the emergence of nomadic identity. See Sparke (2006) *Pol. Geog.* 25, 2 on 'seemingly unbounded global visions of belonging'.

Benguela current The eastern boundary current of the South Atlantic subtropical gyre, which begins as a northward flow off the Cape of Good Hope, then skirts the western African coast equatorwards until around 24–30 °S (Gyory et al. (no date) CIMAS).

Benioff zone An inclined zone of earthquakes, plunging below the Earth's surface at an angle between 30° and 80°, but commonly at around 45°, and extending to a depth of 300–400 km. Benioff zones are associated with the downward movement of a lithospheric *plate at a destructive plate margin. See Scalera (2007) *Geofísica Int.* 46, 1.

benthic Occurring at the base of bodies of water: lakes, oceans, and seas.

Benthos is the life attached to the bottom, or moving in the bottom mud.

Bergeron–Findeisen theory A mechanism by which ice crystals grow from the vapour supplied by cloud droplets. In mixed clouds, where ice crystals and supercooled droplets coexist, the water vapour diffuses from supercooled droplets to ice crystals because the vapour pressure over ice is lower than over liquid water at the same temperature (Castellano et al. (2004) *Atmos. Envir.* 38, 39). See Korolev (2007) *J. Atmos. Scis* 64, 9 on the limitations of the Wegener-Bergeron–Findeisen mechanism in the evolution of mixed-phase clouds.

bergschrund A deep, tensional *crevasse formed around the head of a *cirque glacier. The crevasse forms as ice falls away downslope. Often, a sequence of bergschrunds forms. Johnson's bergschrund hypothesis (1904, *J. Geol.* 12) explained headwall retreat by vigorous frost weathering in a narrow zone; Lewis (1940) *Geog. Rev.* 30 suggested that meltwater forced its way down far below the bergschrund, explaining the great height of some cirque headwalls. For the limitations of bergschrund theory, see D. R. Stoddart (1997).

berm 1. A low embankment or ridge on a sand beach, constructed by *swash or breaking waves. Weir et al. (2006) *Geomorph.* 82, 3–4 examine two models of berm development: vertical growth at spring tides or 'following significant beach cut due to substantial swash overtopping'; and horizontal progradation at neap tides through the formation of a proto-berm, located lower and further seaward of the principal berm.
2. A benchlike remnant of a surface resulting from the interruption of an erosion cycle. Berms may occur naturally in a river channel, or they may be constructed as part of a *river restoration.

(((⊕))) SEE WEB LINKS
• A colourful and useful example of a berm on European Commission's Environment site.

beta index A simple measure of connectivity which can be derived from the formula

$$\beta = \frac{e}{v}$$

where e = number of edges (links), v = number of vertices (nodes). Values of the beta index range from 0.0 for network, which consists just of nodes without any arcs, through 1.0 and greater where networks are well connected. See J. Jenelius (2009) *Transport Geog.* 17, 234 for an illuminating account.

bid-rent theory W. Alonso (1964) notes that when a purchaser acquires land, he acquires two goods (land and location) in one transaction, and a single payment is made for the combination. Thus it is possible to trade off a quantity of land against location. The **residential bid price curve** is 'the set of prices for land the individual could pay at various distances while deriving a constant level of satisfaction'. As individuals consider residential locations at increasing distances from the city centre, they assess the price of land that would allow them to buy enough land (and other goods) to provide as much utility (satisfaction) as a given price, and amount of land, at the city centre. The **bid price function** for the urban firm describes the prices which the firm is willing to pay at different distances from the city centre in order to achieve a certain level of profits. Egan and Nield (2000) *Urb. Studs* 37 develop a model of the Alonso type which casts light upon the intra-urban location decisions of hotels.

Bid-rent

bifurcation ratio In a drainage basin, the ratio of the number of streams of a given order to the number of streams of the next, higher order. 'In general, small streams are tributaries of the next bigger stream in such a way that flow architecture develops from the lowest scale to the highest scale' (Reis (2006) *Geomorph*. 78, 3–4).

Stream ordering after Strahler

Bifurcation ratio

big data A collection of data sets so large and complex that it becomes difficult to process using general database management tools or traditional data processing applications. See Kitchin (2013) *Dialogues Hum. Geog.* 3, 3, 262 on 'big data and human geography: opportunities, challenges, and risks'.

binary distribution A *city-size distribution in which a number of settlements of similar size dominate the upper end of the hierarchy, said to be characteristic of nations with a federal political structure, such as Australia. *See also* RANK-SIZE RULE.

bioaccumulation The build-up of poisonous substances in an organism, occurring when that organism absorbs a toxic substance faster than it sheds it. 'The accumulation of toxins in specific parts of the ecosystem (usually at the higher levels of food chains) due to the greater ability of some chemicals to accumulate in zones where they become bioavailable, and are taken up and stored by producers and consumers. (The materials must be stored in those parts of the individual that will be consumed). It may result in severe adverse effects on an ecosystem once a threshold level of toxin storage is reached' (J. Holden 2005).

bioarchaeology The study of human remains within archaeological sites in order to reconstruct their economy and environment.

biocapacity The capacity of a given area to generate an enduring supply of renewable resources and to absorb its spillover wastes. Earth's biocapacity is determined by the biologically productive area available and the bioproductivity (yield) per hectare. The biocapacity of an area of land is affected by both physical and human factors: Iraq, for example has a low biocapacity, despite its large oil reserves and its position in the once 'fertile crescent'.

The European Union's *Ecological Footprint and Biocapacity* (2006) outlines the world's ability to regenerate resources and absorb waste in a limited time period.

bioclimatology The scientific study of the relationship between organisms and the long-term atmospheric characteristics of a specified area. See Di Filippo et al. (2007) *J. Biogeog.* 34, 11. Global interactions between climate and vegetation may be studied by using *remote sensing.

biocomplexity Those properties resulting from the interaction of the behavioural, biological, chemical, physical, and social factors that impact on living organisms. Colwell (1998) *BioScience 48*, 786 defines biocomplexity as 'the study of the complex chemical biological, and social interactions in our planet's systems'. Duvall (2011) *AAAG* 101, 3, 497 uses this term to indicate a

focus on a biophysical system in which humans are integral. Cadenasso (2006) *Ecol. Complexity* 3, 1, 1 presents a thorough review of definitions.

biodiversity The number and variety of living organisms, from individual parts of communities to ecosystems, regions, and the entire biosphere, including: the genetic diversity of an individual species; the subpopulations of an individual species; the total number of species in a region; the number of endemic species in an area; and the distribution of different ecosystems. Greater plant diversity leads to greater primary productivity, because there is a greater chance that a more productive species would be present at higher diversity, and from the better 'coverage' of habitat heterogeneity caused by the broader range of species traits in a more diverse community (Timan (1999) *Ecology* 80, 5). The effects of global warming on biodiversity are, as yet, unclear; Botkin et al. (2007) *Bioscience* 57, 3 explain why.

The **biodiversity gradient** describes the greater biodiversity of living organisms at the tropics than at the poles in the biomass, the number of individuals, and, in many taxonomic groups, the number of species (Hawkins (2001) *Trends Ecol. & Evo.* 16). Diversity gradients described for the Northern Hemisphere seem to be invalid for the Southern Hemisphere (Platnick (1991) *Natural History* 25) and 'despite this recognition of the generality of latitudinal diversity gradients, our knowledge is biased towards some taxonomic groups, regions and ecosystems' (Boyero (2006) *Ecology Info.* 32). *See* AGROBIODIVERSITY.

biodiversity crisis M. Novacek (2001) reports that Earth is suffering the fastest mass extinction in its history, which is clearly a direct result of human activity, 'yet the public remains unaware of the crisis in sustaining biodiversity'. E. O. Wilson (1992) calculates that if the annual extinction rate is 0.27%, and tropical forests contain 50% of all species, then about 0.14% of all species are committed to extinction each year. Unless immediate action is taken to restrain the rate of tropical habitat loss, the twenty-first century will see another major extinction catastrophe. See Pimm et al. (1995) *Science* 269, 5222, and Gaston and Fuller (2007) *PPG* 31, 2.

(⊕) SEE WEB LINKS
• Website of International Union for Conservation of Nature.

biodiversity hotspot Myers et al. (2000) *Nature* 403 identify 25 global hotspots, which together comprise only 1.4% of the Earth's terrestrial surface, but which, among them, contain 44% of all species of vascular plants, and 35% of all species of four vertebrate groups. Examples include the Mediterranean basin, Madagascar, the Western Ghats, and Sri Lanka.

biofuel Any liquid fuel derived from plant materials, including biodiesel which can be made from vegetable oils such as palm oil, soy, or rapeseed oil. 'There have been two consequences of this change in land use. Firstly, worldwide, less land is being used to produce food at a time when the world population continues to grow. Secondly, as demand and prices have risen for crops such as palm oil, new land has been cleared to start plantations. The land that is being cleared in countries such as Indonesia and Malaysia is often rainforest or other precious environments where plant and animal diversity is high and indigenous people live traditional lives in harmony with the delicate ecosystems' (Royal Geographical Society website).

biogenic Produced or brought about by living organisms. A **biogenous**

sediment is one of biological origin, such as calcarous or siliceous ooze.

biogeochemical cycle The cyclical movement of energy and materials within *ecosystems. See W. Schlesinger (1997).

biogeographic realm A continent- or subcontinent-sized area, separated from other biogeographical realms by physical barriers to plant and animal migration, such as major mountain ranges, and oceans: for example, the Afrotropical Biogeographic Realm, which contains one of the highest levels of biodiversity on Earth (see Udvardy (1975) *IUCN Occ. Paper* 18). Each biogeographic realm may include a number of different biomes. The use of *ecozone* for biogeographic realm is a fairly recent development; see J. Schultz (2005).

biogeography The spatial analysis, and the search for patterns, in any biological feature: demographic, ecological, genetic, morphological, or physiological; see B. Cox and P. Moore (1999). 'Biogeography transcends classical subject areas . . . **ecological biogeography** is concerned with ecological processes occurring over short temporal and short spatial scales, whereas . . . **historical biogeography** is concerned with evolutionary processes over millions of years on a large, sometimes global, scale' (Crisci (2001) *J. Biogeog.* 28, 2). **Comparative biogeography** uses the naturally hierarchical phylogenetic relationships of *clades to discover the biotic area relationships among local and global biogeographic regions; **systematic biogeography** is concerned with the classification of biotic areas based on the naturally hierarchical phylogenetic relationships of clades; and **evolutionary biogeography** seeks to understand evolutionary mechanisms responsible for the distribution of its organisms.

Molecular biogeography is 'that set of approaches that uses genetic data to address the biogeographic structure of lineages and biotas and the evolutionary and Earth history processes that have shaped current population genetic, phylogenetic, and distributional patterns' (Mantooth and Riddle (2011) *Geog. Compass*).

biogeomorphology Any study uniting biota with geomorphic form and process (Osterkamp and Friedman (1997) *Procs USGS Sed. Workshop*). Many biogeomorphological studies concentrate on the processes that link biotic and geomorphic systems, such as bioweathering, bioerosion, bioconstruction, biotransformation, biostabilization, and bioprotection (Naylor et al. (2002) *Geomorph.* 47, 1).

Many 'soft engineering' schemes for environmental management use biogeomorphological systems; see Goodson et al. (2002) *Geomorph.* 47 on *river restoration.

biographical approach An approach which stresses that the behaviour of an individual is made within the context of her or his past. It is 'a way of understanding the relationship between people and place that focuses on an individual's life course' (N. Castree, R. Kitchin, and A. Rogers, 2013). Half-acree and Boyle (1993) *PHG* 17, 3, 333 use in-depth investigation of the biographies of migrants in order to gain an appreciation of the intentions implicated in the migration decision.

biohistory The interplay between human society and the *biosphere (Medley et al. (2003) *AAAG* 93, 1). S. Boyden (1987) gives an account of the stages of civilization, from the primeval and early farming stages to the current high-energy phase, discussing economics, health and disease, and ecology.

biological control The attempt to reduce numbers of pests by the use of predators, either from within the community, or by introduction from outside, rather than by chemicals. See Strobel (1992) *Sci. Amer.* 265 on the biological control of weeds; see also Mannion (1993) *GeoJournal* 31, 4. Determination of the origins of species is important in the search for biological control agents in the source region (Wardill et al. (2005) *J. Biogeog.* 32, 12). Clough et al. (2005) *J. Biogeog.* 32, 11 observe that natural biological control may be more effective in certain regions than in others.

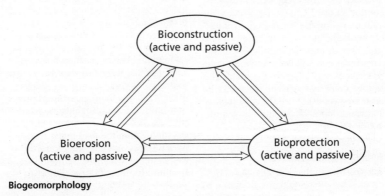

Biogeomorphology

biological invasion The establishment of species in ecosystems to which they are not native. Invasive species cause significant ecological harm: they can alter ecosystem processes, act as vectors of disease, and reduce biodiversity. Worldwide, out of 256 vertebrate extinctions with an identifiable cause, 109 are known to be due to biological invaders (Olsen and Santanou (2002) *Am. J. Agric. Econ.* 84, 5). Simberloff and Alexander (in P. Calow, ed. 1998) estimate that around a quarter of the value of US agricultural output is lost to non-indigenous plant pests or the costs of controlling them.

Case studies abound, among them the UK Environment Agency's material on American signal crayfish and Japanese knotweed, which is also of interest for its use of anthropomorphic phrasing: 'because [Japanese knotweed] does not originate in the UK, it does not compete *fairly* with our native species' (this author's italics). Olson and Roy (*Univ. Maryland, Ag. & Res Econ., W. Paper* 03–06) develop an economic model of a biological invasion, and Peterson (2003) *Q. Rev. of Biol.* 78 claims that the geographic course that invasions are able to take can be predicted.

biological magnification (bio-magnification) The build-up of a toxin as it moves up the *food chain. For example, pesticides wash into waterways, are eaten by aquatic organisms which are in turn eaten by large birds, animals, or humans, and become concentrated in tissues or internal organs. Since aquatic food chains have more links, biological magnification is greater; see, for example, Sutherland et al. (2000) *Arch. Env. Contam. & Toxic.* 38, 4.

biological oxygen demand (BOD) An indicator of the polluting capacity of an effluent where pollution is caused by the take-up of dissolved oxygen by micro-organisms that decompose the organic material present in the effluent. Excess biological oxygen demand loads caused large-scale fish kills in the Cape Fear and Neuse River estuaries (Mallin et al. (1999) *Ecol. Appl.* 9).

biomass The total mass of all the organisms inhabiting a given area, or of a particular population or *trophic level.

biome An ecological zone whose uniformity is defined by the type of plant life that dominates within it, since plant life will often give a strong indication of other ecological features of a zone, such as animal life and soil type. Because biomes are defined by plant life rather than region, some biomes can stretch around the world. For example, the Boreal Forest biome (defined by the predominance of conifers) covers much of northern Europe, Russia, and Canada (Woodward, no date, *Env. Literacy Council*). Other major biomes are: *tundra, temperate (deciduous) forest, *tropical rain forest (selva), tropical grassland (*savanna), temperate grassland (steppe), and *hot deserts. A biome is an idealized type; local variations within a biome are sometimes more significant than variations between biomes.

biometeorology The description and understanding of the impacts of climate and weather on the biosphere and on humans. The International Society for Biometeorology defines it as 'an interdisciplinary science that considers the interactions between atmospheric processes and living organisms (plants, animals, and humans) with the central question within the field being "how does weather and climate impact the well-being of all living creatures"'.
Human biometeorology comprises the study of the influence of weather and

climate on both healthy and diseased persons.

((⊕)) SEE WEB LINKS

- Website of International Society for Biometeorology defining biometeorology. The *International Journal of Biometeorology* is available online.

biophysical diversity The diversity of the natural environment, which controls the intrinsic quality of the natural resource base which is utilized for production. It includes soil characteristics and their productivity, the biodiversity of natural (or spontaneous) plant life, and of the soil biota. It takes account of physical and chemical aspects of the soil, surface and near-surface physical and biological processes, hydrology, microclimate, and also variability and variation in all these elements (Brookfield and Stocking (1999) *Glob. Env. Change* 9, 2). Geertsima and Pojar (2007) *Geomorph.* 89, 1–2 recognize several types of biophysical/ecological diversity: site diversity, soil diversity, and derivative habitat or ecosystem (including aquatic ecosystems) diversity.

biopiracy The exploitation of naturally occurring biological resources: 'searching for economically valuable biochemical and genetic resources from animals, plants, and micro-organisms in nature' (J. Barry and G. Frankland, eds 2001). Other definitions add that individuals or companies then patent such resources for their own benefit. Also known as **ecopiracy**.

biopolitics The term was first used by R. Kjellén (1916) who saw the *state as a quasi-biological organism, a 'super-individual creature denoting social and political power over life'. M. Foucault's (1976) biopolitics is defined as a control apparatus exerted over a population as a whole, citing the ratio of births to deaths, the rate of reproduction, the fertility of a population, and so on. Others have seen biopolitics as the political application of bioethics (J. Hughes 2004); the socio-political consequences of the biotech revolution (Rifkin (2002) *The Nation* 274, 6); the administration and regulation of human and non-human life at the levels of both the population and the individual body (N. Castree, R. Kitchin, and A. Rogers, 2013); and there are more definitions: see Schlosser (2008) *Geog. Compass* 2, 5, 1621.

Morrissey (2011) *Geopolitics* 16, 2, 280 tries to combine a spectrum of definitions into one: biopolitics is 'the spatial politics through which life is constituted and governed; in other words, how life is incorporated into modern forms of governmentality'.

bioregion A region demarcated by the nature of the natural environment, not by man-made divisions. In *ecology, a bioregion is an area that constitutes a natural ecological community (C. Morris, ed. 1992). However, the fact remains that the contours of bioregions are drawn largely by humans; see Meredith (2005) *Ethics, Place & Environ.* 8, 1, 83. Bioregionalists argue that 'modern industrial societies have obscured or lost contact with the extent to which they are reliant on and embedded in ecosystems' and that we should live 'as close to nature as possible . . . letting the rhythms of the land, its soils, winds, patterns, and limits enter into human social life' (J. Barry and E. G. Frankland 2014).

bioscience Any of the life sciences—biology, medicine, anthropology, or ecology—that deal with living organisms and their organization, life processes, and relationships to each other and their environment. **Geographies of bioscience** might analyse the differential impact of developments in relation to inequalities in gender, ethnicity, wealth, and location, and the role of international corporations in these developments; or the development of multinational regulatory endeavours and the distinctive take up of developments in biomedicine and biotechnology in different nations and regions. See Greenhough (2011) *PHG* 35, 2, 153.

biosecurity The protection of people and animals from pests and infectious diseases, notably by managing the movement of agricultural pests and diseases, reducing the effects of invasive species on supposedly indigenous flora and fauna, and preventing the purposeful and inadvertent spreading of biological agents into the human population. See Hinchliffe and Bingham (2008) *Envir. & Plan A* 40, 1534.

biosphere The zone where life is found; the outer portion of the geosphere and the inner portion of the *atmosphere. This extends from 3 m below the ground to some 30 m above it. The biosphere also comprises that region of waters, some 200 m deep, where most marine and freshwater life is found. *Gaia theory postulates biosphere-scale evolution; Free and Barton (2007) *Trends in Ecol. & Evol.* 22, 11 ask whether and why the biosphere might tend towards stability and self-regulation (but don't look for a clear-cut answer). See Phillips (2008) *PPG* 32, 1 on goal functions in ecosystem and biosphere evolution.

biotechnology Using biomolecular and cellular processes to create new products and technological processes. The use of micro-organisms or biological processes for commercial, medical, or social purposes (R. Youngson 2001). As such, biotechnology is one of the earliest technologies used by humans, for example, in cheese-making. Biomedical engineering has, in the last ten years, become a major technology in its own right. Its applications include the development of artificial organs; the design of replacement (prosthetic) devices of all kinds; advances in the medical and biological use of lasers; the development of biological applications of ultrasound; bionics; robotics; and cybernetics (Youngson 2001). F. Bowring (2003) is critical of biotechnology, and raises important issues.

biotic Living or having sustained life. **Biotic interaction** is a mutual or reciprocal interaction between living organisms, in which the distribution of one species is influenced by the distribution of other species. Biotic interaction can be within the same group, for example, competition, facilitation, and parasitism in plants, or between groups, as best exemplified by relationships in food webs: herbivory, predation, and symbiosis (Guisan et al. (2006) *J. Appl. Ecol.* 43). See Gaines in D. Sax et al. (2005) on what species invasions reveal about biotic interaction. **Biotic potential** is the maximum population that an area can support. As the local ecological system approaches the biotic potential of a site, local stability rises, but global sustainability falls; as the system develops feedback loops and complex interactions, it becomes better adapted to local conditions and more stable to local impacts. In the long run, conditions keep changing so that the system becomes sub-optimal to the changed conditions (O'Neill (2001) *Ecology* 82, 12). **Biotic homogenization** describes the process by which species invasions and extinctions increase genetic, taxonomic, or functional similarity; 'evidence is growing that changes in land cover are likely to be accompanied by the ongoing homogenization of biotas, and processes in which native species are replaced by a relatively small set of alien species (Gaston and Fuller (2007)

PPG 31, 2). See Cassey et al. (2008) *J. Biogeog.* 35, 5.

bioturbation The mixing of sediments and soils by animals or plants, for example, when earthworms turn over a soil; see D. R. Butler (1995). These sediments may be marine, lacustrine, or fluvial, where bed-sediment is engineered by plants and algae (Stazner (2012) *Geomorph.* 157–8, 49). Bioturbation is significant in downslope soil transport, particularly in soil formation, soil erosion, and soil creep; see Richards et al. (2012) *ESPL* 36, 9, 1240.

birth rate The crude birth rate is the number of live births per year, per 1 000 population. This does not take the age structure of a population into account, and hence the numbers of women who are capable of giving birth in any year, making accurate comparisons difficult. Because of this, many demographers prefer to use a **standardized birth rate** which indicates what the crude birth rate would have been for a population if the age and sex composition of that population were the same as in a population selected as standard, allowing comparisons to be made between geographical areas, or across social groups within a society (see Fargues (1997) *Pop. & Dev. Rev.* 23, 1 for a demonstration of standardization).

black box view The view that the workings of the human mind cannot be analysed; all that can be observed is the input and output. See Fox (1995) *J. Trans. Geog.* 3, 2.

black earth *See* CHERNOZEM.

black economy The part of the job market which is not reported to the tax authorities and does not appear in official statistics. 'Black' and 'grey' market activities encompass a very diverse set of practices that range from individual bribes and favours to get things done (C. Humphrey 2002), to protection rackets, and to the illegal appropriation of Western aid payments. The trafficking of women and children is a particularly sinister new form of black economy (Smith and Stenson (2006) *PHG* 30, 2). See also Kligman and Limonelli (2004) *Eur. Studs Conf. Chicago.*

black smoker A hydrothermal spring lying along the rift valley of a *mid-ocean ridge. Sea water seeps into fissures in the lava and becomes superheated, chemically interacting with the basaltic rock to create a black precipitate of metal sulphides. (Krom, in J. Holden 2012). One such is located near the Carlsberg Ridge, north of the Seychelle Islands.

blanket bog A continuous covering of bog, needing high humidity and rainfall, mostly comprising peat; only steep slopes and rocky outcrops are dry. Huang (2002) *J. Biogeog.* 29, 2 finds that human impact was the dynamic force responsible for the destruction of the woodland and the formation of blanket bog in upland Ireland. In the UK Peak District blanket bog has suffered a decline in species diversity as a result of 'air pollution, overgrazing, inappropriate or accidental burning, peat extraction and past drainage' (*Blanket Bog Action Plan* 1, section 6. 4).

blind valley A steep-sided karst valley, abruptly terminating in a *streamsink.

blizzards As defined by the US National Weather Service, winds over 16 m s^{-1} together with falling or blowing snow, causing visibility of less than 400 m and lasting for at least 3 hours. See Schmidlin (2002) *J. Climate*, 15, 1765 on the climatology of blizzards in the United States.

bloc An alliance of governments, groups, or parties with a common purpose. Trading blocs, such as the *North America Free Trade Agreement are usually geographically contiguous. Poon et al. (2000) *TIBG* 25, 4 dismiss the notion that international trade is increasingly organized around three economic blocs centred around the United States, European Union, and Japan—the 'triad'—and show that investment intensity patterns do not currently conform to any bloc-like formation, but exhibit instead globally diffused network regions. Badinger and Breuss (2006) *Rev. Int. Econ.* 14, 4 investigate the gains for each member from enlarging a trade bloc, finding that small countries increase in competitiveness by attaining easier access to a larger market, while large countries may have advantages over small countries via relatively more group ties, high market power, and related terms-of-trade effects, advantages in competitiveness due to economies of scale and larger absolute endowments (particularly human capital), larger product varieties (reflecting the greater scope for specialization), and technological advantages.

block 1. A section of the Earth's crust, bounded by faults, that may be tectonically disrupted to form basins and block mountains or complex systems of troughs and ridges. The formation of a **tilt-block** may result in *tectonic deformation along plate margins; see Agarwal and Sharma (2011) *Zeitschrift. Geom.* 55, 2, 194.

2. A generally angular boulder formed by frost-shattering *in situ* on a bedrock surface.

blockbusting A technique, used by estate agents in the USA, of inclining residents of a white neighbourhood to move out because they fear that the district is to be taken over by black families. See Hanlon (2008) *Annual Meeting AAG* Boston.

block field (block stream) A sheet of *block spreads or lines in a *periglacial landscape. Block fields (*felsenmeer*) develop on slopes under 5° (A. L. Washburn 1979), or under 10° (Caine in White (1976) *Quat. Res.* 6). See van Steijn et al. (2002) *PPG* 26, 4. Thompson and Syverson (2006) *GSA Abstracts with Progs*, 38, 4 maintain that this debris is formed *in situ* by *freeze-thaw. Others suggest that the blocks have ridden down on the top of saturated debris during *gelifluction (Creemeens et al. (2005) *Geomorph.* 70, 1–2).

blocking A meteorological condition when the tropospheric circulation takes the form of large-amplitude stationary waves, possibly lasting for weeks. The upper air flow then guides depressions around the edge of the anticyclone(s). Renwick and Revell (1999) *Month. Weath. Rev.* 127, 10 suggest that *Rossby wave propagation links anomalous convection in the tropics and blocking over the south-east Pacific Ocean. **Blocking anticyclones** develop when the Rossby pattern changes from zonal to strongly *meridional, often forming one or two high-level, closed anticyclonic circulations. Athar et al. (2013) *Int. J. Climatology* 33, 3, 585 report that 31% of the Northern Hemisphere mid-latitude blocking anticyclone events occurred over the 10°–70° E longitudes, and of these, the maximum number of mid-latitude blocking anticyclone event onsets are at 30° E (24%); see Schopf et al. (1981) *Dyn. Atmos. Oceans* 5.

block mountain An area of upland identified with an uplifted area bounded by *faults. The Sierra Nevada is a massive block mountain range that broke free along a bounding fault line and has been uplifted and tilted (Huber (1987) *US Geol. Surv. Bull.* 1595).

blowhole A crack in the top of a cliff through which air and sea water blow.

The blowhole is fed from the seaward end via joints and tunnels.

blow-out (blowout) A saucer-, cup-, or trough-shaped depression or hollow formed by *deflation on a pre-existing sand deposit. The adjoining accumulation of sand, the depositional lobe, derived from the depression, and possibly other sources, is normally considered part of the blow-out (Hesp (2002) *Geomorph.* 48, 1–3). Blow-outs usually result from the removal of vegetation; see Dech et al. (2005) *Catena* 60, 2.

bluff A steep, almost vertical, cliffed section of a river bank.

body The body may be seen as: the smallest unit of geography upon which may be inscribed power and resistance (Camp (2002) *J. Southern Hist.* 68); a map of meaning and power (Simonsen (2000) *TIBG* 25); or a cultural representation of masculinity or femininity (R. Ainly 1998). It is also a form of reference by which supposedly 'disembodied' dominant cultures designate certain groups—the elderly, ethnic minorities, females, the obese, the disabled, and so on—as *other (N. Duncan 1996; A. Blunt and G. Rose 1994; and S. Pile and N. Thrift (1995). The figure of the body can be a metaphor for understanding socio-spatial relations in contemporary culture (Dyck and O'Brien (2003) *Canad. Geogr./Géog. canad.* 47, 4). **Sexed bodies** can create spaces, as in a gay pride parade (Johnston (2001) *Annal. Tour. Res.* 28, 1). The notion of the **raced body** tied to the land 'is not novel, having been used to discredit urban Aboriginals, claiming that once they leave the land, they cease to be Aboriginal' (Wazana (2004) *Refuge* 01–03–2004). Every researcher is situated in her or his own body—*embodied*.

body work Labour that involves intimate and often messy contact with the bodies of others through touch or close proximity. Body work has been and continues to be undertaken primarily by women, and is based on a sexual division of labour that assigns to women the care of bodies and the spaces they inhabit. Body work—for example, care work, childcare, and beauty therapy—is generally done by those on the lower rungs of the job ladder; see McDowell, Dyer, and Batnitzky (2008) *PHG* 33, 46. C. Wolkowitz (2006) describes body work as involving the 'care, adornment, pleasure, discipline, or cure' of others' bodies.

bog-burst (bog-slide, bog flow) A peat slide, usually in temperate or cold and wet latitudes, triggered by water pressure beneath a peat surface which causes it to rupture. See M. Evans and J. Warburton (2007) on the geomorphology of upland peat.

bohemian Someone with creative or literary interests, who is socially unconventional. Florida (2002) *J. Econ. Geog.* 2 finds strong relationships between high concentrations of bohemians and concentrations of high human capital, with a particularly strong relationship between the concentrations of bohemians and concentrations of high-technology industry. See Wojan et al. (2007) *J. Econ. Geog.* 7, 6.

bolide A meteoric fireball, or asteroid; see Dawson and Stewart (2007) *PPG* 31, 6.

Bølling The mild Bølling–Allerød interstadials began *c.* 13000, and ended between 12800 and 12300 years BP (Brauer (2002) *Quat. Int.* 88). In Britain the Bølling is not usually distinguished from the *Allerød.

Bond oscillation (Bond cycles, Bond event) Bond et al. (1997) *Science* 278: 1257 established quasi-periodic (*c.*1500 years) climate fluctuations during the

Holocene. They related ice-rafted debris (IRD) with the slowdown of the Atlantic thermohaline circulation (THC) and, later, to variations in solar radiation, as the IRD fluctuations were coherent with the production rates of cosmogenic nuclides of ^{14}C and ^{10}Be (Bond et al. (2001) *Science* 294: 2130). Therefore, Bond cycles suggest that solar radiation must have had a marked influence on Holocene climate variation in all regions influenced by changes in Atlantic ocean circulation (R. W. Battarbee, F. Gasse, and C. E. Stickley (eds) 2004). See also Pèlachs et al. (2011) *Holocene* 21: 95.

Bora (fall-wind) A cold winter wind blowing down from the mountains onto the eastern Adriatic coast. See Pullen et al. (2007) *J. Geophys. Res.* 112.

border A boundary line established by a state, or a region, to define its spatial extent—'central multi-scalar nodes where power, place and identity intersect' (Kaiser and Nikiforova (2006) *Ethnic & Racial Studs* 29, 5). The establishment and maintenance of borders are crucial to the integrity of the modern state. States use a variety of mechanisms to enforce their borders, including checking travel documents, limiting border crossings to select travellers, and in some cases, building obstacles to prohibit unauthorized crossings. While in the European Union borders have become increasingly porous, struggles over the demarcation of borders cause conflicts: there is, for instance, a long ongoing dispute between China and India over borders.

Bauder (2011) *AAAG* 101, 5, 1126 reviews some of the existing literature on the concept of the border and cross-border migration and suggests that various material practices and meanings related to borders can be conceived of as 'aspects' of the border concept.

K. Ohmae (1999) is not alone in identifying a move to a borderless world, brought about by *globalization, but this conclusion is contested by Yeung (1998) *TIBG* 23, 3 and Sparke et al. (2004) *TIBG* 29, 4. See also W. Maley et al. (2000) on refugees and the 'myth of the borderless world'.

(((◉))) SEE WEB LINKS
- The e-journal *borderlands*, published by the University of Adelaide.
- A blog used as an archive for papers and sources regarding the China-India border dispute.

bore A positive surge that may form at flood tide in shallow estuaries with a large tidal range (Wolanski et al. (2004) *Estuarine, Coast. & Shelf Sci.* 60, 4).

Boreal (Period) A division of the *Flandrian, *c.*10000–8000 years BP.

bornhardt A dome-shaped rock outcrop more than 30 m high, and sometimes several hundred metres in width, commonly rising above erosional plains in the tropics, but also found in unglaciated uplands in high latitudes. The origin of these features is problematic: faulting, lithological control, cross folding, and contrasts in fracture density have been cited. C. Twidale and J. R. Vidal Romaní (2005) give an excellent summary.

borral A soil order of the *US soil classification. *See* CHERNOZEM.

Böserup model E. Böserup's (1965) view that increases in population size stimulate agricultural change. 'Pre-industrial farmers, according to the Boserup model, were almost always technically capable of increasing the productivity of their land by applying methods (soil tillage, additional weeding, application of organic fertilisers and irrigation) which allowed it to be cropped more frequently and fallowed for shorter periods. But because such methods typically demanded higher labour inputs per unit of production, they tended to be adopted only when population pressure and the resulting land scarcity forced farmers to work harder in order to maintain existing levels of subsistence' (Henley (2005) *Asia Pac. Viewpt* 46, 2). Brookfield (2001) *Asia Pac. Viewpt* 42 has doubts, observing that agricultural intensification has more often been inspired by price incentives to surplus production, rather than though population pressure.

boulder field As distinct from a *block field, this is an area of boulders which is the result of *spheroidal weathering. See Sevon (1987) *N-E Section GSA* 5.

boundary 1. Of a river, the bed, and banks.

2. A point or limit that distinguishes one social system or group from another and identifies and regulates who may participate in it. Such boundaries may be contested, but Painter, in P. Cloke and R. Johnson (2005) notes that 'It may be impossible to draw a sharp territorial boundary between "British society" and "French society", but social processes operate (somewhat) differently in Britain and France and there are real differences between social structures and social institutions in the two countries'.

3. A line around the edge of an area; a perimeter. J. Nevins (2002) suggests that a boundary is a strict line of separation between two (at least theoretically) distinct territories, where a border is an area of interaction and gradual division between two separate political entities.

Mostly in biogeography and plant ecology, **boundary detection** is the identification of the zones of greatest change in vegetation and associated abiotic/biotic factors; see Jacquez et al.

in T. Poiker and N. Chrisman, eds (1998). Kent et al. (2006) *PPG* 30, 2 are a model of clarity in discussing boundary detection. The extent to which geographically isolated regions support local diversification should predict the placement of biogeographical boundaries (Patten and Smith-Patten (2008) *J. Biogeog.* 35, 3).

boundary layer Any layer of the *atmosphere significantly affected by the Earth's surface. The **laminar boundary layer** is the few millimetres above the surface; the **turbulent boundary layer** denotes the conspicuously turbulent part of the atmosphere.

(⊕) SEE WEB LINKS
• NOAA pages on surface and planetary boundary layer processes.

bounded rationality An understanding of rationality based on the notion that individuals strive to be rational, after first greatly simplifying the choices available, taking into account computational difficulties, the costs of obtaining information, and multivariate preferences, In choosing between a small number, the decision-maker acts as a *satisficer. Salant (2011) *Amer. Econ. Rev.* 101, 2, 724 is clear on this.

Bowen ratio (*B*) The ratio of sensible heat (the heat exchanged by a body or thermodynamic system that has as its sole effect a change of temperature) to *latent heat. The Bowen ratio is the mathematical method generally used to calculate heat lost (or gained) in a substance. When the magnitude of B is less than one, a greater proportion of the available energy at the surface is passed to the atmosphere as latent heat than as sensible heat, and the converse is true for values of B greater than one.

braid bars *See* BAR.

braided channel A river channel in which *bars and islands have been deposited, differing from *anastomosis since the islands are less permanent. Braiding occurs with frequent variations in discharge: when the river cannot carry its load; where the river is wide and shallow; where banks are easily eroded; and where there is a copious bed load. Richardson and Thorne (2001) *Geomorph.* 38, 3–4 show that the division of the flow into multiple threads is a prerequisite for bifurcation. Fotherby (*Geomorph.* 103, 4) argues for valley containment as a causal factor of braided channels. Egozi and Ashmore (2008) *ESPL* 33, 14 commend channel count as a consistent measurement of the complexity of a braided pattern. See Bledoe and Watson (2001) *Geomorph.* 38, 3–4 for an excellent summary.

braunerde *See* BROWN EARTH.

Brazil current A weak boundary current, flowing in the upper 600 m of the water, carrying warm subtropical water along the coast of Brazil—from about 9° S to about 38° S, where it joins the Malvinas Current. The two currents create the strong Brazil–Malvinas Confluence thermohaline frontal region (Bischof et al., no date, *CIMAS*).

break point (breaking point) The point at which the field of influence of one settlement ends, and that of another begins. See Haberkern (2002) *Issues in Shopping Centre Trade Areas* for W. Reilly's (1931) original equation, and a number of modifications to it.

break point bar *See* BAR.

Bretton Woods A shorthand for the system of international finance created in 1944 with the establishment of the *International Monetary Fund, the International Bank for Reconstruction and Development (*World Bank), and the creation of a system of fixed exchange rates. This system lasted until 1971, when it gave way to floating exchange rates.

In 2007, the World Economic Forum and the Reinventing Bretton Woods Committee (RBWC) made specific proposals for improvements in the exchange rate system that could help to prevent large, persistent economic imbalances of the kind that currently threaten world financial stability.

(((●))) **SEE WEB LINKS**
• The International Monetary System, the IMF, and the G–20.

bricolage Initially, a term used in the arts as the process of transforming the meaning of objects or symbols through novel uses or unconventional arrangements of unrelated things. The term bricolage was introduced, in the context of linguistics, by C. Lévi-Strauss (1962–2009), who used the metaphor of the *bricoleur* (handyman) to stand for the practice of creating things out of whatever materials come to hand. From there, it was a short step to using the term as 'an entity that is built out of whatever materials happen to be available' (A. Rogers, N. Castree, R. Kitchin 2013). This entity may be material, for example, a shack in a *squatter settlement, or conceptual, as in *postmodernism.

brown earth (brown forest soil) A free-draining *zonal soil associated with deciduous woodland. Generally, brown earths have a thick litter layer and a *humus-rich A *horizon, containing iron and aluminium sesquioxides in small, crumb-like *peds. The weakly developed, lighter, B horizon contains blocky peds; there is little *leaching. See Storrier and Muir (1962) *Eur. J. Soil Sci.* 13, 2.

brownfield site Any previously used land and/or premises that is not currently fully in use, and may also be vacant, derelict, or contaminated. In an outstanding paper, dealing with both principles and practice, Raco and Henderson (2006) *Local Env.* 11 assess the role of brownfield development in urban and regional policy in the UK.

buffer state A generally neutral state, lying between potentially belligerent neighbours; 'protective devices made obsolete by modern systems of communication' (Spykman (1947) *Geog. Rev.* 32). See V. Pholsena and R. Banomyong (2007) on Laos.

built environment The human-made surroundings that provide the setting for human activity, including buildings, parks, green spaces, neighbourhoods, and cities, and sometimes including their supporting *infrastructure. Very simply, it is defined as 'the human-made space in which people live, work, and recreate on a day-to-day basis' (Roof and Oleru (2008), *J. Environ Health* 71, 24).

buoyancy The quality of an air parcel allowing it to rise through, and remain suspended within, the *atmosphere, due to its relative warmth, and therefore lighter density. Should its water evaporate, the parcel will suffer an *adiabatic heat loss, and lose buoyancy. See Dupilka and Reuter (2006) *Weather & Fcst* 21, 3 on forecasting tornado potential.

bush encroachment The change in vegetation from open *savanna, or mixed grass and woodland, to scrubland. Occurring also in Australia and South Africa, bush encroachment affects the agricultural productivity and biodiversity of 10–20 million ha of South Africa (Ward (2005) *Afr. J. Range & Forage Sci.* 22, 2). Its probable causes are: increased grazing levels; reduced use of burning to create grassland; changes in rainfall regimes; and the interactions between these factors. However, writing in 2005, Ward (op. cit.) writes, 'many people believe that we understand the causes of bush encroachment. We do not.'

bush fallowing A farming system whereby the fallow period is longer than the farming period. Hance (1969, in Omofonmwan and Kadiri (2007) *J. Hum. Ecol.* 22, 2) argues that bush fallowing is a major hindrance to food production in Nigeria, but Chokor (1993) *Env. Manage.* 17, 1 argues that 'modern' methods, which shorten the fallow period, disrupt 'the conservation principles inherent in the traditional man–nature relationship'.

business cycle Economies seem to rise and fall cyclically. Saks and Wozniac (2007) *IZA Discuss. Paper* 2766, link long-distance migration with business cycle fluctuations. Wall (2006) *FRB St Louis, W. Paper* 2006–035A finds regional variations, and an increase in the time period between successive expansions in Japan, 1976–2005.

bustee (basti) In India, a *shanty town; reading S. Mehta (2005) should be mandatory. N. Islam (1996) estimates that almost 50% of Dhaka's population live in bustees, and a major portion of this population is female. To combat negative views of bustees, see Mahmud (2003) *Cities* 20, 5.

butte A small, flat-topped, unvegetated, and very steep-sided hill of layered strata. It may be a volcanic neck or a cone that has resisted erosion, or the residue of a *mesa.

buzz Knowledge transfer via 'in-groups', socialization, and face-to-face contact. 'Individuals in a buzz environment interact and cooperate with other high-ability people, are well-placed to communicate complex ideas with them, and are highly motivated' (Storper and Venables (2004) *J. Econ. Geog.* 4, 351.

C3 and C4 plants C3 plants use the C_3 carbon fixation pathway as the sole mechanism to convert CO_2 into an organic compound. They make up over 95% of Earth's plant species and flourish in cool, wet, and cloudy climates, where light levels may be low, because the metabolic pathway is more efficient there. With enough water, the stomata can stay open and let in more carbon dioxide. However, carbon losses through photorespiration are high.

C4 plants are able to raise the intercellular carbon dioxide concentration at the site of fixation, thus reducing and sometimes eliminating carbon losses by photorespiration. C4 plants first fix CO_2 into a compound containing four carbon atoms. They inhabit hot, dry environments, have very high water-use efficiency, and can double the C3 photosynthesis, but C4 metabolism is inefficient in shady/cool environments. Less than 1% of Earth's plant species can be classified as C4, but they account for around 20% of global gross primary productivity (Wand et al. (1999) *Glob. Change Biol.* 5, 6).

cadastre 1. A record of the area, boundaries, location, value, and ownership of land, achieved by a **cadastral survey**.

2. The term is also used in *geographic information systems (GIS). CARIS (Computer Aided Resource Information System) is a GIS system for hydrography (see Mioc, Anton, and Nickerson, no date, for a web-based GIS model for predicting flood events) and cadastral systems (see Steudler (2004) *GIS@development* for a worldwide comparison of cadastral systems).

Cainozoic *See* CENOZOIC.

calcrete A *duricrust made mostly of calcium carbonate, forming in arid climates via capillary action and prolonged evaporation. See Stokes et al. (2007) *Geomorph.* 85 on groundwater, pedagenic processes, and fan dynamics in calcrete development.

caldera A sunken crater at the centre of a volcano, formed when the magma founders, so that the centre of the volcano collapses. Kennedy and Stix (2007) *GSA Bull.* 119 examine magmatic processes and caldera collapse in New Hampshire; and Furuya (2003) *Earth Planets Space* 55 constructs an optimum source model of caldera formation.

calibration 'The process by which the controlling parameters of a model are adjusted to optimize the . . . degree to which the model's output resembles the reality that the model is designed to simulate . . . While validation may actually be impossible in some cases (see Oreskes et al. (1994) *Science* 263, 641–6), calibration is obviously among the most essential obligation of modelling' (Clarke in P. Atkinson et al., eds. 2005). Lajoie and Hagen-Zanker (2007) *Cybergeo: systs., modél., géostat.* 405 create and calibrate a robot land-use model.

calibre In rivers, the size of the sediment transported. With distance downstream, the calibre of the sediment decreases.

California current A cold *ocean current off the California coast, given to anomalous conditions, that can be termed a Californian *El Niño (Mullin (1998) *Glob. Change Biol.* 4, 1).

calving The breaking away of a mass of ice from an iceberg, ice front, or glacier, caused by: wind and/or wave action; over-extension of an ice shelf; or a collision with an older iceberg. Calving represents the major form of *ablation from a glacial system. See Benn et al. (2006) *PPG* 30, 3 and Skvarca et al. (2002) *Quat. Sci. Revs* 21.

Cambrian The oldest period of *Palaeozoic time, approximately 570–500 million years BP; the time when most of the major groups of animals first appear in the fossil record.

(((⊕))) SEE WEB LINKS
• Provides an overview of the Cambrian period.

canyon A valley with very steep sides and no floor, such as the Grand Canyon of the Colorado River. A canyon differs from a *gorge in that the sides are stepped, reflecting alternating rock resistances. See Spamer (2006) *Annal. Imposs. Res.* 12, 2, who sounds off about the name 'Grand Canyon', but is quite sane in his book of 1984. **Submarine canyons** are deep troughs in the sea bed, sometimes as prolongations of river valleys on land; see Morais et al. (2007) *Marine Ecol.* 28, and suppl. 1. Canyons may also form via earth flows, turbidity currents, submarine springs, or the slipping of sediments.

capability constraint In *time–space geography, a limit to one's actions, due to biological needs, like sleep, and to restricted facilities, like access to transport. See Golob and Regan (2001) *Transp. Res C* 9, 87–91 and Axhausen in D. Ettema and H. Timmermans, eds. (1997).

capacity The maximum amount of debris that a stream can move as *bed load. Capacity depends on *discharge, and on the nature of the load; for the same volume, a stream may be able to carry more gravel than boulders. See Prosser and Rutoni (2000) *PPG* 24.

cap and trade A strategy for lowering the emission of greenhouse gases. The 'cap' is a limit on permitted emissions, while the 'trade' is the market in emissions equal to or above this cap.

capillary In a soil, the fine spaces between soil particles. **Capillary action** is the movement of a liquid in a very narrow space, such as a soil pore, due to surface tension, cohesion, and adhesion. See N. Brady and R. Wei (1998).

capital One of the four *factors of production, capital includes all the items designed by society to further the creation of wealth. Plant, machinery, and buildings are **fixed capital**—they earn profit without circulating further; see Melachroinos and Spence (2001) *PHG* 25, 3. D. Harvey (1982) considers the coercion that fixed capital exerts upon capitalists. **Circulating capital** (**floating capital**) includes raw materials, fuels, components, and labour inputs which are then sold—in the form of the product—at a profit; see Henderson (1998) *Antipode* 30, 2. **Financial capital** is the money necessary for production; although G. Duménil and D. Lévy (2004) argue that financial capital is unproductive.

　Commercial capital mediates in the circulation of commodities for a fee, but has no long-term control over it; 'for Marx, a specialized commercial capital and financial capital drive down the costs of circulation and finance—reducing the deductions they make from the profits of industrial capital and speeding up capital turnover and reinvestment' (Dunford and Pickvance (2007) *PPG* 31, 4).

Knowledge capital encompasses 'talents, skills, know-how, know-what, and relationships' (T. Stewart 2001). Amin and Thrift (2007) *PHG* 31, 2 link urbanism with **fast capitalism**—rapid information and communication technologies together with their impact on self, society, and culture. See P. Taylor (2004).

Capital is seen as a social relation, symbolizing the economic relationships between individuals and/or groups; 'the result of social labour achieved in the creation of goods and services' (Tickell and Peck (1992) *PHG* 16; see also Pasinetti in M. Baranzini and R. Scazzieri, eds. 1986).

Geographers focus on capital as a source of *uneven development; Walker (1978) *Rev. Radical Polit. Econ.* 10 is still worth quoting: '[the] mobility of capital means that capital may use location as a strategy against labour, local development becomes more dependent on outside capital, and development comes and goes over time . . . and capital generates a permanent reserve of stagnant places; a lumpengeography of capital.' *See* SPATIO-TEMPORAL FIX. Grettler (2001) *Env. Hist.* 6 argues that capitalism has been the greatest force for environmental change in America; see also M. W. Lewis (1992).

Capital switching refers to a change in the nature of investment as capitalist production edges towards periodic crises of overaccumulation. The switch is from production into production of the urban built environment as a means to absorb surplus capital and hence avert a crisis, even if only temporarily. Christophers (2011) *AAAG* 101, 6, 1347 finds a clear pattern of capital switching in the

USA into the built environment since the turn of the millennium.

capital intensive With a high ratio of *capital to labour. Barthelmie (2007) *Geog. Compass* 1, 3 notes that, because wind farms are highly capital-intensive, European wind farms are generally owned by utilities, which have access to more attractive financing packages.

capitalism An economic system based on private property and private enterprise, with a major proportion of economic activity carried on by private profit-seeking individuals or organizations, and other material *means of production largely privately owned. Capitalism never stands still. Its central imperative—the search for profit and wealth creation—'drives a perpetual process of economic flux' (Boschma and Martin (2007) *J. Econ. Geog.* 7, 5). This flux is powered by innovation: 'the production, acquisition, absorption, reproduction, and dissemination of knowledge is seen by many as the fundamental characteristic of contemporary competitive dynamics' (Gertler (2003) *J. Econ. Geog.* 3).

'Innovation and growth are at the heart of the capitalist system . . . As new large-scale technologies develop; new forms of firm and factory organisations result in increasingly global value chains producing for global market' (Kaplinsky (2008) *Geog. Compass* 2, 1). See also Metcalfe et al. (2006) *Camb. J. Econ.* 30, 7–32. *Accumulation is at the heart of capitalism—Buck (2007) *Antipode* 39, 4 notes the periodic crises of over-accumulation. A feature of **advanced capitalism** is the possession of capital by fewer and fewer owners.

Capitalism is not uniform: 'nation-states have different trajectories of capitalist development, in which there is considerable variation in the role of markets' (J. Hollingsworth and R. Boyer 1997). Peck and Theodore (2007) *PHG* 31, 6 note 'the persistence of . . . real institutional differences—and the instituted space-economies with which they are associated' and Soskice (1997) *Indust. & Innov.* 4 argues that different national institutional frameworks support different forms of capitalism. See H. Wai-Chung Yeung (2006) and Buck (op. cit.) on Chinese capitalism; E. M. Wood (2003) on English capitalism, M. Wiener (1985) on British capitalism, 'enervated by gentry values'; and Clarke (2007) *TIBG* 32, 1 on British capitalism 'that conceals its destructive effects under the cloak of oldness'. Colas (2005) *Contemp. Polit.* 11, 2–3 explains that capital constantly searches for fresh markets, leading inevitably to 'spatial expansion, as new lands, peoples and re-sources are exposed to the "law of value"'. 'Spatial differentiation is the fundamental building block of capitalism' (Clark (1985) *Econ. Geog.* 61, 3). **Vagabond capitalism** is an increasingly global capitalist production that can discard many of its commitments to place: 'at worst, this disengagement hurls certain people into forms of vagabondage; at best, it leaves people in all parts of the world struggling to secure the material goods and social practices associated with social reproduction' (Katz (2001) *Antipode* 33, 4).

D. Harvey (1985) holds that the inner contradictions of capitalism are expressed through the formation and re-formation of geographical landscapes. Cox, Ohio State U., picks up on this: 'instead of towns organized around cathedrals we have towns organized around "central business districts". Capitalist societies are characterized by extraordinarily differentiated geographic divisions of production.' Similarly, Clark (op. cit.) claims that 'the *longue durée* of a capitalist development trajectory diverted Britain away from the path bequeathing continental European cities their particular flavour in favour of a

voracious, London-centred colonial capitalism'. Brown (2007) *PHG* 32, 4 sees enclosure—the appropriation and privatization of previously common, public or open access resources—as promoted by global capitalism, and commodification and capital.

'Capital reproduces the peaceful civility of the private contract amongst seemingly free and equal individuals so long as the exploited party accedes to its own subordination—once there is significant resistance to the terms and conditions of the contract, the hidden hand of the market becomes the visible fist of the state' (Colas, op. cit.). I. Boal et al. (2005) go further: 'war is virtually *synonymous* with the capitalist state.' 'Urban transformation processes are largely the outcome of space wars, in that deliberate and systematic creative destruction is part of the logic behind current capitalist space economy. These spatial struggles have often been staged in city streets' (Hansen (2008) *Liminalities* 4, 1).

Buck (2007) *Variant* 28 doesn't believe that human-induced global warming will finally destroy capitalism: 'while capitalism may survive, this is not to say that we can safely embrace rosy visions of utopian futures.' See Peck and Theodor (op. cit.) and Faulconbridge (2008) *J. Econ. Geog.* 8, 4 on varieties of capitalism.

capitalist state The state provision of *infrastructure, *external economies, and a regulatory framework facilitates *accumulation. The state can limit or increase the freedom of market actors, raise or lower taxes, intervene in major financial crises, and defend and secure its territory—none of which can be accomplished by *transnational corporations. The **theory of the capitalist state** argues that such functions legitimize the state's existence. B. Jessop (2002) focuses upon the structural coupling and co-evolution of

accumulation regimes and political regimes. It is argued that the state's capacity
to manoeuvre the accumulation regime
has been weakened by international
finance, trading *blocs, and so on, leading
to a loss of national sovereignty; see Zaky
(2005) *Res. Pol. Econ.* 22. Parenti (2014)
New Politics XIV-4 suggests an **environmental theory of the capitalist state**,
arguing that the state is fundamental to
capitalist development and a key part of
what it does for capital is to manage
nature.

capital–labour ratio The amount of
capital per unit of work. The capital-la-
bour ratio is at the heart of the 'one-sector
regional growth model' (McCombie
(1998) *Urb. Studs* 25): as people move
from lower real waged city regions to
higher ones, the capital–labour ratio in
stronger regions decreases, and increases
in weaker regions. The process stops as
soon as the capital–labour ratio is the
same in all regions. P. McCann (2001)
shows how, in time, this development
improves the welfare of society as a whole.

capture (river capture) When a river
is extending its channel upstream by
*headward erosion, it may come into
contact with the headwaters of a less
vigorous river. The headwaters from the
minor river may be diverted into the
more rapidly eroding channel. There is
often a sudden change of stream

direction at the point of capture; this is
the **elbow of capture**. See Mather (2000)
Geomorph. 34, 3 for two examples of river
capture; see also Bishop (1995) *PPG* 19.

carbonation A form of solution where
carbonic acid, formed by the solution of
atmospheric CO_2 in water, dissolves
minerals. Carbonation is best seen in the
solution of calcium-rich rocks such as
limestone, but carbonic acid will also
dissolve silicates.

carbon cycle Carbon is supplied to the
*biosphere as carbon dioxide during vol-
canic eruptions. Most of this is dissolved
in the sea or built into calcareous sediments, which then form limestones and
dolomites. As these rocks are folded and
raised above sea level, they are subjected
to solution by weak carbonic acid and
form sediments once more. This is the
largest and slowest of the carbon cycles.
See Yuan in D. Yuan and Z. Cheng, eds
(2002). The shortest cycle involves respiration by plants and animals whereby
carbon dioxide is expired, and photosynthesis by plants, changing carbon dioxide
and water into organic compounds. Plant
species affect ecosystem carbon uptake
via biomass production, and carbon
release via decomposition. Dorrepaal
(2007) *J. Ecol.* 95, 6 holds that differences
in the way plants respond to climate
change may feed back to the atmospheric
carbon balance, and the climate at a

Headward
erosion

1   2

River capture

global scale. Uncertainty in understanding the carbon cycle has significant cost implications for a climate stabilization policy, 'with cost differences denominated in trillions of dollars' (Smith and Edmons (2006) *Tellus B* 58, 5).

Since 1957, only about half of the CO_2 emissions from fossil-fuel combustion have remained in the atmosphere; the rest being taken up by land and oceans. In the face of increasing fossil-fuel emissions, the rate of carbon absorption by the land and ocean has accelerated over time (Canadell et al. (2007) *Proc. Nat. Acad. Sci. USA* 104). There is no guarantee that the 50% discount will continue; if it disappears we will feel the full climatic brunt of CO_2 emissions from fossil fuels. Climate models that include descriptions of the carbon cycle predict that terrestrial uptake of carbon will decrease in the next century as the climate warms (Miller (2008) *Nature* 451).

carbon footprint The amount of greenhouse gas (GHG) emissions emitted through human activity. It is expressed in terms of tonnes of carbon dioxide equivalent (the mass of carbon dioxide gas that has the same warming effect in the Earth's atmosphere as the given GHG emission) produced in a unit of time.

Carboniferous A period of *Palaeozoic time, stretching approximately from 345 to 280 million years BP, and subdivided into the Mississippian and the Pennsylvanian periods.

(⊕) SEE WEB LINKS
• An overview of the Carboniferous period.

carbon management The monitoring of carbon emissions and their removal through **carbon sequestration**; that is, the storage of carbon dioxide. Smith et al. (2000) *Glob. Change Biol.* 6 believe that energy crops have the

greatest potential for carbon sequestration in Europe, and Bourque et al. (2007) *Mitigation & Adapt. Stats. Glob. Change* 12, 7 provide a methodology for forestry and carbon sequestration.

CO_2 may also be stored underground (see Holt et al. (1995) *Energy Cons. & Manage.* 36) or below the ocean *thermocline. Hoffert et al. (2000) *Science* 298 discuss CO_2 sequestration in deep seas, and Herzog et al. (2000) *Sci. Am.* 282 believe that ocean injections can substantially decrease peak atmospheric CO_2 (although some CO_2 would return to the atmosphere (Kheshgi et al. (1994) *Energy* 19, 967). See Harvey (2003) *Geophys. Res. Lett.* 30, 5 on the impact of deep-ocean carbon sequestration on atmospheric CO_2. The pH impacts of ocean CO_2 disposal could be diminished by accelerating carbonate mineral weathering (Caldeira and Rau (2000) *Geophys. Res. Lett.* 27). Harvey (2004) *Climatic Change* 63, 4 warns that the impact of carbon sequestering on atmospheric CO_2 rapidly decreases over time, as there is a reduction in the rate of absorption of atmospheric CO_2 by the terrestrial biosphere and the oceans.

carbon offset A reduction in emissions of carbon dioxide or greenhouse gases made in order to recompense or to offset for an emission made elsewhere. A **carbon tax** is a compulsory monetary payment imposed by governments. It is designed to reduce the *carbon footprint of commodity producers or consumers.

carbon sink A phenomenon, such as a forest or ocean, which can absorb atmospheric carbon dioxide. Palmroth et al. (2006) *Procs Nat. Acad. Sci. US* 103 report that the above-ground sink strength in forests controls the allocation of carbon below ground.

carceral geography Geographical perspectives on incarceration (imprisonment). This is a fairly new branch of geography; see D. Moran (2013).

care The provision of practical or emotional support. This definition is explored by Milligan and Wiles (2011) *PHG* 34, 6, 737–8, who note that geographers articulate care through the 'differing, and sometimes surprising, social spaces that enable caring interactions'. A **landscape of care** describes places beneficial to one's physical or mental health, such as clinics, gardens, hospitals, holy wells, temples, retreats, and spas.

　Geographies of care study the spatial dimensions of social practices of care. Williams (2002) *Soc. Sci. & Medic.* 55, 141 writes that, in the UK 'changes in health care service delivery have resulted in the transfer of care from formal spaces such as hospitals and institutions towards informal settings such as home. Due to the degree of this transfer, it is increasingly important for geographers to explore the experience and meaning of these changing geographies of care in order to reveal and understand the impact and effect on particular individuals and places'. See also the 2003 issue of *Soc. & Cult. Geog.* 4, 4.

carnivore Any animal which eats only animal flesh. At the top of a *pyramid of numbers, **top carnivores** are usually the rarest, largest, and most complex animals. With their large body size, low abundance, and large home range requirements, top carnivores are particularly vulnerable to changes in habitat (Raffaelli (2004) *Science* 306).

carrying capacity The maximum potential number of inhabitants which can be supported in a given area. The upper limit is set at the point where the environment deteriorates. Carrying capacities are far from being universal constants, but alter with value

judgements and objectives. 'In human society, institutional arrangements are likely to alter the carrying capacities and desired levels of populations, and carrying capacities in the shorter term may differ greatly from those in the longer term' (Seidl and Tisdell (1999) *Ecol. Econ.* 31, 3). See Daily and Ehrlich (1996) *Ecol. Applics* 6, 4 on increasing equality of opportunity and increasing Earth's carrying capacity.

　Seidl and Tisdell (op. cit.) maintain that carrying capacity is a political concept 'generally highlighting that exponential growth, and thus environmental pressures, have to be curbed'. Clarke (2002) *Pop. & Env.* 23, 4 recounts the difficulties in trying to convert the 'vague carrying capacity concept into a functional, qualitative method' and Benjaminsen et al. (2006) *AAAG* 96, 3 decry 'the continued dogged insistence on the application of rigid carrying capacity rules and concepts'.

cartel A system whereby producers divide up the market between themselves, avoiding direct competition and not encroaching on each other's share of the market. See Walker in R. Friman and P. Andreas, eds (1999) on drug cartels.

cartography The production and study of maps and charts. Cartography is a system of information which is used to communicate something of the real world to other people; the map is a *model, to be decoded by the map reader; see Lloyd (1989) *AAAG* 79, 1. All maps are approximations; their clean lines and colours don't reflect the muddled nature of the reality they represent, and they can easily be used to support a point of view; 'a whole arsenal of devices may be used for [persuasive cartography] . . . to influence opinion rather than to inform the reader' (N. Thrower1996). So that the reader is not distracted by 'noise'—anything which stands in the way of understanding—the map has to be encoded using easily understandable signs, symbols, lettering, and lines; see Board and Taylor (1977) *TIBG* 2, 1. F. Farinelli (1989) observed that reason 'is nothing more than the protocol of . . . the cartographic image . . . the history of geography is the history of the confusion between the model and the reality'. See J. Pickles (2004) and C. Jacob (2006).

cascade 1. A sequence of small waterfalls.

2. A configuration in which the output of one component is the input of another. Thus, the catchment hydrological cascade is: bedrock, cascade, step-pool, plane-bed, riffle-pool, and dune-ripple (Branefirun and Roulet (2002) *Hydrol. & Earth. Sys. Sci.* 6, 4). See Couthard and Van Der Wiel (2007) *Geomorph.* 91, 3–4: 'River systems can be viewed at distinctive hierarchical levels that represent a cascade of geomorphological interrelationships'.

((())) SEE WEB LINKS
• AUSRIVAS, Parsons et al. (2002).

cascading systems *See* SYSTEMS.

case hardening The production on a rock outcrop of a resistant *weathering rind mostly made of iron and magnesium hydroxides; sometimes of amorphous silica. Case hardening in limestone occurs when the calcium carbonate dissolved at the surface is re-precipitated internally, most commonly close to the surface from which it was just dissolved (Donovan (2002) *Geol. Today* 18, 4). Case hardening may be related to the production of visors over some *tafoni entrances. See R. Dorn (1998).

cash cropping Growing a crop to produce goods for sale or for barter, rather than for the subsistence of the farmer and family. Tungittiplakorn and

Dearden (2002) *Biodiv. & Conserv.* 11, 11, in a study of increased cash cropping in Thailand, find that, with reduced need for hunting, cash cropping may aid biodiversity conservation. Lim and Douglas (1998) *Asia Pac. Viewpt* 39, 3 describe the results of pressure from the Malaysian government to switch from shifting cultivation to cash cropping: land shortages, increased use of agricultural chemicals, and the failure of sustainable agriculture.

caste A position in society inherited at birth, and from which there is no transfer throughout life. The system is at its strongest in the Indian subcontinent; see Müller-Böker (1988) *Mt. Res. & Dev.* 11, 2 on caste and spatial organization in Nepal, and Ostwald and Baral (2000) *Geograf. Annal. B* 82, 3 on caste, gender, and forest protection in Orissa, India.

catastrophic In geomorphology, any process which is rare, and acts briefly with great force and suddenness. See, for example, Clayton and Knox (2008) *Geomorph.* 93, 3–4 on the catastrophic drainage of Lake Wisconsin, and Guthrie and Evans (2007) *Geomorph.* 88, 3–4. **Catastrophism** is a view that sees major changes in the physical environment as the result of catastrophic events. Although this view fell out of favour, recently it has become accepted as valuable: for example, there have been many explanations of past mass extinctions invoking catastrophic events, and many geomorphological features can be explained by recourse to catastrophic ideas. See, for example, Bintliff (2002) *World Archaeol.* 33 on catastrophism in the study of Mediterranean alluvial history.

catchment The area drained by a river. **Catchment shape** can be expressed by its circularity ratio:

$$Rc = \frac{A}{Ac}$$

where Rc is the circularity ratio, A is the area of the catchment, and Ac is the area of a circle with the same circumference as the catchment. Catchment shape is a major influence on hydrological relationships in a landscape:

$$L = 1.4 \, A^{0.6}$$

where L = the length of the stream from its highest topographical point to the river mouth, and A = the area of the catchment. See K. Fryirs and G. Brierley (2012), for more detail. Sólyom (2011 *ESPL* 36, 4, 533) finds an intimate relationship between catchment shape and channel initiation. The initiation of a channel on an initially unchannelized surface sets up the initial flow-path structure, which consequently defines the boundaries of the channel's first-order catchment.

Catchment Abstraction Management Strategies (CAMS)
A strategy by the *Environment Agency to license water abstraction in the UK.

🌐 SEE WEB LINKS
- Sets out CAMS policy.

catchment area 1. The area served by a city, or some function such as retailers, airports, schools, etc. *See* URBAN FIELD.
2. An alternative term for a *drainage basin/watershed.

categorical data analysis A range of techniques that may be used to analyse *nominal and *ordinal data. The key text is N. Wrigley (2002).

catena A sequence of soil types arising from the same parent rock, but distinct from each other because of the variations—such as *leaching and *mass movement—arising from topography.

De Alba et al. (2004) *Catena* 58, 1 present a model of catena evolution.

cation An ion or group of ions having a positive charge such as the sodium ion in salt (Na^+Cl^-). **Cation exchange** is the interchange between a positively charged cation in the soil water and a negatively charged ion on a clay surface. The most important ions held in this way are the four important plant nutrients sodium, potassium, calcium, and magnesium. **Cation exchange capacity** (CEC) is the overall net negative charge of clay minerals per unit mass of soil, usually expressed as milliequivalents (meq) per kg of oven-dried soil.

causation (causality) The relationship between two phenomena in which the first (the cause) brings about the second (the effect). A **causal relationship** usually requires: the existence of a spatial and temporal contiguity between two events, the occurrence of one event before the other, and the unlikelihood that the second event could have occurred unless the previous event occurred first. A causal relationship is also often thought to exist when a particular type of event always or almost always occurs in a particular way.

The distinction between cause and effect in the development of landforms is a function of time and space (area) because the factors that determine the character of landforms can either be dependent or independent variables as the limits of time and space change (Schumm and Lichty (1965) *Amer. J. Sci.* 263, 110). For example, over a long period, a drainage basin can be seen as an open system, losing potential energy and mass, while over shorter timespans self-regulation is important. Over a very short period, a *steady state may exist. But as Lane and Richards (1997) *ESPL* 22, 949 point out, different scales of form and process are not causally independent of

Catena

each other, as short timescale and small space-scale processes influence processes over longer timescales and larger space-scales. A **causal correlation** between two variables implies that one causes the other, but a correlation between two variables is not necessarily causal.

cave A large, natural, underground hollow, usually with a horizontal opening. **Karst caves** result from solution and *corrasion; see Miller (2006) GSA Special Paper 404.

cavitation The caving-in of bubbles in a liquid, close to a solid surface, causing shock waves. In a river, cavitation usually occurs downstream of an obstruction. It may aid the fluting and potholing of massive, unjointed rocks (Whipple et al. (2000) GSA Bull. 112, 3).

CBD See CENTRAL BUSINESS DISTRICT.

celerity In a river, the square root of the product of the acceleration due to gravity and the mean depth of the river flow. See Richardson and Carling (2006) Geomorph. 82, 1–2.

cellular automata An *agent-based model that works on a two-dimensional

Cellular automata

lattice grid, where the state of any cell is related to that of its neighbours. At its simplest, a grid of elementary cellular automata consists of a one-dimensional row of cells, where each cell can be in one of two states, and the rules for the transformation of a cell are based on the current state of the cell and its two closest neighbours. The neighbourhood thus consists of three cells [6, 7, 8].

In this example, the two states are called 'white' and 'black', and the rules might be that:

- If all three cells are white, the cell remains white
- If all three cells are black, the cell becomes white
- In any other case (that is, if there is a mixture of black and white cells in the neighbourhood), the cell becomes (or remains) black.

The diagram shows four successive generations of cellular automata conforming to these rules: an initial generation consisting of one black cell with all other cells white, and three subsequent generations created by three applications of the rules.

- *Parallelism*: The cells all change state independently and simultaneously (that is, in parallel)
- *Locality*: The new state of a cell is dependent on only the state of the cell itself and the states of its neighbours
- *Homogeneity*: All cells have an identical set of possible states and follow the same rules.

( ) SEE WEB LINKS
- Website with clear and useful explanations on cellular automata and on which the example is based.
- A variety of cellular automata can be found at this website.

cellular clouds Roughly hexagonal patches of clouds. The open cell patterns typically found behind cold fronts can be determined by a large-scale sinking motion of a convectively unstable layer (Krishnamurti (1975) *J. Atmos. Scis* 32).

cellular convection *Convection occurring in semi-regular cells. See Hsu and Sun (1991) *Boundary Layer Met.* 57, 1–2 on mesoscale cellular convection over the East China Sea.

cellular modelling Cellular models in geomorphology can be defined as representing the modelled landscape with a grid of cells, over which the development of the landscape is determined by the interactions between cells (for example, fluxes of water and sediment) using rules based on simplifications of the governing physics. In fluvial geomorphology, cellular models use simplified or 'relaxed' versions of the complex flow equations (Coulthard et al. (2007) *Geomorph.* 90, 3–4; see also Nicholas (2005) *ESPL* 30, 5).

cementation The binding together of particles by adhesive materials, as in sedimentary rocks. See Nash and Smith (2003) *Geomorph.* 50, 1–3 on calcretes.

Cenozoic (Cainozoic) The most recent *era of Earth's history stretching approximately from 65 million years BP to the present day.

((⊕)) SEE WEB LINKS
• An overview of the Cenozoic period.

census An investigation, usually into the size and nature of a population, but occasionally into other things, such as traffic. A **census tract**, known in the UK as an *enumeration district, is a small unit of area used in collecting, recording, and reporting census data. **Census geography** is the spatial organization of conducting a census and the geographical analysis and thematic mapping of census data (N. Castree, R. Kitchin, and A. Rogers 2013). In the UK, **Census Output Areas** are geographical areas consisting of about 125 households, with a population of about 300.

   Census geography design is the delineation of appropriate geographical base units from which to create and use geography; see Vickers and Rees (2007) *JR Statist. Soc. A* 170, 2 and Rees and Martin in P. Rees et al., eds (2002). The UK Office of National Statistics (ONS) offers a range of 2011 Census geography products for England and Wales.

((⊕)) SEE WEB LINKS
• For more detail on census output areas.

Center for Spatially Integrated Social Science (CSISS) A centre focusing on spatial analysis as an integrating force for the social sciences. Their website will lead you to their really useful *GIS cookbook: 'a collection of simple descriptions and illustrations of GIS methods, written with minimal GIS jargon'.

((⊕)) SEE WEB LINKS
• The CSISS website.

central business district (CBD) The heart of an urban area, usually located at the meeting point of the city's transport systems, containing a high percentage of shops and offices. High accessibility leads to high land values, and therefore intensive land use. Consequently, development is often upwards. Within the CBD, specialist areas, such as a jewellery quarter, benefit from *external economies. Vertical land-use zoning is also common, so that retail outlets may be on the ground floor, with commercial users above them and residential users higher up. See Chang (2007) *Urb. Plan. & Dev.* 133, 2 on measuring and assessing the economic activities of CBDs. The CBD is under threat from traffic restrictions, and out-of-town developments, such as superstores. 'Arguments on the future [of the CBD] range from a speculated centre-less form to one of centrality' (Wong (2004) *Land Use Policy* 21).

   Methods of delimiting the CBD include the *central business height index, recording the percentage of floor space given over to CBD functions, charting high level pedestrian flows, and surveying pavement chewing gum.

central business height index (CBHI) A measure of the intensity of land use within the *CBD. The total floor area of all storeys of a building is compared with the ground-floor area of the building under consideration. The higher the index, the more intensive the urban land use. The index can also be used to delimit the CBD. Nahm (1999) *GeoJournal* 49, 3 notes that the CBHI increased fourfold in Seoul, 1953–98.

central flow theory An interlocking network model developed at GaWC. 'In central flow theory, flows generate a network; a space of flows. Central flow theory produces a spatial network of urban settlements' (Taylor (2007) *GaWC Res. Bull.* 238). City networks are constituted by the interlocking of cities by commercial agents in the everyday course of their business practice; central flow theory entails network mutuality, and this suggests that urban policy makers should focus on cooperation with other cities.

centrality Cities dominate their locations economically, politically, and culturally, thus displaying centrality. J. Bird (2013) sees centrality as a mental projection on to space, and discusses the concept in relation to three types of its manifestation in spatial terms: the city as centre of a tributary region; the centres and central areas of cities themselves; and the city considered as a centre or gateway for other distant regions, often overseas. Lefebvre identifies the urban with the sociospatial form of centrality, but as Kipfer et al. (*PHG* (2013) 37, 1, 115) note: 'this is a tricky affair'.

centrally planned economy A synonym for a communist economy, also known as a **command economy**. The *factors of production are owned by the state, and their deployment is planned from the administrative capital. True central planning requires an enormous quantity of information, difficult to gather at the centre. See Zhou (2005) *Env. & Plan. A* 37, 6 on 'an innovative region from a centrally planned economy in Beijing'.

central place A settlement or nodal point which, by its functions, serves an area round about it for goods and services. **Central place theory** is an explanation, advanced by W. Christaller (1933, trans. 1966), of the spatial arrangement, size, and number of settlements. Christaller modelled settlement locations in a hexagonal pattern to examine and define settlement structure and the size of the hinterland. His initial assumptions were of an *isotropic plain proportional to

distance; and *economic man. Economic factors include demand density, economies of scope, localization economies, and agglomeration economies.

While Christaller's theory has waned in popularity, a number of studies use central place theory; see Daniels (2006) *Annal. Tourism Res.* 34, 2 on central place theory and sports tourism; Dennis et al. (2002) *J. Retailing & Consumer Services* 9, 4 on central place theory and UK retail hierarchy; Clarkson et al. (1996) *Int. J. Retail & Distrib. Manage.* 24, 6 on UK supermarket location; and Dean (2003) *China Qly* 174 on hierarchical space in China. See also R. Schiller (2001). Taylor (2007) *GaWC Res. Bull.* 238 takes exception to 'this simple equating of cities with hierarchy and competition'.

central vent eruption A volcanic eruption from a single vent or a cluster of centrally placed vents, fed by a single pipe-like supply channel. Lava from a central vent eruption accumulates to form a conical volcano; see Takada (1997) *Bull. Volcanol.* 58, 7.

centre of calculation The directing and controlling hub of a network; 'the [venue] in which knowledge production builds upon the accumulation of resources through circulatory movements to other places' (Jöns (2011) in J. Agnew and D. Livingstone, eds). See F. Driver (2001).

centrifugal forces In *human geography, those forces which encourage a movement of people, business, and industry away from central urban areas; these include: intra-city location costs (Anas (2004) *J. Econ. Geog.* 4), improvements in transport systems (Alonso-Vilar, *ECINEQ WP* 2006–57), the needs of spatially dispersed clients (Grote (2007) *J. Econ. Geog.* 8, 2); and knowledge spillovers (Baldwin and Forslid

k = 4 network, explanation

- • higher-order centre
- · lower-order centre
- - -→ direction and proportion of custom from lower-order centres to higher-order centres

(a) k = 7 network, explanation; (b) k = 7 network (an alternative orientation)

(a)   (b)

- • higher-order centre
- · lower-order centre
- - -→ direction of custom from lower-order centre

higher order in   higher order in
orientation (a)   this orientation

The marketing principle, k = 3, the G-system. (Top right sector shown in full detail.)

- ◉ G – place
- ⊚ B – place
- ⊙ K – place
- • A – place
- · M – place
- —— boundary of G-region
- —— boundary of B-region
- –·–·– boundary of K-region
- - - - - boundary of A-region
- ·········· boundary of M-region

Central place theory

1.

2.

3.

Central place theory (continued)

(2000) *Economica* 67). Koch (2007) *UN Univ. Res. Paper* 2007/45, lists the dispersion of demand, reputation (a new organization might takes risks that, if successful, would raise their profile), competition effects (where organizations locate away from direct competitors), and high transport costs (when a business might locate near the end-users).

centripetal acceleration In meteorology, the force acting into the centre of a high- or low-pressure system, which makes winds blow along a curved path. It is crucial in *tropical cyclone and *tornado formation; see Wurman and Gill (2000) *Monthly Weather Rev.* 128, 7.

centripetal drainage *See* DRAINAGE PATTERNS.

centripetal forces Those forces which move people, business, and industry towards a centre—accessibility, *functional linkages, *agglomeration economies, and *external economies— thus encouraging the growth of large central places. Porteous in R. Martin, ed. (1999) cites labour market externalities, intermediate services, technological and informational spillovers, and socio-institutional and cultural factors as centripetal forces. Grote (2008) *J. Econ. Geog.* 8, 2 holds that, with ever-falling transaction costs, centripetal forces decline over time.

chain migration A process which depends on a small number of pioneer migrants, who settle in a new place. A feedback process is set up, from the destination to the origin, generating higher than expected levels of migration to these pioneers' neighbourhoods. A metropolis then becomes both segmented and segregated (Skop et al. (2006) *Urb. Geog.* 27, 5). See also Johnston et al. (2007) *AAAG* 97, 4.

channel A watercourse. The nature of any river channel is a result of the interaction of: the material properties of the bed and banks; the flow hydraulics; and sediment transport within the river. Any alteration in any one of them affects the other two. Narrow, stable channel stretches are linked to finer bed material and high levels of river vegetation; wider, unstable channels are associated with larger bed material and less riparian vegetation (Rowntree and Dollar (1996) *S. Afric. Geog. J.* 78, 1).

The Smith and Bretherton (1972, *Water Resources Res.* 8, 1506) explanation of **channel initiation** is based on the continuity equation for sediment transport. Its central result is a criterion for an infinitesimal perturbation on the surface to begin to grow. See also Smith (2000) *Computers and Geoscience* 26, 541.

For more mature landscapes, channel extent is based on competition between diffusion and runoff processes: a channel is initiated once the magnitude of incising runoff processes exceeds the magnitude of infilling diffusion processes, such as creep, rainsplash, and shallow landsliding; see Tucker and Bras (1998) *Water Resources Res.* 34, 2751–64.

The **channel capacity** of a river is its cross-sectional area in square metres, but the borders of a channel are not easy to establish; see Rajaguru et al. (1995) *ESPL* 20. **Channel resistance** slows or impedes the flow; friction with the bed is the major cause. See Righetti and Armanini (2002) *J. Hydrol.* 269, 1 and Wohl (2007) *Phys. Geog.* 28, 3.

channel classification Howard et al. (1994) *J. Geophys. Res.* 99, primarily on the basis of bed morphology, define five

channel types: live bed sand alluvial; live bed gravel alluvial; threshold gravel alluvial; mixed bedrock-alluvial; and bedrock. In a study of mountain drainage basins, Montgomery and Buffington (1997) *GSA Bull.* 109, 5 distinguish seven types: colluvial and bedrock, plus five alluvial channel types (cascade, step-pool, plane-bed, pool-riffle, and dune-ripple). Heritage et al. (2001) *J. Geol.* 109 use cluster and discriminant analyses to classify: bedrock anastomosed, mixed anastomosed, pool-rapid, braided, and alluvial single-thread. These channel types fit on a continuum from bedrock-dominated channels to fully alluvial systems.

channel flow Run-off within the confines of a channel. See L. B. Leopold et al. (1994). **Compound channel flow** occurs in *meandering channels, where the magnitude and directions of the main channel and the upper layer differ, causing sizeable internal *shear stress, and a consequent loss of energy. For meandering compound channels the important parameters affecting the boundary shear distribution are sinuosity (Sr), amplitude (a), relative depth (β), and the width ratio (α) and the aspect ratio (δ); see Khatua and Patra (2007) *Procs 5th Australian Stream Manag. Conf.* Compound channel flow also occurs during floods, when velocity differs between the main channel and the flooded area. See Helmio (2002) *J. Hydrol.* 269 and Fukuoka et al. (1999) *J. Hydrosci. & Eng.* 17, 2.

channelized flow Channels incised at the glacier base, governed by a balance between viscous creep closure of the ice roof and melting due to turbulent heating. Such channels naturally tend to draw in surrounding water and therefore form branching arterial networks, aligned predominantly in the down-glacier direction (Hewitt (2011) *J. Glaciol.*

57, 202). Two types of channels are thought to exist beneath glaciers: Nye (N) channels, which are incised into the bedrock, and Röthlisberger (R) channels, which are melted upwards into the glacier ice; see J. Holden (2012), pp. 475–6.

channel management Any proposals for changing a fluvial channel should: establish past and present morphology, stability, and responsiveness to change; determine the impact of the proposed change on the sedimentary regime; estimate the effects of change in the sedimentary regime on channel morphology, and on bank pore water pressure, and therefore on bank stability. See P. Downs and K. Gregory (2004). Social and environmental conflicts of interest also arise: damming a river will strip out the finer silt, and increase the downstream concentration of coarser silt, leading to increased erosion, for example. Larsen et al. (2007) *Landscape & Urb. Plan.* 79, 3–4 demonstrate the need for taking a landscape-level, long-term view of managing a river channel.

channel order *See* STREAM ORDER.

channel roughness *See* BED ROUGHNESS.

chaos theory 'Chaos Theory deals with simple, deterministic, non-linear, dynamic, closed systems . . . [that] are extremely sensitive to initial conditions, resulting in an unpredictable chaotic response to any minute initial difference or perturbation' (Reitsma (2003) *Geoforum* 34, 1). Chaotic behaviour is not random—current conditions are still linked to future conditions—and chaotic systems do not repeat themselves exactly, but may behave in a loosely recurrent fashion. See Phillips (2003) *PPG* 27, 1.

character displacement The accentuation of differences among similar species whose distributions overlap (or

their minimization in regions where they do not). It results from natural selection; intensifying the aesthetic traits useful for species' discrimination, or for using different parts of a *niche, thus avoiding direct competition. See Montoya and Burns; Russo et al.; and Meiri et al.; all in (2007) *J. Biogeog.* 34, 12.

charter group A group representing the most typical culture of a host community. The charter group creates a system that segregates it from other 'ethno-groups', who may be gradually assimilated, but in such a way that the charter group maintains its dominance. See O. Yiftachel (1999) and Johnston et al. (2007) *AAAG* 97, 4.

chatter mark *See* PERIGLACIAL.

check dam A small, often temporary, dam constructed across a watercourse to reduce flow velocity and/or counter soil erosion. See Mustafa (2005) *AAAG* 95, 3 and Castillo et al. (2007) *Catena* 70, 3 on the effectiveness of check dams for soil erosion control.

chelate A chemical substance formed from the bonding of compounds to metallic ions. Intense podzolization in humic-iron podzols is linked to mineral chelation (Dearing et al. (2001) *Geophys. J. Int.* 144). See Siebner-Freibach et al. (2004) *Soil Sci. Soc. Am. J.* 68 on the interaction of iron chelating agents with clay minerals. The *leaching out of chelates is **cheluviation**. **Chelation**, the formation of chelates, is a process whereby relatively insoluble materials may become soluble, and be released into the soil, increasing nutrient availability to plants.

Chelford interstadial An interstadial during the Devensian glaciation, between 65 000 and 60 000 years BP. *See* PLEISTOCENE.

chemical weathering The breaking down *in situ* of rocks by *carbonation, *hydrolysis, *oxidation, *solution, and attack by organic acids. Bouchard and Jolicoeur (2000) *Geomorph.* 32, 3–4 suggest that hydrology could be the main factor controlling chemical weathering. See Muhs et al. (2001) *Soil Sci. Soc. Amer. J.* 65 on the impact of climate and parent material on chemical weathering; S. Colman and D. Derthier (1986); and Kump et al. (2000) *Ann. Rev. Earth & Planet. Scis* 28 on chemical weathering and climate stability.

chernozem A *zonal soil (US soil classification mollisol, sub-order boroll) with a deep A *horizon, rich in *humus from decomposed grass, and dark in colour. The B horizon is lighter brown, but often absent. The lower horizons are often rich in calcium compounds.

Chernozem development has been associated with temperate continental climates with marked wet and dry seasons, providing enough moisture to permit the decay of the grass litter into humus, but not enough for *leaching to be significant. However, Eckmeier et al. (2007) *Geoderma* 139, 3–4 claim that there is no consensus on the factors controlling the formation, conservation, and degradation of central European chernozems.

chestnut soil A *zonal soil (US soil classification mollisol, sub-order xeroll), found in more arid grasslands. The *xerophytic nature of much of the grassland under which chestnut soils develop retards the development of humus, and there is an accumulation of calcium carbonate in the B *horizon. Liu et al. (2007) *J. Arid Envs.* 68, 4, in a study of soils in the semi-arid steppes of China, find that chestnut soils are the most prone to soil erosion. See also Babayan and Protopopov (1997) *Eurasian Soil Sci.* 30, 10.

chevron A V-shaped, triangular, erosional microform, characteristically developed on the shallow flanks of *cuestas in arid zones. See Baas (2007) *Geomorph.* 91, 3–4 on the effect of different spatial resolutions on the 'visibility' of chevrons.

children, geographies of The study of the socially and spatially uneven impacts of economic, political, cultural, and social processes on children, and the ways in which these are experienced and responded to by individuals and groups. Matthews and Limb (1999) *PHG* 23, 61 propose an agenda highlighting different aspects of children's relationship with their physical and built environment, beyond the home, school, and playground. They emphasize the experiences of children and how they see the world around them. Jeffrey (2009) *PHG* 33, 19 offers a comparative perspective on some of the challenges facing children and young people across the world, focusing especially on young people's practices in the fields of education and employment, as well as discussing conceptual frameworks for analysing children and young people's behaviour.

choice modelling A group of statistical techniques to model the way people choose between alternatives, for example, where to retire. Duncombe et al. (2001) *Int. J. Pop. Geog.* 7 offer a mathematical but clear exposition of the method, together with some software options.

(🌐) **SEE WEB LINKS**
• A full explanation of choice modelling on pdf at Rand Europe's website.

chorology The study of the variation from place to place on the surface of the Earth; as such it is the oldest form of geography.

choropleth A map showing the distribution of a phenomenon by graded shading to indicate the density per unit area of that phenomenon; the greater the density of shading, the greater the density in reality. Choropleths give a clear, but generalized, picture of distribution, but may mask finer details. The choice of values for the classes, and the areal units for which data are available, will affect the visual picture; *see* SCALE. See Goldsberry et al. (2004) *Procs Annual Meeting NACIS* on classification choice and animated choropleths; see also Murray in P. Forer et al., eds (2007) on geographic space and choropleth display.

chronosequence A soil chronosequence is a group of soils whose properties vary primarily as a function of age variability (V. Holliday 2004). See Sauer et al. (2007) *Catena* 71, 3.

chronotope At its most simple, the chronotope is the spatio-temporal 'setting' of a given narrative. Thus, the genre 'Western' in cinema 'does not just indicate a film with a "wild west" setting in the late nineteenth century, a certain cast of characters, typical locations, and a plot following certain conventions. It also inscribes a certain set of morals and mores': Chanan (2000) *Jump Cut*, 43, 56. See Lawson (2011) *Antipode* 43, 2, 384.

c.i.f. pricing (cost, insurance, freight pricing) A form of uniform delivered pricing where prices are quoted with reference to a port; see Fan et al. (2000) *J. Reg. Sci.* 40, 4.

cinder cones Also known as scoria cones, these are composite volcanoes, generally less than 300 m high, with straight, steep sides and very large summit craters. They are made of *tephra shot through with bubbles that were originally filled with volcanic gases. See Cascadden et al. (1997) *New Mexico Bureau of Mines, Bull.* 156.

circuit of capital Money *capital is needed to pay for *factors of production, such as raw materials, plant, energy, and labour. Commodities may then be produced which are more valuable than the cost of all the inputs; surplus value has thus been added. The commodities are sold (turned back into money) so that surplus value leads to reinvestment. *Accumulation occurs with each circuit of capital, and the spatial integration of different productive nodes within society gives concrete substance to the different moments of the money circuit of capital (De Angelis (1998) *Conf. Paper, Eur. Ass. Evol. Pol. Econ.*). As the circuit of capital expands into effective markets, the circuit is diversified (Lee (2006) *TIGB* 31, 4). If the circuit of capital is broken, or if there is over-accumulation, capitalists might 'switch circuits'; see T. Hall (2001). D. Harvey (1996) discusses the different temporalities of short-term financial, medium-term industrial, and long-term infrastructural capital, and the pressures and distortions that these can create within the overall circuit of capital.

circulation 1. The movement of capital, labour, goods, and services through an economy. See Hughes on alternative forms of economic organization (2005, *PHG* 29, 4); on transnational trade and governance (2006, *PHG* 30, 5); and on the circulation of knowledge (2007, *PHG* 31, 4).
   2. In population geography, short-term, repetitive movements of individuals, where there is no intention to change residence permanently. Adepojou in P. Kok (2006) observes that, in southern Africa, poverty has drawn more people into circulation or temporary migration, and notes that 'brain drain' is being transformed into 'brain circulation'. See also Curry and Kobarski (1998) *Asia Pac. Viewpt* 39 on circulation in Papua New Guinea.

circumpolar vortex A vigorous, westerly flow of air in the upper and middle *troposphere over mid-latitudes. Frauenfeld and Davis (2003) *J. Geophys. Res.* 108 think the Northern Hemisphere circumpolar vortex could be a good indicator of climate change. 'During the summer–fall season, the trend towards stronger circumpolar flow [in the Southern Hemisphere] has contributed substantially to the observed warming over the Antarctic Peninsula and Patagonia, and to the cooling over eastern Antarctica and the Antarctic plateau' (Thompson and Solomon (2002) *Science* 296).

cirque Also known as a **corrie** or **cwm**, this is a circular, armchair-shaped, hollow cut into bedrock during glaciation, up to 2 km across. The steep side and back walls are subject to intense physical weathering (Matsuoka and Sakai (1999) *Geomorph.* 74, 1–4); the front opens out downslope.

The formation of cirques remains unclear. Haynes (1968) *Geografiska A* 50, 221–34 concludes that bedrock jointing determines the overall cirque morphology, and Hooke (1991) *GSA Bull.* 103, 1104–80 finds that periglacial freeze–thaw weathering and glacial quarrying facilitate erosion of the headwall. Turnbull and Davies (2006) *ESPL* 31, 9 show that cirques can be initiated by large, deep-seated failures of rock slope. Fluvial erosion and large-scale slope failures can be important for the creation of hollows, which are then deepened during glaciations; see Wilson and Smith (2006) *Geograf. Annal. A* 88, 3.

During glaciation, ice is thought to undergo *rotational slipping, perhaps *overdeepening the cirque floor (Vilborg (1977) *Geografiska A*, 59, 3–4), although Hooke ((1991) *GSA Bull.* 103) suggests that glaciohydraulic supercooling is central to overdeepening. (See also the

classic paper by Lewis (1938) *Geol. Mag.* 75 on a meltwater hypothesis of cirque formation.) Cirques seem to grow by headward extension, biting back into the mountain mass until only *arêtes or *pyramidal peaks remain; see Oskin and Burbank (2005) *Geology* 33, 12 on cirque retreat.

Aspect plays an important role in the orientation of cirques: in the Fuegian Andes, for example, the cirques which are oriented towards the south-east are the most abundant. Glacial ice is preserved in these due to the limited exposure to sunlight, or the effects of humid air masses (Planas et al. (2002) *Geomorph.* 87, 1). Evans (2006) *Geomorph.* 80, 3–4 considers the effects of geology, relief, and region on the *allometric development of cirques in Wales. Gordon (1977) *Geograf. Annal. A* 59 shows that, in general, the larger a cirque, the longer the period of glacial occupancy.

citizenship 'The process of political engagement between diverse groups and individuals . . . urban citizenship involves a politics of connection within, and beyond, particular cities' (Painter (2005) *ICRRDS, U. Durham*). The liberal/individualist approach to citizenship stresses the importance of individual rights; the communitarian approach emphasizes the duties of a citizen towards fellow members of the community they belong to (H. van Gunsteren1998). Radical pluralists view the community critically; hence, they uphold policies that maintain differences; social differentiation without exclusion, variety, and publicity (Ganapati (2008) *J. Plan. Ed. & Res.* 27). Mansvelt (2008) *PHG* 32, 1 has it that citizenship is produced through consumption practice 'with specific commodities and the social practices surrounding them'. See Desforges et al. (2005) *Citizenship Studs* 9, 5 for a wide-ranging review of geographies of citizenship. *See* RIGHT TO THE CITY.

city A large urban centre, functioning as a central place, and providing specialized goods and services. There is no worldwide, or even European, agreement over limiting figures of population size or areal extent for a city. Indeed, A. Amin and N. Thrift (2002) note that 'the traditional divide between the city and the countryside has been perforated'. Geographies of the city include: the politics of public space (D'Arcus (2004) *Space & Polity* 8, 3), and how the city is experienced by different urban dwellers: 'urban space equals difference, not merely coexisting differences and different perspectives, but contesting differences and unequal power relations' (Cupers (2005) *Int. J. Urb. & Reg. Res.* 29, 4). I. Sagan and M. Czepczyński, (2004) discuss the quality of urban life; T. Hall and M. Miles (2003) write on urban futures; see Jones and Evans(2006) *Urb. Studs* 43, 9 and Bell and Jayne (2003) *Local Econ.* 18, 2 on urban regeneration (Atkinson (2003) *Urb. Studs* 40 complains of the blandness of regenerated cities). Glaeser et al. (2001) *J. Econ. Geog.* 1, 1 argue that the critical urban amenities are: a rich variety of services and consumer goods; aesthetics and physical setting; weather; and ease of movement within the city. See Bell (2007) *PHG* 31, 1 on the **hospitable city**, and Hannigan (2007) *Geog. Compass* 1, 4 on **casino cities**, which 'elevate consumption and spectatorship at the expense of economic and social equality'.

A city may drive capital accumulation (Martin and Ottaviano, *CEPII W. Papers* 1996–14); Glückler (2007) *Reg. Studs* 14, 1 sees cities as business development nodes. A city may be a creative field: R. Florida (2005) argues that urban fortunes increasingly turn on the capacity to 'attract, retain and even pamper a mobile and finicky class of "creatives"'. A city may be a multilingual and multi-

ethnic community expressed through its infrastructure, public spaces, and festivals; Podmore (2006) *Soc. Cult. Geog.* 7, 4 observes that gay men produce clearly visible territorial enclaves in inner-city areas. Jouve (2007) *Env. & Plan. C* 25, 3 perceives towns and cities as political spaces, 'in which relations are redefined through a play of confrontation and between social groups'. Amin (2006) *Urb. Studs* 43, 5–6 sees cities as endless inhabited sprawls: 'the contemporary city bears little resemblance to imaginings of the times when urbanism stood for citizenship, the ideal republic, good government, civic behaviour and the ideal public sphere.'

Painter (2005, *ICRRDS Background Paper*) sees access to public space as 'fundamental to the *right to the city'; the core elements of which are 'the promotion of equal access to the potential benefits of the city for all urban dwellers, democratic participation of all inhabitants in decision-making processes and [the] realization of inhabitants' fundamental rights and liberties' (Colin (2005) UNESCO). See Reynolds (2004) *Econ. Geo.* 80, 4 on social justice and the right to the city, and Amin (2006) *Urb. Studs* 43, 5–6 on the 'Good City'.

city marketing The promotion of a city, which aims to attract certain activities. City marketing starts with a thorough analysis of the city's current situation through extensive research on the city's assets, opportunities, and audiences. 'The second step is identifying and choosing a certain vision for the city and the goals that could be achieved with the cooperation and agreement of as broad a range of stakeholders as possible. Then comes the phase of planning specific projects that will collectively achieve the goals set and allocating clear roles for the participating bodies. This is followed by the phase of active implementation of city marketing measures that can be spatial/functional, financial, organizational, and promotional. The process ends with monitoring and regularly evaluating the results of all activities' (Kavaratzis (2007) *Geog. Compass* 1, 3, 695).

city network Taylor (2007) *TIBG* 32, 2 argues that cities, in promoting commerce, efficiency, and innovation, create 'spaces of flows, economic chains, and networks', which are central to economic expansion.

city of refuge San Francisco's City of Refuge ordinance prevents local police from collecting information about a person's immigration status if they have not been charged with any crime (Ridgley (2007) *Urb. Geog.* 29, 1).

city region The area around a city which serves, and is served by, the city (also called *umland or urban field). See Knight (2006) *Gov. Office for NE* on developing the Tyne and Wear city region; for the theoretical background, see Harrison (2007) *J. Econ. Geog.* 7, 3.

city-size distribution The frequency with which the settlements of a country or region occur in certain, arbitrarily defined, population size groups. Generally, large settlements are less frequent than small ones, but when size is graphed against frequency, several patterns may be recognized. Iyer (2003) *Eurasian Geog. & Econ.* 44, 5 documents the change from a relatively uniform city-size distribution 'in accordance to Marxist ideals' to increasing evenness in post-Soviet Russia. See Overman and Ioannides (2001) *J. Urb. Econ.* 49, 3.

city/state relations J. Jacobs (1992) identifies two opposing 'moral syndromes' which give rise to different social spaces, spaces of flows, and spaces of places: cities are seen as commercial, honest, cooperative, enterprising, and

City-size distribution

conscientious; states are seen as guardians, supporting loyalty, tradition, ostentation, exclusion, and, when necessary, vengeance. Taylor (2007) *TIBG* 32, 2 concludes that the **city-state** is not appropriate in a globalizing world, since the guardian functions of the city-state impinge too greatly on economic expansion.

civil society A much-debated term to describe a mix of value-driven organizations—voluntary and community organizations, charities, social enterprises, cooperatives and mutual organizations—which aim to promote social, environmental, or cultural objectives; 'the arena of collective social interaction situated between the state, market, and household' (McIlwaine (2009) in R. Kitchin and N. Thrift).

clade In biogeography, a grouping that includes a common ancestor and all the descendants (living and extinct) of that ancestor.

class Within a society, a set of people who are of the same economic position, and who may share the same tastes and social status. In capitalist societies, class is defined by socio-economic status, but *post-Fordism has created a new division: specialist, skilled workers, in highly paid positions with a large degree of self-determination, and low-paid workers with few skills, working in poor conditions with little or no security. This pattern is common in service employment, but is exacerbated among women (McDowell (2006) *Antipode* 38, 4; Ward et al. (2007) *Geoforum* 38, 2).

Bondi and Christie (2000) *Geoforum* 31, 3 observe that 'the spatial mobility of middle-class lifestyles are taken for granted in ways that sharply remind those struggling to participate in British society of their precarious hold on, if not their exclusion from, the mainstream . . . This self-consciousness about social position operates in relation to gender as much as class.' Lekhi (2000) *New Pol. Sci.* 22, 3 maintains that class is 'a process that is, by necessity, only ever evolving and never fully constituted'. 'Working classness is seen as a dynamic and relational category which is simultaneously economic and cultural. It rests not only on the material and labour market position of individuals, households and communities, but also on symbolic value and cultural practice, intertwining a number of interpretations of class position and class subjectivity' (Stenning (2008) *Antipode* 40, 1).

classification It is important to remember that 'classification is not just a neutral activity' (R. Inkpen, 2005). The classifier looks for groups of like phenomena/objects that can be treated as a single unit for the purpose of making valid generalizations about aspects. Clearly, many different classifications are possible, depending on the purpose of the study being undertaken: 'streams, for example, form a single class, but they can be subdivided on a variety of criteria, such as order, depth, gradient, and so on. The purpose of classification procedures is to provide a grouping which is valid for the scientific activity being undertaken: criteria for classification

must first be determined, and then the objects measured on these must be allocated to classes' (R. J. Johnston, 1976).

Geography uses a range of classificatory systems: **intrinsic classification** depends on natural differences or 'breaks' in the features studied; **extrinsic classification** uses arbitrarily defined class limits; **monothetic classification** uses one criterion; **polythetic classification** uses a number of criteria; **attribute-based classification** is based on 'present' or 'absent' evidence (a climate may or may not have a dry season); **variable-based classification** forms classes on the basis of the degree to which a variable is present.

class-monopoly rent The increases in rents that landlords, as a class, can charge when they have a monopoly over the supply of land. 'Landowners can be treated as a *class* of actors. While these actors—developers, landlords, homeowners, and financial institutions—may act individually, they are compelled, in aggregate, by the same competitive profit maximizing imperatives under capitalism. It is this "coordinating" behaviour that allows these agents to be treated as a *class* based on their collective position as individual landowners. While not necessarily acting in collusion, the profit imperative renders collusion not necessary as similar outcomes are achieved: it is simultaneously individualized and generalized in advanced capitalist societies.' (Anderson (2014) *Geog. Compass* 8, 1, 13).

clast A rock fragment, eroded from a rock mass, and deposited in a new location. See Evans et al. (2007) *Geograf. Annal.* 89, 2 on clast microfabrics; Bourke et al. (2007) *Lunar & Plan. Sci. Conf.* 38 on fluvial clasts; and Clark (1995) *Geology* 19, 2 on **clast pavements**.

clay Mineral particles < 0.002 mm. When dry, clay is hard; when wet it swells and becomes pliable and sticky. **Clay colloids** are finely divided clays, dispersed in water, with a negative surface charge that attracts positively charged ions; they are among the most reactive constituents of a soil (W. van Olphen 1991). **Clay micelles** are individual particles, platey in form, < 2 μm in diameter, having a negative charge, and therefore able to attract cations within a soil; try Murrmann and Koutz in S. Reed, ed. (1972). The **clay-humus complex** is a mixture of clay particles and decaying organic material that attracts and holds the *cations of soluble salts within the soil profile.

clay minerals A group of hydrous aluminium silicates, such as kaolinite, created by intense rock weathering; see Egli et al. (2002) *Clay Minerals* 37, 2. Clay minerals affect the structure of soils, as they expand when wet (Barshad (1952) *Clays & Clay Min.* 1).

clearinghouse In *GIS, a network of the managers, producers, and users of data. The US National Geospatial Data Clearinghouse is a means of publishing descriptions of geospatial data from a variety of governmental and non-governmental sources, helping users determine their fitness for use, and providing tools to access, visualize, or order data as efficiently as possible. It is neither a central repository where data sets are stored nor a set of websites, but a federated system of compatible geospatial data catalogues that can be searched through a common interface—the geodata.gov portal.

🌐 SEE WEB LINKS
• The geodata.gov portal.

cleavage 1. A division in society due to political or partisan allegiance; see Kitschelt (1992) *Pol. & Soc.* 20, 1 on cleavage development in post-communist eastern Europe, and Secor (2001) *Polit. Geog.* 20, 5 on socio-political cleavages in Turkey.
2. The ability of a rock to split along a **cleavage plane**.

cliff A steep rock face, usually facing the sea; see Lee et al. (2001) *Geomorph.* 40, 3–4 on probabilistic models for different cliffs. While an **active cliff** is still subject to the forces of marine erosion, an **abandoned cliff** is protected from wave attack by a *wave-cut platform or by a *barrier beach. As the abandoned cliff is exposed to *subaerial denudation it becomes less steep, and its upper edge more indented. See Hutchinson (1984) *DTIC* ADA144040.

climate A summary of mean weather conditions over a time period, usually based on thirty years of records. Climates are largely determined by location with respect to land- and sea-masses, to large-scale patterns in the *general circulation of the atmosphere, *latitude, altitude, and to local geographical features. Safford provides a summary of the role of radiation in climate (1999, *J. Biogeog.* 26, 4), and of the regional atmospheric circulation (*J. Biogeog.* 26, 3). Feedbacks between land surface and atmosphere are also important determinants of climate (Knorr and Schnitzler (2006) *Clim. Dynam.* 26; G. Bonan 2002). Gallup et al. (1999) *Int. Reg. Sci. Rev.* 22, 2 argue that climate has large effects on income levels and income growth.

climate sensitivity The relationship between the measure of climate forcing and the magnitude of the climate change; see Tomassini et al. (2007) *J. Climate* 20, 7 and Cash (2007) *Tellus A* 59, 2 on the origin of climate sensitivity differences, defined by Denby et al. (2002) *Tellus A* 54, 5 as the change in climate brought about by increasing radiative forcing (radiation). See Voigt et al. (2007) *Glob. Change Biol.* 13, 8 on trophic climate sensitivity.

climatic change During the last 55 million years the Earth has been cooling; during the last million years there have been alternating *glacials and *interglacials. *See* LITTLE ICE AGE.

External causes of climatic change include: changes in solar output (Haigh (2001) *Science* 294, 5549); changes in the number of sunspots, which seem to have an eleven-year cycle (Foukal et al. (2004) *Science* 306, 5693); changes in the ellipticity of the Earth's orbit, which follow a 100000-year cycle (Goldsmith (March 2007) *Natural History*); and changes in the Earth's axis of rotation, which alters the season of perihelion, and which follow roughly a 100000-year cycle (Zachos et al. (2001) *Science* 292, 5517). *See* MILANKOVITCH CYCLES.

Internal causes include changes in the distribution of land and sea, *continental drift, and changes in the atmosphere–surface–ocean system. 'Although vulcanicity is seen as a possible mechanism of climatic change, under present rates of eruptions it is difficult to envisage volcanic activity acting independently of other factors in causing significant and lasting global climate changes' (Wyrwoll and McConchie (1986) *Climate Change* 8, 3).

Nyberg et al. (2007) *Nature* 447 report that the average frequency of major hurricanes decreased from the 1760s until the early 1990s, and that the phase of enhanced hurricane activity since 1995 appears to be a recovery to normal hurricane activity, rather than a direct response to increasing sea surface temperatures. Nicholls and Alexander (2007) *PPG* 31, 1 find a clear pattern of increasing warm extremes and decreasing cold extremes in global climates since the early 1990s. A. E. Dressler and E. A Parson (2006) summarize the key scientific literature on climate change.

See Comrie (2007) *Geog. Compass* 1, 3 on climate change and human health. *See also* GREENHOUSE EFFECT.

climatic climax community *See* CLIMAX COMMUNITY.

climatic geomorphology (climatomorphology) The association of landform types with different climates, such as the evolution of *periglacial landforms in *tundra climates. However, the geomorphic response to climatic changes, fluctuations, and episodes varies spatially and temporally. Some recent landforms result from the long-term and essentially continuous activity of particular processes. Others are surface manifestations of processes that hitherto were active but cryptic and yet others result from the impact of storm or catastrophic events, frequently, though not in every instance, on surfaces previously rendered vulnerable. 'Few, if any, minor landforms can be accepted without question as indicators of climatic change or fluctuation' (Twidale (1997) *Geomorph.* 19). See J. Büdel (1982), M. Gutierrez Elorza (2005), and Gunnell (2000) *Geomorph.* 35, 1–2.

climatology The study of the origins and impacts of climates. For cultural climatology, *see* CULTURAL TURN.

climax community An *ecosystem which experiences a turnover of species, but with an overall steady state. This steady-state ecosystem is characterized by a complex, highly integrated community structure with high species diversity, but relatively low individual population densities not subject to serious fluctuation, and can be seen as the final stage of a *succession. An example is a mature oak woodland. 'Equilibrium or climax communities are rarely affected by intense natural disturbances and are characterised by a predominance of larger, long-lived and often deepburrowing species. Such species can physically modify their surroundings' (Kröger et al. (2006) *Estuarine, Coastal & Shelf Sci.* 67). See Ren et al. (2007) *Acta Ecologica Sinica* 27, 9 on the natural restoration of a degraded rangeland.

climograph A line graph of monthly average temperature against average humidity. The shape and location of line indicates the nature of climate. See R. G. Bailey (1998).

closed system A system marked by clear boundaries that prohibit the movement of energy across them. *Entropy in a closed system can never decrease. See R. Chorley (1964).

closure The delimitation of, or setting boundaries to, an enquiry. Massey (1999) *TIBG* 24 lists the methods of closure commonly used by geographers: oversimplification; bounding a region by process, rather than space; and limiting an enquiry to a specified time period. 'The process of closing research must address the motives behind research topics and questions; the reasons for preferring a particular methodology in addressing a given set of research questions, including the theoretical and practical issues raised by closure; and the importance of place in shaping our understandings of reality' (Brown (2004) *TIBG* 29, 3). The choice of time–space model is also significant in closure; see Lane (2001) *TIBG* 26, 2.

cloud A visible, dense mass of water droplets and/or ice crystals, suspended in the air, and generally forming when air is forced to rise: at a *front, over mountains, or because of *convection. Clouds mirror atmospheric processes; the approach and passage of a warm front, for example, often follows the sequence: cirrus, cirro-stratus, alto-stratus, nimbo-stratus. At active *ana-fronts these clouds may take on a more

cumulus form. Atmospheric *convection currents are generally indicated by the presence of cumulus or cumulo-nimbus clouds. A cumulus cloud will often form over a heated surface and then shift with the wind, so that further cumulus is formed over the same spot; see *cloud street. *Turbulence is a common cause of stratus cloud, which is often trapped beneath an *inversion; it also creates a nearly continuous sheet of strato-cumulus cloud. See C. D. Ahrens (2000).

cloud classification Clouds may be classified by form and by height.

1. Low-level clouds (0–2 km above sea level): stratus, S—extensive, shallow cloud sheet, often yielding *drizzle/light rain; strato-cumulus, Sc—shallow cloud sheet, in roughly recurring *cumuliform masses, often yielding drizzle/snow; cumulus, Cu—separate, hill-shaped clouds with flat, and often level, bases, often at the same height; cumulo-nimbus, Cb—large, high *cumulus, with tops often formed of ice crystals, dark bases, and often showery.

2. Medium-level clouds (2–4 km above mean sea level): alto-cumulus, Ac—shallow cloud sheet broken into roughly regular, rounded clouds; alto-stratus, As—featureless, thin, translucent cloud sheet; nimbus, Ns—extensive, very dark cloud sheet, usually yielding precipitation.

3. High clouds (tropical regions 6–18 km high, temperate regions 5–14 km high, polar regions 3–8 km high): cirrus, Ci—separate, white, feather-like clouds; cirro-cumulus, Cc—shallow, more or less regular patches or ripples of cloud; cirro-stratus, Cs—shallow sheet of largely translucent cloud.

Suffixes may be added: *capillatus*—like a feather, or thread; *congest*—growing rapidly, in cauliflower form; *fractus*—broken or ragged; *humilis*—shallow; *lenticularis*—like a lens, especially of alto-cumulus, cirro-cumulus, and strato-cumulus; *radiates*—banded. See Li et al. (2007) *Remote Sens. Env.* 108 on cloud classification.

cloud droplet *See* PRECIPITATION.

cloud seeding Any technique of inducing rain from a cloud, usually by dropping on it crystals of dry ice (frozen CO_2), or silver nitrate, as *condensation nuclei. Ćurić et al. (2007) *Met. & Atmos. Phys.* 95, 3–4 find that accumulated hail and rain precipitation at the surface are significantly reduced if seeding is at an early stage of cloud development.

cloud street Parallel lines of small cumulus: irregular over land, downwind from a sunny slope, and regular over warm seas. The cloud-free areas come from the mixing of saturated moist cloud air with overlying warmer, drier air: 'an individual cloud is capable of drying out a large area of the surrounding layer through cloud-top entrainment and thus is able to produce significant gaps between the clouds' (Chlond (1992) *Boundary-Layer Met.* 58, 1–2).

Club of Rome A society that aims to understand the workings of the world as a finite system, and to suggest alternative options for meeting critical needs. Concerns have included inequality on global and regional scales, pollution, unemployment, and inflation. Its *Limits to Growth* (1972) argued the necessity of slowing down population growth in order to ward off *Malthusian checks.

(⊕) SEE WEB LINKS
• Club of Rome website.

cluster A geographic concentration of interconnected companies, specialized suppliers, service providers, associated institutions, and firms in related industries (Porter (1998) *Harvard Bus. Rev.* 76, 6). Ketels in (2004) *Innovative City and Business Regions*, 3 identifies the

critical characteristics of clusters as: proximity—firms need to be sufficiently close in space to allow spillovers and the sharing of common resources; linkages; some level of active interaction; and critical mass—there need to be enough participants for the interactions to have a meaningful impact. B. Asheim (2005) adds localization economies and specialization to this list, and Storper and Venables (2004) *J. Econ. Geog.* 4, 4 stress the importance of face-to-face contact. Florida (2003) *City & Community* 2, 1 sees 'concentrations of talented people who power innovation and economic growth' as vital. Glaeser, in G. Clark (2003) argues that places with higher levels of human capital are more innovative and grow more rapidly and robustly over time; see also Cushing (2001) *W. Paper, U. Texas at Austin*. Giuliani (2007) *J. Econ. Geog.* 7, 2 stresses the importance of embeddedness in local business networks. Bathelt (2005) *Reg. Studs* 39 argues that clusters can only create new knowledge if the cluster firms have linkages with external markets and employ a mix of local and non-local transactions. Clustering and globalization are closely intertwined; see M. Fujita et al. (2001)

Maskell and Malmberg (2007) *J. Econ. Geog.* 7, 5 conceive clusters in evolutionary terms: the growing cluster is made up of firms with similar competencies, reducing expenditure on locational search costs, and enjoying *agglomeration economies. Further growth results from *cumulative causation, the presence of support institutions, and the power of the name of the locality as an attraction to a worker: the benefits of being physically located within, for example, Silicon Valley are perceived to be worth the costs of relocation. See Delgado, Porter, and Stern (2012) *NBER* Working paper series, #18250 on clusters, convergence, and economic performance. Sadler (2004)

Reg. Studs 38 argues that an effective regional cluster policy needs to identify clusters on a number of variables, including knowledge flows. The use of the cluster model as a paradigm of economic development has been criticized. Bathelt and Taylor (2002) *Geografiska B* 84, 2 argue that clusters 'depend on specific circumstances in "time–space" and, because of their very transience and specificity, those conditions might be very difficult if not impossible to create through the blunt instruments of policy'; indeed, Huber (2012) *J. Econ. Geog.* 12, 1, 107 casts doubt on the benefits of clustering.

cluster analysis The assignment of a set of objects into groups so that the objects in the same cluster are more similar (in some sense or another) to each other than to those in other clusters. Cluster analysis is used when the researcher does not know the number of groups in advance but wishes to establish groups and then analyse group membership. For example, if the term 'geomorphological processes' is entered onto a search engine, there will be nearly 400000 results. Careful study reveals that these could be clustered into (at least): fluvial, aeolian, hillslope, glacial, tectonic, igneous, and biological processes. Kent (2006) *PPG* 30, 3 reviews recent changing patterns in the use of cluster analysis, and Mohseni Saravi et al. (2010, *PPG* 34, 2) use cluster analysis to delimit homogenous hydrological regions. For similarity analysis/minimum variance, see Ward (1963) *Am. Stat. Ass. J.* 5; for unweighted pairs group average, see Williams et al. (1966) *J. Ecol.* 54. For ordination techniques/ detrenched correspondence analysis, see M. O. Hill (1979); for non-metric multidimensional scaling, see T. Cox and M. Cox (2000). CANOCO software does most multivariate techniques.

coalescence theory *Bergeron–Findeisen's theory cannot explain the formation of all tropical rainfall, since ice crystals are often absent in tropical clouds. Langmuir's coalescence theory (1974) suggests that the small droplets in clouds grow larger by coalescence until they are heavy enough to fall. As they fall, they collide with other droplets, growing more. (Not every collision results in coalescence.) The deeper the cloud, the more the drops grow; up to about 5 mm in diameter. See Xue and Wang (2007) *J. Atmos. Scis* 31, 15.

coastal classification Early coastal classification schemes were based on sea-level variations. Thus, submerged coasts are drowned river and glacial valleys (rias and fjords), while emerged coasts are characterized by coastal plains. A second type of classification distinguishes between primary coasts, which result mainly from non-marine processes and include drowned river valleys and deltas, and secondary coasts, mainly formed from marine processes and/or marine organisms, for example, barrier coasts, coral reefs, and mangrove coasts.

coastal dune A ridge or hill which forms when marine deposits of sand are blown to the back of the beach. 'Dune initiation is often site-specific and dependent on the mechanisms by which sand can be delivered from the shoreface to points where it will be reworked into available accommodation space' (Aagaard et al. (2007) *Geomorph.* 83, 1–2).

coastal flooding The inundation of coastal areas may be caused by *storm surges, *tsunami (Dawson and Stewart (2007) *PPG* 31, 6), or linked to *El Niño (Tingt et al. (2007) *Geol. Today* 23, 6). D. Sobien and C. Paxton (1998) SR/SSD 98-2 provide a primer on coastal flooding. See Carlson et al. (2008) *Ground*

Water 46, 1 on the poisoning of aquifers by the inundation of salt water from storm surges. Nunn (2007) *Geog. Res.* 45, 2 points out that coastal flooding is often destructive at the time, but, 'as many delta dwellers know', not always so thereafter.

coastal geomorphology Stephenson (2006) *Prog. Phys. Geog.* 30, 1 provides an overview of recent coastal research, and M. Schwartz ed. (2005) is recommended reading.

cobble A stone, 60–200 mm, partly or wholly rounded by wave action or running water.

cockpit karst (kegelkarst) A landscape of star-shaped hollows surrounded by steep, rounded hills, found in tropical *karst. The cockpits, up to 100 m deep, usually containing a *streamsink, are the hollows (*dolines) formed by the solution of limestone, and now floored with alluvium. See Fleurant et al. (2006) *J. Geol. Soc.* special issue.

co-constitution The conjoint—joined together—development of two or more phenomena. See Eriksson (2013) *Asian Geogr.* 30, 2 on the co-constitution of high technology and authoritarian politics in Indonesia.

coefficient of localization A measurement of the degree of concentration of a given phenomenon, such as industry, over a set of regions (also called the **index of concentration**). Values of the coefficient, L, lie between 0 (even distribution) and 1 (extreme concentration). Lafourcade and Mion (2007) *Reg. Sci. & Urb. Econ.* 37, 1 use the coefficient to study the geographic distribution of manufacturing and the size of plants.

cognitive dissonance A mismatch between what is perceived and what is (between cognition and reality), so that

an individual may seem to act irrationally. 'Cognitive dissonance theory argues that individuals strive to minimize cognitive discomfort through fostering behavioural consistency' (Hobson (2006) *Area* 38, 3 on recycling). See Lundholm and Malmberg (2006) *Geografiska B* 88, 1.

cognitive geography Cognition is knowledge and knowing by sentient entities (including humans, non-human animals, and artificially intelligent machines), and cognitive geography is the study of cognition, primarily human cognition, about space, place, and environment. The first pages of Butcher (2012) *PHG* 36, 1, 90 review the literature. See also R Kitchin and M. Blades (2001).

cognitive mapping The acquisition, coding, storage, manipulation, and recall of spatial information within the mind (also called mental mapping). Cognitive mapping simplifies the complexity of the landscape, and a mental map influences behaviour. See R. Downs and D. Stea, eds. (2005). **Cognitive trails** represent 'the creation of knowledge by movement through the natural and intellectual environment' (Turnbull (2007) *Geog. Res.* 45, 2).

cohesion Adhesion; the force holding materials together. Cohesion is a measure of strength; a higher cohesion value results from low friction angles (Thermann et al. (2006) *IAEG* 484).

cohort A group of people who experience a significant event, such as birth or leaving school, during the same period of time. **Cohort analysis** traces the subsequent vital history of cohorts; see Frejka and Callow (2001) *Pop. & Dev. Rev.* 27, 1 on cohort reproductive patterns in low-fertility countries. A **cohort effect** is any effect associated with being a member of a group born at roughly the same time and bonded by common life experiences; Cohen and Iams (2007) *Int. Mig. Rev.* 41, 3, for example, see a US **cohort effect** in the fact that foreign-born people have lower retirement incomes than do US-born people. See Franklin and Plane (2004) *Geograf. Anal. B* 60, 1 on the contribution of the cohort effect to fertility decline in Italy. The major problem of cohort analysis is to distinguish between the effects on the cohort of getting older (age effects), of common experiences like National Health orange juice (cohort effects), and particular historical events, like a war (period effects).

Cohort fertility is the total of live births born to a particular birth or marriage group; see Lutz and Skirbekk (2005) *Pop. & Dev. Rev.* 31, 4.

col 1. In the landscape, a pass between two peaks or ridges. Landscape cols may have been formed by: the *headward erosion of a *cirque, *river capture, the beheading of *dip-slope valleys by *scarp retreat, or the localized *differential erosion of a ridge.

2. In meteorology, a narrow belt of relatively low pressure, but not a depression, between two *anticyclones. Here, isobars are few and therefore winds are slack.

cold glacier (cold-based glacier, polar glacier) A glacier with its base well below 0 °C, unlubricated by *meltwater, and therefore frozen to the bedrock. *Accumulation is slow; snow may take 150 years to turn to ice. Cold glaciers move very slowly; rates of 1–2 m per year are not uncommon (Fitzsimmons et al. (2000) *Geol. Soc. London*, special pub. 176), therefore they cause very little erosion through *abrasion, although *plucking may be reinforced. Where subglacial sediments are frozen to a cold glacier base, shear failure will occur at the bottom of the frozen zone, as high pore water pressures develop there

(Owen and Derbyshire (1993) *Zeitschrift* 76). See Glasser et al. (1998) *J. Glaciol.* 44 on Kongsvegen, Svalbard, and its role in landform genesis; see also Vaughn and White (2000) *Geology* 28 on entrainment at cold glacier beds.

cold low (cold pool, polar low) An area of low pressure and temperature in the middle *troposphere. These lows don't appear on *synoptic charts, but are important in *Arctic and *Antarctic meteorology (Serreze et al. (1989) *Meteor. & Atmos. Phys.* 51, 3–4). **Cut-off low** systems are synoptic-scale low pressure systems, formed as a result of meridional shifts of jet streams in the upper troposphere, and playing a role in exchanges between troposphere and stratosphere (Gimeno et al. (2007) *Met. Atmos. Phys.* 96).

collective farm A farming organization identified with socialist regimes. The farm is owned by the state, but leased to the members of the collective, and the workers are shareholders, not state employees. The collective is, in theory, self-governing, although the state may set production targets. See Rausing (2007) *J. Royal Anthr. Inst.* 13, 2 on the end of a collective farm in post-Soviet Estonia.

collectivism A school of thought which maintains that the *factors of production and the means of distribution should be owned by all and not by individuals who might pursue their self-interest at a cost to the state. Huang (2006) *Urb. Geog.* 27, 6 argues that gating and neighbourhood enclosure help to define a sense of collectivism in urban China.

collision margin The boundary between two continental *plates. These are hybrid margins since the two slabs of continental crust that eventually collide

a

Marginal sediments

Continent

Ocean trench

Continent

b

Mountain range

c

Mountain range

Fracture zone

——— Sea level	= = = Marginal sediments
▬▬▬ Oceanic crust	
▓▓▓ Continental crust	∿∿∿ Marginal sediments which have been faulted and metamorphosed
∷∷∷ upper mantle	

Collision margin (after J. F. Deaxy and J. M. Bird 1970) were initially separated by oceanic crust, with passive or destructive margins; when the oceanic crust is consumed, the two continental crusts collide. The lower continental plate forms a double layer of crust together with the upper plate, as in the Himalayas; see D. S. N. Raju et al. (2005).

collision theory This states that raindrops grow by colliding and coalescing with each other, especially in tropical maritime *air masses. *See also* BERGERON–FINDEISEN THEORY and COALESCENCE THEORY.

colluvium The mixture of soil and unconsolidated rock fragments deposited on, or at the foot of, a slope. See Coe et al. (2008) *Geomorph.* 96, 3–4 on colluvium and debris flows. Colluvial boulder deposits are relatively stable colluvium, consisting of linear fields of well-varnished hillslope boulders, ubiquitous on desert hillslopes (Liu and Broecker (2008) *Geomorph.* 93, 3–4).

colonial discourse theory European colonizers tended to construct the identities of colonized peoples and lands as *other: undeveloped, primitive, and immature; as homogeneous objects, rather than sources of knowledge; see Anand (2007) *New Polit. Sci.* 29, 1 on Western colonial representations of the (non-Western) other. The colonizer was represented as having a duty which entailed both financial and emotional cost (and, for many, these costs were very real).

Colonial discourse analysis critically examines the role played by these *representations in colonialism and imperialism, 'rather than focusing on texts, systems of signification, and procedures of knowledge generation . . . a fuller understanding of colonial powers is achieved by explaining colonialism's basic geographical dispossessions of the colonized' (Harris (2004) *AAAG* 96, 1). 'The city-state of Singapore is first

represented in terms of a narrative of masculine heroism' (S. Mathur 1997).

colonialism The acquisition and colonization by a nation of other territories and their peoples. This term took on a more specific meaning in the late 19th century when colonists saw it as the extension of 'civilization' from Europe to the 'inferior' peoples of 'backward' societies: 'the institution of British legal regimes . . . was underpinned by a certain type of violence that served to facilitate colonialism . . . These legal landscapes were founded . . . upon a mythological hierarchy wherein white peoples were accorded the label of civilized in contrasting them from savage, non-white (read: non-European) peoples' (Rossiter (2007) *J. Hist. Geog.* 33, 4). Colonialism also entailed colonial philanthropy: 'a new doctrine of responsibility toward the unprivileged, a doctrine which received its chief impulse from the Evangelical emphasis on the value of the human soul, and hence, of the individual' (E. Howse 1953). Colonialism is also seen as a search for raw materials, new markets, and new fields of investment; see Giordano and Matzke, and Redclift (both in (2001) *PHG* 25, 4). Characteristics of colonialism include inequality between rulers and subjects; political and legal domination by the imperial power; and the exploitation of the subject people. Many commentators link colonialism with *uneven development; see, among others, Dutt (1992) *Am. Econ. Dev.* 82, 2 on India, and K. Cox (2002).

Sometimes, but not always, colonialism was accompanied by **colonization**; that is, the physical settling of people from the imperial country. Lester (1998) *TIBG* 23, 4 argues that 'in their new colonial environment, settlers, who were initially divided by prevailing British social boundaries of class, gender, religion and ethnicity, soon faced an imperative for solidarity'. Although independence from former colonization has been achieved almost everywhere, most accept that it has been replaced by *neo-colonialism; see Dutt (1992) *World Dev.* 20, 8.

command economy A communist economic system where the state controls macroeconomic policy and entrepreneurial activity, but allows some local freedom for decisions on employment and consumption (also called a **centrally planned economy**). See D. Shaw (1999) on the Soviet command economy and its transition to capitalism.

commercial agriculture The production of agricultural goods for sale. See Keys (2004) *Land Degrad. & Dev.* 15 on the transition to commercial farming in southern Yucatán.

commercial spaces (commercial hospitality spaces) Commercial hospitality is becoming increasingly important for the branding and promoting of cities. Latham (2003) *Urb. Studs* 40 detects a new form of public culture, based around cafés, bars, and restaurants, and Laurier and Philo (2004) *Env. & Plan. A* 36, 3 represent cafés as 'sites where economic, political and cultural matters come into contact and are mutually transformed'. These joyful judgements are shared by Anderson (2004) *Annals Am. Acad. Polit. Soc. Sci.* 595 and Esperdy (2002) *Architectural Design* 72.

D. Hobbs (2003) takes a more cynical view, seeing the creation of hospitality as a means of generating customer loyalty, but Bell (2007) *PHG* 31, 1 argues that it shouldn't be seen as merely calculative. Jayne et al. (2006) *PHG* 30 stress the downside of pub culture, and Hobbs (op. cit.) illustrates the affects and effects of 'the alcohol-fuelled night-time economy'—from boozy matinées to random drunken brawls.

comminution Changing rock debris to fine powder via *abrasion and *attrition. See Hiemstra et al. (2007) *Boreas* 36, 4 on comminution in subglacial tills. *See also* ROCK FLOUR.

commodification The process of turning something—an artefact, symbol, or idea—into a *commodity that can be exchanged, usually for money. Ioris (2007) *Capital Nature Socialism* 18, 1 shows how the commodification of water in Brazil has created social and environmental inequities; Naidoo (2007) *Rev. Afr. Polit. Econ.* 34, 111 describes the negative impacts of the commodification of service delivery in South Africa; and Thornes and Randalls (2007) *Geografiska A* 86, 9 describe the commodification of the atmosphere. R. Sack (1992) sees commodification as a feature of *modernity. See Marks and Bezzoli (2001) *Urb. Forum* 12, 1 on **commodified culture**.

commodity An object of economic value, intended for exchange in a *capitalist system, such as labour and money. **Commodity chains (commodity networks)** are the connections between the production, circulation, and consumption of goods; for example, the chains which bring cut flowers from Kenya to the UK. See A. Hughes and S. Reimer (2004) and J.-P. Rodrigue et al. (2006).

commodity fetishism The collective belief that the value of things is measured by money; 'the true source of their value—human labour—is not visible. Consumers perceive commodities as natural, as having a life of their own, rather than as socially produced' (Leslie (2009) in J. Agnew and D. Livingstone, eds). Kosoy and Corbera (2010) *Ecol. Econs.* 69, 1229 provide an excellent overview of this term, and S. Aitken and G. Valentine (2006), p. 63, claim that 'commodity fetishism has itself become a commodity'.

Common Agricultural Policy (CAP) Since the early 1960s, the

European Union's CAP has been highly protective, encouraging production through high price support, underpinned by import levies, state intervention and trading, and export subsidies. This helped to transform the European Union from a net importer of major temperate agricultural products to a major net exporter. EU exports needed heavy subsidization to be saleable internationally, and consequently depressed world prices. The EU has also maintained preferential access to regulated amounts of imports from members' former colonies and traditional trading partners.

communication geography The study of 'the role of communication in exerting and legitimating political and economic power at various scales; the influence of distance, space, and accessibility on communication technologies; the geography of messages and messengers' (Thrift in R. J. Johnston 2000). The study of information and communication technologies (ICT) by geographers has evolved from a concentration on friction of distance and spatial organization toward a set of fout interrelated social approaches: 'ICT as a set of contested terrains, ICT as a means of perception, ICT as a form of embodiment, and ICT as virtual places or spaces—distanciated social contexts' (Adams and Ghose (2003) *PHG* 27, 4). Adams and Janssen (2012) *Communication Theory* 22, 3, 299 is helpful

communism Historically, the principle of communal ownership of all property; basic economic resources are held

in common. See Shaw and Oldfield (2008) *Pol. Geog.* 27, 1.

community 1. In ecology, a naturally occurring, non-random, collection of plant and animal life within a specified environment. Badano and Cavieres (2006) *J. Biogeog.* 33, 2 discuss the view that increases in habitat complexity due to the presence of ecosystem engineers lead to higher **community diversity**.
  2. In human geography, the population and the interconnections of that population in a particular area, town, village, suburb, or *neighbourhood; a set of shared values, practices, and ways of being in the world (A. Latham et al. 2009). Ryan-Nicholls and Racher (2004) *Rural & Remote Health* 4, 244 see community as a subject with its own construction of reality, unique needs, values, and assets, while Lacy (2000) *Rural Soc.* 65 sees community as the container for the development of social cohesion, human endeavour, empowerment, and place formation. 'In taking up . . . the classical distinction between Gemeinschaft and Gesellschaft, networking practices in the transient word of projects might be differentiated into network communality, network sociality and network connectivity' (Grabner (2004) *PHG* 24, 4). A. Amin (with N. Thrift, 2004), uses the concept of *communities of practice: the process of social learning that occurs when people who have a common interest in some subject or problem collaborate over an extended period to share ideas, find solutions, and build innovations. Communities of practice act as performative spaces for learning by homogeneous groups of employees, not only via spatial proximity, but also at a distance (J. Peck and H. Yeung, eds. 2003).

community of practice A group of people, not necessarily physically near to each other, who form common values, norms, and understanding through common training, experience, and expertise. See Amin and Roberts (2008) *Res. Policy* 37, 353.

community-supported agriculture (community-shared agriculture) (CSA) A CSA is organized around a contract between a farmer, commonly an organic producer, and a set of local residents, who share the risks of the farming enterprise by contributing money up front for a 'share' of the harvest prior to the farming season; see O'Hara and Stagl (2001) *Pop. & Env.* 22. Wells et al. (1999) *Community Develop. J.* 34 hold CSA brings a growing circle of people into a closer relationship with place—'farming, nature, each other'.

commuting The movement from suburban or rural locations to the place of work and back. In a Montreal-based study, Shearmur (2006) *Urb. Geog.* 27, 4 concludes that differences in commuting length between different places of work are, by and large, independent of possible explanatory factors (such as residential location, economic sector, occupation, and income). Despite their shorter average overall commutes, women travel farther than men to reach jobs in the *CBD (Shearmur op. cit.). Murata and Thisse (2005) *CEPR Disc. Paper* 4936 find that low transportation costs lead to the dispersion of economic activities. Elias et al. (2003) *Reg. Studs* 37, 8 find that accessibility to job openings in surrounding regions significantly increases the likelihood of commuting, but that labour mobility decreases with access to job opportunities in neighbouring regions.

comparative advantage The advantage of some nations or regions to produce goods better and more cheaply than less favoured nations or regions; an important concept in understanding regional specialization. Llang (2004) *RURDS* 16, 3 finds that comparative advantages from improved technical

efficiency, scale economies, and growing economic openness contribute positively and significantly to regional specialization in China.

comparative biogeography The study of biotic area relationships among local and global biogeographic regions: 'the comparative study of *biotic areas and their relationships through time' (L. R. Parenti and M. C. Ebach 2009).

comparative cost analysis An analysis of costs in order to establish a least-cost location; see R. Hayter (2004). Guyatt et al. (2002) *Health Policy & Plan.* 17, 2 apply the method to combating malaria in Kenya.

comparative urbanism The systematic study of similarity and difference among cities or urban processes. 'It should not surprise us that each and every place is different or even unique in some ways—this is the idiosyncratic nature of place. What begs our attention is why separate places can be very similar in certain respects' (Nijman (2007) *Urb. Geog.* 28, 1).

compatibility 1. Being able to accommodate more than one form of activity; for example, 'to achieve compatible physical and economic coexistence of rural residences, agriculture and cities, through preservation of resources, clustering of development contiguous to cities, buffering uses, and adequate infrastructure' (*Leavenworth County Plan*). **2.** To harmonize existing land-use datasets in order to make comparisons within and between countries (Jensen (2006) *J. Land Use Sci.* 1, 2–4).

competence The largest size of particle that a river can carry. Competence varies with water depth, but is reduced with channel widening (Cheetham (1979) *GSA Bull.* 90, 9). The sixth power law suggests that a doubling, for example, of river velocity would raise competence by 2^6, i.e. by a factor of 64; see Houbrechts et al. (2011, *Geomorph.* 126). Baker and Ritter (1975) *GSA Bull.* 86, 7 demonstrate the analysis of shear stress as an estimate of competence. *See also* HJULSTRÖM DIAGRAM.

competition 1. Rivalry between suppliers providing goods or services for a market (Jonathan Law, ed. 2006). Essletzbichler and Rigby (2007) *J. Econ. Geog.* 7, 5 seek to show how competition may produce distinct economic spaces: 'as competition unfolds over time, new spaces of economic activity are created.' See H. Garretsen and S. Brakman (2005) on competition between decision-makers. **Competitive advantage** describes the relative economic gains a firm (or a group of similar firms) enjoys over its business rivals at home or overseas.

2. In biogeography, competition occurs when a necessary resource is sought by a number of organisms. 'Vegetation dynamics, in a modelling context, comprise the processes of competition for resources among individuals or plant functional types' (Smith et al. (2001) *Glob. Ecol. Biogeog.* 10, 6).

complementarity A mutual dependency based on an ability to produce goods in one area which are needed in another, as with the import of Japanese manufactured goods to Australia, and the export of Australian agricultural goods to Japan (D. Rumley 1999). Nazara et al. (2006) *J. Geog. Sys.* 8, 3 explore the degree to which complementarity and competitive interaction at one level in the hierarchy persist at lower or higher levels.

complexity In *geomorphology, a **complex system** is characterized by nonlinear feedbacks, rather than straightforward relationships between forcing and response. Complex systems spring from nonlinear dynamics and chaos theory and include: self-similarity and (multi) fractals; emergence and self-organization; network analysis; and self-organized criticality. A complex system when interfered with or modified is unable to adjust in a progressive and systematic fashion. Phillips (2003) *PPG* 27, 1, 3 is a good place to start.

complexity science 'Complexity theory . . . takes on new meanings as it circulates in and through a number of actor-networks, and, specifically global science, global business and global New Age' (Thrift (1999) *Theory Cult. & Soc.* 16, 3; bizarrely, this is the only use of the term in this paper). At the heart of all definitions, however, is the realization that, in general, the reductionist methods used in science cannot be used in studying complex systems: 'the inter-action among constituents of the system, and the interaction between the system and its environment, are such that the system as a whole cannot be fully understood simply by analysing its components' (P. Cilliers 1998). 'Each entity has different relations to others, and, *where* an entity is in the system has significance for the unfolding behaviour of entities individually, and of the system collectively' (O'Sullivan (2004) *TIBG* 29, 1). This means that scaling up from local to global behaviour is not straightforward. As P. W. Anderson (1972) *Science* 177, 393 shows, 'more is different'; the whole is very much more than the sum of the parts. See Trudghill (in S. Trudghill and A. Roy, eds. 2003) for a wonderfully user-friendly discussion of models, narratives, and constructs in complexity theory. 'The resulting geography . . . is expected to reach the public sooner or later, even if the effect—according to complexity theory itself—cannot be foreseen in detail' (O'Sullivan (2004) *TIBG* 29, 3).

At the time of writing, the last comment must come from Martin and Sunley (2007) *J. Econ. Geog.* 7, 5, who, in a model of exegesis, write that 'to refer to complexity "theory" is perhaps to exaggerate the degree of conceptual coherence and explanatory power associated with the notion'. For **complexity in geomorphology** see (2007) *Geomorph.* 91, 3–4, special issue.

compositional data Population data based on percentages or proportions. In geography, the vast majority of analyses of such data make use of standard statistical approaches, which, according to Aitchison (1986) should not be employed directly in the analysis of such data. Lloyd et al. (2012) *AAAG* 102, 6, 1251 highlight some problems associated with using standard approaches in the analysis of compositional data in population studies and outline some key log-ratio-based approaches for the transformation and the proper analysis of compositional data.

compressive flow (compressional flow) Where a glacier is slowing up, the flow is compressive; in zones of compression, the ice may erode the valley floor. Compressional flow on the *stoss side of hills causes till sheets to shear, and debris bands to stack (Gore et al. (2003) *Antarc. Sci.* 15, 2). See also Swift et al. (2006) *Q. Sci. Revs* 25, 13–14.

compulsory purchase order A directive to enforce the sale of land from a private landowner to national or local governments (Gallent and Shaw (2007) *J. Env. Plan. & Manage.* 50, 5).

concave slope A slope which declines in steepness with movement downslope, also known as a *waning slope. Claessens et al. (2007) *Geomorph.* 74, 1–4 find that steep concave slopes are one of the main factors responsible for landslides in Uganda. Most often, the concave element

of a slope occurs as a basal concavity. Where the scarp is not being undercut, the base and the crest make straight-line intersections with the mid-section to form a sharp basal concavity and crestal convexity (Nash and Beaujon (2006) *Geomorph.* 75, 3–4). The basal concavity is poorly developed in humid temperate environments (J. Gerrard 1988).

concealed unemployment A situation whereby individuals know that there are no jobs available and don't register as unemployed, even though they would like a job; official figures thus under-represent the true situation. P. Alcock (2003) uncovers extensive concealed unemployment in the UK among officially 'economically inactive' men.

concentration and centralization The tendency of economic activity to congregate in a restricted number of central places, encouraged by *functional linkages, and *external and *agglomeration economies. See Hanson in G. L. Clark et al., eds (2000). Falcioğlu and Akgüngör (2008) *Eur. Plan. Studs* 16, 2, in a study of centralization in Turkey, confirm the Krugman hypothesis (1991, *J. Pol. Econ.* 99) that industries become more concentrated with economic integration. Rodríguez-Pose and Zademach (2003) *Urb. Studs* 40, 10 find that proximity appears to play a distinctive role in the geography of mergers and acquisitions in Germany.

concentration dynamics Changes in the concentrations of suspended sediment in a river. Concentrations usually rise with increased river velocity. See Wang et al. (2014) *Sci. Total Env.* 470–71, 907.

concentric zone theory R. Park and E. Burgess (1925) suggested that the struggle for scarce urban resources, especially land, led to competition for land

and resources, which ultimately led to the spatial differentiation of urban space into zones. They predicted that cities would take the form of five concentric rings with areas of social and physical deterioration concentrated near the city centre and more prosperous areas nearer the periphery. In the post-war period, Park and Burgess's model fell out of favour as critics suggested that the models were overly simplistic. However, M. Davis (1992) used the concentric rings model to describe Los Angeles as a city with an inner core of 'urban decay metastasizing in the heart of suburbia' (Brown, no date, *USBC Spatial Rev.*).

concordant Complying with. In geomorphology, relief and drainage may be concordant with geological structure. A **concordant coast** is parallel to the ridges and valleys of the country.

condensation The change of a vapour or gas into liquid form, accompanied by the release of *latent heat, altering the *adiabatic temperature change in rising air. See Wang and Wang (1999) *J. Atmos. Scis* 56, 18 on the classic turbulence condensation theory. Condensation in meteorology can be caused by the cooling of a constant volume of air to *dew point; the expansion of a parcel of air without heating; the evaporation of extra moisture into the air; the fall in the moisture-holding capacity of the air due to changes in volume and/or temperature; and contact with a colder material or *air mass. The likelihood of water vapour condensing will depend on the *saturation vapour pressures of water and ice at a given temperature, and/or the presence of *condensation nuclei, since water vapour can cool well below 0 °C before condensation occurs. *See* CONDENSATION NUCLEI; BERGERON–FINDEISEN THEORY. The **condensation level** is the height at which rising air will cool, condense, and form clouds.

condensation nuclei Microscopic *atmospheric particles that attract water droplets, that may then *coalesce to form a raindrop. Condensation nuclei come from pollen, salt from sea spray, dust from volcanic eruptions and soil erosion, and particulate air pollution (C. Park 2001). Many small condensation nuclei cause many small droplets to remain liquid well below normal freezing point, and small ice nuclei may increase the rainfall rate from single clouds, which is the principle of *cloud seeding (Graf (2004) *Science* 303, 5662). *See* HYGROSCOPIC NUCLEI.

conditional instability *See* INSTABILITY.

cone volcano A volcanic peak with a roughly circular base, tapering to a point. Cones may be built solely of *lava or of *scoria, or of both. Parasitic cones form round smaller vents on the flanks of the volcano. See Frazzetta et al. (1983) *J. Volc. & Geotherm. Res.* 17.

congelifluction (congelifluxion) *See* GELIFLUCTION.

congelifraction *See* FREEZE–THAW.

congeliturbation *See* GELITURBATION.

congestion 'The nation's principal means of allocating scarce road space among competing users during periods when too many people want to use that limited space at the same time' (A. Downs 2004). 'Combinations of public transport/cycling/traffic management measures (traffic calming, park and ride), are not enough to reduce traffic in city centres; restraint measures, such as traffic calming, parking controls or road pricing are needed as well. To achieve public/political acceptability for such policies, improved public transport, providing alternative means of travel, needs to be in place before the restraints are introduced. The danger is that once the situation improves, the cars will immediately return.

A reduction in congestion will have the potential to induce extra trips and traffic, which would not otherwise have occurred; all assessments of measures to reduce congestion should take account of the potential for induced traffic' (*House of Commons, Res. Paper* 98/16). This term can apply to an excessive demand for any facility.

conglomerate. A *sedimentary rock composed of rounded, water-borne pebbles which have been naturally cemented together.

coniferous forest (boreal forest) Natural forest, found between 55° and 66°N, where winters are long and very cold, summers short and warm, and annual precipitation is around 600 mm. Pure stands are common and species relatively few. Trees are evergreen, and the needle-shaped leaves restrict surface area, thus preventing loss of water via transpiration. Undergrowth is sparse. Animal species are dominated by insects and seed-eating rodents; larger animals include deer, bear, and wolves. See Väisänen (1995) *IUFRO XX World Congr.*

connectivity The degree to which the *nodes of a network are directly connected with each other. A graph is connected if every node can be reached from some other node. The higher the ratio of the *edges to the nodes in a network, the greater the connectivity. See Urban and Keitt (2001) *Ecology* 82 on graph theory as applied to connectivity in heterogeneous landscapes. Connectivity in a network is said to increase as economic development proceeds; see G. Vigar (2002). Taylor (2002) *Urb. Studs* 39 measures a city's connectivity by summing the products

of every firm's service value in the city with their service values in all other cities. US cities are generally less globally connected than their European Union and Pacific Asian counterparts—New York is the only US city with more non-US cities than US cities in its top ten of global connections (C. Taylor 2005, Brookings Institution).

Landscape connectivity is the degree to which the landscape facilitates movement across its existing resources. It can be considered as an emergent property of the landscape that results from the interaction between landscape structure and landscape function; see García-Fecedet et al. (2011) *Forest Ecol. & Manage.* 261, 154.

consequent stream Any stream whose course is controlled by the initial slope of a land surface (A. Strahler and A. Strahler 1979).

conservation The protection of natural or man-made resources and landscapes for later use. A distinction is made between conservation and preservation; while a **conservationist** recognizes that people will use some of the fish in a lake, a preservationist would ban fishing entirely. 'If the twentieth century question was how to stop extinction, then perhaps the twenty-first-century challenge is how to avoid the total collapse of the biosphere, our life support system' (W. Adams 2004).

Conservation is a highly political enterprise: McKenna et al. (2005) *Area* 37 argue that a failure to conserve coastal dunes in Ireland results 'more from management deficiencies than from shortcomings in scientific understanding'. Duffy (2006) *Pol. Geog.* 25, 1 casts light on the struggles encountered in environmental politics over access to key natural resources; and Sundberg (2003) *Pol. Geog.* 22, 7 observes that the existence of democratic regimes and formal institutions does not guarantee that environmental projects will be implemented through demographic processes. C. Hambler (2004) favours establishment of an environmental fund into which polluters pay and from which societies can draw, such that those whose development opportunities are curtailed for global conservation objectives can access compensatory development funds. See Vogiatzakis et al. (2006) *PPG* 30 on using *GIS for conservation.

Conservation biology is a sub-discipline of biology concerned with the impacts of people on the environment and the conservation of biological diversity (see Murray et al. (2002) *Austral. Ecol.* 27; see also Klinkenberg (2001) *Canad. Geogr./Géog. canad.* 45, 3). Van Kleunen and Richardson (2007) *PPG* 31, 4 argue for more integration between conservation biology and invasion biology.

conservation of angular momentum The momentum of a body taking a curved path is *angular momentum*. A body's angular momentum will remain constant unless changed by some force. This is important in understanding *jet streams; a body of air at the equator will move at the same speed as the Earth. When this body of air moves polewards, it must increase its velocity to maintain its angular momentum, since Earth's radius decreases polewards. At 30°, for example, the air moves very much faster than the Earth below (which is much slower than at equatorial speed). This very fast-moving stream of air is a jet stream.

conservative margin A plate margin where the movement of the plates is parallel to the margin, such as the San Andreas Fault. See Chéry (2008) *Earth & Pl. Sci. Lett.* 269, 3–4.

consolidation The reform and reorganization of land ownership in order to redress *fragmentation. Crecente et al. (2002) *Land Use Policy* 19, 2 review land consolidation in Galicia, Spain. Ramutsindela (2007) *Tijdschrift* 98, 4 argues that the South African government's approach to land restitution accounts in part for the former bantustans. See Gonzalez et al. (2004) *Ag. Systs.* 82, 1 on the practicalities of land consolidation.

constant slope A straight, sloping element of a hillslope, located either in the middle of a slope profile, or at the base of the *free face. Burt (2003) *PPG* 27, 4 finds that a constant slope is characteristic of valley incision, but that it cannot survive at its original angle once downcutting stops. Calvache et al. (1997) *Geomorph.* 21, 1 note that particle size influences the longitudinal constant slope profiles of alluvial fans.

constructive margin A boundary where two *plates are spreading apart from one another and new oceanic crust is being created and added to the outer shell of the planet. See Jackson and Gunnarsson (1990) *Tectonophys.* 172 and Edmonds et al. (2003) *Nature* 421, 6920, on the slow-spreading Gakkel Ridge.

constructive wave A low-frequency (6–8 per minute) spilling wave, with a long wavelength and a low crest, running gently up the beach. *Swash greatly exceeds *backwash (which is reduced by *percolation), leading to deposition. Spilling waves usually occur on gently sloping beaches, and may form beach ridges and *berms (Hughes (2004) *Coastal Eng.* 51, 11–12; Coelho and Veloso-Gomes (2004) *J. Coast. Res.* SI-39). See Cooper et al. (2000) *Water & Env. J.* 14, 2 on beach profiling.

constructivism The argument that each of us constructs her or his own understanding by relating new information on what they already know. It is the assertion that all forms of knowledge are shaped by the social and material cultures within which they are produced and consumed. In relation to climatology, Demerritt (2001) *AAAG* 91, 307 argues that the reality studied by physical geographers is socially constructed. This has not been universally accepted; see Scheider (2001) *AAAG* 91, 2, 338.

consumer 1. Organisms in all the *trophic levels, (herbivores, *carnivores, omnivores, and parasites) except for *producers. **Primary consumers** feed only on plant material; **secondary consumers** feed on primary consumers, and so on. Sullivan and Rohde (2002) *J. Biogeog.* 29, 12 see tight consumer-resource relations as 'pulling a system towards equilibrium'.

2. A buyer of goods and services for her/his/others' personal satisfaction, as opposed to income generation. Jackson (2002) *PHG* 26, 1, sees garment labels that one is looking at as a confirmation of one's moral superiority, seeming to say to one 'you're an "alternative" person'; 'you'd like to help the Indian economy'. Jackson sees these labels as rounds of *commodification in an all-encompassing consumer culture. Slocum (2004) *Env. & Plan. A* 36 is more upbeat: 'citizens . . . cannot be wholly embodied by constructions such as the consumer . . . [and] can position themselves as reasoning *publics'.

consumption The utilization of consumer goods and services for the satisfaction of the present needs of individuals, organizations, and governments. Paolucci (2001) *Antipode* 33, 4 illustrates the way that the contemporary city is 'a concrete manifestation of the centrality of consumption'. Jackson (2004) *TIBG* 29, 2 examines the way that producers have 'customized' their products for different markets, and demonstrates the resilience of local consumption cultures in the face of *globalization. **'Alternative' consumption** is a practice in the North that is underpinned by concern about social injustice and unfair labour practices in the South, and the practices adopted as part of a 'radical' lifestyle (Bryant and Goodman (2001) *TIBG* 29, 3). Barnett et al. (2005) *Antipode* 3 see ethical consumption as a significant moment in consumer activism; 'one in which large numbers of people are explicitly mobilised in support of various political causes through a shared identity as consumers, but where the spatiality of this mobilisation exceeds the scale of the nation-state'. **Spaces of consumption** are purpose-built spaces, characterized by the provision of consumption-related services, visual consumption, and cultural products. They are 'the physical manifestation of consumerism as a way of life . . . play[ing] a key role in the re-invention of the post-industrial city' (S. Miles 2010).

consumption, geographies of
Mansvelt (2008) *PHG* 32, 1 identifies the themes that draw attention to shaping of consumption in place: consumption as an arena in which governance, regulation, and citizenship are produced; relationships between consumption and forms of urban space; and the way in which practice is embedded in social and spatial contexts which extend beyond the act of purchase. For example, in India, alcohol is prohibited in Gujarat, Manipur, Mizoram, Nagaland, and Lakshadweep, so the spatial pattern of alcohol use varies over India; see Subramanian et al. (2005) *Bull. WHO* 83, 11, who also find variations by economic status and caste. McNeill (2008) *PHG* 32, 3 argues that the hotel as hybrid of public and private space breaks down the artificial

analytical distinction between 'public' (the high street, the mall, the department store) and 'private' consumption spaces (the home, the garden). 'To see private urban malls as iconic of contemporary retailing is to simplify the complexity of consumption spaces and practices' (Crewe (2000) *TIBG* 24, 2).

content analysis A methodology in social science for making sense of recorded human communication including news media, policy documents, letters, and video or novels—particularly written texts. Jackson et al. (2006) *Geoforum* 37, 447 use content analysis to chart the course of human geography from the 1950s.

contested landscape A site of competing and often contrasting discourses (Rofe and Oakley (2006) *Geog. Res.* 44, 3). T. Marsden et al. (2003) see the productivist rural landscape as a contested landscape 'where decision making reflects local economic priorities and development projects are likely to cause less contention'.

context The setting of an event, statement, or idea. Geographical context 'is often thought about in terms of national or political territories, physical landscapes or exotic places' (J. Anderson 2010).

contextual effect The effect of their local community on the decisions that people make, so that they may be influenced by the views of others in the community; 'converted', in other words, by their neighbours. Mondak in D. C. Mutz et al., eds. (1996) reports that some contextual effects emerge when citizens use their local environment as a default source of information in the absence of national information. Cutler (2007) *Pol. Geog.* 26, 5 thinks that local conditions will be more influential on issues that

are: clearly related to a particular political judgement; easy to observe; with localized policy impacts; and where national-level information is scant or open to multiple interpretations. Pearce et al. (2008) *J. Epidemiol. & Comm. Health* 62 study the contextual effect on diet. *See also* EMBEDDEDNESS.

contiguous zone A zone of the sea beyond the *territorial seas of a nation, over which it claims exclusive rights.

( ) **SEE WEB LINKS**
• UN discussion of the Treaty on the Law of the Sea.

continental climate A climatic type associated with the interior of large land masses in mid-latitudes. Without the moderating influence of the sea, summer and winter temperatures are extreme. Precipitation is low, as the region is distant from moisture-bearing winds. The European Union PESETA project predicts that the greatest temperature rises related to global warming will be in continental climates; Klingbjer et al. (2005) *Geografiska A* 87, 1 consider the effect of global warming on continental climate and glacier dynamics. Bergh et al. (2003) *For. Ecol. & Manage.* 183, 1–3 compare net primary productivity in continental and maritime climates.

continental crust The outer, rigid surface of the Earth which forms the continents. Wilde et al. (2001) *Nature* 409, 175–8 find evidence for the existence of continental crust on the Earth 4.4 gigayears ago. Continental crust is rich in feldspars and granitic rocks and about 35 km thick (Gao et al. (1998) *Geochimica et Cosmochimica Acta* 62, 11)—thicker, but less dense, on average, than *oceanic crust. Collins (2002) *Geology* 30, 6 explains the role of cycles of tectonic switching in the production of continental crust. Crust-mantle interactions occur in the lower continental crust (Yang et al. (2007) *Prog. Nat. Sci.* 17, 2).

continental drift The theory that continents which are now separate were united in a supercontinent, suggested by Alfred A. Wegener (1916). Wegener's ideas were vindicated in the mid-1960s when the development of plate-tectonic theory provided a new framework for planetary-scale tectonics, explicable only in terms of continental drift (Harper (2001) *Geol. Today* 17, 4): Hess (1962), Buddington memorial volume *Geol. Soc. Am.* 599–620, identified sea-floor spreading as a geologic mechanism to account for Wegener's moving continents. See Jackson (1993) *AAAG* 83, 2 on the role played by palaeomagnetic studies in continental drift research.

Dobson (1992) *AAAG* 82, 187–206 suggests that convection drives circular plate motion while gravity drives lateral motion. Convection currents well up at high-pressure centres, spiral outward, transfer to low-pressure cells, spiral inward, and descend at low-pressure centres. In most instances, upwelling and descent of the asthenosphere occur at opposing plate centres rather than at plate margins. Cells migrate laterally in a global pattern driven by gravity. Sea-floor spreading and subduction occur because of differential rates of lateral plate motion. Bokelmann (2001) *Geology* 30, 11 believes that movements of the mantle play an important role in driving the plates.

In 2006, NASA scientists released the first direct measurements of continental drift, showing that the Atlantic is gradually widening, and that Australia is receding from South America and heading for Hawaii (Anonymous (2006) *New Scientist* 192, 2578; p. S24). Silver and Behn (Carnegie Institution) suggest that plate tectonics may have halted at least once, and may do so again. See Pratt (2000)

J. Sci. Exploration 14, 4 on the challenges confronting plate tectonics.

continental shelf The gently sloping submarine fringe of a continent, ended by a steep continental slope which occurs at around 150 m below sea level. Mossop (2007) *Ocean Dev. & Int. Law* 38, 3 identifies potential threats to marine biodiversity on the continental shelf, and considers the legal rules that coastal states may use to protect marine biodiversity.

contingency An event or state of affairs dependent on another, uncertain event or occurrence. **Contingence** is the influence of the history of a system in determining its response to an input (Lane (2001) *TIBG* 26). In the context of a fluvial system, for example, the response of a length of bank to a hydrological event will be influenced by the bank morphology resulting from a previous event. *Hysteresis—the dependence of the equilibrium position of a system on what happens during the process of dynamic adjustment—seems to be the human geographer's term for contingence (J. Rogers Hollingsworth and R. Boyer, eds. (1997); Vandenberghe and Woo (2002) *PPG* 26, 4; Jordaan (2005) *PHG* 29).

continuity equation All landscapes must obey an equation for the conservation of mass: mass, energy, or momentum are conserved in a system, so that for any part of the system the net increase in storage is equal to the excess of inflow over outflow of the quantity conserved. In rectangular coordinates and in differential form, the continuity equation is:

$$\frac{\partial \rho}{\partial t} = -\left[\frac{\partial \rho u}{\partial x} + \frac{\partial \rho v}{\partial y} + \frac{\partial \rho w}{\partial z}\right]$$

where ρ is fluid density and u, v, and w are velocity components in the x, y, and z directions. Geomorphologists apply this equation to all cases of mass balance. Applied to mass transactions on a hillslope, the continuity equation states that if more material enters a slope section than leaves it, then the difference must be represented by aggradation; conversely, if less material enters than leaves, then the difference must represent net erosion. The same principle of continuity of sediment transport applies to other geomorphic systems including rivers; see Dietrich and Perron, 2006 *Nature* 439, 411.

contour A line on a map joining places of equal heights, and sometimes equal depths, above and below sea level. The **contour interval** is the vertical change between consecutive contours. **Contour ploughing** is a method of ploughing parallel to the contours rather than up or down a slope. It is used to check soil erosion and the formation of *gullies.

conurbation A group of towns forming a continuous built-up area as a result of *urban sprawl (also called **metropolitan area**). See the classification outlined by Raybould et al. (2000) *Update* 25, 2–4S; see also S. Brunn (2003).

convection The process whereby heat is transferred from one part of a liquid or gas to another, by movement of the fluid itself. Convection carries excess heat from the Earth's surface and distributes it through the *troposphere (Fanaki (1971) *Boundary-Layer Met.* 1, 3; Takayabu et al. (2006) *J. Met. Soc. Japan* 84A). **Convective storms** include thunderstorms, hail, tornadoes, heavy rains, and high winds; Combardo and Colle (2011) *Weath. Forecasting*, 26, 940 document the convective storm structures over the northeastern United States.

Thermal convection (**free convection**) is propelled by buoyancy (R. Barry and R. Chorley 2003). **Mechanical**

convection (**forced convection**) is the upward movement of an air parcel over mountains, at fronts, or because of *turbulence (Sorbjan (2006) *Boundary-Layer Met.* 119, 1). Kijazi and Reason (2005) *Theoret. & Appl. Climatol.* 82, 3–4 observe that wet conditions during *El Niño years are associated with **enhanced convection** (convection with a higher Bowen ratio). See Ludlam and Scorer (2006) *Qly. J. Royal Met. Soc.* 79, 341.

convection rain When upward *convection occurs in a parcel of moist air, the rising air will cool. Further cooling will cause condensation of the water vapour in the air, and rain may result. If the air is very moist, the cooling results in condensation, and hence the release of *latent heat. This causes the rising air to accelerate, and very tall *cumulo-nimbus clouds form; see Sui et al. (1997) *J. Atmos. Scis* 54, 5.

convective condensation level The point on a thermodynamic diagram where a sounding curve (representing the vertical distribution of temperature in an atmospheric column) intersects the saturation mixing ratio line corresponding to the average mixing ratio in the surface layer (approximately the lowest 500 m). The dry adiabat through this point determines, approximately, the lowest temperature to which the surface air must be heated before a parcel can rise dry-adiabatically to its lifting condensation level without ever being colder than the environment. This temperature, the convective temperature, is a useful parameter in forecasting the onset of convection (AMS online).

convective instability Instability that is caused by the rising of very dry air over warm, moist air below. C.-G. Rossby (1932) argued that it exists when one of these conditions is met over a layer of atmosphere: the lapse rate of wet-bulb temperature exceeds the moist-adiabatic lapse rate; the equivalent potential temperature decreases with height; the wet-bulb potential temperature decreases with height. Convective instability of the terrestrial atmosphere is largely determined by water vapour. See Gutzler (1996) *J. Atmos Scis* 53, 19.

convective turbulence Turbulence generated when warm fluid parcels are accelerated upward by their buoyancy, normally initiated by daytime solar radiation.

convenience distance The ease, or otherwise, of travel. A town 50 km away may be well served by transport and thus 'nearer' than one 20 km away which is less accessible. **Convenience goods** are *low-order goods like milk, and occasional groceries, which are frequently bought locally, with little consideration of the prices.

convergence 1. The meeting of *tectonic plates. See Rosenbaum (2002) *Tectonophys.* 359, 1–2 on the convergence of Africa with Europe, and Gelabert et al. (2004) *Geologica Acta* 2, 3 on the convergence of the Indoaustralian and Eurasian plates.

2. In meteorology, air streams flowing to meet each other. An area of **convergence** is an area of rising air. Convergence can occur aloft over dense, cold air and is not necessarily confined to a layer bounded by the surface. Byers and Rodebush (1948) *J. Meteor.* 5 were the first to show a correlation between convective rainfall and convergence of low-level winds. These areas of low-level convergence were often restricted to well-defined lines of convergence with widths of the order of 1–2 km (Crook and Klemp (2000) *J. Atmos. Scis* 57, 6). In the upper troposphere, convergence causes air to subside, creating *anticyclonic

conditions at ground level (Hastenrath (2007) *Dynam. Atmos. & Oceans* 43, 1–2).

3. Within human geography, the use of the same production methods or practices by firms operating in different national-institutional spaces. Gertler (2001) *J. Econ. Geog.* 1, 1 describes the convergence between Anglo-American, German, and Japanese models of economic growth, and Domosh (2004) *TIBG* 24, 4 explores the convergence of economic and cultural approaches to *imperialism. See also A. Leyshon and J. Pollard (2000).

convergence space Spaces for actors, movements, and struggles at particular moments in time; 'a heterogeneous affinity between various social formations, such as social movements' (Routledge (2003) *TIBG* 28, 3); 'a world made of many worlds' (Marcos in K. Abramsky, ed. 2001). 'Convergence spaces are comprised of place-based, but not necessarily, place-bound movements . . . relational achievements involving a practical relational politics of solidarity. Such solidarity takes place in the form of changing and overlapping circuits of relations that are enacted both virtually through the internet and materially in particular forums where connections are grounded in place- and face-to-face based moments of articulation' (Cumbers et al. (2008) *PHG* 32, 2). These places of protest enter the collective memory of the participants of convergence spaces, thereby reinforcing the moral commitment of activists to their struggles (Routledge, op. cit.).

converging margin *See* DESTRUCTIVE MARGIN.

convex slope A slope, or slope element, that gets progressively steeper downhill. It may be determined by structure (*exfoliation domes), by lithology (Allison and Higgitt (1998) *Catena*

33, 1), or by landslides (Di Luzio et al. (2004) *Geomorph.* 60, 3). Convex slopes may result from weathering and debris transport—see J. Wilson and J. Gallant (2000). See Lin and Oguchi (2004) *Geomorph.* 63, 3 on drainage density and convex slopes.

conveyor belt 1. An expansive upwards, polewards flow of air. As a mid-latitude depression develops, a warm conveyor belt rises at *c.*20 cm s⁻¹, ahead of the cold front, and a reciprocal cold conveyor belt descends from medium/upper levels, below the warm conveyor, well ahead of the surface warm front; see Pomroy and Thorpe (2000) *Monthly Weather Rev.* 128, for a useful diagram. Schultz (2001) *Monthly Weather Rev.* 129 suggests that cyclones with well-defined warm fronts show a sharp break between the cyclonic and anticyclonic paths of the cold conveyor belt, while cyclones with weaker warm fronts have a broad transition zone between the two.

2. In the oceans of the world, a global system of currents. The **Atlantic conveyor** is a vast ocean current, which is part of the global ocean circulation. Warm water travels northwards close to the surface of the ocean. Beneath it there is a slow, deep-water flow carrying cold dense Arctic waters to Antarctica along the ocean bottom. As it travels north, water in the upper current becomes colder and denser. Winter sea ice forming in the Arctic Ocean leaves salt in the water making it denser still. At certain places in the North Atlantic the dense water sinks down towards the sea floor and begins to flow south. In this way the circulation 'overturns' to form the loop in the conveyor belt between the warm shallow current and the cold deep current (PEEP, *What Drives the Atlantic Conveyor?*). Broecker et al. (1985) *Nature* 315, 21 were the first to suggest a link between the Atlantic conveyor belt and

possible climate change, and Alley (2007) *Ann. Rev. Earth & Plan. Scis* 35: 241 gives an impressive review of the acceptance of this link. *See also* THERMOHALINE CIRCULATION.

(((●))) **SEE WEB LINKS**
- Website of Physics & Ethics Education Project (PEEP).

co-occurrence An event or situation that happens at the same time as or in connection with another. For example, 'any map which shows the impact of having a number of resources in the same geographic area is a co-occurrence map. This was once accomplished with actual transparent overlays, but with the development of computer mapping, a summary map can be generated showing the locations where different numbers of factors are co-located' (from the excellent Natural Resources Inventory on the Natural Resources Conservation Service website).

(((●))) **SEE WEB LINKS**
- Natural Resources Conservation Service website.

cooperative A type of business organization owned by its employees or customers. **Agricultural cooperatives** are collective projects initiated by local producers which aim to organize the implementation, collection, and transformation of the members' production, and to ensure their reputation through collective actions. This means that they have to reconcile economic performance (marketing) with their social responsibilities towards their members (Filippi and Torre (2003) *Int. J. Technol. Manage.* 26, 2–4).

Agricultural cooperatives have been encouraged in the developing countries, where several farms pool resources to jointly purchase and use agricultural machinery; see anon. (2005) *Rev. Int. Cooperation* 98, 2 on cooperatives in Colombia. Rice and Lavoie (2005)

Canad. Geogr. 49, 4 evaluate **business cooperatives** in Canadian economic growth.

coral reef An offshore ridge, mainly of calcium carbonate, formed by the secretions of small marine animals. Corals flourish in shallow waters over 21 °C and need abundant sunlight, so the water must be mud free, and shallow. Fringing reefs lie close to the shore, while barrier reefs are found further from the shore, in deeper water. Reef health is mapped by: live coral cover (Mumby et al. (2004) *Marine Pollut. Bull.* 48), diversity (Hodgson et al. 2004), and remote sensing (Kutser et al. (2006) *Limnol. & Oceanog.* 48).

Coral reefs are the most biodiverse marine ecosystems on the planet; M. L. Reaka-Kudla (1997) estimates that they harbour nearly one million species globally. However, the health of coral reefs is declining (see C. Wilkinson 2004), due to overfishing, nutrient enrichment, and coral diseases, at the local scale, and ocean warming, acidification, and sea-level rise at the global scale; see also Hoegh-Guldberg (1999) *Marine & Freshw. Res.* 50, Hempel and Morozova (2001) *Bull. Marine Sci.* 69, 2 on coral reef management from the benthos up, and the US NOAA Coral Reef Conservation Programs.

A **coral atoll** is a horseshoe-shaped ring of coral which almost encircles a calm lagoon. Many coral reefs are hundreds of metres deep; Charles Darwin's theory (1842) is that deep reefs formed during a long period of subsidence. Thus, coral forms in shallow waters and then sinks. Woodfine et al. (2008) *Sedimentol.* 52 seem to support the subsidence theory. See Braithwaite et al. (2000) *Int. J. Earth Scis* 89, 2 on revisiting Darwin's model. Woodruffe et al. (1999) *Marine Geol.* 160, 1 dispute Darwin's theory. There is also the possibility of a positive feedback loop: sea level rises, providing more habitat for coral reefs, the reefs flourish and produce more CO_2, the additional CO_2 causes additional temperature increase, and sea levels rise even more; see G. Camoin and P. J. Davies (1998).

core 1. The central part of the Earth, the core is the point within the Earth where S-waves cannot penetrate, marked by an abrupt increase in pressure. It is believed to be composed primarily of a nickel-iron alloy (along with abundant platinum-group elements), with a liquid outer zone, 2900–5000 km deep, and a solid inner zone, 5000–6370 km deep.

2. *See* CORE REGION.

core–frame concept A model of the central area of the city that shows a core of intensive land use indicated by high-rise buildings (the *CBD), beyond which is the frame, with less intensive land use: warehousing, wholesaling, car parking, medical facilities, and so on.

core–periphery The core—a central region in an economy, with good communications and high population density, which conduce to its prosperity—is contrasted with the periphery—outlying regions with poor communications and sparse population (for examples, *see* UNEMPLOYMENT). 'Either defined in geographical or sociological terms, the center represents the locus of power and dominance and importantly, the source of prestige, while the periphery is sub-ordinate. Simply put, a center–periphery relationship is about hierarchy' (Azaryahu (2008) *Soc. & Cult. Geog.* 9, 4).

Cores are associated with high wages, high technology, and high profit inputs and outcomes. Geographically, these processes have tended to concentrate and segregate—this produces places where core processes dominate and places where peripheral processes dominate. 'For short-hand purposes these may be designated as "core" and "periphery" but they must never be seen as purely one or the other: so-called "core countries" encompass numerous, if minority, peripheral processes; and the opposite is so for "peripheral countries"' (Brown et al. (2002) *GaWC Res. Bull.* 236). 'When transport costs fall below a critical value, a core–periphery spontaneously forms, and nations that find themselves in the periphery suffer a decline in real income' (Krugman and Venables (1995) *Qly J. Econ.* 110, 4). Lanaspa and Sanz (2001) *Papers Reg. Sci.* 80 add congestion costs and infrastructure to Krugman's model, and Baldwin and Forslid (2000) *Economica* 67, 267 introduce Romerian product innovation growth into the model. M. Fujita and J.-F. Thisse (2002) develop a core–periphery model in which the agglomeration effects from concentrating R&D activity in the core, combined with relatively low transportation costs, generate sufficient value added to more than compensate the periphery for the loss of R&D activity.

The model has been criticized—it has been argued that *uneven development is not the inevitable consequence of development, but of the particular *mode of production used to bring about that development (Harris in L. Blume and S. Durlauf 2006). Copus (2001) *Eur. Plan. Studs* 9, 4 finds the validity of conventional (spatial) models of peripherality 'increasingly questionable'.

core region In a nation, a centre of power where innovation, technology, and employment are at a high level, such as Madrid or south-east England. See Harrison (2007) *J. Econ. Geog.* 7, 3 on England's Core Cities Group, and Pain, no date, *GaWC Res. Bull.* 220 on the London core and the global 'mega-city region'.

(⊕) SEE WEB LINKS

- Brand and Bhatti (2006) on the core region of Madrid.

corestone *See* BOULDER FIELD.

CORINE The Coordination of Information on the Environment is an ecological commission of the European Communities located in Brussels, Belgium. Its main goals are to coordinate information and action within the EU, to define and protect biotopes, to combat air pollution, and to preserve the ecology of the Mediterranean region.

Coriolis force An apparent, rather than real, force which causes the deflection of moving objects, especially of air streams, through the rotation of the Earth on its axis. It shows up, for example, in the movement of an air stream, relative to the rotating Earth beneath it. The Coriolis force is proportional to the wind speed U (z), and the **Coriolis parameter**, $f = 2\,\Omega \sin\Phi$ —where Φ is the latitude and Ω is the Earth's rate of rotation—acts perpendicular to the wind direction (Orr et al. (2005) *Weather* 60, 10).

corporate governance The system by which companies are directed and controlled; it is concerned with systems, processes, controls, accountabilities, and decision-making at the heart of and at the highest level of an organization. Issues of corporate governance are central to corporate social responsibility (Gouldson (2006) *Area* 38, 4). See also A. Bongenaar (2001) and G. Clark and D. Wójcik (2007). There is a growing European market for corporate governance; large continental companies appear to be adopting shareholder-friendly practices more consistent with global financial market imperatives than their national traditions; see Bauer et al. (2008) *J. Econ. Geog.* 8, 4.

corporate social responsibility The belief that a business or corporation is responsible for the economic, political, or social impacts of its practice; 'the attempt by large firms to make a wider

contribution to society, often in recognition of the impact their business activities have on communities, localities, and citizens. It can be described as a form of ethical capitalism', N. Castree, R. Kitchin, and A. Rogers, (2013).
See A. Barry 2004 in W. Larner and W. Walters, eds (2004).

corrasion The erosive action of particles carried by ice, water, or wind (although the term is most often used in a fluvial context). Corrasion is another term for *abrasion.

corridor A limb of one state's territory cutting through another state, usually for access. The most famous is the Polish corridor (Howkins (2006) *J. Transp. Geog.* 4, 4). A **corridor of commerce** extends along a line of communication, such as the Mekong River (Furlong (2006) *Pol. Geog.* 25, 4), or the UK 'M4 corridor' (Coe and Townsend (1998) *TIBG* 23, 3).

corrie *See* CIRQUE.

cosmopolitanism 'The ethic and practice of being open to, and respecting, ways of life that may vary significantly from one's own. To be "cosmopolitan" is to be "well travelled" and genuinely curious about the world beyond one's doorstep. It is to have experienced, and been enriched by, all manner of other people, places, customs, and norms' (N. Castree, R. Kitchin, and A. Rogers 2013). Datta (2012) *Antipode* 44, 3, 745 describes and analyses cosmopolitanism in a Delhi squatter settlement, while Mith and Jenkyns (2012) *Antipode* 44, 3, 640 consider the relationship between cosmopolitanism, development, and civil society on Indian NGOs. A **cosmopolitan distribution** is when 'everything is everywhere' (Smith (2007) *J. Biogeog.* 34, 10).

cost–benefit analysis The attempt to compare the total social costs and

benefits of implementing a decision, usually expressed in money terms; 'the decision-making tool of choice in contested planning decisions' (J. Adams 1995). Costs and benefits include money, externalities—for example pollution, noise, and disturbance to wildlife—and external benefits, such as reductions in travelling time or traffic accidents. Rouwendal and van der Straaten (2008) *CPB Disc. Papers* 98 note that 'when costs and benefits of actual projects are listed and have to be traded off against each other as to determine the net result, consensus often disappears'. They discuss a hedonic method as an attempt to compute the monetary value of the benefits objectively. Hanley et al. (2003) *J. Env. Manag.* 68, 3 observe that decisions over the size of the population affected are crucial in the calculation of aggregate benefits and costs. Castree (2003) *PHG* 27, 3 argues that cost–benefit analysis is one mechanism of 'achieving proxy commodification and can serve as [a precursor] to the creation of real markets'. See A. Boardman et al. (2006).

cost–space convergence Especially in the last 50 years, innovations in air, rail, road, and sea transport technology, communications, and handling technology have caused cost/space (and time/space) to shrink, but differentially. These differential spatial effects largely stem from unequal investment in modal capacity, routes, and terminals at international, national, and local scales, and tend to enhance the importance of the largest demand centres in developed countries (Knowles (2006) *J. Trans. Geog.* 14, 6). *See* CORE–PERIPHERY.

cost surface A three-dimensional 'contour model' representing the variation in costs over an area. See Martin et al. (2002) *Health & Place* 8 for a *GIS-based example.

Cost surface | Cost surface with revenue surface

75
50
25
O

75
50
25
O
revenue surface

spatial margin to profitability
•O

Cost surface a | Cost surface b

Cost surface

cottage industry The production of finished goods by a worker at home, sometimes together with her or his family, such as the production of wool and woollen goods in the Himalayan district of Uttarkashi. See Kristiansen (2002) *Sing. J. Trop. Geog.* 23 on Indonesia.

counter-mapping 1. Mapping running counter to existing mapping, or mapping for conservation 'to overcome predominant power hierarchies, interspecies injustices' (Harris and Hazen (2006) *ACME* 4); to protect indigenous land rights, demarcate and protect 'traditional' territories, manage community lands, protect biodiversity, raise awareness about conflict and mediate resolutions, empower communities, and promote cultural diversity.

2. Bauer (2009) *Cult. Geogs.* 16, 229 uses the term to describe the process and rhetoric of participatory mapping.

counter-radiation The long-wave radiation emitted from the Earth to the *atmosphere after it has absorbed the shorter-wave radiation of the sun.

countertrade A reciprocal trading system between two countries, useful for nations wishing to import high technology but lacking foreign exchange, or exporting goods for which world demand is low. Choi et al. (1999) *J. Int. Bus. Studs* 30, 1 argue that international countertrade is an organizational remedy for uncertainty and institutional deficiency.

counter-urbanization The movement of population and economic activity away from *urban areas—see Mitchell (2004) *J. Rur. Studs* 20, 1 for an exhaustive dissertation on the meaning of this term, and Escribano (2007) *Tijdschrift* 98, 1 for a more nuanced view.

Increased mobility and improved communications, higher incomes, changing household composition, changing technology in manufacturing, and growth of the service sector have led to increased commuting, counter-urbanization, and the urban–rural employment shift (Hodge and Monk (2004) *J. Rur. Studs* 20, 3). Between 1970 and 2000, the shift of employment and population from the urban core to the suburban fringe carried with it huge inequalities in the distribution of income among Detroit area municipalities. The lowest-income communities, located in the areas suffering the greatest population and employment decline, concentrate in and around the central city; the most affluent are in Detroit's outlying north-west suburbs (Jacobs (2003) *Urb. Studs* 40).

country rock A pre-existing rock which has suffered later igneous *intrusion.

Countryside Stewardship Scheme (CSS) Introduced in the UK in 1991, the CSS encourages land managers, in practice mostly farmers, to restore and maintain traditional features of the countryside for wildlife conservation and the enjoyment of the public. See Wilson and Hart (2001) *Sociologia Ruralis* 41, 2 for an evaluation.

coupled human-environment system (CHANS) Any system displaying interactions between humans and their environments. The study of these systems considers stability and change between human and environment systems and the management of these interconnections. E. Sheppard and R. B. McMaster, eds (2004) note that social and natural systems are inseparable.

coupling (geomorphic coupling) The connectivity within the components of a system. For example, the physics, chemistry, and biology of the ocean are coupled since the oceanic CO_2 cycle is controlled by the ocean circulation, the supply of sunlight, and major and trace nutrients delivered by the oceanic and atmospheric circulation. Strong couplings link human dynamics, biology, biochemistry, geochemistry, geology, hydrology, and geomorphology; see Barry et al. (2009) *Geomorph* 103, 3, 496. Coupling behaviour conditions the way a system responds to disturbance, and is therefore important in determining the response to human-induced, climatically induced, or tectonically induced environmental change. See Harvey, (2002) *Geomorph.* 44, 3-4, 175 on coupling

within a fluvial system, and Tulaczyk in D. M. Mickelson and J. W. Attig, eds (1999) considers coupling between ice and till at the glacier bed.

coupling constraint In *time–space geography a limit to an individual's actions because of the necessity of being in the same space and time as other individuals. Fixed activities dictate strict coupling constraints while flexible activities allow more fluid coupling in space and time (Miller (2005) *Geog. Analys.* 37).

coverage Used in GIS, coverage refers to a single layer, such as soil or a stream. K.-T. Chang (2014) defines a coverage as a 'special kind of geographic feature, with the distinguishing characteristics that other features have one particular value associated (such as a road number, which remains constant over all the road's extent) whereas a coverage typically conveys different values at different locations'.

cover crop Any seasonal crop that is planted to protect and/or improve the soil. 'Investing money, time, and management in a crop that grows during the period between cash crops is a bold, but often highly productive strategy for a farming system that is healthy over the long term—both economically and ecologically' (APAARI *2012–2013 Cover Crop Survey*).

(((⊕))) SEE WEB LINKS

• APAARI website with *2012–2013 Cover Crop Survey*.

crag and tail A large rock outcrop, with a rampart of *till on its lee side, formed by erosion on the up-glacier side of the outcrop and deposition of debris in the low-pressure lee area. Glasser and Bennett (2004) *Prog. Phys. Geog.* 28, 1 define **micro crag and tails** as small tails of rock that are preferentially protected from glacial abrasion in the lee of resistant grains or mineral crystals on the surface of a rock.

crater A circular depression around the vent of a volcano. Craters form the summit of most volcanoes. They occur where lava overflows and hardens or where the walls collapse as the *magma sinks down the vent after an eruption. Funnel-shaped craters are typical of *stratovolcanoes, while kettle-shaped craters are characteristic of *shield volcanoes.

craton A core of stable continental crust within a continent and composed wholly or largely of *Precambrian rocks with complex structures. Two types are recognized: platforms, which are parts of cratons on which largely undeformed sedimentary rocks lie, and *shields. See Zheng et al. (2004) *Geology* 32, 3 on the North China Craton.

creative cluster Florida (2004) *Next American City* 5 identifies as crucial to creativity: creative human capital; innovative activity associated with high-tech agglomerations; and the quality of a place as 'liberal, tolerant, and capable of attracting skilled people able to generate new ideas'. Florida also envisages the creative class as a mobile section of the population, willing and able to uproot in order to find the best jobs and lifestyles, but this thesis is attacked by J. Peck (2011) and by Borén and Young (2013) *AAAG* 103, 1, 195. Storper and Venables (2004) *J. Econ. Geog.* 4, 4 argue that face-to-face contacts are the most fundamental aspect of proximity, particularly important in environments where information is imperfect, rapidly changing, and not easily codified.

creative industries Those industries that focus on creating and exploiting intellectual property; for example, fashion designs, computer games, or providing creative services for business, such as advertising (also called the **creative economy**).

In 1998, the cultural industries employed 5% of the total UK workforce at the time, making up 4% of UK GDP—a larger share than for any traditional manufacturing industry (UK Department of Media, Culture and Sport, *Creative Industries Mapping Document*). Peck (2005) *Int. J. Urb. & Reg. Res.* 29, 4 argues that 'an unobserved creative milieu, that attracts artists, increases local economic dynamism', while Wojan et al. (2007) *J. Econ. Geog.* 7, 6 feel that '**creative milieu**' remains an amorphous construct.

creep The slow, gradual movement downslope of soil, *scree, or glacier ice. Most creep involves a deformation of the material, i.e. *plastic flow. **Soil creep** is more frequent in slopes of calcareous rock with surface deposits and slope angles of approximately 25–35°. The rate of **ice creep** is a function of the *shear stress applied and is expressed in the Glen–Nye flow law:

$$\Sigma = k\tau^n$$

where:

Σ = shear strain(flow) rate

τ = stress

n = a constant between 2–4 (typically 3 for most glaciers) that increases with lower temperature

k = a temperature-dependent constant.

creolization The hybridization of a culture, as it absorbs and transforms forces from outside; the production of new local forms in response to *globalization; 'a term now widely used to refer to processes of cultural change in the Caribbean and elsewhere, that contributed to the development of culturally hybrid social forms' (Lambert (2004) *PHG*

28). Carter (2005) *Area* 37, 1 thinks that creolization overcomes 'the fixed and essentialized assumptions regarding both identity and territory'. Adams and Ghose (2003) *PHG* 27, 4 discuss **cultural creolization**—tandoori pizza, anyone?

Cretaceous The youngest period of *Mesozoic time from *c.*136 to 65 million years BP.

(((•))) **SEE WEB LINKS**
• An overview of the Cretaceous period.

crevasse A vertical or wedge-shaped crack in a glacier, varying in width from centimetres to tens of metres. Its maximum depth is about 40 m, as, at that depth, ice becomes plastic and any cracks merge together. **Transverse crevasses** occur when the ice extends down a steep slope. **Longitudinal crevasses** form parallel with the direction of flow as the ice extends laterally. **Marginal crevasses** occur across the sides of a glacier as friction occurs between the ice and the valley walls. **Radial crevasses** fan out when the ice spreads out into a lobe.

crime, geography of The analysis of crime—its effects, the offences, and the offenders—to understand the interactions between crime, society, and space. In the UK, Australia, New Zealand, and North America, the *inner city is a crime hot spot; Canada's Research and Statistics Division notes that 'several indicators were found to have a significant effect on crime levels [in Saskatoon] including higher proportions of single people and youth not attending school as well as lower average household incomes'. The residential pattern of the offenders resembles the pattern of offences. See Monmonier (2006) *PHG* 30, specifically 375-8, on mapping crime.

Feminist geographers stress that, while most girls are brought up to fear violence by strangers in public places (thus hugely limiting female spatial mobility), most violence against women is located within the home. It's difficult for many women to acknowledge that their homes are not a place of safety; see Meth (2003) *Geoforum* 34, 3. Brunel (2005) *Gender Place & Cult.* 12, 3 shows that organizations devoted to reducing domestic violence are still rare in urban environments and virtually non-existent in rural ones. On the perception of crime, Stenning (2005) *TIBG* 30, 1 addresses the 'wider tendencies to characterize and stigmatize working-class communities'. Pain (2000) *PHG* 24 reviews work on fear of crime, social identity, and exclusion, and 'local ethnographies of fear'.

Herbert and Brown (2006) *Antipode* 38, 4 note that the onset of *neoliberalism in the United States coincided with 'an unprecedented expansion of punishment practices that intensify social divisions rooted in class and race'. Pallot (2005) *TIBG* 30, 1 argues that the use of the peripheries as a place of exile and detention in the USSR was driven by the pursuit of national economic goals by a powerful central state. Post-1992, the continued expulsion of people to the peripheries may be seen as Moscow cleaning itself of 'unwelcome elements' at the expense of provincial Russia. S. Herbert (2006) finds that the community is 'unbearably light' as a basis for political action or a foil against violent crime—'only rarely can people rally around a sense of common place'—but Martin (2003) *AAAG* 93, 3 is more sanguine about the role of place in motivating activism. See also Martin (2000) *Geography* 21, 5.

critical In a *geomorphological system, the magnitude of an input—object or process—required to cross a *threshold. Thus, **critical shear stress** in an overland flow is that point where the *shear stress is great enough for incision to occur; or within a channel, for mobilizing sediment (*entrainment). This point occurs at the **critical distance** from the *drainage divide, where the flow reaches the **critical depth**. When the velocity of a flow is less than critical, it is considered a **subcritical flow**. Subcritical flows are characterized by a water surface that is smooth and on which a wave produced at the water surface is quickly damped out; subcritical flows are also termed 'tranquil' (K. J. Gregory and A. S. Goudie 2011). **Supercritical flows** are highly erosive, moving rapidly and efficiently through the channel, often over-shooting bends. Turbulent mixing is less intense (M. Kay 1998).

As quantified by Bagnold (1980, *Proc. R. Soc. Lond.* A 372, 453), **critical stream power** is the amount of stream power needed to transport the average sediment load. Ferguson (2005) *Geomorph.* 70, 33 amends Bagnold's equation, and Parker et al. (2011) *Geomorph.* 126, 51–65 present an interesting review on critical stream power. This paper is critiqued and supplemented by Camenen (2012) *Geomorph.* 139–40, 34. Gob et al. (2010) *ESPL* 35 investigate the relationships between sediment size and critical stream power.

critical cartography A cartography grounded on *critical theory. The MIT SLAB website describes critical cartography as 'mapping the unmapped; [noticing] people and things that are overlooked or denied, and usually [debunking] professionalism with an anti-sleek visual vocabulary and techniques'. J. Crampton (2010) writes that 'critical cartography assumes that maps *make* reality as much as they represent it'.

⊕ SEE WEB LINKS
- MIT SLAB website.

critical geography (critical human geography) The umbrella term for a group of geographical concepts and procedures that are centred on opposition to repressive and inequitable power relations in: *capitalism, *class, *colonialism, disability, ethnicity, gender, race, and sexuality. The keystone to all of these is the contention that 'knowledge can never come in an unpoliticized form' (J. W. Crampton 2007). Critical geographers stress the role of dominance and confrontation in the production and reproduction of *landscape, *place, and *space: see, for example, Thompson-Fawcett and Bond (2003) *Prog. Plan.* 60, 2 on the urbanist discourse. *ACME* (an online international journal for critical and radical analyses of the social, the spatial and the political) sees critical geographies as 'part of the praxis of social and political change aimed at challenging, dismantling, and transforming prevalent relations, systems, and structures of capitalist exploitation, oppression, imperialism, neo-liberalism, national aggression, and environmental destruction'.

Although Blomley (2007) *PHG* 31, 1 judges that critical geography has become deeply entrenched within the academy, many share Cloke's frustration with 'our apparent inability to retain a critical political edge in human geography' (2002, *PHG* 26). For Baeten (2002, *Geografiskal B* 3–4), urban geography 'fails to crystallize in a convincing political project that would provide a credible alternative for the poverty-generating capitalist shaping of today's city'. However, Peck and Wills (2000) *Antipode* 32, 1–3 note that being radical in the late 1960s 'involved a very different cluster of beliefs, ideas and affiliations than might be expected today'. Mitchell (2000) *2nd Int. Crit. Geog. Conf.* 2 argues that critical geography should be 'for people, not just ourselves'. Oberhauser (2007) *AAAG* 97,

2 judges C. Katz (2004) as 'a model for research in critical geography'. See Simonsen (2004) *Geoforum* 35, 5 and Blomley (2008) *PHG* 32, 2 for reviews of this sub-discipline; see also Oswin (2008) *PHG* 32, 1. Unwin's words (2000, *TIBG* 25, 1) are inspirational: 'A critical geography needs to engage with the everyday practices of all of us who live in the places that we do; it needs to focus on the needs and interests of the poor and the underprivileged; it remains a very modern enterprise, retaining a belief that it is possible to make the world a "better" place.' See Glassman (2011) *PHG* 35, 5, 669 on **critical development geography**.

critical geopolitics *See* GEOPOLITICS.

critical legal geography The study of the connections between critical legal studies and *critical geography scholarship. Legal geography scholarship stresses that nearly every aspect of law has some spatial frame of reference, and that social spaces, lived places, and landscapes are inscribed with legal significance. Distinctively legal forms of meaning are projected on to every segment of the physical world. Such fragments of a socially segmented world are not simply the inert sites of law, but are inextricably implicated in how law happens (I. Braverman, N. Blomley, D. Delaney, and A. Kedar, eds, 2013). See N. Blomley, D. Delaney, and R. Ford, eds (2001).

critical race theory A theoretical approach used to deconstruct racial ideologies, and analyse the politics of 'race' and the links between nationalism—or localism—and racism (Dunn (2006) *Geog. Res.* 4, 4). The interdisciplinary field of critical race studies pursues the claim that race has no necessary epistemological status in itself, but depends on the context and organization of its

production for its political effects (Anderson (2008) *Cult. Geogs* 15, 2).

critique A process of looking carefully at the assumptions made about any form of knowledge; a critique is not necessarily about finding fault; see J. W. Crampton (2010). The fundamental questions are: 'what is an authority?' and 'who shall have authority?', which means that critique is a political project.

crop rotation The practice of planting a succession of crops in a field over a period of years. Rotations can maintain field fertility since different crops use different soil nutrients, so excessive demands are not made of one nutrient. In certain rotations, plants like legumes (peas and beans) are grown to restore fertility. See D. Stone (2005) on medieval agriculture, and Sainju et al. (2006) *J. Env. Qual.* 35 on contemporary practice.

(((⊕))) SEE WEB LINKS

• Wikipedia entry about crop rotation in India.

cross-bedding (current bedding) In a sedimentary rock, the arrangement of beds at an angle to the main bedding plane.

cross-correlation analysis A way of identifying any changes in land use that have occurred in a previously mapped area. See Butson and Fraser (2005) *3rd Int. Wkshp. Analysis Multi-Temp. Rem. Sens.* for an evaluation of the technique.

crowd-sourcing (crowdsourcing) Getting data or ideas (user-generated content) by consulting large numbers of people, especially from the 'online' community; Wikipedia is a good example. The OpenStreetMap project is a crowd-sourced project, where detailed map data are produced by people pooling GPS and metadata. Matt Ball's blog of October 2010 explains the distinction between crowd-sourcing, volunteered geographic information, and authoritative data.

(⊕) SEE WEB LINKS
• OpenStreetMap site.
• Matt Ball's blog.

crude rate A *vital rate which is not adjusted for the age and sex structure of a population.

crumb A spheroidal cluster of soil particles, i.e. a type of *ped.

crust The outer shell of the Earth including the continents and the ocean floor; the *lithosphere. *Sial overlies *sima in the continental crust; the oceanic crust is mostly sima.

cryofront *See* CRYOTIC.

cryogenic (cryergic) Processes carried out by ground ice. Needle ice and frost heaving are connected with the development of cryogenic phenomena (Grab et al. (2004) *Geografiska. B* 86, 2). *See* FROST HEAVING; FROST WEDGING; CRYOPLANATION.

cryopediment In *periglacial environments, a low-angle, concave, piedmont footslope developed by slope retreat, brought about through frost *weathering and the sapping of slopes, and by *nivation surfaces. Cryopediments may coalesce to form a **cryopediplain** (Tzudek (1993) *Permaf. & Perig. Procs* 4, 1—with useful photographs).

cryoplanation The lowering and smoothing of a landscape by *cryogenic processes. 'To deny the fairly widespread existence of features commonly called cryoplanation benches, terraces, and pediments would be foolish: to claim that there is anything approaching an adequate explanation of their origin(s) would be even more foolish . . . there is little to say about cryoplanation other than

the concept needs to be challenged directly' (Thorn and Hall (2002) *PPG* 26, 4).

cryosphere The ice at or below the Earth's surface, including *glaciers, *ice-caps and *ice sheets, *pack ice and *permafrost. Climatic cooling increases the cryosphere (Lewis et al. (2007) *GSA Bull.* 119, 11). The mass balance of the cryosphere is most likely contributing to sea-level rise (J. Bamber and A. Payne 2004).

cryostatic pressure The pressure exerted on rocks and soil when freezing occurs. As the *freezing front advances, the pressure of the trapped soil moisture increases (Mackay and Mackay (1976) *Can. J. Earth Scis* 13). This pressure can separate individual grains of soil, forming a mass of fluid mud, which may dome up the ground, or form mud blisters.

cryotic Having temperatures below 0 °C. Van Everdingen (1976) *Can. J. Earth Scis* 13 proposes that 'cryotic' be used to describe permafrost on the basis of temperature, and that 'frozen' be restricted to ground in which part or all of the pore water is ice.

cryoturbation A general term describing all frost-based movements of the *regolith, including *frost heaving and *gelifluction. Grab (2005) *PPG* 29, 2 analyses cryogenic mounds.

cuesta A ridge with a *dip slope and a *scarp slope. *See* ESCARPMENT. For **cuesta topography**, see Pánek et al. (2008) *Geomorph.* 95, 3–4.

cultural capital The symbols, ideas, tastes, and preferences that can be strategically used as resources in social action, and serve as markers of collective identity and social difference. Cultural capital includes cultural goods: pictures, books, dictionaries, instruments, machines, and so on. See Hubbard (2002)

Env. & Plan. A 34. The commodification of cultural capital is carried out by the **cultural industries** which convert cultural capital to economic capital. R. Florida (2002) argues that urban fortunes increasingly turn on the capacity to 'attract, retain and even pamper a mobile and finicky class of "creatives"'.

It's possible to be high in cultural capital, but low in economic capital: S. Fernandes (2006), for example, notes that Cuban arts and popular culture have been commercialized to entice foreign investment. Ottaviano and Peri (2006) *J. Econ. Geog.* 6, 1 find that US-born citizens living in cities of increasing **cultural diversity** earned significantly higher wages. Khawaja and Mowafi (2007) *Scandi. J. Pub. Health* 35, 5 find that cultural capital has significant associations with general and mental health status in Lebanese women.

cultural distance A gap between the culture of two different groups, as between Vietnamese immigrants and the Czech majority in the Czech Republic (Drbohlav and Dzúrová (2007) *Int. Mig.* 45, 2).

cultural ecology An investigation that focuses on the dynamic interactions between human societies and their environments with an understanding of key ecological concepts such as resilience, stability, and biodiversity. Culture is seen as the primary adaptive mechanism used by human societies to deal with, understand, give meaning to, and generally cope with their environment (T. Darvill 2002). The landscape cannot be understood without attention to the behaviour of land managers, who, after all, are the critical agents modifying or transforming the landscape; see Chowdhury and Turner (2006) *AAAG* 96, 2.

cultural economy At its most simple, that part of the economy dealing with

cultural goods and services: film, TV, radio, photography, advertising, and so on. However, with admirable clarity, Gibson and Kong (2005) *PHG* 29, 5, 54 explain why this term is so difficult to define.

cultural geography The study of the impact of human culture on the landscape; 'the ways in which place and identity are embedded in a range of cultural landscapes, and the ways in which those social and material landscapes have reflected and influenced various experiences and notions of movement' (Mains (2004) *J. Cult. Geog.* 22, 1). Cultural geographers emphasize the symbolic dimension of human activities and the relevance of historical understanding of societal processes, and are committed to an interpretative epistemology, that is, a philosophy that investigates the origin, nature, methods, and limits of human knowledge as perceived and experienced by people and organizations, rather than as perceived only by social scientists. As things travel, their meanings and material nature can change; 'things themselves therefore have complex cultural geographies' (Crang in P. Cloke et al., eds 2006). Kirsch (2012) *PHG* 1 believes that cultural geography is currently suffering 'something of an identity crisis', a theme also expressed by Wylie (2010) *Cult. Geogs.* 17, 211.

Mitchell (2002) *Antipode* 34, 2 gives a breezy account of cultural geography, later writing that 'inequality, domination, oppression, exclusion, and power (and still the possibility for good lives despite all this)—at all scales from the household to the globe—are the true object of cultural geographic study' (2004, *J. Cult. Geog.* 22, 1). Lorimer (2005) *PHG* 29, 1 writes, somewhat diffusely, on the tendency for cultural analyses to cleave towards a conservative, categorical politics of identity and textual meaning, which can, he argues, be overcome 'by allowing in much more of the excessive and transient aspects of living'. See Lorimer (2007) *PHG* 31, 1 and K. Anderson et al., eds (2004).

cultural hearth The location in which a particular culture has evolved. Voigt-Graf (2004) *Glob. Networks* 4, 1 uses the term as 'the country, region or place of origin of migrants and their descendants'.

cultural landscape The impact of cultural groups in shaping and changing the natural landscape. 'Landscape itself is a cultural image; a way of symbolizing, representing, and structuring our surroundings' (P. Cloke et al., eds 2006). Meinig 1979 (in D. Meinig and J. Jackson, eds 1979) sees that individuals interpret the landscape in many ways, 'all of them dependent on the viewer's mental ideas evoked by . . . their previous experiences'. Röhring (2006) *IASCP* argues for the cultural landscape as a regional common good.

cultural turn The increased focus, from the 1990s onwards, on culture as a geographical agent and product: 'a reorientation of human geography's interdisciplinary concerns toward the wide field of cultural studies' (Barnett (1998) *Antipode* 30, 4). 'It is because all human realities are expressed through cultures that a cultural turn was needed in geography' (Claval (2007) *Tijdschrift* 98, 2). 'With the cultural turn in economic geography, the emphasis on economic necessity . . . has been contested by an agency-orientated cultural reading' that views engagement in such spaces as about the search for fun, sociality, distinction, discernment, the spectacular, and so forth, and more recently by a geographically sensitive approach that ascribes agency to affluent populations and economic rationales to deprived

populations (Williams and Paddock (2003) *Geografiska B* 85, 3).

Thornes and McGregor in S. Trudgill and A. Roy (2003) propose a **cultural climatology**: 'the study of the processes of, and the interactions and feedbacks between, the physical and human components of the climate system at a variety of temporal and spatial scales' while Gregory (2006), *Geomorph.* 79, 172, argues for a **cultural geomorphology**; for example, 'with knowledge of processes and changes of river channels it seems expedient to consider cultural reactions and perceptions'.

culture 'The stuff of culture [is] meaning, symbols, signification' argues Barnes (2003) *Urb. Geog.* 24, 6, but J. Anderson (2009) is kinder to the student: 'culture is what people do. Different cultures are produced when individuals get together to live their lives differently'. In a study of MTV, Rosati (2007) *TIBG* 32, 4 argues that 'cultural forms—and the authority of their production—are always implicated in a wider field of struggle over how the practice and meaning of daily life will be organised and in whose interests'. 'Popular culture . . . is one of the ways in which people come to understand their position both within a larger collective identity and within an even broader geopolitical narrative' (Dittmer (2005) *AAAG* 95, 3). Brons (2006) *Tijdschrift* 97, 5 sees culture as a 'map for behaviour'. See Hedberg and Kepsu (2003) *Geografiska B* 85, 2 on migration to Sweden as an expression of culture.

cultures of work The attitudes, values, beliefs, modes of perception, and habits of thought and activity in the working environment. Cowen (2005) *Antipode* 37, 4 documents changing cultures of work in Canada, associated with the shift towards higher education: increasing mobility and individualism in career choices. Faulconbridge (2008) *J. Econ. Geog.* 8, 4 shows how managers in TNCs act as **cultural entrepreneurs**, driving change in institutionalized cultures of work through strategies that alter the cognitive frames of workers. See Jarvis (2003) *Area* 34, 4.

cumec A measurement of *discharge. One cumec is one *cu*bic *m*etre of water per second.

cumulative causation The unfolding of events connected with a change in the economy, as a consequence of the *multiplier effect. Cumulative causation can be set in motion where the expansion of the cluster, through attraction of firms with complementary competencies, adds to the initial attractiveness of the cluster (Maskers and Malmberg (2007) *J. Econ. Geog.* 7, 5).

Kaldor (1981) *Econ. appliquée* 34, 4 argues that any initial differences in productivity between regions are sufficient to produce progressive, self-reinforcing divergence—but this demonstrates how the nature of 'historicity' in the new economic geography models is very circumscribed (Martin and Sunley (2006) *J. Econ. Geog.* 6, 4). Improvements made through cumulative causation are made at a cost to some other part of the economy; although underdeveloped regions offer the advantage of low-wage labour, these benefits tend to be offset by the agglomeration economies in the industrialized regions. 'Therefore in order to overcome the [negative] effect of cumulative causation intervention on the part of government is necessary' (Cibulskiené and Butkus (2007) *Jahrbuch für Regionalwissenschaft* 27). The process is not limitless and can be reversed through negative externalities, such as congestion costs, or reductions in trade costs. Simulation models tend to show that large changes in trade costs may be required to cause deconcentration and that at intermediate levels of trade costs concentration remains high (Weiss (2005) *ADB Institute Discuss. Paper* 2005/04).

cumuliform In the shape of a *cumulus cloud. **Cumuliform convection** is vertical convection.

cumulo-nimbus A low-based, rain-bearing *cumulus cloud, dark grey at the base and white at the crown, which spreads into an anvil shape, as it is levelled by strong upper-air winds.

cumulus An immense, heaped cloud with a rounded, white crown and a low, flat, horizontal base, extending as high as

Cumulative causation

5000 m. Updraughts within the cloud are strong—up to 10 m s^{-1}, and cloud growth is often rapid, through *entrainment and the saturation of surrounding air. **Cumulus congestus** is a swelling, small cumulus which often becomes *cumulo-nimbus.

current A horizontal movement of water, as along a stream or through an ocean. The rate of flow of a river current varies with depth because friction operates along the bed and sides (R. Chorley et al. 1985). **Tidal currents** are associated with the rise and fall of the sea, and the velocities of ebb and flow vary with the morphology of the coast and any outflow of fresh water; see FitzGerald et al. (2002) *Geomorph.* 48, 1–2. **Rip currents** are generally strong, shore-normal, jet-like flows that originate within the surf zone and are directed seaward through the breakers. They form in the nearshore zone and balance the inflow of seashore currents, influence the morphology of the shoreline, and may be important for transporting fine sediments offshore (MacMahan et al. (2005) *Marine Geol.* 214, 1–4). *Ocean currents are driven by the planetary winds; see Hidaka (1952) *J. Oceanograph. Soc. Japan* (available online, captivating English, many equations).

current bedding In a sedimentary rock, bedding which is oblique to the 'lie' of the formation as a whole. The structure is not due to tilting or folding, but develops when sandbanks are built up in shallow water, or where sand dunes accumulate from wind-blown sands, where the pattern of bedding reproduces all or part of the outline of the dunes.

curvature In meteorology, the wind speed divided by the radius of curvature of the bending air stream. Conventionally, *cyclonic curvatures have positive, and anticyclonic curvatures negative, radii of curvature in the Northern Hemisphere. See W. Saucier (2003).

cusp A small hollow in a beach, U-shaped in plan; the arms pointing seawards. Beaches tend to have a series of cusps, formed when outgoing *rip currents and incoming waves set up nearly circular water movements. They form by self-organization, wherein positive feedback between swash flow and the developing morphology initiates the development of the pattern, and negative feedback—from the circulation of flow within the beach cusps—causes pattern stabilization (Coco et al. (2003) *J. Geophys. Res.* 108, C3). Ciriano et al. (2005) *J. Geophys. Res. Oceans* 110, C02018 suggest that beach cusp evolution might control low-mode edge wave dynamics.

cuspate foreland An accretion of sand and shingle, shaped by *longshore drift and constructive waves from two directions. See Roberts and Plater (2007) *Holocene* 17, 4.

customs union A common market encompassing two or more states within whose boundaries there is free trade; with no tariffs or barriers to the movement of goods. In 1996, Turkey established a customs union with the EU; Falcioğlu and Akgüngör (2008) *Eur. Plan. Studs* 16, 2 find that Turkey's economic integration with the EU is a significant factor in explaining the change in industry concentration.

cut-off low *See* COLD LOW.

cwm *See* CIRQUE.

cyberspace The information 'space' created by information technologies, notably the internet, the worldwide web, and virtual reality. Kitchin (1998) *PHG* 22, 3 argues that spatiality is central to understanding cyberspace. 'Cyberspace is firmly embedded in the social spaces seen in the "real" world, and this relationship is reflected in how cyberspace is created, conceptualized and studied' (M. Dodge and R. Kitchin 2001). Valentine and Holloway (2002) *AAAG* 92, 2, showing how on-line spaces are used, encountered, and interpreted within young people's off-line, everyday lives, demonstrate the way the real and the virtual are mutually constituted. M.-P. Kwan (2001) proposes a behavioural model of cyber-accessibility and examines the way notions underlying conventional accessibility measures such as impedance and opportunity set can be extended for measuring individual cyber-accessibility, and suggests that behavioural theories and models may provide a theoretical foundation for cybergeography. N. Lin (2001) contends that social capital has been on the ascent in the past decade—in the form of networks in cyberspace. 'The tales of total cyberspace are contradicted even by its own, very material, necessities: the world of "friction-free" money-flows is organised from places like the City of London and Canary Wharf, where the physicality of location, the implacable materiality of those pompous buildings, even the symbolism of the precise address, is of fundamental importance' (D. Massey, Open2Net).

cyborg An android whose artificial intelligence and sensitivity is comparable with that of their human creators. As such, 'the cyborg metaphor has been deployed to challenge disembodied, dualistic, masculinist, and teleological bodies of knowledge' (Gandy (2005) *Int. J. Urb. & Reg. Res.* 29, 1, 26). Gandy describes the way that urban environments have become cyborgs through a re-*materialization of the city, away from a purely mediated/virtual/digital urban experience, to one that privileges the negotiated in-betweenness of urban space, virtuality, and the human body.

cycle of poverty A vicious spiral of poverty and deprivation passing from one generation to the next. Poverty leads very often to inadequate schooling and then to poorly paid employment. As a result, the affordable housing is substandard—low housing costs in poor neighbourhoods attract migrants from rural areas—leads to overcrowding, overuse of facilities and services, and can also contribute to the perpetuation of the cycle of poverty (Okoko (2004) *Habitat Int.* 28, 3; *US Dept. Housing & Urb. Dev. (HUD)* 1995). Children growing up in such areas start off at a disadvantage, and so the cycle continues; see Chilton et al. (2007) *Indian J. Med. Res.* 126. J. Lin and C. Mele (2005) argue the need for class-specific policies in America, designed to raise educational levels, improve the quality of public schools, create employment, reduce crime, and strengthen the family: 'only a simultaneous attack along all fronts has any hope of breaking the cycle of poverty.'

cyclogenesis The formation of *cyclones, especially mid-latitude depressions. Cyclogenesis occurs in specific areas, such as the western North Atlantic, western North Pacific, and the Mediterranean Sea, and is favoured where thermal contrasts between air masses are greatest. Cyclogenesis occurs when *divergence in the upper *troposphere removes air more quickly than it can be replaced by convergence at ground level. The net result is low pressure. The significance of upper-air movements in cyclogenesis is also indicated by the link with *Rossby waves; see Molinari et al. (forthcoming) *J. Atmos. Sci.* Surface depressions develop below the downstream, or eastern, limbs of Rossby waves, where the airflow is divergent. The routes of mid-latitude cyclones (*depression tracks) closely parallel the movements of the upper-air *jet stream.

cyclomatic number In *network analysis, the number of circuits in the network. It is given by:

$$\mu = e - v + p$$

where e = number of edges, v = number of vertices (nodes), p = number of graphs or sub-graphs. A high value of the cyclomatic number indicates a highly connected network. Gorman and Maleki (2002) *Telecomm. Policy* 2 note the relationship between the level of economic development of a region and the cyclomatic number of its major transport networks.

cyclone A *synoptic-scale area of low *atmospheric pressure with winds spiralling about a central *low. As air near ground level flows into a cyclone, its absolute *vorticity increases, and it is therefore subject to horizontal *convergence, causing air to ascend. This rising causes cooling, which often leads to *condensation, so that *precipitation is associated with cyclones. See

Kristjánsson and McInnes (1999) *Q'ly J. Royal Met. Soc.* 25, 560 on Greenland and North Atlantic cyclones; see McTaggart-Cowan et al. (2007) *Month. Weather Rev.* 135, 12 on Hurricane Katrina. See also Löptien et al. (2008) *Clim. Dynamic.*, online.

cyclone wave The wave-like distortion of flow in the middle and upper troposphere associated with a mid-latitude depression. See Orlanski and Katzfey (1991) *J. Atmos. Scis* 48, 17.

cyclostrophic Referring to the balance of forces in a horizontal, tightly circular flow of air. **Cyclostrophic flow** is a form of gradient flow parallel to the isobars where the centripetal acceleration exactly offsets the horizontal *pressure gradient. The **cyclostrophic wind** is the horizontal wind velocity producing such a centripetal acceleration; see J. Dutton (2002). It equals the real wind only where the *Coriolis force is small, or where wind speed and curvature, and hence centripetal acceleration, are great.

daily urban system The commuting area around a city. Bretagnolle and Pumain (2001) *Cybergeo* 335 show that this area varies according to the definition of 'city'.

dam A barrier for holding back water, usually in the form of a reservoir, which may be used to generate electricity and/or to supply irrigation water. Csiki and Rhodes (2010) *PPG* 34, 6 review current knowledge related to the influence of run-of-river dams on the hydraulics and geomorphology of rivers and suggest types of studies that need to be undertaken to address gaps in current knowledge. See Graf (1999) *Water Res. Research* 35, 4, 1305 on the large-scale hydrologic impacts of the 75000 dams in the United States.

dambo A gently sloping, shallow *savanna wetland, with no permanent river channel, often wider upstream than downstream (perhaps through chemical sapping), and often infilled with sands and clays. Core boulders indicate *etchplanation; see von der Heyden (2004) *PPG* 28, 4.

Dansgaard–Oeschger (D–O) cycles Very rapid climate changes—up to 7 °C in some fifty years—during the *Quaternary period, especially during the most recent glacial cycle. The cause of D–O cycles has yet to be comprehensively established. See Genty et al. (2003) *Nature* 421 on the precise dating of Dansgaard–Oeschger climate oscillations in western Europe.

Darcy's law When the *Reynolds number is very low, the velocity of flow of a fluid through a saturated porous medium is directly proportional to the difference in water pressure at the two sites:

$$V = \frac{h}{P_l}$$

where h is the height difference between the highest point of the water-table and the point at which flow is being calculated, V the velocity of flow, P the coefficient of permeability for the medium in question, and l the length of flow. Darcy's law does not hold good for well-jointed limestone which has numerous channels and fissures. **Darcy's equation** can be written in several ways: see GroundwaterSoftwater.com (March 2004). See also Buchan and Cameron in S W. Trimble et al., eds (2003).

data acquisition The process of sampling signals that measure real-world physical conditions and convert the resulting samples into digitized data that can be used by a computer. Data acquisition systems (abbreviated to **DAS** or **DAQ**) typically convert analog waveforms into digital values for processing.

(⊕) SEE WEB LINKS
• Source of free data acquisition applications.

data mining A process of extracting information from a data set in order to transform it into an understandable structure. Multivariate statistical methods include multiple linear regression, *principal component analysis, *cluster analysis, and the parallel coordinate method; see Sun and Wang (2011) *IEEE International Conference* DOI: 10.1109/ICSDM.2011.5969006.

data model In *geographic information systems, a construct used in the representation of geographic features or surfaces. Thus the *vector data model uses points, lines, and polygons, and the *raster data model uses cell matrices that store numeric values, and the *triangular irregular network represents geography as sets of contiguous, non-overlapping triangles. See K.-T. Chang (2014).

data structure In computer science, the way data are organized and stored in a computer.

datum In *geodesy and *GIS, a geographic coordinate system covering the Earth's surface, such as latitude and longitude, by which phenomena may be located. When loading a GIS, the correct datum for each map source should be used. When working with more than one map source at a time, all data must be converted to a common datum. The most frequently used datum is the Universal Transverse Mercator (UTM).

(⊕) SEE WEB LINKS
• Website providing an exhaustive list of datums.

dead cliff A sea cliff no longer subject to wave attack, with an emerged strandline, often overlain with *shingle at its foot. Dead cliffs result either through a fall in sea level or because they are protected by a broad beach. See A. Guilcher (1958).

dead ice (stagnant ice) Static glacier or ice-sheet material. Ice may stagnate after a surging extension is abandoned, or at the end of a glacial period. As stagnant ice melts, the remnants are covered in *ablation moraine; see Kjær et al. (2001) *Sedimentol.* 48, 5 and Ebert and Kleman (2004) *Geomorph.* 62, 3. Dead ice in valleys may generate *kame terraces.

death rate The number of deaths in a year per 1000 of the population, as measured at mid-year. This is a crude rate—no allowance is made for different distributions of age and sex. For example, in 2003, Sri Lanka had a crude death rate of 5.9%, compared with 10.2% in the UK, 2004. The low rate for Sri Lanka is a reflection of the youth of its population—24.3% are under 14, as opposed to 17.2% in the UK. A **standardized death rate** compares the rate with a real or assumed population which is chosen as standard.

debris Material such as scree, gravel, sand, or clay formed by *weathering. Through the air, debris is transported by *saltation and *deflation; in water, debris moves by rolling, by saltation, in solution, and in suspension. Debris can be carried on a glacier—supraglacially—within a glacier—englacially—or subglacially. **Debris entrainment** is the process whereby ice picks up material: by the freezing of basal ice to the bed as the ice flows forward (Cuffy et al. (2000) *Geology* 28, 4); through the drag between ice and bedrock particles; via the closure of debris-filled basal cavities in the ice; by the refreezing of meltwater; and on the glacier surface. **Debris flow** is the very rapid downslope movement of saturated material, guided by stream channels, and both less deep seated and rarer than a landslide. See Blikra and Nemec (1998) *Sedimentol.* 45 on debris flow, deposits, and controls on debris-flow activity, Tunusluoglu et al. (2007) *Nat. Haz. Earth Sys. Sci.* 7 on artificial neural networks and debris flows, and Hartshorn and Lewkowicz (2000) *Zeitschrift* 44.

debt-for-nature swap A transaction whereby some of a developing nation's foreign debt is written off with the proviso that the debtor nation invests in environmental conservation measures. For example, in 2010 the United States agreed to convert $21m of Brazilian debt into a fund to protect tropical ecosystems.

decalcification The *leaching of calcium carbonate from a soil *horizon by the downward movement of soil water. See van den Berg and Loch (2000) *Eur. J. Soil Sci.* 51, 1.

decentralization (deconcentration) The devolution of decision-making powers to the lowest levels of government authority. Bryld (2001) *Democratization* 8, 3 questions the use of decentralization as a development strategy. However, Faguet (2003) *Crisis States, W. Paper* 29 argues that decentralization in Bolivia 'put real power over public resources in the hands of ordinary citizens'. National governments may use decentralization to regenerate declining regions; see K. Altes (2002) *Urb. Studs* 39, 8 on decentralization in the Netherlands. In 2003, the government of the Republic of Ireland began the transfer of complete government departments to provincial locations.

(((()))) SEE WEB LINKS
• Decentralization in the Republic of Ireland.

deciduous forest In the cold season of temperate latitudes, a tree's water supply is restricted when the temperature falls below 0 °C. To lessen water loss, deciduous trees shed their leaves until the spring brings more available moisture. See Bradshaw et al. (2005) *Ecography* 28, 2.

decision-making Some concern themselves with the *political nature* of

the decision-making unit. Thus Archibugi (2000) *New Left Rev.* 4 proposes replacing the state as decision-maker with 'new flexible frameworks based on the rights of the global citizen, freed from territorial restrictions'. Meanwhile, Hinchcliffe (2001) *TIBG* 26, 2 observes that precautionary and *participatory forms of decision-making predominate in environmental geography. Pickerill and Chatterton (2006) *PHG* 30, 6 value the **consensus-based decision-making** characteristic of alter-globalization (globalization which puts democracy, environmental justice, and environmental protection before economic considerations) activists; Herbert (2007) *Pol. Geog.* 26, 5 writes that, in contrast to 'the anti-democratic tendencies within the WTO', these activists hoped that their consensus decision-making would illustrate the merits of their cause. See Dengler (2007) *Pol. Geog.* 26, 4 on the formation and evolution of ad hoc organizational spaces of decision-making.

As to *who* makes the decisions, Trauger (2007) *Tijdschrift* 98, 1 reports on giving decision-making to the workers, which 'enhances the employees' job satisfaction and makes it more likely that opportunities will be spotted' (Stam (2007) *Economic Geog.* 83, 1). Aitken in W. Craig et al., eds. (2002) explores the potential of Public Participation *GIS to enable local issues to 'jump'. See Viehe et al. (2006) *Geografiska B* 88, 2 on the DaisyGIS decision-making model.

It is argued that 'satisficing' (as opposed to 'maximizing') is a much more faithful account of the way people make choices: 'people pick the alternative that satisfies a given level of aspirations without looking necessarily for the best possible option' (Landemore (2004) *J. Moral Philos.* 1, 177). Medda and Nijkamp (2003) *Integr. Assess.* 4, 3 observe a tendency to favour simpler, model-free modes of planning based on satisficing or compromise. Moreover, 'decision support models tend to be complex and difficult to understand, often resulting in their neglect by policy practitioners' (Lootsma (1999) IIASA/ DAS workshop); thus, Strauss (2008) *J. Econ. Geog.* 8 advocates economic decision-making that 'combines insights from behavioural economics and cognitive science with both quantitative and qualitative methods, and the theorisation of context, embeddedness and the role of institutions'.

declining region A region suffering the economic decline associated with factory closures, outmoded industry, and high unemployment. For government action in declining areas, see Charles et al. (2005) *OECD Territorial Review* (Newcastle and the North-East). See also Klosterman et al. (2006) *Int. J. Env. Tec. & Manage.* 6, 1–2, on the What if?™, *GIS system, as used in evaluating remedial strategies in declining regions.

decommodification Given that a commodity is an artefact, symbol, or idea that can be exchanged, usually for money, *de*commodification would indicate that a commodity no longer has a status as such. This might result when it is thrown away, or when it is no longer deemed by ethical consumers to be exchangeable: if the commodity is produced by child labour, if its production or transportation pollutes the environment, and so on. As an example of this sort of decommodification, in 2006 the Dalai Lama stated that Tibetans should cease wearing clothing lined with skins of endangered animals (Yeh (2013) *J. Asian Studs* 72, 2, 319).

Decommodification is also conceived as 'any political, social, or cultural process that reduces the scope and influence of the market in everyday life' (Vail, *Politics & Soc.* (2010) 38, 3, 310). In a decommodified system, welfare

services such as education and healthcare are provided to all and are not linked to market processes because they are seen as an entitlement rather than as a commodity that must be paid for or traded.

decomposer *See* DETRITIVORE.

deconcentration *See* DECENTRALIZATION.

deconstruction Used for a text, a type of *critical analysis that considers the internal workings of the language and modes of thought used: 'a theory, with methodological implications, that emphasizes the incompleteness of the apparently complete categories and concepts used in the West to make sense of the world' (N. Castree, R. Kitchin, and A. Rogers, 2013). J. Derrida (1989) sees deconstruction as bringing alternative layers of significance in a *discourse to light, pointing out contradictions and unspoken assumptions. (Doel (2005) *Antipode* 37, 2 provides an entertaining, instructive account. Colling (2006) *TIBG* 31, 2 and Jones et al. (2007) *TIBG* 32, 2 make very heavy weather of deconstruction in geography. See Dixon and Jones in J. S. Duncan et al., eds (2004).

deepening A decrease in the central *atmospheric pressure, usually of a low pressure system. See Killer and Petty (1998) *Monthly Weather Rev.* 126, 9.

deep weathering The *weathering of rocks, up to hundreds of metres in depth, by air and water, forming *saprolite—see Smith et al. (2002) *Catena* 49, 1–2 on the premises of deep weathering. Migon and Lidmar-Bergstrom (2001) *Earth-Sci. Revs* 56, 1 hold that deep weathering has played a major part in the geomorphological evolution of Europe.

defensible space A community managed and controlled by its inhabitants (O. Newman 1972). This involves physical and design changes to maximize the potential for surveillance by residents: if public space can be segmented into small, controllable areas, this would encourage residents to exercise control over these areas, and this would discourage crime; see Cozens et al. (2001) *Property Manage.* 19, 3.

deflation The *denudation and lowering of a surface by wind action. Washington et al. (2006) *Geophys. Res. Lett.* 33, L09401 show that contemporary deflation from the Bodélé Depression, Chad, is controlled by topography and wind stress. See French and Demitroff (2001) *Permaf. Perig. Procs* 12 on deflation hollows in periglacial environments.

deforestation The complete clearance of forests by cutting and/or burning. Between 1995 and 2005, the world's top deforesters were Brazil, Indonesia, and Burma. Grainger (2008) *Procs Nat. Acad. Scis* 105, 2 shows that it is difficult to demonstrate tropical deforestation convincingly, and that a 'forest return' effect may be operating, in which forest regeneration in some areas offsets deforestation (but not biodiversity loss) elsewhere.

deglaciation The process by which glaciers thin and recede—through climatic change (Turner et al. (2005) *Geografiska A* 87, 363), or by increased *calving when sea levels rise. Deglaciation is generally accompanied by the release of *meltwater (Tarasov and Peltier (2006) *Q. Sci. Revs* 25, 7–8), and the formation of recessional *moraines (Barrows et al. (2002) *Q. Sci. Revs* 21, 3).

deglobalization The process, unwitting or deliberate, of reversing *globalization, of reducing the volume, scale, and velocity of social and environmental interactions. This might entail: producing for domestic markets only;

protecting local industries through tariffs; and the replacement of global banks by regional ones. However, Evans (2010) *Glob. Economy J.* 10, 2, 3 argues that deglobalization in the form of reduced trade interdependence, reduced capital flows, and reduced migration has few positive economic effects, and that politically, deglobalization makes for less stable domestic politics and a greater likelihood of war.

⊕ SEE WEB LINKS

• Transnational Institute website listing 11 principles of deglobalization.

deglomeration The movement of activity, usually industry, away from *agglomerations, perhaps when congestion makes further agglomeration in a region difficult and expensive. See Werker and Athreye (2004) *J. Evol. Econ.* 14 on deglomeration, agglomeration, and evolutionary economic geography. **Deglomeration economies**, such as lower transport costs, or longer commuting times and congestion externalities (Y. Kanemoto et al. 2005), lead to *decentralization.

degradation The lowering and flattening of a surface through erosion. **Land degradation** is the long-term loss of ecosystem function and productivity caused by disturbances from which the land cannot recover unaided (Bai et al. (2008) FAO/ISRIC–Rome/Wageningen). Lal (1990) *Advances Soil Sci.* 11, 129 estimates that globally, 2000 million hectares of once biologically productive land had been degraded or destroyed. See Kriage (2013) *PPG* 37, 5, 664.

degree-day Generally, a measure of the departure of the mean daily temperature from a given standard; one degree-day for each °C or °F departure above (or below) the standard during one day. See Roeber (1998) *Weather & Fcst.* 13, 3 on degree-day and forecasting.

degrowth The opposite of growth. It entails a reduction of production and consumption—the contraction of local, national, and global economies. Degrowth devotees argue that this does not require individual self-sacrifice or a decrease in well-being; instead, their aim is to maximize happiness and well-being through non-consumptive means—'sharing work, consuming less, while devoting more time to art, music, family, culture and community'.

(()) SEE WEB LINKS
• Degrowth magazine.

deindustrialization The decreasing significance of the manufacturing sector, both in terms of employment, and national production. Bowen (2006) *Prof. Geogr.* 58, 3 distinguishes between **positive deindustrialization**—a normal feature of economic development, since, as manufacturing increases, workers are increasingly employed in services—and **negative deindustrialization**—'characterized by a vicious cycle linking falling incomes, rising unemployment, the erosion of competitiveness', and one result of the 'credit crunch'. In a study of South Africa, Barchiesi and Kenny (2002) *Int. Inst. Sociale Geschiedenis* 47 observe that it is also marked by low-wage economies, the casualization of employment, and increasing *social exclusion.

delta A low-lying area at the mouth of a river, formed from *alluvial deposition, that occurs as the river's silt-carrying

capacity is checked when it slows in the more tranquil waters of a lake or sea. Deltas develop through the progradation of river mouths and delta shorelines producing a subaerial deltaic plain over the seaward delta-front deposits. See Bellotti et al. (2011) *Holocene* 21, 1105 for a detailed study of the Tiber delta. Gao (2007) *Geomorph.* 85, 3–4 thinks that river deltas may be limited when sediment retention approaches zero, when the delta front advances into deep waters, or by the combination of sea-level rise and subsidence.

Inland deltas form in hot, arid inland drainage areas, where water dynamics and sediment transport are very different, especially deposition associated with expansion–contraction dynamics at the channel head. As a result, the mean slope of the delta decreases with distance from the apex, while in coastal deltas it increases towards; see Seybold et al. (2010) *Geophys. Res. Lett.* DOI:10.1029/2009GL041605.

demographic transition (DTM) An account, but not a complete explanation, of changing rates of fertility, mortality, and *natural increase. Four stages may be recognized: in the High Stationary Stage, birth and death rates are high and the death rate fluctuates from year to year; the Early Expanding Stage is characterized by high birth rates, falling death rates, and increases in population; in the Late Expanding Stage death rates are low and fertility is declining, but population is still increasing; and, lastly,

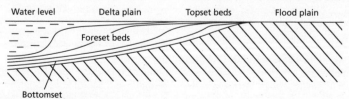

Water level   Delta plain   Topset beds   Flood plain

Foreset beds

Bottomset
Delta shore composition

in the Low Stationary Stage birth and death rates are low and the birth rate fluctuates. There seems to be a fifth stage where birth rates fall below death rates so population levels fall.

Death rates fall because of improved conditions and health care. The reasons for falling fertility are less clear, but include: anti-child labour laws; pensions (which remove the need for children as a support in old age); a higher proportion of women in work; and increasing costs of bringing up children, as standards of living rise. Everywhere, mortality decline appears to have played a central role for fertility decline. 'Female literacy and aggregate income do not seem to matter as much' (Conley et al. (2007) *NBER W. Paper* 2892). Reher (2004) *Pop. Space Place* 10, 19 notes that, with falling child death rates, fertility falls rapidly in the developed world, and more slowly in more recent transitions. In an analysis of developing countries, Zlidar et al. (2003) *Pop. Reports*, M 17 note that fertility has fallen by an average of 1% per year, due to increased contraception use, better child survival rates, and better maternal health care. Because of their late start to the DTM, developing countries will not catch up demographically with more developed regions for over 200 years: 'in fact, less developed regions never seem fully to catch up' (UN Dept. Econ. & Soc. Affairs, Pop. Division). See McNicoll (2006) *Pop. & Dev. Rev.* 32, 1 on politics and the East Asian demographic transition. 'The end of the transition is also the end of the explanatory paradigm, and no other paradigm is, at present, capable of enlightening us on the future' (Vallin (2002) *Pop. & Dev. Rev.* 28, 1).

The **second demographic transition** is marked by: a later age at marriage; an increase in the proportion of adults living alone or cohabiting; slower rates of remarriage; increased fertility outside of marriage; and delaying or forgoing childbearing. 'With respect to *timing*, *process*, and *explanation*, the second demographic transition is characterized by no less diversity than the historical first transition' (Lesthaeghe and Moors (2000) *Rev. Pop. & Soc. Policy* 9).

demography The observed, statistical, and mathematical study of human populations, concerned with the size, distribution, and composition of such populations.

dendritic drainage *See* DRAINAGE PATTERNS.

dendrochronology An absolute dating technique using the growth rings of trees. The combined oak and pine tree-ring chronologies of Hohenheim University are the backbone of the Holocene radiocarbon calibration for central Europe; see Friedrich et al. *Radiocarbon* 46, 3, 1111. Tree rings represent a very valuable natural archive on past debris-flow occurrence as they may record the impact of events in their tree-ring series; see Bollschweiler and Stoffel (2010) *PPG* 34, 5, Vandekerckhove et al. (2001) *Catena* 45, and Malik (2008) *Geomorph.* 93, 3–4. **Dendrogeomorphology** is the dendrochronological interpretation of geomorphic processes.

density dependent factors The checks to population growth, such as *competition, which result from overcrowding; factors which connect the dynamics of human-dominated landscapes, and 'natural' ecological processes (Fagan et al. (2001) *Landsc. Ecol.* 16).

density gradient The rate at which the intensity of land use or the density of population falls with distance from a central point. See Tannier and Pumain (2005) *Cybergeo* (*systèmes, modélisation, géostatistiques*) 307, online 20 April 2005.

denudation A general name for the processes of weathering, transport, and erosion. See Twidale (2007) *Phys. Geog.* 28, 1 on inequalities in denudation.

dependency ratio The ratio between the number of people in a population between the ages of 15 and 64 and the **dependent population**: children (0–14) and elderly people (65 and over). It is used as a rough way of quantifying the ratio between the *economically active population and those they must support, but the age limits are somewhat arbitrary as, in the UK, for example, 30% of over-16-year-olds go on to higher education, and are therefore still dependent, either on the state or, increasingly, on their parents or bank managers.

dependency theory The view that the development of the advanced economies has caused the underdevelopment of less economically developed nations; see A. G. Frank (1967). D. Slater (2004) finds that while dependency theory possesses a continuing relevance, owing to its concern with the marginalized and oppressed, a post-colonial version would overcome a 'hyper' focus on class conditions by introducing concerns with agency, discourse, and knowledge-making. Shepherd (2005) *TIBG* 30, 2 thinks dependency theory has been ignored because 'it fails to conform to the neoclassical hard core'.

depopulation The decline, in absolute terms, of the total population of an area, more often brought about by out-migration than by a fall in fertility or excessive mortality. Depopulation typically occurs in more remote areas; it is initiated by loss of employment, effected through out-migration, and exacerbated by the 'vicious cycle' of decline depicted in the Drudy–Gilg model (P. J. Drudy 1978; A. W. Gilg 1983; Millward (2005) *Canad. Geogr./Géog. canad.* 49, 2). See Zaslavsky (2007) *J. Reg.* *Sci.* 40, 4 on depopulation and Russia's demographic crisis.

depression An area of low pressure (roughly, below 1000 mb); *see* MID-LATITUDE DEPRESSION. **Depression tracks** are influenced by the courses of *jet streams, energy sources—such as warm seas—and mountain barriers.

deprivation Loss; lacking in desired objects or aims. Within the less developed countries deprivation can be acute: water, housing, or food may be lacking. Within the developed world basic provisions may be supplied but, in comparison with the better-off, the poor and the old may well feel a sense of deprivation. This is **relative deprivation**, which entails comparison, and is usually defined subjectively. In the UK, inequalities in health seem to be more marked in deprived areas than in more affluent ones (Norman et al. (2005) *Soc. Sci. & Medicine* 60), and in more affluent Western societies relative deprivation seems to be more important than absolute deprivation in determining population health (R. G. Wilkinson 2005). 'The impetus for social change comes not so much from changes in absolute deprivation, but from relative deprivation and it is this which perhaps above all threatens the sustainability of globalisation' (Kaplinsky (2008) *Geog. Compass* 2, 2). See M. Pacione (2009) on poverty and deprivation in the Western city.

The **cycle of deprivation** is the transmission of deprivation from one generation to the next: lack of knowledge and skills all contribute to a cycle of deprivation, leaving the poor with few resources and many barriers to overcome.

(⊕) SEE WEB LINKS
- Describes circle of deprivation.
- UK indices of deprivation.

deregulation Rolling back state 'interference' in social and environmental life; 'state regulation has a "light touch", and more and more actors become self-governing within centrally prescribed frameworks and rules' (Castree (2008) *Env. & Plan. A* 40).

desakota region A region characterized by an intense mix of agricultural and non-agricultural activities that often stretch along corridors between large city cores. See Sui and Zeng (2001) *Landscape & Urb. Plan.* 53, 1 on urbanization in Asian desakota.

desalination (desalinization, desalting) The removal of dissolved salts and minerals from saline water to produce fresh, drinkable water. Commercial desalination processes include: reverse osmosis, electrodialysis, multi-effect distillation, and vapour compression. The key issues associated with desalination are cost, energy use, and the environmental impacts of brine disposal and feedwater intake (Trimble et al. (2005) *Encyc. Water Sci.*).

desegregation The phasing out of the separation of humans, usually with reference to different ethnic, caste, or national groups. Bolt, Phillips, and van Kempen (2010) *Housing Studs* 25, 2, 129 conclude that 'policies prompting desegregation and social mix rarely meet policy expectations. First, residential mixing cannot be assumed to enhance community cohesion or people's social capital. On the contrary, urban renewal policies have been found to disrupt communities. Displaced households experience difficulties in establishing new social ties. At least in the short term, dispersal often leads to a decline in bonding capital, and thus social support, whilst there are no gains in terms of more bridging capital. The idea that desegregation leads to more social mobility

for poor and minority ethnic households is overly optimistic'. India has passed multiple laws to penalize the caste system present in the society. Estevan et al. (2010) in a paper published online by MIT Sloan School of Management note that while the use of quotas in India's public service has been implemented, forcing school integration in terms of caste appears to be difficult, since lower caste children tend to drop out early.

((⊕)) SEE WEB LINKS

- Estevan et al. (2010) paper 'Dynamic Desegregation'.

desert An area of very low precipitation, sometimes specified as below 250 mm/yr^{-1} or where *evapotranspiration exceeds precipitation. A **desert pavement** is a continuous mantle of flat-lying, densely packed, partially overlapping pebbles, typically overlying a soft, silty layer (Quaid (2001) *Geology* 29, 9). Most **desert soils** lack clear soil *horizons, as the climate is too dry for chemical *weathering or humus formation. *Leaching occurs only after occasional rain, and is soon reversed by evaporation. **Desert varnish** is a film of iron oxide or quartz on desert rocks. Silica, slowly dissolved from other minerals, gels together to form a glaze on the rock face. Detrital grains, organic compounds, and pollutants from local environments become entombed in the coating as it forms (Perry (2006) *Geology* 34, 7).

desertification The meaning of this word is much debated. It is defined by the UN as the degradition or destruction of land, which results in desert-like conditions. Thomas in J. Holden (2012) defines desertification as 'an often misunderstood and abused term used to describe land degradation in drylands. Misunderstanding often arises because natural environmental (especially plant system) responses to drought, from which recovery usually occurs, have been confused with longer-term and more persistent negative changes'. The causes and consequences of desertification are complex and little understood. 'Occasional droughts, due to the seasonal factors or inter-annual variations of rains, and severe droughts of long periods can be caused or aggravated by human influence on the environment, reduction of vegetation covering, change of the effect of albedo, local climate changes, greenhouse effect, etc.' (UNCCD 2004). However, J. Letera et al. (2006) challenge negative perceptions of pastoralism. 'Herd mobility is now recognised as a rational strategy in unstable environments, and some countries have taken steps to facilitate movement of herds' (see Touré (2004) *IIED issue paper* 126). Rasmussen et al. (2001) *Glob. Env. Change* 11, 4 conclude that 'broad generalisations on land degradation processes, based on local-scale studies, are risky. Significant variations exist at the landscape level, and trends in ecosystem dynamics are sometimes totally reversed even within small regions . . . conclusions concerning "irreversible" degradation, based on few years or decades of observations may be premature'.

desilication (desilification) The removal of silica from a *soil profile by intense weathering and *leaching. It occurs when water that is undersaturated with silica percolates through a soil at the correct pH, and is facilitated by freely draining conditions in a hot, humid environment (R. Schaetzl and S. Anderson 2005).

deskilling Breaking down jobs into smaller units, each to be tackled separately, so that low levels of skill—and therefore cheaper labour—are required for restricted tasks. 'When combined with the separation of residential ownership from direct control, deskilling meant industrial capital could draw on a much larger pool of labour over much broader geographic space' (Moore (2002) *Antipode* 34, 2). Man (2004) *Women's Studs & Int. Forum* 27, 2 suggests that deskilling is also an inherent part of Canadian immigration policy. Although L. Grossman (1998) questions the assumption that contract farming necessarily brings about deskilling, Hendrickson and Heffernan (2002) *Sociologia Ruralis* 42 claim that the deskilling of farmers is well advanced in parts of the USA. Epp in R. Epp and D. Whitson, eds. (2001) refers to the **political deskilling** of rural communities through neoliberal political reforms.

destructive margin The zone where two *plates meet, and oceanic crust is subducted beneath either oceanic or continental lithosphere, changing to the more ductile, fluid-like, denser rocks of the aesthenosphere via increases in temperature and pressure. Subduction creates strong compression at the plate margins, piling up unconsolidated segments and pieces of rock slabs into a wedge (R. Krebs 2003).

destructive wave A plunging wave, with a short wavelength, a high frequency (13–15 per minute), and a high crest. *Backwash greatly exceeds *swash. Destructive waves comb beach material seawards (Joliffe (1978) *PPG* 2).

determinism The view that human actions are stimulated and governed by some outside agency. **Environmental determinism** assumes that the physical environment is the primary determinant of cultural forms. During the late 19th and early 20th century, some geographers claimed a link between imperialism and environmental determinism (see Smith in J. N. Entrikin et al., eds

1989)—a view that has long been rejected; 'if political ecology has taught us anything it is that we can do better than environmental determinism' (P. Robbins 2004). Barnes (2001) *PHG* 25, 4 argues that proponents of economic geography tended towards technological, rather than environmental, determinism.

deterritorialization At its simplest, deterritorialization may mean the takeover of an area or place (territory) that is already established. G. Deleuze and F. Guattari (1972) use the term to represent the freeing of labour-power from specific means of production, as when the English Enclosure Acts (1709–1869) enclosed previously common land for private landlords. As such, deterritorialization is the severance of social, political, or cultural practices from their native places and populations. Connell (2004) *PHG*, 28, 3, 342 uses this concept of separation to describe world music as a deterritorialization of place and identity, undoing what has already been done. From this, we can see the decline in the role of the state as deterritorialization; 'certain *transnational corporations have more financial power, within a state, than the state itself' (Kapferer (2005) *Anthrop. Theory* 5, 3). 'The nation-state may be in decline but it is giving way to a relatively original state order or political/economic formation with multiple state-like effects that is able to act in ways systemic with deterritorializing global processes' (Hudson (2002) *PHG* 26, 3). Globalization has been explicitly seen as deterritorialization by J. A. Scholte (2000), but Eldon (2005, *TIBG* 30, 1) is reluctant to accept Scholte's straightforward understanding of territory: 'work proposing an idea of deterritorialization requires an explicit theorization of what territory is'. Akinwumi (2006) *Pol. Geog.* 25, 7 seems to see the euro as a deterritorialization

strategy. For tax havens, 'the legal advantages of non-sovereignty are significant' (R. T. Naylor 2002).

Deterritorialization makes it increasingly difficult to maintain a convincing nationalistic discourse that involves 'cultural purity' (Gjerde and Onishi (2000) *Hum. Dev.* 43). Connell and Gibson (2004) *PHG* 28, 3 claim that the expansion of world music exemplifies the deterritorialization of cultures.

detritivore An animal which feeds on fragments of dead and decaying plant and animal material. Detritivores have a vital part to play in *food webs and nutrient recycling.

detritus Fragments of weathered rock which have been transported from the place of origin. See Doyle (2006) *Geomorph.* 77, 3–4 on detritus loss rates; try Valdiya (2002) *PHG* 26, 3.

development The use of resources to relieve poverty and raise living standards; the means by which a traditional, low-technology society is changed into a modern, high-technology society, with a corresponding increase in incomes. This can be done through mechanization, improvements in infrastructure and financial systems, and the intensification of agriculture. This definition is based on the more obvious distinctions in living standards between developed and less developed countries. Narrowly economic definitions of development have been criticized, and many believe that true development includes social justice; 'a focus on "global social justice" could provide a framework to think interesting and possibly dangerous thoughts about how to take the international development agenda beyond the relatively instrumental approach of the Millennium Development Goals'. Maxwell (2008) *ODI Opin.* reflect on the means and ends of development in its various forms.

For example, Smith (2002) *Area* 34 argues against development as 'some neutral concept of progress' and Andolina et al. (2005) *Pol. Geog.* 24, 6, argue that as 'indigenous movement platforms and concepts are increasingly institutionalized . . . their demands for culturally appropriate government and development are in practice implemented as governmentally and developmentally appropriate culture'. Potter (2001) *Area* 33 argues that development is a non-linear process, and 'one should not mimic the experience of "developed" countries'. 'For many local communities, the maintenance of social and cultural practices is central to **participatory development** and just as important as income gains and poverty reduction' (Connell (2007) *Sing. J. Trop. Geog.* 28, 2). O'Reilly (2007) *AAAG* 97, 3 observes that women's participation in development is a dialogic process that includes negotiating how women fieldworkers will participate within their own organizations; see also Sharpe et al. (2003) *TIBG* 28, 3. **Development indicators** as measurements are simply concerned with statistics and do not indicate social structures and patterns of behaviour; there is no definitive definition of what development should be for each society, and no blueprint for how to achieve it. See D. Gasper (2004). **Ethnodevelopment** interprets culture and indigenous identity in ways that make it compatible with neoliberal social policy, but potentially restrict indigenous agency (Laurie et al. (2005) *Antipode* 37, 3).

(⊕) SEE WEB LINKS
• World Bank development indicators.

developmental state A state that derives political legitimacy from its record in economic development, which results mainly from selective industrial policy. The classic model of the developmental state is derived from Japan between the 1950s and the 1980s. O. Edigheji ed. (2010), on South Africa as a developmental state is useful and comprehensive.

Devensian The last full glacial series in Britain, lasting from about 70000 to about 10000 years BP. During this time, sea levels were 130–160 m below current levels; Ireland was joined to Britain, and Britain to northern Europe and Scandinavia by a broad dry plain. The later part of the Devensian embraces pollen zones 1–111 as defined by Godwin (1940).

devolution The transfer of powers from central government to regional and local governments. UK devolution has been based on the principle that decision-making powers be devolved to the appropriate level of *governance, creating new assemblies in Wales and Scotland for the purpose. Wood et al. (2005) *Pol. Geog.* 24, 3 find that UK businesses have been reconfigured to match these new institutions more closely. 'Devolution of state responsibility only occurs with localization if the state is not also decentralized' (Hayes-Conroy (2008) *Pol. Geog.* 27, 1). In response to *nationalism, devolution may be seen as one way to avoid ethnic unrest, but neither the devolution of powers to the Basque Autonomous Community in Spain nor more than a decade of power-sharing within this region has led to a peaceful settlement of the Basque conflict (Mansvelt-Beck (2008) *Nations & Nationalism* 14, 1). **Devolution *discourses** draw on nationalist identities to create new systems of governance: see Raco (2003) *TIBG* 28.

Devonian A period of *Palaeozoic time stretching approximately from 395 to 345 million years BP.

 SEE WEB LINKS
• An overview of the Devonian period.

de Vries cycle (Suess cycle) This is an approximately 200-year solar cycle, believed to be one of the most intense. Raspopov et al. (2008) *Pal., Pal., Pal.* 259, 6 find that palaeoclimatic reconstructions correlate well with these variations in solar activity.

dew A type of condensation where water droplets form on the ground, or on objects close to it. Dew forms when strong night-time *terrestrial radiation cools the ground to chill the overlying air to *dew point. See Gidron (1999) *Agric. & Met.* 96, 1–3.

dew point The temperature that a body must be chilled to for it to become *saturated with respect to water, so that condensation can begin; also seen as the temperature of a chilled surface just low enough to attract *dew from the *ambient air. The dew point of an *air mass varies with its initial temperature and humidity; see Dodd (1965) *Monthly Weather Rev.* 93, 2.

dialect A variant of a language in terms of vocabulary grammar, syntax, and pronunciation. Dialect differences arise and persist because of geographical barriers (the Ohio River, for instance, helps to define the division between the dialect areas of the North and the South USA), political boundaries, settlement patterns, migration and immigration routes, territorial conquest, and language contact (Rickford (2002) *Discover*). See also Langevelde and Pellenbar (2001) *Tijdschrift* 92, 3. **Dialectology** is the study of social and linguistic variations within a language; **dialect geography** is the study of local differentiations in a speech area: see J. K. Chambers and P. Trudgill (1998). **Dialectometric** techniques analyse linguistic variation quantitatively, to map geographic patterns of individual linguistic variants, such as which word is used for a particular concept in a language area, or which sounds are used in particular words (Kretzschmar (2006) *Literary & Ling. Comp.* 21, 4).

dialectics Arriving at the truth by the exchange of opposing logical arguments. Especially associated with Hegel, this process states a thesis, then develops a contradictory antithesis, and resolves these into a coherent synthesis. Sheppard (2008) *Env. & Plan. A* (2008), 40, 2603 is not for the fainthearted, but rewards study.

diapir An upward-directed, dome-like intrusion of a lighter rock mass into a denser cover. See Petford and Clemens (2001) *Geology Today* 16, 1.

diaspora The dispersion of people from their homelands. 'Those in diaspora may be seen as subject to power relations, tensions, disconnections and . . . notions of belonging, identity and community' (Mavroudi (2007) *Geog. Compass* 1/3). 'Diaspora involves feeling "at home", in the area of settlement while retaining significant identification outside it' (B. Walter 2001). 'Citizenship is inherently geographic in that it is used to determine who can participate in (trans)national affairs and, more broadly, who is seen as "belonging". The bifurcation of this legal and cultural sense of belonging and representation is even more intensified for diaspora populations' (Mains (2007) *Focus on Geog.* 50, 1). Drozdzewski (2007) *Soc. & Cult. Geog.* 8, 6 discusses the way that different groups of Polish people contextualize their diasporic identity, and attribute meaning to place. Mercer and Page (2010) *African Diaspora* 3, 1 consider the ways in which home associations provide space for debate about what is an intrinsically *good* way to live together in the diaspora. See also Blunt (2007) *PHG* 31, 5. Writing on India's **diaspora strategy**, Dickinson

(2012) *TIBG* 37, 4, 609 argues that 'specifically, incorporating the diaspora into a narrative of Indian sovereignty is an act that marks out India's own reinvention from that of a post-colonial developmentalist nation (and the concomitant excision of its colonial diaspora from its anti-colonial nationalist rhetoric) to an emergent power in the international economy'. New Zealand's diaspora strategy is to engage expatriates to extend international marketing opportunities, without requiring them to return home (Larner (2007) *TIBG* 32, 3).

diatoms Single-celled algae with a resistant siliceous outer shell that enables them to be preserved in marine and freshwater sediments. For the use of diatoms in physical geography, see Mannion (1982) *PPG* 6: 233.

difference Difference exists only with reference to another; in the case of the social sciences, this may be patriarchy, 'whiteness', 'colour', *feminism, *colonialism, or *post-colonialism, all of which may be defined by what they are *not*. 'The differentiation of an owning (capitalist) class from . . . the working class is foundational for capitalism in ways that gender, racial or sexual differences, however violent and life-destroying, are not' (Malden (2008) *PHG* 31, 1, paraphrasing N. Castree and D. Gregory 2006). 'Civil wars in Sri Lanka and sectarian fighting in Iraq are violent expressions of the difficulties of incorporating difference in the nation state' (Staeheli (2008) *PHG* 32, 4). 'Walls to limit the movement of people between two territories—whether between Israel and the Palestinian territories or between Mexico and the United States— suggest that difference may no longer be accommodated within the liberal state' (Staeheli, op. cit.). I. Rivera-Bonilla (2000) analyses the way gating

residential areas recreates spatial markers of class and racial differentiation. See also Valentine (2008) *PPG* 32, 3, 323.

differential erosion The selective erosion of surfaces, so that softer rocks, such as clays and shales, or lines of weakness, such as joints and faults, are eroded more rapidly than resistant, competent, and unjointed materials.

diffuse pollution A widespread pollution that cannot be attributed to a specific source. Hodgson et al. (2009) *PPG* 33, 3 critically evaluate the complexities and uncertainties attributed to the development of models for assessing the agriculturally derived microbial pollution of watercourses. See D'Arcy et al. (2000) Chartered Institution of Water and Environmental Management.

diffusion The widespread dispersal of an innovation from a centre or centres. T. Hägerstrand's model of diffusion (1968) implies the existence of a *mean information field which regulates the flows of information around a regional system; flows that are moderated by barriers which can obstruct the evolution of information into innovation. 'All the basic ideas that have come to be associated with diffusion research can be found in Lawrence Brown, 1981' (Lutz (2006) *PHG* 30, 4).

'To the extent that a high proportion of personal contacts are local, the diffusion of information must also be spatially constrained: the further people are from early adopters the later they are likely to be to adopt the innovation. To the extent that people's non-local contacts are hierarchical (more likely to be with people in big cities than in small), the diffusion of information must also be hierarchically constrained: the lower down the urban hierarchy people live, the later they are likely to be to adopt the

digital divide The differential access to computers, information, the internet, and telecommunications, globally, regionally, nationally, and locally, that also comprises unequal access to job opportunities, resources, and training. 'To characterise the digital divide solely as a technological divide is a (common) misconception . . . To some extent it is true that the digital divide is a new name for an old phenomenon: uneven spatial distribution of resources' (Crampton (2001) *Pres. to Ass. Internet Researchers*). 'Advanced technologies have a pro-rich bias: they are essentially designed for developed countries' (J. James 2002). For less economically developed countries, strategies for closing the divide aim to leapfrog the earlier stages of economic development, bringing advanced communication technologies to areas of deprivation, in a way suited to local socio-economic conditions; for example, the 'phoneshop' system (James (2000) *Development and Change* 31, 4). Dasgupta et al. (2001 *W. Bank Working Paper; WPS 2567*) highlight the importance of government policy: 'simulations based on the econometric results, suggest that feasible reforms could sharply narrow the digital divide during the next decade for many countries in Africa, Asia, and Latin America'. See the World Bank's annual *Little Data Book* for data on technology and infrastructure from over 200 countries.

In England and Wales, the areas with the highest proportion of households with broadband tend to be in dense, middle-class suburbs. For UK government initiatives, see the Dept. Trade and Industry *Policy Action Team 15 Report.*

innovation . . . the diffusion of innovations reflects supplier behaviour as well as adopter behaviour' (Webber (2006) *PHG* 30, 4). See Belhedi (2007) *Cybergeo* 372 and E. Rogers (2003).

digital elevation model (DEM) An array of regularly spaced elevation values referenced horizontally either to a Universal Transverse Mercator (UTM) projection, or to a geographic coordinate system. See USGS (2000) *Fact Sheet 040-00* for a thorough and helpful summary.

digital terrain analysis (geomorphological analysis, landform parametrization, land surface analysis) The quanitification of terrain. A **digital terrain model** (*digital elevation model) is a digital relief map. See Deng (2007) *PPG* 31, 4; Lacroix et al. (2002) *Env. Model. & Software* 17, 2; and Thwaites (2002) *New Forests* 24, 2.

dike *See* DYKE.

dilatancy theory Emphasizes the role played by variations of stress in *till found at the base of moving ice masses. A local reduction in stress causes the till to settle and become more compact. See Crozier (1975) *Canad. Geogr.* 19. Dilatancy theory has also been used in *earthquake prediction—see F. Naeim (2001). **Dilatation** is *pressure release.

diminishing returns, law of The principle that further inputs into a system produce ever lower increases in outputs. Radelet, Clemens, and Bhavnani (2005) *Finance and Dev.* 42, 3 find that *aid has a positive relationship with growth, but with diminishing returns.

dip The angle of inclination of a rock down its steepest slope, that is to say, the direction at right angles to the strike. Dip is the angle between the maximum slope and the horizontal. A **dip slope** occurs where the slope of land mirrors the slope of the underlying *strata.

direct cell *See* ATMOSPHERIC CELLS.

dirt cone A cone of ice, up to 2 m high, covered with a thin layer of debris, found in the *ablation zone of a glacier.

disability; geographies of disability, welfare, and social exclusion Hunt (1966) *Union of the Physically Impaired against Segregation* distinguishes between *impairment*, which relates to individually based bio-physical conditions, and *disability*, 'which is about the exclusion of disabled people from "normal" or mainstream society'. 'The dominant societal stereo-type of disability as a "pitiful" state reinforces the view that people with disabilities are somehow "less than human"' (R. Imrie 1996). 'The biggest obstacle to disabled people's meaningful inclusion into mainstream community life is negative public attitudes' (Barnes in L. Barton and M. Oliver, eds 1997). Wilton (2004) *TIBG* 29, 4 outlines the multiple strategies of control of the disabled in the workplace; notably the denial of access to accommodation. B. Gleeson (1998) sees space as 'a social artefact that can be used to disable, rather than enable people with physical impairments'. I. Kawachi and L. F. Berkman, eds (2003) explore the way physical and cognitive disability affect and are affected by a neighbourhood's physical and social environment. V. Chouinard, E. Hall, and R. Wilton, eds (2010) consider the 'spatial interpreta-tion of disability' in a strongly recom-mended text.

(( )) SEE WEB LINKS
- Disability awareness in action.

disaster Usually, a sudden catastrophe with serious consequences. While the causes and nature of a disaster are largely the preserve of physical geogra-phers, human geographers also consider the causes and impact of disasters. Le Billon and Waizenegger (2007) *TIBG* 32,

3, 411 note that 'disaster risk has "political roots", notably (unequal) power relations and (under-) develop-ment processes'. Hurricane Katrina, for example, renewed attention to inter-racial and class-based inequalities, prejudices, and tensions in the United States. Politics, rather than the strict "needs" of disaster victims, also influence responses to disasters. Interpretations of disasters represent political choices with political impacts, particularly from a gender perspective, and for disaster recovery or future risk mitigation.

discharge The quantity of water flow-ing through a cross-section of a stream or river in unit time. Discharge (Q) is usually measured in *cumecs, and calculated as $A \times V$ where A is the cross-sectional area of the channel and V is the mean velocity. See R. Charlton (2007) pp. 57–9 for instructions on measuring river discharge. Definitions of **dominant discharge** (also known as **effective discharge**) vary; at its simplest it is the flood discharge that achieves the great-est total geomorphic work, but see Ferro and Porto (2012) *Geomorph.* 139–40, 313 for a thorough review of definitions. For those rivers in which the relationships of bedload discharge to water discharge are not dominated by very large-sized parti-cles, and where daily mean discharges are an adequate description of stream-flow characteristics, dominant discharge is often considered to equate to bankfull discharge (Emmett and Wolman (2001) *ESPL* 26, 1369).

discordant Cutting across the geolog-ical grain, as in a stream cutting across an *anticline.

discourse A specific assembly of categorizations, concepts, and ideas that is produced, reproduced, *performed*, and transformed in a particular set of practices. Schott (2007) *Tijdschrift* 98, 2 defines discourse as 'a social process which can be called "inclination of sense" or, in a more sophisticated manner, "genesis of meaning" . . . It is appealing to use the word discourse in Foucault's sense of a system of exclusion.'

Discourses create their own 'regimes of truth'—the acceptable formulation of problems, and solutions to those problems (M. Foucault 1980). 'Although many of the urban studies that undertake discourse analysis are interested in urban social justice . . . some critics charge that talking about social justice is not enough, and as critical urban researchers we need to create it' (Lees (2004) *PHG* 28, 1). See Nijman (2007) *Tijdschrift* 98, 2 on 'the prevailing political and ideological discourse [that] seeks to maintain the Haitian identity'.

discrete choice modelling *See* CHOICE MODELLING.

discrimination The exclusion of specific groups from certain sectors of society. 'This type of exclusion is to the disadvantage of the immigrant, but to differentiate it from xenophobic discrimination or racism . . . the concept *systemic discrimination* would cover the situation where exclusion is connected to the ethnic background of individuals, but is not a result primarily of hostility towards "the other"' (Bäcklund (2003) *Geografiska B* 85, 1). Hyndman (2007) *AAAG* 97, 2 shows how, in post-tsunami Sri Lanka, the implementation of no-build setback areas along the coastline stirred feelings of discrimination. See also C. Flint, ed. (2004).

diseconomies Financial drawbacks. **Diseconomies of scale** may occur when an enterprise becomes too large, sites become constricted, the flow of goods is congested, or the workforce is alienated. See Barcena-Ruiz and Begona Garzon (2005) *Eur. J. Pol. Econ.* 21, 1 on economic integration and privatization under diseconomies of scale. *See* ECONOMIES OF SCALE.

dispersed settlement A settlement pattern of scattered, isolated dwellings. See Cruickshank (2006) *Norw. J. Geog.* 60, 3 on the fight in rural Norway for a dispersed settlement pattern.

dissection The incision of valleys by river erosion. A **dissected plateau** is a level surface which has been deeply cut into by rivers. See Stanford et al. (2000) *ESPL* 25, 9.

dissimilarity index A measurement of the overall difference between two percentage distributions. Dissimilarity indices are often used in studies of occupational, or racial, *segregation. Jones (2008) *Urb. Geog.* 29, 3 develops a typology to explain the different pathways by which the dissimilarity index may increase or decrease in a metropolis.

dissipative Describing a *non-linear system where energy is lost at positions far from equilibrium. A **dissipative beach** is gently sloping, running from the seaward edge of the forezone to the surf zone. See Ruggiero et al. (2004) *J. Geophys. Res.* 109, C06025.

dissolved load Material carried in solution by a river; capable of passing through a 0.45-μm membrane filter (S. Trimble et al., eds (2006) *Encycl. Water Sci.*).

dissolved oxygen Oxygen from the atmosphere and from photosynthesis, dissolved in the upper levels of bodies of water. Dissolved oxygen decreases with depth, rising temperatures, and the oxidation of organic matter. See Kung and Ying (1991) *Geog. J.* 157, 1.

distance Absolute distance is expressed in physical units such as kilometres and is unchangeable. 'The geography of the world as we experience it is being twisted and contorted so as to bear very little relation to the physical distances that are involved. The human aspect of this is sometimes called **social distance** or **cultural distance**—those gulfs that can exist in the understanding and in the history of experiences that we bring to the world. Physical proximity is not necessarily a good measure of social and cultural distance' (D. Massey, BBC/ OU Open2.net—Free Thinking).

Relative distance includes any other kind of distance such as **time-distance**, measured in hours and minutes, and changing with varying technology. 'It's the relative distance that counts' (Chauvin and Gaulier (2002) *TIPS Annual Forum*). Breschi and Lissoni (2006) *CESPRI W. Papers* 184 show that spatial distance and technological distance are both proxies of **social distance**. Carrere and Schiff (2004) *World Bank W. Paper Series* 3206 find that 'distance has become more important over time for a majority of countries'.

distance decay The lessening in force of a phenomenon or interaction with increasing distance from the location of maximum intensity; the **inverse distance effect**. Seidl et al. (2006) *BMC Health Serv. Rev.* 6, 13 find an inverse relationship between the geographical distance from a patient's home to a clinic, and the likelihood of the patient actually turning up there. Gimpel et al. (2008) *Pol. Geog.* 27, 2 find that technology has lessened, but not eradicated, the **friction of distance**. See Farhan and Murray (2006) *Annals Reg. Sci.* 40, 2.

distributary A branch of a river or glacier which flows away from the mainstream and does not return to it.

distributed mapping Sometimes called 'online GIS', this is the sharing of cartographic data and tools via the Internet. *GIS software and spatial databases are on a server which is accessible online, enabling users to interact, manipulate, query, and map spatial data.

(SEE WEB LINKS)

• A good website to start looking at distributed mapping.

disturbance In *ecology, any event, such as fire, flood, or earthquake, that disrupts the everyday running of an *ecosystem. A **disturbance regime** is marked by short, but recurring episodes of high magnitude, or at critical sites. **Disturbance-related landforms** tend to be polygenetic and heterogeneous, and are long-term/permanent adjustments to fires, volcanoes, etc. Church (2002) *Freshw. Biol.* 47, 4 argues for a disturbance-related interpretation of mountain rivers.

divergence The spreading out of an air mass into paths of different directions, linked with the vertical shrinking of the atmosphere (see R. Barry and R. Chorley (1992) fig. 3.6). Upper-air divergence is associated with ground level depressions (Sutcliffe and Forsdyke (1950) *Quart. J. RMS* 76, 328), and closely related to vertical *vorticity; see Chen and Bosart (1979) *J. Atmos. Scis* 36, 2.

diverging margin See CONSTRUCTIVE MARGIN.

diverse economies Non-capitalist economies, and non-market economies. See Gibson-Graham (2008) *PHG*, 36, 16, 616.

diversification A spread of the activities of a firm or a country between different types of products or different markets, in order to reduce over-dependence; as in the Arab Gulf (Malecki and Ewers (2007) *PHG* 31, 4). **Horizontal**

diversification is diversification between same-type investments; see
Achilladelis and Antonakis (2001) *Res.
Policy* 30, 4. **Vertical diversification**
is adding different investments to a
portfolio. Diversification for a small
open economy is inevitably protracted,
involving a change in economic structure. Supportive conditions include low
inflation, stable exchange rates, and a
labour-market environment with few
worker–employer conflicts. Diversification is not an inescapable route to output
stability (Chan (2006) *Monetary Authority of Macao*). Herzer et al. (2006) *Appl.
Econ.* 38, 15 find that export diversification plays an important role in economic
growth.

diversity The abundance of species
within an ecosystem. **Alpha diversity** is
the diversity of species within a particular area or ecosystem, expressed by its
species richness (Moreno et al. (2006)
INCI 31, 1). **Beta diversity** is the difference in diversity of species between two
or more ecosystems, expressed as the
total number of species unique to each
of the ecosystems compared; see Vellend
(2001) *J. Veg. Sci.* 12, 4 on beta diversity
and species turnover.

divestment (divestiture) In business,
the sale of an asset; the opposite of an
investment. Geographical analyses of
divestment focus on: the conduits
through which surplus or excess effects
are discarded; the practices of divestment; disinvestment as practice; and the
connections between divestment practices. Gregson et al. (2007) *TIBG* 32, 2
show how surplus/excess things are
routinely divested through specific conduits; throwing something *out* is actually
throwing it *in* to another location—a bin,
a skip, a charity shop. In international
business, divestment is the closure or
sell-off of units in foreign locations,
or, conversely, units owned by foreign

firms (Benito (2004) *J. Econ. Geog.*
5, 2, 235).

division of labour The partitioning
of a production process into separate
elements. **Gender divisions of labour**
separate tasks into male and female; see
Padmanabhan (2008) *J. Int. Dev.* 20.
A **social division of labour** divides
workers by product: Mitsui (2003) *30th
ISBC, Singapore* sees social divisions of
labour as crucial to small and medium
enterprises in manufacturing, unable to
survive if they simply rely on own limited
resources and specialized skills. Meijers
(2007) *Reg. Studs* 41, 7 detects **spatial
divisions of labour** at the intra-urban
level in the Randstad, the Flemish Diamond, and the Rhein–Ruhr Area,
'whereas at the inter-urban level this division of labour is diminishing'. Within
the 'new' international division of labour, capital and control functions are
concentrated in the advanced economies or global core, while labour has
been relocated to developing economies
on the global periphery (Tonkiss (2008)
Sociol. Compass 2, 2). May et al. (2007,
TIBG 32, 2) explore the emergence of a
new **migrant division of labour** in
London.

doldrums Regions of light, variable
winds, low pressure, and high temperature and humidity, occurring over the
East Pacific, the East Atlantic, and
from the Indian Ocean to the West
Pacific; see Graham et al. (1988)
J. Atmos. Scis 46.

doline (dolina) A closed, steep-sided,
and flat-floored depression in *karst
country. The sides are 2–10 m deep; the
floor 10–1000 m wide. **Solution dolines**
form when solution enlarges a point of
weakness in the rock into a hollow.
Subsidence dolines form when limestone caves develop below insoluble

deposits. The majority of dolines are corrosion forms.

dome An uplifted section of rocks, highest at the centre, from which the rocks dip all around. **Volcanic domes** may be formed from slow-moving, *viscous lava, rounded as the result of pressure from lava below (Jousset and Okada (1999) *J. Volc. & Geotherm. Res.* 89, 1). A **plug dome** is a small, irregular dome within a crater, possibly with projecting, spiny extrusions.

domestic labour Work performed in the home. B. Anderson (2000) observes that 'there has been a tendency amongst feminists to see domestic work as the great leveller, a common burden imposed on all women equally by patriarchy', and looks at the racialization of paid domestic labour in the north—'a phenomenon which challenges feminism and political theory at a fundamental level'.

domestic space The home as a distinct, private, space. It may be produced through the inhabitants' possessions (see Rose (2003) *TIBG* 28, 1 on postcards, and Brace (2007) *AAAG* 97, 1 on the 'temperance household'), or through *capitalist and patriarchal reproduction. Lau (2006) *Mod. Asian Studs* 40 argues that the division of domestic space within South Asian households reflects the social status of women. Halford (2006) *Gender, Work, Organiz.* 13, 4 argues that, in the modern Western world, the construction of fatherhood has been underpinned by the spatial separations of work/home, of public/private. Jepson (2005) *Antipode* 37, 4 chronicles how Mexican-American women in South Texas changed a domesticated space into one of empowerment. 'The apartheid system dichotomized physical space into masculine and feminine categories, marginalizing the feminine.

Urban and public space is masculine while rural and domestic space is feminine. Only men who were economically productive were permitted in urban work spaces, and the associated housing system. Women were relegated to the inferior physical and social space of the homelands where they were expected to farm, raise children, and care for the sick and elderly' (G. Elder 2003).

dominant discharge Usually, *bankfull discharge, but dominant discharge has also been equated to *effective discharge; see Simon et al. (2004) *Geomorph.* 58, 1–4.

donga In South Africa, a steep-sided gully resulting from severe *soil erosion.

dormant volcano An inactive, but not extinct, volcano.

dormitory town A settlement made up largely of daily commuters, and with relatively few retail outlets, as the commuters will use city-centre services, or out-of-town shopping centres.

doughnut city A city with empty buildings, crime, and vandalism in the centre, and wealth, happiness, and family life in the suburbs. See Priemus (2004) *Housing Theory Soc.* 21, 1.

downward transition region In John Friedmann's *core–periphery model, a peripheral region marked by depleted resources, low agricultural productivity, or outdated industry.

downwearing, declining slope retreat A model of hillslope retreat where the slope angle decreases over time due to a combination of soil *creep, rain splash, and *sheet wash that causes slope convexities and concavities at the expense of straight hillslope segments.

drag The braking effect of friction, imparted by a fluid on bodies passing through it.

drainage The naturally occurring channelled flow formed by streams and rivers. A **drainage basin** is the area of land drained by a river and its tributaries (the term is synonymous with river basin, and termed *watershed* by North Americans). Stream channel length and equilibrium conditions within the system are revealed by the statistical relationships between the various parameters; see Gardiner (1978) *TIBG* 3, 4.

drainage basin geometry Drainage basin geometry postulates that, as the *stream order increases, (a) the number of streams decreases, (b) the average stream length increases, and (c) the average slope decreases. See Strahler (1952) *GSA Bull.* 63 and Sólyom and Tucker (2004) *J. Geophys. Res.* 109, F 03012.

drainage density The total length of streams per unit area. Any attempt to calculate drainage density is impeded by the difficulty of calculating total stream length, as the exact point at which a stream starts is problematical. See Lin and Oguchi (2004) *Geomorph.* 63, 3–4. Since water flow in channels is faster than *overland flow and *throughflow, the higher the drainage density, the faster the hydrograph rise, and the greater the peak discharge.

drainage network evolution The drainage basin can increase its area and extend its channels by landslides at the edge of the network; by *headward erosion; by the extension upslope of underground *pipes; and by rill formation. See Coulthard and Van De Wiel (2007) *Geomorph.* 91, 3–4 on *self-organization in drainage basin evolution, and Sagri

et al. (2007) *Geomorph.* 94, 1–2 on tectonics and drainage basins.

drainage patterns The pattern of a drainage network is strongly influenced by geological structure. **Anastomotic drainage**, developing on nearly horizontal, coarse sediments, is the division of a river into several channels, and **annular** or **radial drainage**—where the major rivers radiate from a centre—develops on domes, particularly where belts of resistant rock are separated by weaker belts. **Centripetal drainage** moves into a centre created by a crater or depression. **Dendritic drainage**—a branching, tree-like network—is most common on horizontally bedded or crystalline rocks, with uniform geology. **Rectangular drainage** has tributaries running at right angles to the major river and occurs on rocks with intersecting, rectangular joints and faults; **contorted drainage** is rectangular drainage on complex metamorphosed rocks. **Trellised drainage** resembles a trained fruit tree, usually found on dipping or folded sedimentary or weakly metamorphosed rocks. See Riedel et al. (2007) *Geomorph.* 91, 1–2 on structural controls on drainage patterns.

dreikanter A stone with three clearly cut faces, like a Brazil nut, formed by sand-blasting in desert environments.

drift-aligned beach A beach which has developed parallel to the line of *longshore drift. If the wave direction becomes more right-angular, the beach realigns itself.

dripstone Secondary calcite deposits, precipitated as *vadose water drips from/flows over cave roofs, walls, and floors (Tabarosi and Stafford (2003) *Int. J. Speleol.* 32, 1–4).

driving force A propulsive force; for example, the three main driving forces for the movement of the Earth's tectonic plates are mantle convection, gravity, and the Earth's rotation. The opposing forces are **resisting forces**.

drizzle Rain droplets, up to 0.2 mm across, and with a fall speed of around 0.8 m s^{-1}.

drought A long, continuous period of dry weather. Drought over Africa has increased, especially over West Africa. Northern and East Asia show positive trends and central Asia and the Tibetan Plateau show decreasing trends. In South Asia all trends are insignificant. Drought extent over Australia has decreased. Over the Americas, trends are uniformly negative and mostly significant (Sheffield and Wood (2008) *J. Climate* 21, 3).

drumlin A long hummock or hill deposited and shaped under an ice sheet or very broad glacier, while the ice was still moving. The classic view sees the drumlin longitudinal profile as asymmetric, with a steeper stoss (upstream) side and a gentler lee (downstream) side. However, in a study of 29000 British drumlins, Spagnolo et al. (2011) *ESPL* 36, 6, 790 found that the average profile built from all mapped drumlins appears almost symmetric. Furthermore, they found that drumlin shape is not a good indicator of the palaeo ice flow direction. Drumlins have been formed via a variety of mechanisms; see Murray (in J. Holden, ed. (2012) pp. 496–7), for a very accessible summary of drumlin formation theories. Phillips et al. (2010) *Sediment. Geol.* 232, 98 highlight the role played by bedrock geology on the distribution of drumlins beneath an ice stream (this entire issue of *Sediment. Geol.* is devoted to drumlins. A large group of drumlins is a **drumlin swarm**, or **drumlin field**. Briner (2007) *Boreas* 36, 2 argues that the

bedforms in the New York drumlin field indicate fast ice flow. **Rock drumlins** are more commonly known as roches moutonnées.

dry adiabatic lapse rate (DALR) *See* ADIABAT.

Dryas Part of the characteristic threefold late-glacial sequence of climatic change and associated deposits following the last, Devensian, ice advance and prior to the current, Flandrian, interglacial. The colder Dryas phases mark times of cold, *tundra conditions throughout what is now temperate Europe (Britain did not experience the Older Dryas). The **Oldest Dryas** (Dryas I/Pollen Zone Ia) lasted from about 16000 to 15000 years BP, with tundra-grassland characterized by *Dryas octopetala* (mountain avens); the **Older Dryas** (Dryas II/Pollen Zone Ic) from 12300 to 11800 years BP; and the **Younger Dryas** (Dryas III/Pollen Zone III) from about 11000 to 10000 years BP. *See* PLEISTOCENE.

dry farming Farming without irrigation, using mulching, frequent fallowing, working the soil to a fine tilth, and frequent weeding to conserve water. See R. Shetto and M. Owenya (2007).

dry snow zone An area of a glacier where there is no surface melt, even in summer. Most alpine glaciers do not have a dry snow zone as some melting occurs, but much of Antarctica and the central region of the Greenland ice sheet are characterized by no melting (J. Holden, ed. 2012).

dry valley A valley, usually in chalk or *karst, with no permanent watercourse. These valleys may have been cut during *periglacial phases, when the *permafrost would stop *meltwater from soaking through chalk or limestone, and thus permit *dissection.

dual economy A. Lewis (1954) identified two sectors in an economy: the capitalist sector—'that part of the economy which uses *reproducible capital* and pays capitalists for the use thereof'—and the subsistence sector—'all that part of the economy which is not using reproducible capital'. Per capita output is lower in the subsistence sector than in the capitalist sector, through a lack of capital. With additional capital, more workers can move from the subsistence sector to the capitalist sector. The process of economic development then depends on the transfer of surplus labour from the subsistence to the capitalist sector. The nature of risk in a peasant-dominated subsistence sector tends to hinder the creation of employment and the mobilization of surplus labour in the capitalist sector. 'For this reason the path of industrialization is not smooth and trouble-free, or susceptible to analysis using the tools of equilibrium economics' (Ghosh (2007) *J. Econ. Methodol.* 14, 1).

dumping The off-loading of goods at below cost, usually as exports.

dune *See* SAND DUNE; COASTAL DUNE.

duricrust A hard surface layer on tropical upland soils. See Clarke (2006) *Geomorph.* 73, 1.

dust (atmospheric dust) The effect of dust aerosols on climate depends on both dust distribution in the atmosphere and the physical and chemical properties of the dust particles themselves (Tegen (2003) *Quat. Sci. Revs* 22). Changes in atmospheric dust, derived mostly from deserts, can trigger large-scale climate responses (Zhang et al. (2002) *Earth Planet. Sc. Lett.* 202). Yang et al. (2008) *Atmos. Chem. Phys.* 8 demonstrate the modulation of the monsoon climate by dust. Dust particles contribute to the *greenhouse effect by absorbing and emitting radiation; see K. Kohfeld and I. Tegen (2007).

Dust Bowl Parts of Colorado, Kansas, New Mexico, Texas, and Oklahoma, which were severely afflicted by *drought and dust storms in the 1930s. One preconception is that the Dust Bowl was created by farmers ploughing the prairies, but G. Cunfer (2005) thinks it was probably induced over many years by unusually low rainfall and unusually high air temperatures: 'the Dust Bowl would have happened even without farmers ever being there'.

dust storm Washington et al. (2003) *AAAG* 93, 2 highlight the primacy of the Sahara as a source of dust, and Qian et al. (2002) *J. Climate* 15, 10 find that winter temperatures and spring cyclones trigger dust storms.

duyoda In a *periglacial landscape, a steep-sided, shallow, and often circular depression formed as baydzharakhs collapse. A duyoda is smaller than, but can develop into, an *alas.

dyke (dike) A discordant vertical or semi-vertical wall-like igneous intrusion; see McDonnell et al. (2004) *Mineralogical Mag.* 68, 5 on the Slieve Gullion ring dyke, Ireland.

dynamic contributing areas When the soil is saturated, the water table rises to the surface, and the saturated area acts as an extension to the channel, transferring a significant volume of water in a short space of time. These saturated areas are known as variable source areas or dynamic contributing areas (R. Charlton, 2008).

dynamic equilibrium In a landform, a state of balance, in spite of changes taking place within it. For example, dynamic equilibrium in an alluvial stream would be 'delicately adjusted to provide, with available discharge and with prevailing channel characteristics, just the velocity required for the transportation of the load supplied from the drainage basin' (Biedenharn et al. (2000) *Geomorph.* 34, 3–4).

dynamic meteorology The study of those motions of the atmosphere that are associated with weather and climate. For all such motions, the discrete molecular nature of the atmosphere can be ignored, and the atmosphere can be regarded as a continuous fluid medium (J. Holton 1992). See Weber and Névir (2008) *Tellus A* 60, 1 on DSI (Dynamic State Index) analysis in dynamic meteorology.

earth flow A form of *mass movement, where water-saturated, weak-slope material flows under the action of gravity at speeds between 10 cm s^{-1} to 10 cm day^{-1}. A rapid earth flow generally quickly follows the initial fall-slide phase (Di Crescenzo and Santo (1999) *Geogr. Fis. Din. Quat.* 22). Di Crescenzo and Santo (2005) *Geomorph.* 66, 1–4 find relationships between: the height of the landslide crown zone and the slope relief energy; the slope relief energy and the sliding zone area; and the landslide area and the basin area.

earth pillar An upstanding column of soil that has been sheltered from erosion by a natural capstone on top, common where boulder-rich moraines have been subject to gully erosion.

Earth Summit A major conference that focuses on human-environment relationships and aims to promote *sustainable development via inter-governmental action.

SEE WEB LINKS
• Website of Earth Summit 2012.

Earth system science (ESS) A science concerned with how the Earth works today; its natural systems and how these have been perturbed by human actions; how the components of the system have evolved over time in response to changes in other parts of the system; and predicting what will happen in the future.

SEE WEB LINKS
• Online *Journal of Earth System Science.*

earthquake A sudden and violent movement, or fracture, within the Earth, followed by a resultant series of shocks. Earthquakes occur in narrow, continuous belts of activity which correspond with *plate junctions; see Witter et al. (2003) *GSA Bull.* 115, 10. The scale of the shock of an earthquake is known as the **magnitude**; the most commonly used scale is the *Richter scale, while the **intensity** of an earthquake is measured by the *Mercalli scale.

Earthquake waves are of three basic types: P, primary, push waves travel from the focus by the displacement of surrounding particles and are transmitted though solids, liquids, and gases; S, secondary or shake waves travel through solids; and L, long or surface waves travel on the Earth's surface. Computers have been used in earthquake wave analysis since the 1980s; see Tan et al. (2008) *Turkish J. Earth Sci.* 17. For earthquake prediction, see Brace et al. (1966) *J. Geophys. Res.* 71, 16 on *dilatancy theory; Kuo et al. (2006) *Ground Water* 44, 5 on radon-222; and Chen et al. (2002) *Geology* 30 on older fault scarps. Rhoades et al. (2011) *Acta Geophysica* 59, 4, 728 investigate the efficiency of earthquake forecasting models.

easterlies Winds from the east. Easterlies blow in low latitudes in both high and low *tropospheres, and in high latitudes in the lower troposphere (Ichigawa and Yasunari (2006) *J. Climate*

19, 7). The downward propagation of easterly winds occurs throughout the equatorial zone; see Baldwin et al. (2001) *Rev. Geophys.* 39. **Easterly waves** are weak troughs of low pressure in *tropical areas. See Fuller and Stensrud (2000) *Monthly Weath. Rev.* 128, 8 on easterly waves and the North American monsoon, and Ross and Krishnamurti (2007) *Monthly Weath. Rev.* 135, 12 on easterly waves and tropical cyclones.

ecocompensation In China, eco-compensation describes any kind of payment that reflects the environmental impact by those who manage, maintain, or conserve a watershed that provides valuable environmental services. For example, Han et al. (2012) *Chin. Geogra. Sci.* 22, 1, 119–26 calculate an economic value for the loss of wetlands using remote sensing, field investigations, and department visits to survey wetland types, assess wetland area changes, and calculate wetland economic value. Market-value loss and ecological function-value loss caused by reduction of the wetland area and environmental pollution are calculated using market valuations, ecological valuation, environmental protection investment cost analysis, and outcome parameters.

ecodevelopment 'The ideal of eco-development action [places] its national emphasis upon economic equity, social harmony and environmental balance in the local pursuit of individual fulfilment, household self-sufficiency and community self-reliance' (R. Riddell 1981). Mahanty (2002) *World Dev.* 30, 8 thinks the current emphasis on *what* conservation programmes do should be balanced with seeing *how* conservation programmes are undertaken.

ecofeminism A marriage of ecology and feminism. In a pleasingly frank paragraph, Mary Mellor (1992) *Capitalism*

Nature Socialism, 3, 2, 43 writes that ecofeminism cannot be easily defined, reflecting the fact that it draws on many feminisms, from New Age thinkers to socialists. Nightingale in B. Warf (2010) is bolder: 'ecofeminism is the study of women's connections to nature and how they inspire particular forms of environmental activism, stewardship, and spiritual attachments to Earth'.

ecohydrology When vegetation types and patterns change, so do the ways in which water flows. Thus, ecohydrology is the study of the influence of dynamic hydrology, *hydraulics, and geomorphology on stream ecosystems, particularly nutrient cycling, primary productivity, and trophic interactions. See Wilcox et al. (2011) *Geog. Compass* DOI: 10.1111/j.1749-8198.2011.00413.x, on dryland ecohydrology, giving clear and practical insights into this field.

ecological balance The equilibrium between, and amicable coexistence of, organisms and their environment. Changes in air temperature, precipitation, and carbon dioxide levels, plus lesser ecological changes, may alter the ecological balance between communities: 'slight increases in individual mortality of one community, coupled with increased success rates of another, could eventually cause changes in the geographical distribution of forests, grasslands, and deserts' (Masek (2001) *J. Biogeog.* 28, 8). See Ding et al. (1998) *PPG* 22 on the concept of ecological balance and the definition of *desertification.

ecological climatology (**eclimatology**) An interdisciplinary approach to the role of terrestrial ecosystems in the climate system. The central theme is that terrestrial ecosystems, through their cycling of energy, water, chemical elements, and trace gases, are important determinants of climate (G. Bonan 2008; but see Beerling (2003) *Trends Ecol. Evol.* 18, 2).

ecological crisis A state of human-induced ecological disorder that could lead to the destruction of this planet's ecosystem to the extent that human life will at least be seriously impaired for generations, if not destroyed. Evidence to suggest that this crisis has already been reached includes *deforestation, increasing levels of atmospheric carbon dioxide, and current rates of energy use; see C. Merchant (1989).

ecological efficiency The ability of the organisms at one *trophic level to convert to their own use the potential energy supplied by their foodstuff at the trophic level directly beneath them; the energy content of prey consumed by a predator population, divided by the energy content of the food consumed by the prey population (R. Wootton 1998). Ecological efficiency tends to be inversely correlated with primary production (P. Herring 2002).

ecological energetics The study of the flow of energy from the sun through and up the *trophic levels, expressed in calories. With movement up each trophic level, there is a very great loss of the energy available as food. 'A more appropriate name for ecological energetics could refer to limitations by a single component of the resource' (Brooks et al. (1996) *Ecoscience* 3, 3). See also Kooijman et al. (1999) *J. Theor. Biol.* 197.

ecological fallacy A wrong assumption about an individual based on aggregate data for a group; the presence of a relationship between two variables at an aggregated level that is due simply to aggregation, rather than to any real association (Openshaw (1984) *Env. & Plan. A* 16). See Dark and Bram (2007) *PPG* 31, 5 on the *modifiable areal unit problem

and the ecological fallacy; see G. King (1997) on a solution.

ecological footprint The demand of a society on biocapacity (Rees in J. Dewulf and H. Van Langenhove, eds. 2006); the area of land functionally required to support a human society. In more economically developed countries, this will include land beyond the territory inhabited by that society; that is, the land from which its imports are sourced. Chen and Lin (2008) *Energy Policy* 36, 5 identify the factors that affect the annual growth of China's ecological footprint. Wackernagel et al. (2004) *Land Use Policy* 21, 3 calculate national and global ecological footprints, and Bagliani et al. (2008) *Ecol. Econ.* 65, 3 disaggregate an ecological footprint into its energy and non-energy components.

ecological imperialism A process of biological transformation brought about, intentionally or unintentionally, by colonizers in the developing world. The classic text is A. Crosby (2004), which is well reviewed by Hody (1991) *Pols & Life Scis* 10, 1, 81, although she adds the rider 'to acknowledge the impact of ecological imperialism should in no way diminish the importance and relevance of other types of imperialism'.

ecological modernization The argument that the economy benefits from moves towards *environmentalism. Ecological modernization indicates the possibility of overcoming environmental crises without leaving the path of modernization (Mol and Spaargaren (1993) *Intl. Sociol.* 8, 43). See Gibbs (2006) *Econ. Geog.* 82, 2, 193 for a critique of this argument.

ecological overshoot The shortfall in Earth's biological capacity to meet the demands of its human population. Wackernagel et al. (2002) *Procs Nat. Acad. Sci.* 99, 14 argue that human demand may well have exceeded the biosphere's regenerative capacity; in 1961 humanity's load was 70% of the capacity of the global biosphere, rising to 120% in 1999.

ecological politics Advocacy for, or work towards, protecting the natural environment from degradation or pollution. See Philip (2001) *Capitalism, Nature, Socialism* 12, 2.

ecological restoration The re-creation of entire *communities of organisms, closely modelled on those occurring naturally, together with long-term maintenance and management to ensure integrity, and stability. See Abella et al. (2007) *Natural Areas J.* 227, 1.

ecology The study of the interrelationships between organisms and their surrounding, outer world; the study of animals and plants in relation to each other and to their habitats. **Production ecology (community ecology)** is the study of *communities in terms of the throughput of energy and chemical compounds; key concepts are *primary production, *trophic levels, and *nutrient cycles. See Chesson and Huntly (1997) *Am. Naturalist* 150 and Shea and Chesson (2002) *Trends Ecol. & Evol.* 17, 4.

economically active population The total population between the ages of 15 and 65 in any country, used in the calculation of dependent population.

economic determinism The thesis, as advanced by Marx and Engels, that economic factors underlie all of society's decisions. The social relations specific to a particular *mode of production are said to structure social relations between classes, underpinning the legal and political systems. This implies that all political, cultural, and social life can be predicted from the prevailing mode of production. This extreme view has been severely criticized as it denies the existence of free will and individual independence. A more moderate view sees the relations of production as a constraint on the ways in which individuals and superstructures can develop, 'but once the chains of causality anchoring consciousness and value to collective production of material goods are broken, as they are in "humanist" Marxism, the theory loses much of its effective explanatory edge' (Duncan (2005) *PHG* 29, 479).

economic development Taylor (2007) *GaWC Res. Bull.* 238 distinguishes between a growth in the economy, but with the same division of labour— *economic growth*—and the production of new commodities—*economic expansion*—where the division of labour is altered, creating a more complex economy. Taylor calls the former 'old work': 'State economic planning has produced such through increasing old work. But this has its limitations.' The second, he associates with cities 'where cycles of new work generate vibrant, cosmopolitan city economies'. Frenken and Boschma (2007) *J. Econ. Geog.* 7, 5 view the process of economic development as an evolutionary branching process of product innovations. Essletzbichler in K. Franken, ed. (2007) finds significant positive relationships between diversity and stability on the one hand, and growth and instability on the other. Park (2003) *Pol. Geog.* 22, 8 argues that the politics of local economic development in South Korea engendered regionalism and uneven development.

economic distance The distance a commodity may travel before transport costs exceed the value of the freight. Economic distance can be reduced by better physical infrastructure, improvements in the institutional framework, better cost recovery and maintenance, simpler transit procedures, better information, and the reduction of unofficial payments/corruption (Crochet et al. (2004) *IBRD*, World Bank).

economic dynamism Vigorous activity, progress, and strength in an economy. The Ewing Marion Kauffman Foundation measures economic dynamism by: the share of jobs in fast-growing firms; the degree of job turnover (a product of new business start-ups and existing business failures); and the value of companies' initial public offering (the event of a firm's first sale of stock shares). J. Jacobs (1984) sees the economic dynamism of cities as the outcome of the closely related processes of innovation and import replacement. Wojan et al. (2007) *J. Econ. Geog.* 7, 6 find that a relative abundance of *bohemians positively influences local economic dynamism.

(((()))) SEE WEB LINKS

• Website of the Ewing Marion Kauffman Foundation.

economic geography The analysis of the spatial distribution of the transportation and consumption of resources, goods, and services, and their effects on the landscape; 'taking seriously the relations between economic and other social and bio-physical processes, rather than analyzing the economic as either separable from or foundational to such other processes' (Sheppard in S. Bagchi-Sen and H. Lawton Smith 2006).

There are two distinct new economic geographies in the Anglo-American literature. The first uses sophisticated spatial modelling to explain uneven development and the emergence of industrial clusters—by considering centripetal and centrifugal forces, especially *economies of scale and transport costs. See Krugman (1998) *Oxford Review of Economic Policy* 14, 2; for a

critical review, see Martin (1999) *Camb. J. Econ.* 23.

The second approach also attempts to explain the emergence of industrial clusters, but emphasizes relational, social, and contextual aspects of economic behaviour, particularly the importance of knowledge and learning—which takes place most effectively through personal contacts at the local–regional level (Amin (1999) *International Journal of Urban and Regional Research* 23). This economic geography emphasizes aspects of economic behaviour that are considered intangible by the first; see Perrons (2004) *Econ. Geog.* 80, 1. See Gertler (2003) *J. Econ. Geog.* 3 on tacit knowledge and the economic geography of context.

Bathelt and Glückler (2003) *J. Econ. Geog.* 3 argue for a **relational economic geography**, wherein economic actors are situated in contexts of social and institutional relations, and economic processes are contingent in these agents' strategies. Their key words are: organization, evolution, innovation, and interaction. Crevoisier (2004) *Econ. Geog.* 80, 4 in an innovative milieux approach to economic geography, holds that space—or, more precisely, territory—is the matrix of economic development, and that economic mechanisms transform space. N. Coe et al. (2007) is very favourably reviewed. For a review of the 'institutional turn' in economic geography see Peck (2005) *Econ. Geog.* 81.

Evolutionary economic geography seeks to integrate growth and innovation theories, and looks at *endogenous reasons for regional economic development. 'Evolutionary Economic Geography explains the spatial evolution of firms, industries, networks, cities and regions from elementary processes of the entry, growth, exit and (re-)location of firms' (Frenken and Boschma (2007) *J. Econ. Geog.* 7, 5). Lee and Saxenian (2008) *J. Econ. Geog.* 8, 2 claim that the 'evolutionary turn' in economic geography has led to increasing emphasis on coevolution among technologies, organizations, and territories. 'The weakness of this approach, however, is a focus on broad coevolutionary pictures that pays little attention to coordination processes that guide interdependent actions on the ground'.

(⊕) SEE WEB LINKS

• Online *Journal of Evolutionary Economics* providing more about the aims and scope of this discipline.

economic growth The growth in wealth of a nation, as measured by an increase in *gross national product, or in national income. Kang and Meernik (2005) *J. Politics* 67, 1 and Capasso (2004) *J. Econ. Surveys* 18, 3 suggest that government fiscal and monetary policies are critical in improving economic growth; Gupta and Mitra (2004) *Dev. Policy Rev.* 22, 2 stress the crucial role of health care. 'The critical driving force of economic growth is not the super-normal profits that technological change generates but rather the continuous creation of opportunities for further technological development' (Carlaw and Lipsey (2003) *J. Econ. Surveys*, 17 3).

Regulation theorists explore the way 'economic and non-economic procedures can be articulated to produce a relatively stable, coherent and dynamic framework, which can in turn secure the expanded reproduction of capitalism' (Jessop (1990) *Econ. & Soc.* 19, 2). R. Boyer (1990) sees the variability of economic and social dynamics, in both time and space, as 'the central question' for regulation theory; see B. Jessop and N.-L. Sum (2006).

economic integration This includes regional cooperation and integration in the areas of infrastructure and software,

trade and investment, money and finance, and regional public goods. B. Dluhosch (2000) has been reviewed by Bennett (2002) *PHG* 26, 2, 268 as 'an important and readable book that deserves to make [a] significant impact on the European economic development debate'.

economic man A theoretical being who has perfect knowledge of an economy and the ability to act in his or her interests to maximize profits; the concept of the *satisficer is more realistic.

economic rent A measurement of land profitability, economic rent is a payment (rent) over and above that which is necessary to stay in business. Economic rent R, is given as:

$$R = Q\,(p - c)$$

where Q is the quantity of production (as in kilos of grain per hectare), p is market price per unit, and c is the cost per unit of production (E. Sheppard 2008). Economic rent is not synonymous with profit, since opportunity cost (the cost of choosing one alternative good or service in terms of the sacrifice of the next best alternative) is built into the concept. See Coe et al. (2004) *TIBG* 29, 4.

economic system Economic systems mainly deal with the relationships between production/supply and consumption/demand: manufacturing, regulation, circulation, and distribution. P. Hall and D. Soskice (2001) argue that national economic systems often experience external shocks, but adjust, and restore equilibrium. W. Eucken (1952) holds that 19th-century laissez-faire shows that 'the economic system cannot be left to organize itself', a view confirmed by the economic crises of recent years, and the calls for better regulation. N. Brenner and R. Kiel (2006) highlight the interplay between the global and the local within the economic system.

economies of scale The benefits of producing on a large scale: as the volume of production increases, cost per unit article decreases. After a certain volume, this fall in cost will be halted as diseconomies arise, but only at very high levels of production, if at all. Pinch and Henry (1999) *Reg. Studs* 33 argue that regional specialization is sustained by economies of scale that 'lock' regions into particular development paths. Parr (2002) *Env. & Plan. A* 34 has a classification of internal and external economies of scale.

economies of scope The economies or benefits that firms derive by producing particular combinations of output. These exist if one firm can produce two separate products more efficiently than two firms can independently produce them separately. 'Economies of scope resemble economies of scale but operate in a different manner. Whereas economies of scale refer to the lower costs involved in producing larger quantities of a single type of good, economies of scope refer to the benefits generated from producing a mix of goods and thus are common in multiproduct firms' (B. Warf 2007).

economism The belief that economic values are all that is real or important; human beings are motivated mainly by economic drives. Thornton (2007) *Antipode* 39, 5 thinks that American-style economism 'poses a grave threat to the core values of Islam and the West alike'. 'Economism is the claim that decision makers and theorists have overestimated the contribution that the economic realm can make to policy making' (Wolff and Haubrich in M. Moran et al., eds 2006).

economy The state of a country or region in terms of the production and consumption of goods and services and the supply of money; a particular system or stage of an economy.

ecoregion A large area (smaller than a *bioregion, but larger than an ecozone) with a distinctive assemblage of communities, environment, and species. Splinter (2013) *Phs. Geog.* 34, 6 observes that a certain amount of physical variability exists in ecoregions, which makes them spatially diverse in their characteristics.

ecosystem A community of plants and animals within a particular physical environment that is linked by a flow of materials through the non-living (abiotic) as well as the living (biotic) sections of the system. In practice, the hierarchy is: cell, organism, population, community, ecosystem, landscape, biome, and biosphere, although T. Allen and T. Hoekstra (1992) point out that this hierarchy incorporates conceptionally different, non-homogeneous levels of organization, and should therefore be regarded with suspicion. Seastedt et al. (2008) *ESA Online J.*, DOI: 10.1890/070046, suggest that the focus in **ecosystem management** should be less on restoring ecosystems to their original state and more on sustaining new, healthy ecosystems that are resilient to further environmental change.

ecosystem services The benefits to humans of one or more ecosystems. According to the United Nations 2005 Millennium Ecosystem Assessment (MEA), there are four broad categories of ecosystem services: *provisioning*—the production of food and water; *regulating*—the control of climate and disease; *supporting*—nutrient cycles and crop pollination; and *cultural*—spiritual and recreational benefits.

ecotaxation (eco taxation) An economic instrument using taxes or subsidies to encourage behaviour that reduces pollution. Also known as pollution taxation, pigovian taxation, or green taxation. See Bazin et al. (2007) *Ecol. Econs.* 63, 4, 732.

ecotone A region of rapidly changing species between two *ecosystems. Within the ecotone, local factors, such as soil and groundwater conditions, determine species (Walker et al. (2003) *J. Veg. Sci.* 14. See DeDeckker and Forester in C. DeDeckker and J. P. Peypouquet (1988) on the **aquatic ecotone**.

ecotope A defined *niche or niche space within a habitat.

ecotourism The development and management of tourism such that the environment is preserved. The income from tourism adds to the investment into landscape conservation. Lake and Hanson (2000) *Urb. Geog.* 21, 1–4 argue that most of the key concepts in ecotourism are chaotic; 'their uses imply or signify a great range of things, to the point where nearly anything can be deemed sustainable ecotourism' (D. Fennell 2003). Klack (2007) *Geog. Compass* 1, 5 thinks that **sustainable ecotourism** should give equal weighting to ecological integrity, economic viability, and social justice.

ecozone *See* BIOGEOGRAPHIC REALM.

ecumene (oecumene) The inhabited areas of the world, as opposed to the non-ecumene which is sparsely or not at all inhabited. The ecumene of a nation is its more densely inhabited core. These very simplistic classifications pose difficulties of delimitation.

edaphic Of the soil; produced or influenced by the soil.

eddy A roughly circular movement within a current of air or water. An eddy

may be a *vortex (Thompson and Young (2006) *J. Phys. Oceanog.* 36, 4), show *helicoidal flow, or the cylindrical motion of rollers, or appear as a surge phenomenon (a short-lived outbreak of greater velocity in a flow; see Mass and Allbright (1989) *Monthly Weather Rev.* 117, 11).

edge In *network analysis, another term for the *link between two *nodes.

edge city A self-contained employment, shopping, and entertainment *node, located at the periphery of a pre-existing city, that allows its inhabitants to live, work, and consume in the same place. J. Garreau (1991) lists the criteria for an edge city: at least 465 000 m^2 office space; at least 56 000 m^2 retail space; more jobs than bedrooms; identification as a distinct 'place'; and a total dissimilarity to a 'city' of thirty years ago. Edge cities have developed with mass car ownership and a 'whole system of supportive infrastructures, from highways to service stations, to drive-through fast food centres, out-of-town malls and auto-access leisure and retail complexes . . . which thereby renders it at least *functionally* a city' (MacLeod (2003) ICRRDS *Rept Office Dep. PM*).

edge effect The effect on an area of its boundary or edge. This underlines the notion of a *boundary as a transition zone, a porous area rather than a clear, linear demarcation. Van Meter et al. (2010) *Int. J. Health Geographics* 9, 40 is more than competent.

education and learning, geographies of Geographies of education and learning stress the important influence of *spatiality in making and using education systems, whether formal—as in kindergarten and higher education—or informal as in homes and work spaces. 'Between them, these geographies foreground the wider political, economic, social, and cultural processes shaping and being reshaped through formal and informal spaces of education across the globe, and the ways they are experienced, embraced, and contested by educators and diverse subjects of education, including children, young people, parents, and workers' (Holloway and Jöns (2012) *TIBG* 37, 4, 482).

effective discharge In a channel, the discharge that, on average, transports the largest proportion of the annual sediment load, calculated by multiplying the frequency of different discharges by the corresponding sediment transport rate. See Doyle et al. (2005) *Water Resour. Res.* 41, W11411 on effective discharge analysis.

effective normal pressure The force per unit area imposed vertically by a glacier on its bed. For a cold-based glacier, it is effectively equal to the weight of the overlying ice. This is expressed as:

$$N = pgh$$

where N is the effective water pressure, p is the density of the ice, g is the acceleration due to gravity, and h is the ice thickness. However, if water is present at the glacier bed, the effective normal pressure is reduced by an amount equal to the subglacial water pressure. The higher the subglacial water pressure, the more it can counteract the pressure of the ice. The equation then becomes:

$$N = pgh - wp$$

where N is the effective water pressure, p is the density of the ice, g is the acceleration due to gravity, h is the ice thickness, and wp is the subglacial water pressure. This is true for flat-bedded glaciers. Effective normal pressure is modified by the flow of ice over obstacles, increasing upstream of the obstacle and decreasing to its lee.

efficiency The ratio of the work done to the effort used; usually the relation of output to input. **Spatial efficiency**—the organization of space in order to minimize costs—is central to *location theory; see Borrás and Pastor (1998) *Location Sci.* 6, 1–4.

EFTA Current members of the **European Free Trade Association**, established in 1960, are Iceland, Liechtenstein, Norway, and Switzerland.

(((⊕))) SEE WEB LINKS
• The EFTA website.

Ekman layer In the *atmosphere, the transition stratum between the surface *boundary layer and the free atmosphere. **Ekman layer rectification** describes the way the Ekman layer velocity profile differs with variations in surface winds (McWilliams and Huckle (2006) *J. Phys. Oceanog.* 36, 8). The **Ekman spiral** describes how the horizontal wind sets surface waters in motion. As represented by horizontal vectors, the speed and direction of water motion change with increasing depth.

electoral geography The geographical analysis of elections; the study of the spatial patterns of voting and power. Johnston et al. (2004) *Pol. Geog.* 23, 4 find that 'similar people vote differently in different places' (see Pattie and Johnston (1998) *Area* 30 on the *contextual effect). Other concerns include the influence of voting decisions upon the environment, and the drawing of constituency boundaries; in India, for example, constituency boundaries had to be redrawn frequently to take into account the latest census figures, reorganization of states, graduation of some of the union territories into states, and bifurcation or trifurcation of some states,

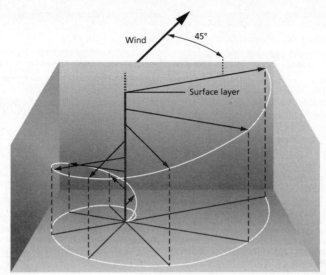

Wind   45°

Surface layer

Ekman spiral

e

to fulfil the demands and aspirations of the people within a new federal structure (Singh (2000) *Pol. Geog.* 19, 4).

elevation The elevation of a geographic location is its height above a fixed reference point, most commonly a reference geoid (a mathematical model of the Earth's sea level as an equipotential gravitational surface) (K.-T. Chang 2014).

(SEE WEB LINKS)
• An interface to query locations on the Earth for elevation data.

elites Powerful and/or very rich persons who disproportionately influence public and private policy. Ahmed (2009) *Hum. Geog.* 2, 3, 37 shows how the coercive power of global governance institutions has worked in tandem with the interests of local elites to produce neoliberal changes in India. Ahmed also

points out that class elites are not a homogeneous group in India.

El Niño A disruption of the equatorial Pacific, east–west *Walker circulatory cell. The gradient between the low pressure of the warm, western tropical Pacific and the high pressure of the cold, eastern tropical Pacific decreases, weakening the easterly trade winds, and allowing warm surface water to move eastwards. The rising branch of the circulation cell, and therefore the *precipitation associated with it, moves with the water. An El Niño is only one element of a dual-phase oscillating ocean–atmosphere system; when the system reverts to its 'normal' phase, Pacific waters cool off the coast of Ecuador and Peru, and become warm again in the western Pacific. Sometimes the eastern Pacific becomes unusually cool; this is a *La Niña event. This entire ocean–atmosphere system is the El Niño, and the 'normal' phase (sometimes developing into La Niña) is the **El Niño–Southern Oscillation**. Dawson and O'Hare (2000) *Geog.* 85, 3 give an exceptionally clear explanation. See Moy et al. (2002) *Nature* 420 on the variability of the El Niño–Southern Oscillation.

El Niño events have been *teleconnected to abnormally heavy rain in the southern USA (*Science Daily*, 2008/01/080128113104) and western South America, a decrease in tropical cyclones in the Atlantic, and to droughts in Indonesia and Australia; see M. E. Mann, R. S. Bradley, and M. K. Hughes (2000). See Nicholls (1988) *Bull. Am. Met. Soc.* 69, 2 on El Niño–Southern Oscillation Impact Prediction.

eluviation The movement of material in true solution or colloidal suspension from one place to another within the soil; soil horizons that have lost material through eluviation are **eluvial**.

embeddedness The state of being located or secured within a larger entity or *context. The economic life of a firm or market is territorially embedded in its peculiar social and cultural relations; in place-specific characteristics, infrastructure, operating environments; and conditions of production. *Transnational corporations in the USA are 'highly constrained by dynamic and deep capital . . . [while] their Japanese counterparts are effectively bound by complex but reliable networks of domestic relationships' (Yeung (2000) *TIBG* 23, 3).

Hess (2004, *PHG* 28, 1) argues that there are interconnected societal, network, and territorial dimensions of embeddedness: **territorial embeddedness**—the extent to which an 'actor' is anchored in particular territories or places; **societal embeddedness**—the cultural, political, institutional, and regulatory framework the actor is located in; and **network embeddedness**—the

structure of relationships among a set of individuals/organizations. Granovetter in N. Nohria and R. Eccles, eds (1992) defines **structural embeddedness** as 'the connectedness of not only two parties, but the extent of interconnection among third parties or mutual contacts of dyadic partners'. Orderud (2007) *Geografiska B* 89, 4 refers to an embeddedness 'where trust acts as a reinforcement . . . as a capacity restraint and a socially constructed need for face-to-face meetings'. See Feagan (2007) *PHG* 31, 1 on the 'quality turn' and embeddedness in food supply; see also Jones (2008) *PHG* 32, 1. **Ecological embeddedness** is embedded in 'natural' or ecological processes, to which Morris and Kirwan (2011) *J. Rural Studs* 27, 322 would add that 'it must reflect a change in the relations between economic actors and the natural environment that produces a benefit to both'.

embodiment A perspective that roots social existence and experience through the body and the cognition of the self. The body is symbolically constructed as one incorporates socially constructed labels and assumptions relating to the body; 'these take on the weight of unquestionable fact once applied to self and others' (Adams and Ghose (2003) *PHG* 27, 4). 'Intellectual production is always materialized through human bodies, and nonhuman objects . . . knowledge never arrives from pure brainpower, from only sparking synapses. It is the outcome of embodied practice' (Barnes (2004) *PHG* 28, 5).

Perception is seen as a practical bodily involvement; an active process relating to our ongoing projects and practices. 'This means that the human body is unique in playing a dual role both as the vehicle of perception and the object perceived' (Simonsen (2005) *Geografiska B* 87, 1). H. Nast and S. Pile (1998) consider the relationship between space and embodiment: bodies are relational and territorialized—they are woven together with space in 'intricate webs of social and spatial relations that are made by, and make, embodied subjects'. 'Our understanding of spaces and places must incorporate the ways in which bodies are constituted and contested, and acknowledge their challenge to conventional boundaries' (Little and Leyshon (2003) *PHG* 27, 3). S. Franklin and S. McKinnon (2001) list the 'vastly different scales of embodiment', from the gene, to the body, to the family or species, to the nation, to the commodity form, and cyberspace.

emergence The creation of new phenomena, requiring new laws and principles, at each level of organization of a complex, and often non-linear, system. This concept from physics has been applied to geomorphology; see Harrison (2001) *TIBG* 26, 3.

emigration The movement of people from one place to another, usually from one country to another. Green et al. (2008) *NZ Geogr.* 64, 1 observe that the 'push–pull' model incompletely describes the motivations of those who migrate, and who are not necessarily either pushed or pulled, but motivated by things like a desire for a change, or for more adventure. Hugo (2006), *Geogr. Res.* 44, 2 shows how increasing gradients of difference between nations in the pattern of growth in the workforce, in income and poverty levels, and in governance have been important drivers of emigration. In a study of emigration from rural Albania, Vullnetari and King (2008) *Glob. Netwks* 8, 1 remind us that migration statistics are about real people.

emissions credits Through photosynthesis, green plants remove carbon dioxide from the atmosphere. Mindful of rising levels of atmospheric carbon dioxide and *global warming, the *Kyoto Protocol proposed that, if nations increase their forested areas, they should be granted emissions credits to count in their CO_2 **emission standards** (their permissible, and legally binding, pollution maxima); see Depledge and Lamb (2005) *UNFCCC*. A further proposal, **emissions trading**, would allow a nation with *greenhouse gas emissions below its agreed target to trade its 'credit' to over-target nations. The impact of emissions trading will depend on the structure of the system (Voon (2000) *J. Int. Law & Policy* 10, 1).

emotional geographies The study of the location of emotions in both bodies and places; the emotional relations between people and environments; and the task of representing emotional geographies; see L. L. Bondi et al. (2007). Emotional geography has 'a common concern with the spatiality and temporality of emotions, with the way they coalesce around and within certain places' (J. Davidson et al. 2007). One of its central themes is the examination of how feelings mediate both our conscious and unconscious behaviours in the places where we live out, or 'enact', our lives (Hallman et al. (2007) *Soc. & Cult. Geog.* 8, 6). See Thien (2005) *Area* 37 for consideration of emotion in geography. See also Anderson and Smith (2000) *TIBG* 26, 1.

enclave A small area within one country administered by another country. Enclaves often appear to become the regions of international tensions and wars: East Bengal, West Berlin, Cabinda, Northern Ireland, and the Palestinian territories (E. Vinokurov 2007). Kaliningrad is situated simultaneously on the European Union's border with Russia as well as physically separated from its home country Russia by the surrounding land boundaries of EU states, thus resembling an enclave (Diener and Hagan (2011) *Euras. Geog. & Econs.* 22, 4, 567.

enclosure The conversion of common or open fields into private, exclusive parcels on which sole proprietorship gave rents and/or management decisions to individual owners (Williamson in J. Eatwell, M. Milgate, and P. Newman, eds 1987).

encoding Digitizing data into a computer format, as happens when an image is scanned. Encoding may be used to improve security, reduce storage needs, or to transfer data between different file formats.

encounter, geographies of
'Social behaviours draw upon previous encounters' (Hitchings (2007) *TIBG* 32, 3). Valentine (2008) *PHG* 32, 3 observes that 'encounters in public space . . . always carry with them a set of contextual expectations about appropriate ways of behaving which regulate our coexistence. These serve as an implicit regulatory framework for our performances and practices.' R. Fincher and K. Iveson (2008) argue that libraries are spaces of encounter that have a redistributive function, offering free and equal access and a safe space for individuals and groups. See also Laurier and Philo (2006) *Area* 38, 4; Tolia-Kelly (2006) *Area* 38, 2.

endemic Occurring within a specified locality; not introduced. **Endemism** describes the restriction of a species (or other taxonomic group) to a particular geographic region, and is due to factors such as isolation or response to soil or climatic conditions. Giokas et al. (2008) *J. Biogeog.* 35, 1 outline a technique to identify areas of endemism.

endogenetic (endogenic) 'From within'. In geomorphology, this means those forces operating below the crust which are involved in the formation of surface features. In human geography, it is those forces acting from within, for example, a society. An **endogenous variable** in economics is a variable explained within the theory under consideration.

endogenous growth theory An economic theory claiming that economic growth is generated from within a system as a direct result of internal processes, underlining the way some regions create an internal mechanism for promoting or perpetuating their growth. Firms operating in places with pools of skilled professionals, for example, are more innovative, leading to faster technological and productivity growth (Ó Huallacháin and Leslie (2007) *J. Econ. Geog.* 7, 6). Frenken and Boschma (2007) *J. Econ. Geog.* 7, 5 observe that the more product varieties are already present in a firm or city, the higher the probability that new product varieties can be generated through recombination of old routines—'this feedback relationship is non-linear in that the potential for new ideas rises more than proportional with the stock of existing ideas'. See also D. Felsenstein et al., eds (2001).

energy The physical capacity for doing work. Nearly all our energy derives from the sun, and technical progress has reflected more and more sophisticated uses of energy, from wind and water, through *fossil fuels, to *nuclear power.

In the early stages of industrialization, energy consumption is closely related to levels of economic development, and per capita *GNP—see Zeng et al. (2008) *Science* 319, 5864 on China. Mature economies tend to be more energy efficient, perhaps because technology improves, and the emphasis shifts to service industries (M. Carr 1997). Even so, the *advanced economies still account for most of the world's energy consumption.

World demand for energy has increased so much that an **energy crisis** (a potential shortage of energy) has been identified; see the Roosevelt Institution's 25 ways of solving the energy crisis. This crisis, together with the adverse environmental effects associated with the burning of fossil fuels (*greenhouse effect, *acid rain) has led to increased emphasis on energy conservation. **Energy intensity** is energy consumption per unit GDP. *AAAG* 101 (2011), issue 4 is devoted to new geographies of energy.

(🌐 SEE WEB LINKS)

• Energy Administration Information.

englacial Within a glacier.

ENGO Environmental NGO.

ENSO *See* EL NIÑO–SOUTHERN OSCILLATION.

enterprise zones 'Place-oriented policies that are a tool of regional development, and refer to geographically targeted areas chosen for development that are designated on the basis of unemployment, poverty, population, age of housing stock, and other criteria. Firms that locate in the area and create jobs are given tax credits, abatements, and exemptions. The underlying assumption is that firms and employees in the zone area benefit because of a reduction in the price of capital and/or labour, and there is expanded investment and employment generation through deregulation' (S. Sridhar (2004) *Nat. Inst. Public Finance & Prop. N. Delhi W. Paper* 19). In a study of US enterprise zones, A. Peters and P. Fisher (2002) find that while early econometric studies of the effects of state enterprise zones usually found positive results, more recent results have been much less favourable.

entisol In *US soil classification, young soils, high in mineral content and

without developed soil *horizons. *See* RANKER.

entitlement A guarantee of access to benefits through rights, or through legal agreement. By entitlements, A. Sen (1981) means one's ability to achieve the commodities necessary for basic needs (including food and shelter), either through income-generating work or from direct production of these goods. The size of one's entitlement determines the amount of food (and other commodities) one can command. Thus Smith (2001) *PHG* 25, 140 writes that access to food in urban areas is affected by a wide range of entitlements and blockages, most of which revolve around the fact that unlike rural areas where self-sufficiency or exchange is possible, cities operate on the basis of a cash economy. Sen uses the **entitlement approach** to explain famines. Watts (2000) *Zeitschrift* 88 comments that 'one of the great strengths of Sen's approach to hunger is that its entitlements are part of a larger architecture of thinking about development as a state of well-being and choice or freedom'. See also G. Robinson (2004).

entrainment 1. The picking up and setting into motion of particles, either by wind, water, or *ice. The main entrainment forces are provided by impact, *lift force, and *turbulence. Resistance to entrainment is provided by packing and by the submerged weight of the particle.

2. In *meteorology, the incorporation of buoyant air into a cloud; see Moeng (2000) *J. Atmos. Scis* 57, 21.

entrepôt A transhipment point where goods are held without incurring customs duties. See Dobbs (2002) *Sing. J. Trop. Geog.* 23, 3 on Singapore as an entrepôt.

entrepreneur An organizer, singly or in partnership, who takes risks in

UK Enterprise zones

- • Enterprise Zones (26)
- Development Areas
- Intermediate areas
- Special status for Northern Ireland

creating, investing in, and developing a *firm from its inception through to hoped-for profitability as goods and services are marketed. In recent decades, most industrialized receiving countries have experienced a rise of

immigrant and ethnic entrepreneurship: see R. Kloosterman and J. Rath (2003). Jessop and Sum (2000) *Urb. Studs* 12 hold that an entrepreneurial city pursues innovative strategies intended to maintain or enhance its economic competitiveness *vis-à-vis* other cities and economic spaces. A. Cronin and K. Hetherington (2008) see an **entre-preneurial city** as characterized by business-led urban development, technological innovation, social capital featuring skilled labour, and social and *environmental *sustainability, and supported by local authorities, educational and training institutions, and its own heritage and cultural industries.

entropy A measure of disorder in a system; the higher the entropy, the greater the disorder. 'Entropy maximization is a method used to derive the most probable state of a spatial interaction system, given information on total inflows and total outflows and capacity constraints' (Sheppard and Plummer in N. Thrift and R. Kitchin, eds 2009). For an exceptionally clear explanation of the role of entropy in the atmosphere, see Lockwood in J. Holden (2012), p. 84. Wilson (2010) *Geogr. Analysis* 42, 4, 364 provides a useful summary of entropy in human geography.

enumeration district A unit of census survey; in the UK, about 150 households.

environment The surroundings. The **natural environment** includes the nature of the living space (sea or land, soil or water), the chemical constituents and physical properties of the living space, the climate, and the assortment of other organisms present. The **phenomenal environment** includes changes and modifications of the natural environment made by man. The effect of the environment on man is modified, in part, by the way the environment is perceived, and human geographers distinguish this—the **subjective environment**—from the **objective environment**—the real world as it is. Eden (2001) *PHG* 25, 1 shows that nature

and the environment are not only differently perceived and classified by different groups but also used differently in association with policies and actions which can then materially influence what is being perceived. Simmons (2000) *Geog.* 85 describes the 'many layers' of environment as 'something more like a double helix of mind and matter, whose gyre seems to be ever-widening, spinning unpredictable combinations of society and economy into the main space, but also throwing off minor gyres, which might be short-lived but might equally be the germs of the main arms of the future'.

Environment Agency An independent public body, created to protect and improve the environment in England and Wales. Its remit includes tackling flooding and pollution incidents, reducing industry's impacts on the environment, cleaning up rivers, coastal waters, and contaminated land, and improving wildlife habitats.

(∰) SEE WEB LINKS

• The Environment Agency website.

environmental accounting The inclusion of economic and environmental information in a common framework to measure the contribution of the environment to the economy and the impact of the economy on the environment. Environmental accounts enable governments to set priorities, monitor economic policies more precisely, enact more effective environmental regulations and resource management strategies, and design more efficient market instruments for environmental policies (Lange et al. (2003) *Nat. Resources Forum* 27, 1). Pulselli et al. (2006) *Ecol. Econ.* 60, 1 use environmental accounting in calculating an index of sustainable economic welfare. See C. Barrow (2005).

An **environmental audit** is the practice of assessing, checking, testing, and verifying a process of environmental management, and/or the product(s) arising from it; see Alvarez-Larrauri and Fogel (2008) *J. Cleaner Prod.* 16, 1 on environmental audits in Mexico.

environmental archaeology The study of the relationship between climatic/environmental change and human cultural evolution during the *Holocene. The entire issue of *The Holocene* (2012) 22, 6 is devoted to multidisciplinary studies in environmental archaeology.

environmental determinism The view that human activities are governed by the environment, primarily the physical environment. J. M. Blaut (2005) challenges the central thesis of environmental determinism, while J. Sachs (2005) reconsiders the role of the environment in shaping human history. Judkins et al. (2008) *Geog. J.* 174, 1 think that Jared Diamond's books are pure environmental determinism: 'this resurgence threatens negative consequences if uncritically adopted by policymakers.'

environmental dynamics The behaviour of an environmental system in space and through time. 'Modelling environmental dynamics aids in the understanding and anticipation of future evolutions, and their prospective simulation supports decision making for environmental management' (M. Paegelow and M Olmedo 2008).

environmental economics 'Understanding why environmental issues exist, identifying them if it is worth remedying the situation, and developing solutions' (Rolfe (2008) *Australian J. Agric. & Resources Econ.* 52, 1).

environmental gradient A gradual and continuous change in communities and environmental condition. The gradients can be related to environmental factors such as altitude, temperature, and moisture supply. See Damgaard (2003) *Ecol. Modelling* 170 on modelling plant competition along an environmental gradient.

environmental impact A change in the make-up, working, or appearance of the environment. An **Environmental Impact Assessment** considers the probable consequences of human intervention in the environment, seeking to restrict environmental damage. In most developed countries, the methodology includes provision for the responsible authority to produce a draft, release this for public comment, and gather responses from a variety of perspectives: other agencies and levels of government, environmental and community groups, interested corporations, resource users, and ordinary citizens. This generates information that might otherwise be excluded from administrative decision making, and although this information may not directly influence the decision, it makes environmental and democratic values more visible and legitimate than before (Turnbull (2004) *Geog. J.* 170, 1). See S. Atapattu (2006).

environmentalism A concern for the environment, and especially with the bond between humans and the environment; not solely in terms of technology but also in ethical terms. In 2004, M. Shellenberger and E. Nordhaus savaged modern environmentalism as 'no longer capable of dealing with the world's most serious ecological crisis'. 'If we are to understand modern conservation of nature, we need to see environmentalism not as a thing, but as a process' (L. Vivanco 2006). Wade-Benzoni et al. (2007) *Presidential Studs Qly* 37, 4 find that self-perceptions of environmentalism are changed by subtle manipulations of context and, in turn, affect environmental behaviour. **Market environmentalism** is defined by T. Anderson and D. Leal (2001) as a mode of resource regulation that promises both economic and environmental health via market means, offering a 'virtuous fusion of economic growth, efficiency, and environmental conservation'. See Bailey (2007) *AAAG* 97, 3 on territorializing market environmentalism, and Davis (2006) *Geog. J.* 172, 2.

environmental justice 'The fair treatment and meaningful involvement of all people regardless of race, colour, national origin, culture, education, or income with respect to the development, implementation, and enforcement of environmental laws, regulations, and policies' (US Environmental Protection Agency). See Ranco and Suagee (2007) *Antipode* 39, 4. It is a social movement focusing on the fair distribution of environmental benefits and burdens, and is also an interdisciplinary group of scholars interested in the linkages between social justice and environmental change, with a particular focus on the global dimensions of (in)justice. The **environmental justice foundation** campaigns on a number of issues, but is not an academic source. See Buckingham (2004) *Geog. J.* 170, 2 on environmental justice and ecofeminism.

environmental lapse rate The fall in temperature of stationary air with height, averaging 6 °C per 1000 m. This fall of temperature with height in stable air is due to a fall in the density of the air, varying greatly with the time of day, and with the nature of the *air mass concerned. See Schultz et al. (2000) *Monthly Weather Rev.* 128, 4143–8, and Houston and Niyogi (2007) Monthly Weather Rev. 135, 3013–32.

Environmentally Sensitive Area (ESA) A fragile *ecosystem area where the conservation or preservation of the natural environment is sustained by state controls and/or grants, such as the eastern South Downs, UK (Boardman (2003) *TIBG* 28, 2).

environmental perception The way in which an individual perceives the environment; the process of evaluating and storing information received about the environment. Decision-makers cannot but base their judgements on the environment as they perceive it. See J. Hobbs and C. Salter (2006) and Y.-F. Tuan (1974).

environmental policy The statement by a supranational, national, or regional government of its approach to environmental protection. Pezzey (2004) *Scand. J. Econ.* 106, 2 contrasts environmental policy, 'which internalises externalised environmental values', with sustainability policy, 'which achieves some form of intergenerational equity'. See C. Barrow (2005); try Tews et al. (2003) *Eur. J. Pol. Res.* 42, 4.

environmental realism The notion that the environment is a 'real thing'; environmental idealism posits a set of underlying, stable, and consistent values that relate to the character, sense, and quality of nature. See P. Macnaghten and J. Urry (1998).

environmental science The study of environments; traditionally the physical environment, or, more widely, social and cultural environments. Major themes include: environmental accounting; environmental performance evaluation; reporting; participative processes in environmental policy-making; sustainable management; and the effectiveness of Local Agenda 21. See the special 2007 issue of *Environment and Urbanization* 19, 1. O'Riordan (2004) *TIBG* 29, 2 notes that environmental science 'has become highly political . . .

the antagonistic pressures of established power and economic hegemony are never far away'.

Environment Directorate An agency of the OECD, the environment directorate provides governments with the analytical basis to develop policies that are effective and economically efficient, including country performance reviews, data collection, policy analysis, projections and modelling, and the development of common approaches.

((⊕)) SEE WEB LINKS

- The Environment Directorate on the OECD website.

eolian *See* AEOLIAN.

epeirogeny Broad, and generally large-scale, limited, or slow vertical movements of the Earth's crust (Chardon et al. (2006) *Geomorph.* 82, 3–4) which do not involve much alteration in the structure of the rock; hence **epeirogenic**—caused by the relatively gentle raising or lowering of the Earth's crust.

ephemeral Short-lived. R-strategist plants are ephemeral in that they grow and reproduce rapidly when conditions are favourable, dying within a short space of time; see Boulos (2003) *J. Biogeog.* 30, 12. **Ephemeral streams** flow only during and after intense rain, and are typical of arid and semi-arid areas; see Billi (2007) *Geomorph.* 85, 1–2.

epicentre The point of the Earth's surface which is directly above the *focus of an *earthquake, usually 0–50 km below it. See Bondár et al. (2004) *Geophys. J. Int.* 156, 3 on epicentre accuracy based on seismic network criteria.

epilimnion The upper layer of a body of water, penetrated by light, thus enabling photosynthesis. This zone is warmer, and contains more oxygen, than the layers below.

epiphyte A plant growing on another plant but using it only for support and not for food, most commonly found in *tropical rain forests. See Zotz et al. (1999) *J. Biogeog.* 26, 4 on the community composition, structure, and dynamics of epiphyte vegetation.

epistemology The philosophical theory of knowledge which considers how we know what we know, and establishes just what ought to be defined as knowledge. In *geography, the term is used to indicate the examination of geographical knowledge—how it is gained, sent, changed, and absorbed. O'Sullivan (2004) *TIBG* 29, 3 distinguishes between epistemology (1) 'experimenting on theories' and (2) epistemology learning from models.

epochs An interval of geological time; several epochs form a period, several periods an era. An epoch is ranked as a third-order time unit (M. Allaby 2006).

equal area map A map so drawn that a square kilometre in one portion of the map is equal in size to a square kilometre in any other portion. Equal area maps of the whole globe tend to be elliptical in shape, and severely distort the shapes of regions far from the equator as in Peters' projection below.

equatorial rain forest *See* TROPICAL RAIN FOREST.

equatorial trough A narrow zone of low pressure, between the two belts of *trade winds, arising from high *insolation, especially in the centre of continents in summer; see Quinn and Burt (1967) *J. Appl. Met.* 6, 6. Also known as the *inter-tropical convergence zone, the equatorial trough is not constant in

Peters' projection

Interrupted equal area

position, breadth, or intensity; see Chao (2000) *J. Atmos. Scis* 57, 641–51.

equifinality The principle that a given end state can be reached by many potential means. Thus, in geomorphology, this term indicates that similar landforms might arise as a result of quite different sets of processes. For example, Cruslock et al. (2010) *Geomorph.* 114 note that different erosion mechanisms—in this case, wave action and ice scour—have produced similar *shore platforms in Sweden and South Wales in the UK. Thus from the landform alone, without additional evidence, it might be difficult to identify the particular set of causes or to differentiate different feasible causes.

equilibrium In physical geography, the state of an open system where the inputs, throughputs, and outputs of energy or matter are in balance. Equilibrium relies on the presence of negative feedback loops within the system (R. Inkpen 2005) and J. Holden (2012) recognizes: 'static equilibrium (no change over time) or a steady-state equilibrium (short-term fluctuations about a longer-term mean value) while over longer time periods the equilibrium might be dynamic (shorter-term fluctuations with a longer-term mean value that is changing)'. Holden adds 'the concept of equilibrium has always been somewhat confusing because different people have chosen to identify different types of equilibrium and because the precise meaning is time dependent ... Often it depends on where and when you measure something as to whether it will show equilibrium'. Within a system, the **equilibrium time** is the time required for a system to establish a new steady state after the application of a stimulus or a change in external conditions. Within human geography **spatial equilibrium** is regional equality, brought about by the interaction of labour moving to high-wage areas and capital moving to regions where wage rates are low, and usually expressed in terms of wages: 'workers and firms are indifferent among alternative locations as they have eliminated any systematic difference in indirect utility and profits through migration' (Ottoviano and Perri (2005) *J. Econ. Geog.* 6, 1.

equilibrium line In glaciology, the point at which expansion of the glacier by accumulation is balanced by ice losses through *ablation. See Carrivick and Brewer (2004) *Geografiska A* 86, 1.

equilibrium species Species that show characteristics consonant with a stable *niche. Dispersal is less important, perseverance is more significant than recovery from adverse conditions, and the survival of the young is more important than high fecundity.

equinox Equinoxes, those times when day and night are of equal length, occur twice a year. The **spring**, or **vernal**, **equinox** is on 21 March and the **autumn equinox** is on 22 September. On these dates, the sun is directly overhead at the equator. The changes in day length result from the changes in the tilt of the Earth with respect to its axis.

era The largest unit of geological time. The approximate datings of the eras are:

era	duration in millions of years before present
Precambrian	4600–570
Palaeozoic	570–225
Mesozoic	225–65
Cenozoic	65–0

erg Arid, sandy desert, particularly within the Sahara. The Rub Al Khali aeolian system of the Arabian Peninsula is the world's largest erg (Bray and Stokes (2004) *Geomorph.* 59, 1–4).

ergodic A form of reasoning whereby data are selected as examples of a landform at different stages of evolution, since we cannot travel back in time to see all the stages through which it has evolved. Ergodic reasoning is, in effect, a special case of the axiom that 'the present is the key to the past'. See Paine (1985) *PPG* 9, 1.

erodibility The vulnerability of a material to erosion. This will vary according to the nature of the material: degree of consolidation, organic content, particle size, hardness, and strength, among others; on the extent and type of ground vegetation; thus on *land use. See Darboux and Le Bissonnais (2007) *Eur. J. Soil Sci.* 58, 5 on measuring soil erodibility. Quantifying soil erodibility is crucial for forecasting dust events (Zender et al. (2003) *J. Geophys. Res.-Atmos.* 108, 4543).

erosion The removal of part of the land surface by wind, water, gravity, or ice. These agents can only transport matter if the material has first been broken up by weathering. Some writers use a very narrow interpretation of the word, claiming that erosion refers only to the transport of debris; see Stallard in P. Hancock and B. J. Skinner, eds (2000). If so, *denudation would be the correct term for the weathering as well as the transport of rocks.

Primary erosion is the initial *in situ* erosion of rock, *regolith, or soil.

erosion surface A relatively level surface produced by erosion. Much of Africa is composed of extensive plains, often cutting *discordantly across varied geological structures, and these have therefore been identified as erosion surfaces as such (Lister (1965) *Rec. Geol. Surv. Malawi* 7). See also Ojany (1978) *GeoJournal* 2, 5. The whole concept is bound up with the theory of the cycle of erosion.

erosivity The ability of rainfall to cause erosion, which depends on rainfall intensity and drop size, as well as the material it lands on. Rainfall intensity is used as the primary erosivity indicator at shorter times and smaller areal scales. Larger areas and greater time scales generally require use of run-off as the measure of erosivity. See Cerro et al. (1998) *Soil Sci. Soc. Am. J.* 62.

erratic A large boulder of rock which has been transported by a glacier so that it has come to rest on country rock of different *lithology; all erratic blocks in northern Germany, for example, originated in Scandinavia (Bernhardi (1832) *Jahrbuch für Mineralogie, Geognosie, und Petrefaktenkunde*, **3**). Agassiz (1840–1) *Procs Geo. Soc.* 3, 2 used the presence of erratics as evidence of glaciation in the British Isles.

ESA *See* ENVIRONMENTALLY SENSITIVE AREA.

escarpment A more or less continuous line of steep slopes, facing in the same direction, and caused by the erosion of folded rock (A. Bowen and J. Pallister 2001).

esker A long ridge of material deposited from *meltwater streams running *subglacially, roughly parallel to the direction of ice flow, and ranging from tens of metres to several hundred kilometres, as in Finland. Eskers wind up and down hill because subglacial streams are under great hydrostatic pressure, and can flow uphill. Brennan (2002) *Geomorph.* 32, 3–4 identifies and investigates five types of esker. Thomas and Montague (1997) *Quat. Sci. Rev.* 16 argue that glacial lake systems are crucial for the development of esker systems.

estuary That area of a river mouth which is affected by sea tides. A **wave-dominated estuary** has waves as the dominant shaping force; in a **tide-dominated estuary**, tidal currents dominate geomorphologically (Komatsu (1999) *Cretaceous Res.* 20, 3).

eta index **(η)** An expression of the relationship between a network as a whole and its *edges. For a worked example, see Alberto and Manly (2006) *J. Biogeog.* 33, 4.

etchplain A tropical *planation surface where deep weathering has etched into the bedrock. Fluvial denudation may lay the etchplain bare. See Stengel and Busche in X. Yang, ed. (2002) on a Late Proterozoic etchplain in Namibia.

ethical geography Ethics are the principles of right and wrong conduct; the basis for doing what is right, and the discernment of what is right (W. R. F. Browning 1997); morals are the accepted norms and standards of conduct of a society, community, or nation (J. Last 2007). Smith (2001) *PHG* 25, 2 seeks to clarify the difference between these two concepts: 'moral philosophers would doubtless agree that writing about certain things being right or wrong counts as ethics, but would wish to consider how such normative claims are made'. Cloke (2002) *PHG* 26 asks geographers to live ethically and act politically, but Evanoff (2007) *Eth. Place. Env.* 10, 2 argues that a single environmental ethic that can be universally applied in all geographic settings and across cultures 'cannot be formulated'. Madge (2007) *PHG* 31, 5 provides a wide-ranging review of research ethics in geography.

ethical trade Trade based on social and environmental responsibility in existing large-scale businesses. Freidberg (2003) *Soc. & Cult. Geog.* 4, 1 notes that supermarket supply networks are characterized by highly unequal power relations and fetishized standards: 'cleaning up down South comes cheap'. Hughes (2001) *J. Econ. Soc.* 1 argues that for ethical trading strategy to make a more significant difference, attention needs to be paid to worker welfare at sites of production, and the terms of trade between retailers and suppliers.

ethnicity A cultural and geographic entity that emerges when a group shares a common ancestry, origin, and tradition. Ethnicity may relate to a geographical territory, world view, custom, ritual, and language (J. Anderson 2010). Ethnicity in a group may become pronounced as a result of migration: 'in Trinidad and Guyana, Hindus were at the bottom of the social structure . . . When they migrated to Britain, they were still in an ethnic quandary, with the white British population thinking of them derogatorily as "Paki" (subcontinental Indian)' (Kong (2001) *PHG* 25, 2). However, Peach (2000) *PHG* 24, 4 observes that the minorities in the 1960s to the 1980s which were sometimes compressed into a single 'black race' category are now being teased out into their constituent ethnicities, classes, and genders. Whites are also being unpacked from their homogenized categorization: see Shaw (2006) *Antipode* 38, 4. 'Transnationalism is transforming the ethnic enclave from being a feature of deprivation in the inner city into affluent closure in the suburbs; ethnic villages of the audible rather than the visible minorities are appearing' (Simpson et al. (2008) *Urb. Studs* 45, 1).

ethnic segregation The evolution of distinct neighbourhoods recognizable by their characteristic ethnic identity. Massey and Denton (1988) *Social Forces* 67 distinguish five dimensions of residential segregation: evenness, exposure, concentration, centralization, and clustering. Phillips et al. (2007) *TIBG* 32, 2 discuss multiple readings of ethnic segregation.

External causes (imposed by the *charter group) of ethnic segregation include discrimination, low incomes—which direct them towards inner-city locations—and the need for minorities to locate near the *CBD since much of their employment is located there. Internal causes (springing from the ethnic group) include a desire to locate near facilities serving the group, such as specialized shops and places of worship, desire for proximity to kin, and protection against attack; see Fossett (2006) *J. Math. Soc.* 30, 3–4.

Clark and Morrison (2008) *Pop. Res. & Policy Rev.* 27, 3 find that an upgrading of housing stock may promote ethnically homogeneous apartment-building 'neighbourhoods' within an otherwise unsegregated ethnically diverse area, and Fossett (2006) *J. Math. Soc.* 30, 3–4 reports that reductions in housing discrimination may not necessarily lead to large declines in ethnic segregation in the short run. Ellis et al. (2004) *AAAG* 940 reveal that segregation by work tract is considerably lower than by residential tract. Lobo et al. (2007) *Urb. Geog.* 28, 7 argue that ethnicity itself is wielded as a potent force as a group moves into a neighbourhood and eventually dominates it. Phillips (2007) *Geog. Compass* 1, 5 explores the politics of data collection, categorization, and representation in ethnic segregation research.

ethnocentricity Making assumptions about other societies, based on the norms of one's own. This may lead to the development of global theories based, for example, on Western conditions. 'Ethnocentric accounts of "hate" and "love" are continually disingenuous to the flows and rhythms of the development and experience of affective capacities of individual bodies, and thus are negligent in their theoretical project' (Tolia-Kelly (2006) *Area* 38, 2).

ethnography The study of the customs, habits, and behaviour of specific groups of people (usually people in non-literate societies). Herbert (2000) *PHG* 24, 4 believes that ethnography is an underused methodology in geography. 'This neglect is especially injurious to the discipline because ethnography provides unreplicable insight into the processes and meanings that sustain and motivate social groups. These processes and meanings vary across space, and are central to the construction and transformation of landscapes; they are both place-bound and place-making. Ethnography's potential contribution to geography is thus profound.' Hart (2006) *Antipode* 38, 5 holds that using critical conceptions of spatiality can extend and enrich global ethnography. Hörschelman and Stenning (2008) *PHG* 32, 2 argue that ethnographic research is most valuable when it is employed 'within an explicitly spatial imaginary, placed within a multi-sited context and critically intervening in broader theoretical debates'. Would it not then be geography?

euphotic zone The upper layer of a body of water receiving light and thus where photosynthesis is possible. Deep-sea benthic communities are largely controlled by the organic carbon flux from the euphotic zone (McClain et al. (2006) *J. Biogeog.* 33, 9).

European Economic Area A single, free-trade grouping, comprising the member states of the *European Union and *EFTA.

(⊕) SEE WEB LINKS
• The European Economic Area defined on the European Commission website.

European Neighbourhood Policy (ENP) Through its ENP, the EU works with its southern and eastern neighbours to achieve the closest possible political association and the greatest possible degree of economic integration. This goal builds on common interests and on values—democracy, the rule of law, respect for human rights, and social cohesion. Partner countries agree an ENP action plan with the EU, demonstrating their commitment to democracy, human rights, rule of law, good governance, market economy principles, and sustainable development. In early 2014, participating partners were: Armenia, Azerbaijan, Egypt, Georgia, Israel, Jordan, Lebanon, Moldova, Morocco, Palestine, Tunisia, and Ukraine.

European Union (EU) A *free trade area comprising Austria, Belgium, Bulgaria, Cyprus, the Czech Republic, Denmark, Estonia, Finland, Germany, Greece, Hungary, Ireland, Italy, Latvia, Lithuania, Luxembourg, Malta, the Netherlands, Poland, Portugal, Romania, Slovakia, Slovenia, Spain, Sweden, and the UK. Designed initially as an economic unit, the European Union now attempts uniformity in social policies.

(⊕) SEE WEB LINKS
• The European Union website.

European Water Framework This directive's key aims are: expanding the scope of water protection to all waters, surface waters, and groundwater; combining emission limit values and quality standards; getting the citizen involved more closely; and streamlining legislation. See Carter (2007) *Geog. J.* 173, 4 and Kay et al. (2007) *PPG* 31, 1.

(⊕) SEE WEB LINKS
• The EU Water Framework Directive website.

eustasy A worldwide change of sea level, which may be caused by the growth and decay of ice sheets, by the deposition of sediment, or by a change in the volume of the oceanic basins. 'It is commonly assumed that melting of polar ice sheets induced, for example, by climate warming yields a nearly eustatic trend in sea level, although it has long been recognized that self-gravitation and loading effects will lead to significant departures from eustasy' (Mitrovica et al. (2001) *Geophys. J. Int.* 147, 3).

eutrophication The process by which *ecosystems, usually lakes, become more fertile environments as detergents, sewage, and agricultural fertilizers flow in; the enrichment of nutrients (De Anda et al. (2001) *Lakes & Reservoirs: Res. & Manage.* 6, 4). For eutrophication in the UK, see National Expert Group on Transboundary Air Pollution (NEGTAP) 2001.

evaporite A deposit formed when mineral-rich water evaporates; the most common are of gypsum (hydrated calcium sulphate) and halite (sodium chloride). *Playas are **evaporite basins**; see T. M. Peryt, ed. (1988); see also Ayora et al. (2001) *Geology* 29, 3 on Mesozoic and Tertiary evaporite basins.

evapotranspiration The release of water vapour from the Earth's surface by evaporation and transpiration. Since evapotranspiration is so variable, physical geographers prefer to use the concept of **potential evapotranspiration** (**PE**). This is the greatest amount of water vapour which could be diffused into the atmosphere given unlimited supplies of water.

evolutionary economic geography The evolutionary economic geography approach aims to understand actions of economic actors and paths of change in the context of time and space. It explains how behaviour of agents is situated and conditioned, but not determined, by structures accumulated at the level of organizations and the environment (Boschma (2004) *Reg. Studs* 38, 9). However, as Martin and Sunley, in a vital summary of the sub-discipline (2007 *J. Econ. Geog.* 7), point out, just as 'there is as yet no single, coherent, or widely agreed body of theory or methodology that defines evolutionary economics', so there is a wide variety of approaches to evolutionary economic geography. Key approaches include using the evolutionary principles of variety, selection, and retention (Essletzbichler and

Rigby (2007) *J. Econ. Geog.* 7). See also Maskell and Malmberg (2007) *J. Econ. Geog.* 7.

exclave A portion of a nation which lies beyond national boundaries. The Kaliningrad Oblast has become a Russian exclave within the enlarged EU (Prozorov (2007) *Pol. Geog.* 26, 3). Newman (2006) *PHG* 30, 2 argues that transition zones are exclaves demarcated by the hard lines separating them from the external world around them.

exclusion, geographies of The study of the way different social groups are geographically marginalized and discriminated against along lines of *gender, *race, *class, *sexuality, *age, and *disability; how some groups of people are deemed 'out of place' in some locales and are pushed to the margins of society. For example, in the Indian state of Kerala, certain groups such as fisherfolk, scheduled castes, and scheduled tribes have been marginalized, a situation that is now being addressed; see the seminal work of D. Sibley (1995). *See also* SOCIAL EXCLUSION.

exfoliation The *sheeting of rocks and their disintegration, thought to be due to *thermal expansion, at least on small structures. Waragai (1998) *Arct. Alp. Res.* 30 finds that high rates of change in rock temperature may produce exfoliation sheets in *periglacial environments. Le Pera and Sorriso-Valvo (2000) *Geomorph.* 34, 3–4 think exfoliation boulders, now found at the ground surface, may have been formed at depth.

exhumation The removal of young deposits to reveal the underlying structure of older rocks.

exogenetic 1. Applying to processes which occur at or near the Earth's surface.

2. In human geography, as a result of outside, environmental influences.

expatriates, migrants, diaspora strategies New Zealand is a nation of 4 million people, plus 1 million expatriates: 'whereas early efforts involved encouraging highly skilled migrants to return to their home countries, today a range of new techniques [is] being used to engage expatriates in activities in their countries of origin without requiring them to return' (Larner (2007) *TIBG* 32, 3).

extended family A family unit of relatives by blood and by marriage as well as two parents and their children. Kovacheva in C. Wallace, ed (2002) concludes that extended family and household are 'an indispensable means for survival in the difficult economic situation'. Smith and Stenning (2006) *PHG* 30, 2 find that remittance economies from migrants to their extended family back home link households with wider circuits of value.

extending flow The extension and thinning of a glacier, often marked by an *ice fall. Extending flow occurs near the *equilibrium line, and where the velocity of the glacier increases, for example, down a rock *step. Extending flow can transmit material from the surface of a glacier to its base, thus increasing its powers of *abrasion. It is also responsible for *crevasses and is typical of the zone of *accumulation of ice. King (1970) *Geografiska A* 52, 3–4 notes that, in zones of extending flow, ice reaches the bed tangentially, diminishing the ice's erosive capacity, and thus diminishing erosion.

external economies The cost-saving benefits of locating near factors which are external to a firm, such as locally available skilled labour, training, and research and development facilities.

Moran et al. (2002) *PHG* 26, 4 writes that regional specialization is sustained by external economies of scale that impart certain advantages and 'lock' regions into particular development paths. See also Crescenzi et al. (2007) *J. Econ. Geog.* 7, 6.

externality A side effect on others following from the actions of an individual or group. This effect is not brought by those affected and may be unwished for. Two types of externality are recognized: **public behaviour externalities** covering property, maintenance, crime, and public behaviour, and **status externalities** resulting from the social and ethnic standing of the household. McCarthy (2004) *Geoforum* 35 argues that capitalists today can profit from resource use irrespective of the externalities this use generates. See also Helmsing (2001) *Dev. & Change* 32 on externalities, learning, and governance.

extinction The end of the existence of a species or group of taxa, or the end of their ability to reproduce. Gaston (2008) *PPG* 32, 1 observes that rare species are disproportionately vulnerable to extinction in the short term. Loss of a species to chance extinction leaves an area of resource space unoccupied; extinctions could trigger behavioural or evolutionary adaptations (Phillips (2008) *PPG* 32, 1). Van Kleunen and Richardson (2007) *PPG* 31, 4 consider the links between species traits and extinction risk and invasiveness. See Kiesling and Aberhan (2007) *J. Biogeog.* 34, 9 on geographical distribution and extinction risk. Losos

and Schluter (2000) *Nature* 408 explain the relationship as the outcome of the effect of area on immigration and extinction rates. 'If the twentieth-century question was how to stop extinction, then perhaps the twenty-first-century challenge is how to avoid the total collapse of the biosphere, our life support system' (W. Adams 2004). See Sanderson et al. (2002) *Bioscience* 52, 10 on human activity and extinction rates. *See also* BIODIVERSITY CRISIS.

extra-tropical cyclone A cyclonic disturbance outside the tropics; for example, a *mid-latitude depression. See Knox et al. (2011) *Geog. Compass* DOI: 10.1111/j.1749–8198.2010.00395.x on the development of high winds in extra-tropical cyclones.

extrusion A formation of rock made of *magma which has erupted onto the Earth's surface as lava and has then solidified. The crystals in extrusive rocks are small, since the lava solidifies rapidly, giving little time for crystal growth. Extrusions emerge from *fissure eruptions and *volcanoes. See Redfield et al. (2007) *Geology* 35, 11.

exurb American for *dormitory settlement. In many cases, amenity migration and the 'urbanization of the rural' it produces can be understood as the first signs of exurban development (McCarthy (2008) *PHG* 32, 1).

eye The calm area at the centre of a *tropical cyclone.

fabric The physical make-up of a rock or sediment. **Fabric analysis** can determine the *dip and orientation of particles in a sediment: see Carr and Goddard (2007) *Boreas* on fabric analysis and glacier dynamics; Cifelli et al. (2006) *Eos Trans. AGU* 87 on magnetic/mineral fabric analysis and regional deformation; and Brown and Bell (2007) *Bull. Volcan.* 69, 8 on fabric analysis and mass wasting. Millar (2005) *Geomorph.* 72, 1–4 seriously doubts the time/cost effectiveness of fabric analysis to identify periglacial landforms.

facies The characteristics of a rock, such as fossil content, or chemical composition, which distinguish it from other formations and give some indication of its formation. Torsvik et al. (2004) *J. Geol. Soc.* 161, 4 present a facies review of Earth geography from 400 to 250 Ma.

factor analysis A way to get a small set of (preferably uncorrelated) variables from a large set of variables, most of which are correlated to each other. This will reduce the number of variables and detect structure in the relationships between variables. For an example of its use in geography, see Brown and Raymond (2007) *Appl. Geog.* 27, 2.

⊕ SEE WEB LINKS
• Website with online tutorial on factor analysis using Microsoft Excel.

factorial ecology The investigation of urban spatial structure by *factor analysis. The classic study is Murdie's

(1969) *U. Chicago Res. Paper* 116. For the problems associated with factorial ecology, see Martinez-Martin (2005) *ITC Diss. Series* 127, 44–5.

factors of production The resources required to produce economic goods. They are land (including all natural resources), labour (including all human work and skill), capital (including all money, assets, machinery, raw materials, etc.), and entrepreneurial ability (including organizational and management skills, inventiveness, and the willingness to take risks). For each of these factors there is a price, i.e. rent for land, wages for labour, interest for capital, and profit for the entrepreneur. Desmet and Fafchamps (2005) *J. Econ. Geog.* 5, 3 state that perfect mobility of capital and labour tends to bring about an even distribution of economic activity across space, since the use of land leads to decreasing returns to the mobile factors of production—but Puga (2002) *J. Econ. Geog.* 2, 4 disagrees. Jiménez (2003) *Growth & Change* 34, 2 observes that public capital is believed to increase the productivity of the private factors of production whereas human capital is thought to contribute to the production process as an additional input. Ettinger (2008) *PHG* 32, 1 refers to **intangible factors of production** such as trust, cooperation, and the sharing of knowledge.

Maskell and Malmberg (1999) *Eur. Urb. Reg. Studs* 6, 1 point out that, historically, while some factors of production were ubiquitous, the cost of certain others varied between locations. They go on to discuss the process whereby some previously important locational factors are actively converted into ubiquities, which they label, with apologies, 'ubiquitification': 'ubiquitification tends to undermine the competitiveness of firms in the high-cost areas of the world. When international markets are opened up and when knowledge of the latest production technologies and organizational designs become globally available, firms in low-cost areas become more competitive.' Usher (2003) *Canad. Geogr./Géog. canad.* 47, 4 shows how an aboriginal household allocates the factors of production to optimize income flows from both the market and subsistence economies.

factory farming A system of livestock farming in which animals—most commonly, pigs, laying hens, broiler chickens, and veal calves—are kept indoors, with very restricted mobility, for most of their lives. Factory farming has created its own set of environmental damage: raised output of greenhouse gases (farm animals account for 15–20% of methane emissions), and 'manure surpluses' (J. Wolch and J. Emel, eds. 1998).

factory system A concentration of the processes of manufacturing—fixed capital, raw material, and labour—under one roof, in order to provide the mass production of a standardized product or products. Ngai and Smith (2007) *Work Employment Soc.* 21 comment on China's 'unique factory system—a workforce of permanently young workers'.

fact-value distinction The distinction between what *is*—the realm of facts, and what *ought to be*—the realm of values; 'the supposed *ontological

difference between how the world is and how people evaluate the world' (N. Castree, R. Kitchin, and A. Rogers, 2013). However, as Levine (2008) *A J. of Philosophy & Geog.* 11, 1, 91 observes, in many cases considerations of fact and value cannot be neatly separated. Hansson (2010) *J. Risk Res.* 13, 3, 231 tackles the fact–value distinction with relation to risk.

failed state A *state may be said to have failed when it cannot carry out basic functions such as education, governance, or security; a state where the government is not effectively in a position to exert its authority over its entire territory. 'In many cases, the authority of the state is challenged by internal factions resulting in bloody civil conflicts and a scenario wherein the effective control of different parts of the state territory lies in the hands of different regional warring factions' (NUIM's *2013 Failed States Index*).

(⊕) SEE WEB LINKS
• Website of NUIM with failed state index.

fair trade A *commodity chain set up to promote worker welfare, minimize environmental damage, and allow producers to increase their living standards, and control within the supply chain, through *development projects. See Hughes in D. Burch and G. Lawrence, eds. (2007) on supermarkets and ethical/fair trade, and Hughes (2006) *Geoforum* 37, 6 on UK food and clothing retailers, and ethical trading.

fall A form of *mass movement in which fractured rock and soil separates into blocks and falls away from the parent slope. A fall starts with the detachment of soil or rock from a steep slope along a surface on which little or no shear displacement takes place. The material then descends mainly through the air by falling, bouncing, or rolling (Varnes in R. Schuster and R. Krizek, eds 1996).

fallowing An agricultural practice where the land is ploughed, but not seeded. Fallowing is based on the rotation of land between different uses rather than a single permanent use, to regenerate soils through a vegetation-soil nutrient cycling. See Awanyo in B. Warf, ed. (2010) on **bush fallowing**.

family In geolinguistics, a genetic group of languages with numerous cognates and regular correspondences, arising as a language diffuses and diverges over time. See Millar (2003) *J. Linguistics* 7, 1 and Sharma (2005) *J. Linguistics* 9, 2 on new varieties of English.

famine A relatively sudden flare-up of mass death by starvation, usually relatively localized, and usually associated with a sharp rise in food prices, the sale of household goods, begging, the consumption of wild foods, and out-migration. P. Walker (1989) describes famine as a 'socio-economic process which causes the accelerated destitution of the most vulnerable, marginal and least-powerful groups in a community, to a point where they can no longer, as a group, maintain a sustainable livelihood'. See Woo-Cumings (2002) *ADB Institute Research Paper* 31 on famine in North Korea.

A. Sen (1982) sees famine as distinguishable from chronic hunger and deprivation, in that speedy intervention can prevent it. Sen developed the concept of *entitlement, identifying declining wages, unemployment, rising food prices, and poor food-distribution systems, as causes of starvation. Lin and Dennis (2000) *Econ. J.* 110, 1 endorse Sen's approach, adding a decline in food availability as a further factor. Pluemper and Neumayer (2007) http://ssrn.com/abstract=920852 suggest that it can be politically rational for a government to remain inactive in the face of severe famine threat. Howe and Devereux (2004) *Disasters* 28, 4 note that definitions of famine (including this one) tend to be vague, and propose new famine scales based on magnitude and/or intensity.

(((●))) SEE WEB LINKS
• The famine early warning systems network.

farm fragmentation The division of a farmer's land into a collection of scattered lots. Fragmentation usually results from inheritance but may also reflect *bush fallowing. See Pan et al. (2005) *IGU Commission Land Use* on farm fragmentation in Ecuador.

fascism A political system characterized by dictatorship, strong socioeconomic controls, suppression of the opposition through terror and censorship, and often militant nationalism and racism. N. Castree et al. (2012) describe fascism as 'a radical and authoritarian political ideology', but Ingham (2001) *Pol. Geog.* 20, 1029 finds it a 'difficult and contentious term'. Heffernan (2005) *Pol. Geog.* 24, 731 provides a useful analysis.

fault A fractured surface in the Earth's crust along which rocks have travelled relative to each other. Usually, faults occur together in large numbers, parallel to each other or crossing each other at different angles; these are then described as a **fault system**. The slope of the fault is known as the *dip. Where rocks have moved down the dip there is a **normal fault**; where rocks have moved up the dip, there is a **reverse fault**. A **thrust fault** is a reverse fault where the angle of dip is very shallow and an **overthrust fault** has a nearly horizontal dip. A **fault plane** is the surface against which the movement takes place. A **tear fault** is where movement along the fault plane is lateral. This latter

Fault-guided valley

Horst

Graben

- - - Fault
⇒ Direction of movement
▨ Stratum of rock

Normal

Normal

Normal

Reverse

Transcurrent

Fault

type of fault may be termed a **strike-slip fault**. Regions, such as the Harz of Germany, that are split by faults into upland *horsts or depressed *rift valleys are said to be **block faulted**.

SEE WEB LINKS

• The USGS Visual Glossary provides definitions and outstanding photographs.

fault block A section of country rock demarcated by faults, which has usually been affected by *tectonic movement. See Young et al. (2002) *Basin Res.* 4, 1 on the Hammam Faraun Fault Block, Egypt.

fault breccia A zone of angular rock fragments, located along a fault-line, and formed by the grinding action associated with movement either side of the fault. Killick (2003) *S. African J. Geol.* 106 outlines a system of fault rock classification.

fault scarp A steep slope resulting from the movement of rock strata down the *dip of a normal fault; Locke et al. (1992) *Bull. Seismological Soc. America* 82 describe the Teton Range fault scarps.

A **fault-line scarp** is formed when faulting juxtaposes stronger and weaker rocks, and the latter are eroded.

faunal realms The simplest groupings of the world's animals. The *holarctic* realm covers the *nearctic* (most of North America, plus Greenland) and the *palearctic* (extra-tropical Asia, Europe, and North Africa). The *neotropical* realm covers Central and South America; the *Ethiopian* Africa south of the Sahara and Arabia. The *oriental* realm is tropical Asia, with an ill-defined boundary between it and the *Australian* realm (New Zealand, Australia, Oceania, and some of South-East Asia). See Smith (2004) *Complexity* 10, 2 for an evolutionary understanding of faunal realms, albeit in very ornate language.

fear, geographies of The study of those places that people indicate and understand as fear-filled or dangerous (Modly (2000) *PHG* 24, 3, 365 is worth reading). In a commendable paper, Wilson and Little (2008) *Current Issues Tourism* 11, 2, 167 observe that 'fear itself is complexly tied to social expectations and messages, our individual expectations and messages, our gender/class and our capacity, or belief in our capacity, to manage fearful and/or dangerous situations'. See also R. Pain and S. Smith, eds (2008). Paul (2011) *Gender Tech. & Develop.* 15, 3 411 argues that women's restricted access to public space in India is a manifestation of socially produced fear.

fecundity The potential of a woman/women to bear live children. Fecundity in a population is, of course, closely linked to the proportion of women of childbearing age. See Bhalotra and van Soest (2006) *IZA Dis. Paper* 2163 on fertility and neonatal mortality in India.

federalism A two-tier system of government. The higher, central government is usually concerned with matters which affect the whole nation such as defence and foreign policy, while a lower, regional authority generally takes responsibility for local concerns such as education, housing, and planning, although, of course, the division of responsibilities varies from case to case. Cutler (2004) *J. Federalism* 34, 2 argues that federalism may reduce voters' ability to hold their governments accountable.

Federations were designed to preserve regional characteristics within a united nation, or to constrain *nationalist elements. Thus, the Nigerian federation was substantially amended after the 1967 Biafran war; see Philips (2005) *African Studs Rev.* 48, 2. In 1935, India was given a federal form of government by the British; see Parich and Weingarst (1997) *Virginia Law Rev.* 83, 7, 1593.

feedback The response within a *system to an action or process. **Negative feedback** causes the situation to revert to the original; in a river system, erosion is a negative feedback mechanism that restores stream stability by lowering channel gradient and increasing bed material size (Nunally (1985) *Env. Manage.* 9, 5). **Positive feedback** causes further change: positive feedbacks involve a move away from an equilibrium state. Perry and Enright (2002) *J. Biogeog.* 29 note that positive feedback loops between *fire and vegetation have degraded tropical and humid forest ecosystems.

felsenmeer A surface of broken rock fragments found in *periglacial environments. See Ebert and Kleman (2004) *Geomorph.* 62, 3–4 on circular moraines on felsenmeer.

feminist geography Feminist geographies are concerned with the relationships between gender relations and space—how space impacts upon gender relations, and how gender relations express themselves spatially. Dixon and Jones in S. Aitken and G. Valentine eds (2006) are more vehement: 'feminist geography is concerned, first and foremost, with improving women's lives, by understanding the sources, dynamics, and spatiality of women's oppression, and with documenting strategies of resistance'. The defining characteristic of feminist geography is a concern with *patriarchy—feminist geographies question the patriarchal and hierarchical assumptions on which geography is based, and emphasize the oppression of women and the gender inequality between men and women. G.-W. Falah and C. Nagel (2005) examine gender relations as they are mediated by Islamic practices: 'Muslim women are not only subject to patriarchal forces within their own communities and societies, but are also subjected to patriarchal structures of the west.' Gender can be understood as an organizing principle of societies; a social relation that shapes the forms, functions, structures, and governance of cities (Bondi (2006) *estudos feministas/études feministes*). Women's subjectivities, relationships, and symbolic productions occur within processes and practices of development, generating multifaceted experiences and diverse challenges to exclusion (Radcliffe (2006) *PHG* 30, 4). M. Domosh and J. Seager (2001) discuss the gendering of space, and the ways that geographies restrict women's access to, and movements within, space. J. K. Gibson-Graham (2006) and Pratt and Yeoh (2003) *Gender, Place & Cult.* 10 provide accounts of the entanglement of gender in the global workings of the economy; Nagar et al. (2002) *Econ. Geog.* 78 examine the 'double marginalization' of subjects and space characteristic of dominant accounts of globalization: 'Women are sidelined, as is gender analysis more broadly, and southern countries are positioned as the feminized other to advanced economies.' Hyndman (2003) *ACME* 2, 1 argues that feminist geopolitics 'provides more accountable, embodied ways of seeing and understanding the intersection of power and space'; that it 'refers to an analytic that is contingent on context, place, and time, rather than a new theory of geopolitics or a new ordering of space' (Hyndman (2007) *Prof. Geogr.* 59, 1). See also Sharp (2007) *PHG* 31, 3.

L. Staeheli and E. Kofman, eds. (2004) argue that a **feminist political geography** recognizes a different outline of the political from other forms of *political geography: 'the political is not just about differences—either between people or between perspectives; it is also about the webs of power and social relationships that are the basis of connections.'

Dias and Blecha (2007) *Prof. Geogr.* 59, 1, believe that feminist geographies should be seen not solely as a separate sub-discipline in geography, 'but as a

critical perspective useful to all sub disciplines'; Longhurst (2001) *PHG* 25, 4 claims that feminist geography can be seen as a 'strong protagonist in a complex and mature politics of geographical knowledge'. L. Nelson and J. Seager (2004) offer a useful, concise, and densely referenced overview of the history of feminist geography.

feminist standpoint theory A theory concerned with the way that women's perceptions are shaped by their daily activities and realities. The central tenet is that all knowledge is *situated knowledge. Feminist standpoint theory has proven controversial because of its implication that women form a coherent, homogeneous group (Thien in R. Kitchin and N. Thrift 2009).

feminization The social process which has resulted in certain occupations, such as nursing, being regarded as women's work. Watson and Stratford (2008) *Soc. & Cult. Geog.* 9, 4 suggest that the feminization of HIV/AIDS invokes an ethic of care for women and their dependants in situations where risk means both hazard and vulnerability.

This term is also used to describe the increasing presence and influence of women in previously male-dominated occupations or institutions (see D. Perrons 2004). Gallo (2006) *Glob. Netwks* 6, 4 explores the relationship between a 'feminization of migration' and the construction of masculine identities among migrants from Kerala. See Kodras and Jones (1991) *Geoforum* 22, 2, 159 on the **feminization of poverty**.

ferrallitization The effect on a soil of strong, tropical *leaching and intense weathering. Organic matter is rapidly destroyed (so there is little *humus), and bases and silica are leached.

Sesquioxides in the B2 *horizon form a hardpan that may be revealed by the erosion of the upper horizons. In *US soil classification, ferrallitic soils are oxisols.

Ferrel cell A weak *atmospheric cell lying between the Polar cell and the Hadley cell, and travelling in the direction opposite to that of the Hadley cell. It transfers warm air to high latitudes and shifts cold air back to the subtropics, where it is warmed. The Ferrel cell is a consequence of the stationary and transient eddy circulations in the mid-latitudes. See Salustri and Stone (1983) *J. Atmos. Scis* 40, 5.

ferricrete A soil *horizon made from the *cementation of iron oxides at or near the land surface. See Phillips (2000) *Geografiska A.* 82, 1 on groundwater and ferricrete formation.

ferruginous Of or containing iron or iron rust. A **ferruginous soil** is a very deep, *zonal soil found in warm temperate climates without a dry season, or in tropical *savannas/bushlands. The A *horizon is dark red-brown with a weak crumbling structure; the B horizon is stained red by ferruginous gravel. In *US soil classification, a ferruginous soil is an ultisol. See B. Warkentin (2006).

fertility The level of childbearing; in an individual, but more often in a society or nation. Crude *birth rate is the simplest measure of fertility, but does not relate the number of births to the number of women of childbearing age, while the **general fertility rate**, or **fertility ratio**, shows the number of births in a year per 1000 women of reproductive age (generally 15–45, sometimes 15–49). **Cohort fertility rates** show the number of births to women grouped according to either their year of birth or their year of marriage. The **total fertility rate** is the average number of children that would be born per woman, if women experienced the age-specific fertility rates of the year in question throughout their childbearing lifespan. Globally, fertility rates vary widely; in 2006, the rate for the UK was 1.84; for Malawi, 5.7. Abernethy (1999) *Pop. & Env.* 21, 2 argues for the **fertility opportunity hypothesis**—that humans are alert to environmental signs that indicate whether conditions for childbearing and nurture are more or less optimal.

There are very strong global correlations between fertility rates and per capita *GNP, fertility rates and women's education, and fertility and infant mortality rates (Dasgupta in K.-G. Mäler and J. Vincent, eds 2001). Abernethy (2004) *AAAS Annual Meeting* proposes a link between energy and fertility. Fertility has been declining in industrial societies since the late nineteenth century; a decline which preceded easily available artificial contraception.

(⊕) SEE WEB LINKS
- UN website with tables on estimated and projected total fertility for the world, development groups, and major areas (p. 11).

fetch The distance that a sea wave has travelled, from its initiation to the coast where it breaks. The fetch controls the energy and height of a wave, and the longest fetch, and hence the dominant wave direction, will affect beach orientation and the direction of *longshore drift. See Stephenson and Brander (2004) *PPG* 28 on the effect of fetch on a beach, and Davidson-Arnott et al. (2005) *Geomorph.* 68, 1–2 on sediment transport and fetch.

feudalism A system, common in Europe in the Middle Ages, where access to farm land was gained by service to the owner: the **feudal lord**. Initially, no money was involved in transactions between the serf and the lord, although the payment of cash in lieu of service became common in the later Middle Ages;

see T. Aston, ed (1987). In India, the feudal *zamindari* system divided rural society into three broad classes: landlords (*zamindars*), tenant farmers, and landless labourers. Women, Untouchables, and tribal peoples were excluded from decision-making at both regional and national levels. See N. Mohammad (1992); also on feudalism in India, see Sharma (1985) *J. Peasant Studs* 12: 2–3.

fiard (fjärd) An inlet of the sea with low banks on either side, common along the Gulf of Finland, and formed by the postglacial drowning of the Fenno-Scandian shield.

fictitious capital Money that is thrown into circulation as capital without any material backing in the form of commodities or productive activity. D. Harvey (2006) explains: 'consider the case of a producer who receives credit against the collateral of an unsold commodity. The money equivalent of the commodity is acquired *before an actual sale*. This money can then be used to purchase fresh means of production and labour power. The lender, however, holds a piece of paper, the value of which is backed by an unsold commodity. This piece of paper may be characterized as fictitious value'. A **fictitious commodity** is one with an ecological and/or social value exceeding its monetary value functions.

field capacity The volume of water which is the maximum that a soil can hold in its *pores after excess water has been drained away. Field capacity generally increases with finer soil textures, but Gabrielle and Bories (1999) *Transport in Porous Media* 35, 1 note that field capacity hardly applies for sandy soil, because of its high hydraulic conductivity.

field system The layout and use of fields. The extent and use of fields varies with the natural environment, the nature of the crops and livestock produced, and aspects of the culture of the farming community such as inheritance rights and available technology. See M. Aston (1983) for medieval field systems in Britain; M. Widgren and J. Sutton, eds. (1999) on the Engaruka field system, Tanzania; Xiong (1999) *Toung Pao* 85, 4–5 on the highly political equal-field system in China and Japan; and McCoy (2005) *J. Polynesian Sociol.* 116, 4.

filière *See* GLOBAL COMMODITY CHAIN.

film Cinema is seen as 'an essentially geographic art, a way of "writing the world"', according to D. Strauch (Geography of Film, Geog. 425 syllabus), for film contributes to the production of meaning, culture, economy, and identity. Film is important to geography because it mediates social knowledge, reinforces ideological constructions of the status quo, and is an active agent of hegemony (Lukinbeal in R. Kitchin and N. Thrift, eds 2009).

((⊕)) SEE WEB LINKS
- Website for Geog. 425 syllabus—a really useful resource.

filtering down The movement of progressively poorer people into housing stock. It is suggested that when the rich move away from the city to newly built houses, the next social and occupational class moves in. While there are reservations to this theory (not only the rich get new homes, many higher-status housing areas withstand infiltration, and the well-off *gentrify run-down areas), Harris and Wahba (2002) *Int. J. Urb. & Reg. Res.* 26, 1 illustrate filtering in low-income settlements in cities in the developing world.

financial exclusion The disbarring of the disadvantaged and the poor from financial services. Financially excluded people typically: lack a bank account; rely on alternative forms of credit (such as doorstep lenders and pawnbrokers); and lack key financial products such as savings, insurance, and pensions. An estimated 8% of UK households (around 2.8 million adults) lacked a bank account of any kind during 2002–3 (HM Treasury 2004).

The socio-spatial characteristics commonly associated with financial exclusion include low-income households, single- and lone-parent households, social housing tenants, and residence within a deprived urban area (E. Kempson and C. Whyley 1999). 'The concept of exclusion is problematic, not only because of its predominant association with economic (labour market) participation but also because of its highly moral undertones' (Midgley (2005) *Area* 37, 3). Financial exclusion often reinforces other aspects of social exclusion: see Chakravarty (2006) *Reg. Studs* 40. Pollard and Samers (2007) *TIBG* 32, 3 observe that the 'excluded' may also comprise a whole range of individuals who prefer, on religious, moral, or other grounds, not to engage with Western financial institutions. See Li et al. (2001) *Env. & Plan. A* 33 on ethnic minorities, banks, and overcoming financial exclusion in Los Angeles.

financial services, geography of
Mergers, acquisitions, and takeovers, online and phone banking databases of customers' credit rating, and the central processing of cheques have radically altered the spatial distribution of providers of financial services. The 1990s saw a 28% fall in the number of UK bank branches (French et al. (2008) *Antipode* 40, 1). The spatial implications of these changes have been: disproportionately more bank closures in less affluent areas;

a shift in *central business district land use from financial services to restaurants and bars; and the development of call centres. Bristow et al. (2000) *Env. & Plan. A* 32, 3 note the propensity to site call centres close to existing concentrations of allied activity, with preferences for densely populated areas mediated by needs to maintain employee access and avoid staff turnover problems: 'this has important implications for the spatial division of labour, with call centre growth likely to reinforce existing spatial unevenness in employment in key service activities'. See Sokol (2007) *Growth and Change* 38, 2 on 'spaces of flows' and 'flows of value' in the geography of financial services.

Interestingly, Argent (2003) *Bureau Rural Sci., Canberra*, on financial services in rural and remote areas of Australia, reports that a diverse range of service modes, from face-to-face branches to virtual banking, has helped to reduce potential *financial exclusion. Wojcik (2007) *Soc. Sci. Res. Network* confirms the leading position of the UK in international finance. London, for example, has four distinct clusters of financial services: a very cohesive City of London cluster featuring banks, insurance, auxiliary finance, law, and recruitment firms; a less cohesive West End cluster, with distinctive cluster zones, such as banks near Mayfair, and advertising in Soho; an incipient general cluster north of the City of London featuring architecture, and business support; and the law cluster that straddles the City of London and the West End (Corporation of London 2003). See Faulconbridge (2004) *Area* 36, 3 on Frankfurt and London.

fines Small stone particles, whose poor thermal conductivity may aid the development of *patterned ground in *periglacial environments; see Haugland

(2006) *Arct. Antarct. & Alp. Res.* 38, 1. For fines in desert landforms, see A. Goudie and N. Middleton (2006).

finger lake A long, narrow lake occupying an over-deepened basin in a glacial trough.

finger plan The development of new towns or suburbs along routes, road or rail, radiating from the city centre, since planners see commuting as a fact of life which should be made as efficient as possible. Copenhagen, for example, has a finger plan: leisure areas are close to the urbanized areas, which are located along public transport axes (Østergård (2007) *Ministry Env. + Plan 09*).

fiord (fjord) A long, narrow arm of the sea which is the result of the 'drowning' of a glaciated valley; the classic text is J. P. Syvitski et al. (1987).

Fiords are distinctive because of their great depth, and the overdeepening of their middle sections which are deeper than the water at the mouth. Søgnefjord, for example, is 1200 m deep, but its mouth is only 150 m below sea level. The shallow bar at the seaward end of the fiord is thought to represent the spreading and thinning of ice as it was released from its narrow valley and spread out over the lowland; see Nesje et al. (1992) *Geomorph.* 5, 6 on Søgnefjord. Norway's 2004 nomination for its fjords to be added to the UNESCO World Heritage List makes interesting reading: the geomorphology is outlined, together with issues of management, developmental pressures, and environmental pressures. Glaciation may not be the only fiord-forming factor; see Nesje (op. cit.) on fracture systems and Søgnefjord.

fire ecology The study of the impacts of fire on ecosystems, fire ecology can be considered a sub-discipline of landscape ecology as both span the temporal, spatial, and social dimensions of landscapes (Bowman and Franklin (2005) *PPG* 29, 2). The primary focus has been to understand the direct and indirect effects of fire disturbance on plants and, to a lesser extent, animals (Bowman and Boggs (2005) *PPG* 30, 2). See also Bowman (2007) *PPG* 32, 2; and Roberts (2001) *PPG* 25, 2 on tropical fire ecology. See Mistry (2003) *J. Biogeog.* 30, 7 for a collection of papers on fire ecology. Kalabokidis et al. (2007) *Area* 39, 3 find that natural and/or agricultural vegetation, grazing, and topography are the major factors in landscape wildfire dynamics.

firm In *neoclassical economics, an ordered, autonomous, rational independent unit which utilizes the factors of production to produce goods and services. Revenue is kept high enough to cover costs and to generate profit. In Marxist analysis, a firm is a complex organization embodied in a logic of *accumulation, driven to increase profits by avoiding the costs of community or environmental degradation, exploiting labour, and manipulating interactions with governments and unions (see Susman and Schutz (1983) *Econ. Geog.* 59, 2).

Economic geographers are interested in the nature of a firm: its dimensions of gender, sexuality, corporate culture, and cultural embeddedness, and its locating factors, including *institutional thickness. See Taylor and Asheim (2001) *Econ. Geog.* 77, 4 on the concept of the firm in economic geography, and Pollard (2003) *Econ. Geog.* 3, 4 on firm finance.

firn Ice formed when falls of snow fail to melt from one season to another. As further snow accumulates, its weight presses on earlier snow, compacting and melting it to a mass of globular particles of ice with interconnecting air spaces. Where temperatures are around 0 °C, snow can turn to firn within five years, but the process takes much longer in very cold conditions. See Parry et al. (2007) *Annals Glaciol.* 46 on firn within the Greenland ice sheet. The **firn line** is the line, close to the *equilibrium line, at which firn forms, and varies with *aspect (Menzies (1951) *J. Glaciol.* 1, 9). See König et al. (2000) *Procs EARSeL-SIG-Workshop* on equilibrium and firn-line detection.

First World A misleading term (since 'first' does not mean better) for western Europe, Japan, Australia, New Zealand, and North America; the first areas to *industrialize. Widely used synonyms include 'the developed world', 'the North', 'the more economically developed countries' (MEDCs), and 'the advanced economies'.

fish farming The rearing of fish in pools or tanks; Kapetsky (1994) *CIFA Technical Paper* No. 27 gives a strategic assessment of warm water fish farming potential in Africa. In China, manuring has traditionally taken place by raising ducks and allowing their droppings to fall into the water. In a somewhat left-field approach to the topic, Tejeda and Townsend (2006) *Gender, Tech. & Dev.* 10, 1 discuss the different meanings of fish farming for different family members as a key issue in understanding relative successes and failures in fish farming; look, too, for Ho's atmospheric account of the Chinese dyke-pond system (2006, *Institute of Science in Society*).

fissure eruption A volcanic eruption where *lava—usually very fluid and basic—wells up through fissures in the Earth's crust and spreads over a large area. See Walker (1995) *Geol. Soc. Memoirs* 16 on the Antrim plateau, Northern Ireland; see also Mege (2004) *J. Volcan. & Geothermal Res.* 131, 1 on fissure eruptions and dyke fracture length.

Flandrian The present *interglacial, during which there has been a global rise in sea level—the **Flandrian transgression**, caused by the melting of *ice sheets and *glaciers (Smith et al. (2006) *ESPL* 8, 5). The Flandrian is divided into the *Pre-Boreal, *Boreal, and *Atlantic.

flashy In hydrology, applied to a natural watercourse which responds rapidly to a storm event, hence **flash flood**; a very sudden, brief, and dramatic flood event—rising and falling limbs are steep, and the period of peak flow is short. Flash floods are major hazards in deserts, for, although rainfall is rare, the hard-baked ground may be impermeable, and there is little or no vegetation to intercept the rainfall or slow the floodwater. For the prediction of flash floods, see Mukhopadhyay et al. (2003) *Hydrol. Procs* 17, 4 and Foody (2004) *J. Hydrol.* 292, 1–4. Flash flooding is the major process in the formation of *wadis—which, incidentally, should therefore be avoided as camp sites. See Chhopel (2006) *Procs Int. Workshop Flash Flood Forecasting* on flash floods and glacial lake outbursts, and Chan (1997) *Disaster Prevention & Management* 6, 2 on urbanization and flash flooding in Kuala Lumpur; see also Gupta in O. Hassan et al., eds (2000).

(((⊕))) SEE WEB LINKS
- USGS description and depiction of effects of flash flooding precipitation on desert geomorphology during El Niño years.

flattening 1. Of a data set, a reduction in the divergence of statistical data over a geographic area; for example, trade union membership.
  2. In *cartography and *GIS, representing the shape of the Earth as a spheroid—that is to say, not a perfect sphere—rather than a sphere. (The Earth is slightly flattened in the direction of its axis.)

flexible accumulation The use of innovative industrial technologies, adaptable inter-firm relations, variable organizational structures, and flexible consumption, in response to *competition from *newly industrializing and *less developed countries, and to the saturation and fragmentation of markets within *more economically developed countries. Wong (2004) *Land Use Policy* 21, 1 argues that digitization has created new spatial dynamics underlain by flexible specialization and flexible accumulation. 'Flexible accumulation under neoliberal globalisation produces a highly labile space-time most fully achieved through the spatially and temporally unmoored labouring body of the unauthorised immigrant' (Smith and Winders (2008) *TIBG* 33, 1).

flexible spaces Space which admits of, or is revealed as having, multiple uses and meanings, which overlap in time and space. Crang (2000) *City* 4, 3 outlines the case for seeing space as fluid, folded in complex dimensions, and eventful. Wilson (*AAG Annual Meeting* 2007) contends that flexible spaces are thoroughly integrated into global circuits of capital, while Hsu and Lin (*AAG Annual Meeting* 2007) demonstrate the contradictory and conflictual nature of the socio-spatial relations which underlie the production of flexible space, questioning the possibility of successfully creating such spaces. ICT and the developing knowledge-based economy are driving forces for a more flexible space. Thus, cyberspace creates and sustains an increasingly ambient virtual space 'that is still flexible enough for individuals and communities to appropriate their share—especially as the opportunities to acquire technologies and the skills to develop and use them broaden' (R. Murthy (2006) *First Monday, Spec. Issue* 4).

flexible specialization A strategy at company level of permanent innovation,

and accommodation to, rather than control of, ceaseless change. This strategy is based on innovation, multi-use equipment, flexible, skilled workers, or on subcontracting to major firms in sweatshops and homeworking (S. Sassen 2006). Locke and Romis (2007) *MIT Sloan Manag. Rev.* 48, 2 show that suppliers that adopt flexible specialization, through multi-skilled workers operating in groups, working with incentive bonuses and empowered to stop the production line to ensure quality, had better outcomes for workers—in terms of wages and working conditions—and for firms—in terms of productivity gains, efficiency, and compliance. See also Essletzbichler (2003) *Reg. Studs* 37, 8. M. Dunford and L. Greco (2006) criticize theorists who see flexible specialization as a replicable alternative to the mass production system and a way out of the crisis of Fordism. Press (2008) *J. Econ. Geog.* 8, 4 finds that instability for firms is substantially higher under flexible specialization.

flocculation The process whereby very small soil particles, usually clays, aggregate to form *crumbs. Flocculation plays an important role in the transport and behaviour of fine sediment in aquatic systems—see Dyer and Manning (1999) *J. Sea Res.* 41 and Curran et al. (2004) *Marine Geol.* 203, 1–2. In some subsoils of arid areas, downward translocation of soluble salts leads to the breakdown of these crumbs in the process of **deflocculation**.

flood Floods occur when peak discharge exceeds channel capacity, and this may be brought about naturally by intense *precipitation (Wilson and Rashid (2005) *Canad. Geogr./Géog. canad.* 49, 1), snow- and ice-melt (Kaczmarek (2003) *Risk Analys.* 23, 3), the rifting of barriers, such as ice dams; the failure of man-made structures (Cioffi and

Gallerana (2003) *Rivers Res. & Applications* 19, 1); deforestation and urbanization, which reduces *infiltration and *interception (Boardman (2003) *TIBG* 28); and by land drainage and the straightening and embankment of rivers (Gilvear and Black (1999) *Hydrol. Scis J.* 44, 3). Macklin and Rumsby (2007, *TIBG* 32, 2) show that the incidence and size of **extreme floods** have markedly decreased over the last 50 years. Extreme upland flooding appears to be associated with negative *North Atlantic Oscillation index values.

Flood prevention and **flood control measures** include afforestation, the construction of relief channels and reservoirs, water meadow areas in which to divert flood water, and a ban on building in flood-prone environments, such as *flood plains. These measures may increase with per capita income, individual preparedness, and/or experience with flooding, but may decrease with distance from a river, acceptability of flood risk, and provision of environmental information (Zhai et al. (2006) *J. Am. Water Resources Ass.* 42, 4). In *more economically developed countries, there is a move for flood costs to be borne by the private citizen (Salthouse (2002) *Insurance Res. & Practices* 17, 1; Penning-Rowsell and Wilson (2006) *TIBG* 31). See Ono (2002) *Australian Geogr. Studs* 40, 2 on problems and conflicts arising from flood control in Hokkaido. Chen and Hou (2004) *J. Am. Water Resources Ass.* 40, 1 develop a multi-criterion, fuzzy recognition model for flood control.

flood frequency analysis (FFA)

The calculation of the statistical probability that a flood of a certain magnitude for a given river will occur in a certain period of time. Each flood of the river is recorded and ranked in order of magnitude with the highest rank being assigned to the largest flood. The *return period here is the likely time interval between floods of a given magnitude and can be calculated as:

$$\frac{\text{number of years of river record} + 1}{\text{rank of a given flood}}$$

All methods of FFA are methods of extrapolation, which requires the fitting of a model, and this, in turn, needs an assumption about the underlying distribution generating flood events: 'not only is this not known for extreme hydrological events beyond the observed record, but it is untestable within human timescales' (Kidson and Richards (2005) *PPG* 29, 3).

flood plain

The relatively flat land stretching from either side of a river to the bottom of the valley walls. Flood plains are periodically inundated by the river water; hence the name. Flood plains are often ill drained and marshy, and characteristic *fluvial features include meanders, levées, and ox-bow lakes.

The **flood-plain hydrology system** is dependent on interactions among dynamic, non-linear physical and biological processes linking water, heat, and materials (biota, sediment, plant-growth nutrients), flux, and retention to fluvial landscape change. The key processes driving biogeochemical patterns and cycles include flood-caused scour and sedimentation (cut and fill alluviation), routing of river water and nutrients above and below ground, channel movement (avulsion). Groundwater routing through the flood plain and upwelling back to the surface involves penetration of river water into zones of high hydraulic conductivity (subsurface palaeochannels) within the bed sediments that are created by channel scour and subsequent filling with sorted gravel and cobbles. Strong interactions between short-duration, high stream-power floods, channel and sediment movement, increased roughness due to presence of vegetation and dead wood, and upwelling of groundwater creates a complex, dynamic distribution of resource patches, referred to as the shifting habitat mosaic (*University of Montana Floodplain biocomplexity site*).

See Poole et al. (2002) *Freshwater Biol.* 48, 10 on geomorphic controls on flood-plain hydrology and connectivity. For flood-plain management, see Evers and Urban (2003) *COFIS abstracts*.

floral (floristic) realms Areas characterized by the indigenous plant species and not by the *biome. Moreira-Muñoz (2007) *J. Biogeog*. 34, 10 presents a detailed revision of 19th- and 20th-century floralistic realm classifications.

flow 1. A 'river' of rock, earth, and other debris saturated with water; differing from *slides in that no shear plane exists. Flows are classified by the size of the particles: **debris flow** refers to coarse material; earth flow to soil; and mud flow to clay. See Quinta-Ferreira (2007) online *Bull. Engineer. Geol. & Env.* 66, 1 on the natural and man-made causes of an **earth flow** in Portugal, 2000; and Dowdeswell et al. (1987) *ESPL* 13, 8 reconstruct an active debris flow from mudlines. See also Stoffel and Bollschweiler (2007) *Geophys. Res. Abs.* 9 on the dendrogeomorphological reconstruction of a past debris flow.

2. The movement of goods, people, services, and information along a *network. P. Dicken (2003) observes that flows of goods, people, and information are rising, based on changes in scale, and driven by innovations in information and communication technologies. Janelle in Hanson et al., eds. (2004) illustrates newly emerging systems of flows and nodes, and Hughes (2007) *PHG* 31, 4 reviews the geographical literature on the flows and networks of

knowledgeable capitalism. Jensen (2006) *Mobilities* 1, 2 offers a 'new' perspective on flows in the contemporary city, and Carr et al. (2005) *J. World Business* 40, 4 write about **talent flow**—whereby economically valuable individuals migrate between countries. **Spaces of flows** are the networks that bind a world system of cities (M. Castells 1989).

flowage The movement of solids such as ice or rock without fracturing. The term is also used in tectonics (Dickinson (2004) *PPP* 213, 3–4).

flow regime The 'climate' of a river, which is defined by average flow conditions over the course of a year (R. Charlton 2007). Characteristics of the flow regime include: seasonal variations in *discharge; the size and frequency of floods; and frequency and duration of droughts. See Poff et al. (1997) *BioScience* 47, 11, 769.

fluid stressing The erosion of weak, cohesive rocks by the force of water in a river. The effect of this force depends, among other factors, on the strength of the bed, the percentage of clay in the bed, and the velocity and *turbulence of the water. **Fluid-stressing** is effective for weakly cohesive muds, 'where the current may initiate its own defects' (Gibling and Tandon (1997) *Sedimentol*. 44). Evorsion is an alternative term.

flume In hydrology, an apparatus placed across a watercourse to measure *discharge.

fluvial Of or referring to a river, including the organisms within a river or the landforms produced by river action. **Fluvial processes** include erosion, flow processes, and sediment and solute transport in rivers; see Pitty, ed. (1979) *Geoabstracts* 127–47 and Reid et al. in A. Brown and T. Quine, eds (1999). **Fluvial erosion** is the destruction of bedrock on the sides and bottom of the river; the erosion of channel banks; and the breaking down of rock fragments into smaller fragments; see Powell et al. (2001) *Water Resources Res.* 37 and Wijdenes et al. (2001) *ESPL* 26. **Fluvial flux** refers to the material carried by a river; see McCarney-Castle et al. (2012) *Holocene* 22, 91 on sediment transfers on the Danube. **Fluvial deposition** dumps material worked or deposited by rivers; 'at current sea levels, the locus of fluvial deposition is not necessarily the ocean, estuary, or delta, but floodplains in and upstream of the fluvial-estuarine transition zone' (Phillips and Slattery (2006) *PPG* 30, 4. **Fluvial dunes** are trains of highly ordered sediment waves along the sediment on a river bed. Coleman and Nikora (2011) *ESPL* 36, 1, 39 review current understanding of these bedforms.

fluvial geomorphology The study of the processes and pressures operating on river systems. Changes in the independent variables of *discharge, *sediment *load supplied to reach, and valley slope, give rise to adjustments in the dependent variables of sediment load and particle size, hydraulic characteristics, and morphologies, all of which interact with each other. Dollar (2004) *PPG* 8, 3 provides a progress report.

fluvial system A river can be viewed as an *open system; see G. M. Kondolf and H. Piégay, eds (2003). The fluvial system is a process-response system that includes the morphologic component of channels, floodplains, hill slopes, and the cascading component of water and sediment. The system changes progressively through geologic time as a result of normal erosional and depositional processes, and it responds to changes of climate, base level, and tectonics.

A fluvial system can be divided into three zones:

Zone 1   is the drainage basin, watershed, or sediment-source area. This is the area from which water and sediment are derived. It is primarily a zone of sediment production, although sediment storage does occur there in important ways.

Zone 2   is the transfer zone, where, for a stable channel, input of sediment can equal output.

Zone 3   is the sediment sink or area of deposition (delta, alluvial fan).

These three subdivisions of the fluvial system may appear artificial because sediments are stored, eroded, and transported in all the zones; nevertheless, within each zone one process is dominant. From this viewpoint, a river system is supplied with water, sediment of specified sizes, and potential energy. The principles that govern the fluvial system include hydraulic principles that govern the transport of water and sediment, constraints imposed by external variables, and time (S. A. Schumm, 1977).

fluvio-glacial Of, or concerned with, watercourses from melting glaciers. See Price in K. Clayton, ed. (1973) and Gretener and Strömquist (1981) *Geografiska A* 63.

flux In geomorphology, the amount of matter that flows through a unit area per unit time. **Solute flux** (sometimes called the **solute load**) is the solute concentration of the water at a point in time multiplied by the discharge occurring at the same time:

solute flux, mgs^{-1} = solute concentration
$$(mg\ L^{-1}) \times discharge\ (Ls^{-1})$$

See Heal in J. Holden (2012) pp. 372–3 for techniques in estimating solute flux.

flying geese theory An East Asian variety of *modernization theory, which argues that East Asian economies have followed Japan's economic development pathway of import substitution-cum-export promotion and the progression from crude and simple goods to complex and refined goods. See Schoeppel and Mariko (2002) *Japanstudien* 14, 203.

(⊕) SEE WEB LINKS

• Ozawa (2010) provides a thorough critique of the theory at Columbia University website.

focal area In geolinguistics, an area of relative uniformity, as indicated by sets of shared linguistic features, which acts as the central area of a *dialect: for example, the central area of industrial Scotland. See J. Chambers et al. (2008). Britain (2005) *Linguistics* 43, 5 on Estuary English looks interesting.

focus The point of origin of an earthquake. Akira (2000) *J. Geog.* 109, 6 explains how multi-beam echo sounding and seafloor imaging provides information on an earthquake focus.

fog A cloud of water droplets suspended in the air, limiting visibility to less than 1000 m. Fog forms when a layer of air close to a surface becomes slightly supersaturated and produces a layer of cloud, that is, when vapour-laden air is cooled below *dew point. In **advection fog**, this cooling is brought about as warm, moist air passes over cold sea currents. See Nakanishi and Niino (2006) *Boundary-Layer Met.* 119, 2, in a highly technical paper, on predicting advection fog.

  Radiation fog forms during cloudless autumn nights when strong *terrestrial radiation causes ground temperatures to fall. Moist air is chilled by contact with the ground surface. See Sachweh and Koepke (1995) *Geophys. Res. Lett.* 22, 9 on radiation fog and urban climate, and Underwood et al.

(2004) *J. Appl. Met.* 43, 2 on radiation fog in California. Where cold air streams across warm waters, **steam fog** forms. This is common when relatively warm surface air over lakes in *frost hollows convects into the cold *katabatic airflow above it, and is also the mechanism behind *Arctic sea smoke; see Walker (2003) *Weather* 58, 5 on radiation and steam fog. **Frontal fog** forms when fine rain falling at a warm front is chilled to dew point as it falls through cold air at ground level. See also Roach (1995) *Weather* 50 on land fog.

föhn When moist air rises over a mountain barrier, it cools at the slow saturated *adiabatic lapse rate; precipitation is common. Once past the mountains, the air, now much drier, descends, warming at the dry adiabatic lapse rate, higher than the saturated rate by some 3 °C/1000 m. A dry, warm, gusty wind, which can reach gale force, results. In summer, desiccation brings a serious risk of bush fires; in winter, snow melt can be rapid. See T. McKnight and D. Hess (2000) on föhn/Chinook winds, and Chan (2005) *Croatian Met. J.* 40 on föhn winds in Hong Kong.

fold A buckled, bent, or contorted rock. Folds result from complex processes including fracture, sliding, *shearing, and *flowage. Delcaillau et al. (2006) *Geomorph.* 76, 3–4 report on recent fold growth in India. An arch-like upfold is an **anticline**, a downfold is a **syncline**. A complex anticline is an **anticlinorium** (see J. Conley 1973 on the Blue Ridge anticlinorium) and a complex syncline is a **synclinorium** (see Röhlich (2007) *Bull. Geosci.* 82, 2 on the Prague synclinorium). In an **overturned fold** the upper limb of the syncline and the lower limb of the anticline dip in the same direction. In **recumbent folds** the beds in the lower limb of the anticline and the upper limb of the syncline are upside down. G. Bennison

Anticline   Syncline   Anticlinorium   Overturned fold

Synclinorium

Recumbent fold

Fold

and K. Moseley (2003) have excellent block diagrams of various fold types.

fold mountain An upland area, such as the Alps or Andes, formed by the buckling of the Earth's crust. Many fold mountains are associated with destructive or collision margins of *plates. **Young fold mountains**, such as the Caucasus and Alps, were formed by the Alpine orogeny of 65 million years BP, and reach elevations of 10 000 m. **Old fold mountains**, such as the Grampian mountains of Scotland, were created by earth movements predating the Alpine orogeny but have been extensively eroded. Some old fold mountains have been uplifted and re-eroded.

food chain Plants (primary *producers) and *consumers at various *trophic levels are interconnected in their diet and in their role as sources of food. A food chain is a linear sequence representing the nutrition of various species from the simplest plant through to top *carnivores, as in: rose→greenfly→ladybird→sparrow→ sparrowhawk. (This direct pathway is too simplified, and plants and animals are usually linked together in a *food web.) Schoener (1989) *Ecology* 70 shows that **food-chain length** is governed by the total energy available to a given trophic level. See Post and Takimoto (2007) *Oikos* 116 on variations in food-chain length, and Steinberg et al. (1995) *Procs Nat. Acad.*

Sci. USA 92 on the evolutionary consequences of food-chain length.

food conversion ratio The ratio of the number of calories of a prey required to produce one calorie for a predator. Many strategies for sustainable agricultural development strive to improve the food conversion ratio while minimizing negative environmental impacts; see Phillips (2005) Pacific States Marine Fisheries Commission.

food production, industrialization of The increasing intensification and capitalization of the production, transport, storage, and retailing of foods, especially of 'Westernized' products, associated with *agribusiness and *transnational corporations. The Indian broiler industry, for example, has grown from 31 million birds a year in 1981 to 800 million two decades later (M. Gold 2003). 'Very powerful corporations dominate many sectors. Primary producers are locked into tight specifications and contracts. Consumers may benefit from cheaper food but there are quality implications and health externalities. As consumer confidence has been shaken, new quality agencies have been created. Tensions have emerged about the state's role as facilitator of industrial efficiencies. Food policy is thus torn between the pursuit of productivity and reduced prices and the demand for higher quality, with implications for both producers and consumers in the developing world' (Lang (2003) *Dev. Policy Rev.* 21, 5/6).

food security The status of people who live without hunger, fear of hunger, or starvation. 'Food security depends on robust food systems that encompass issues of availability, access and utilization—not merely production alone' (Gregory et al. (2005) *Phil. Trans. R. Soc. B* 360, 21). 'A key component of these newer definitions of food security is attention to building local capacity to produce and distribute food and control food supplies . . . [and] to keep decision-making power within the community rather than losing it through dependence on external sources of food' (Anderson and Cook in J. Harris 2000).

India is a unique nutrition case study—despite enormous growth in economic and political power, in 2011 46% of Indian children were malnourished. 'Food justice will not happen in India without recognizing the harsh reality facing the agriculture sector . . . the land rights of many are restrained by corporate interests and smallholder farmers play a crucial but forgotten role. A fair food system for India requires sound knowledge, transparent institutions, and, most importantly, the political will of leaders within India and globally' (*IDS Bulletin* 17 July 2012).

Nally (2010) *TIBG* 36, 1, 37 explores the way a moral economy of hunger (viewing hunger as immoral and requiring intervention by governments and NGOs) is gradually being replaced by a political economy of food security that promotes market mechanisms as a better protection against scarcity. 'The spectre of hunger in a world of plenty seems set to continue into the 21st century. It bears repeating that this is *not* the failure of the modern food regime (J. Edkins 2000), but the logical expression of its central paradoxes, particularly its reliance on over-production in some

places and under-production in others. To think seriously about global hunger means addressing the legal, institutional, and biotechnical mechanisms— including trade tariffs, agricultural subsidies, enforcement of intellectual property rights, and the privatization of public provisioning systems—that directly restrict certain people's ability to subsist.' Slocum (2006) *Antipode* 38, 2, in a study of a community food coalition in New York, reveals that 'people of colour disproportionately experience food insecurity, lose their farms, and face the dangerous work of food processing and agricultural labor'. See Hyman et al. (2005) *Food Policy* 30 on poverty and food security mapping; and P. Porter (2006).

foodshed The flow of food from the area where it is grown into the place where it is consumed; recently, a way of looking at and thinking about local, sustainable food systems: 'a socio-geographic space: human activity embedded in the natural integument of a particular place' (Kloppenburg et al. (1996) *Agric. & Hum. Values* 13). Starr et al. (2003) *Agric. & Hum. Values* 20 think foodsheds are embedded in the moral economy of a particular place, 'just as watersheds reattach water systems to a natural ecology'. See P. Ackerman-Leist (2013).

food web A series of interconnected and overlapping *food chains in an *ecosystem. A small change in the number of species in a food web can have consequences both for community structure and ecosystem processes (Emmerson et al. (2005) *Glob. Change Biol.* 11, 3).

footloose industry An industry whose location is not influenced strongly by access either to materials or markets, and which can therefore operate within

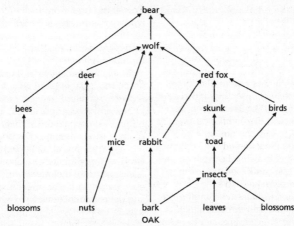

Food web

a very wide range of locations. Any form of 'direct line' business, operated almost entirely through telephone and fax lines, would be an example; see T. Friedman (2005) and J. R. Bryson and P. W. Daniels (2004). Such industries are also liberated from locational constraints by **footloose capital**.

Stichele and de Haan (2007) *SOMO*, look at the consequences of investment in garment making—a classic footloose industry. The negative impacts they detail include the problems of companies leaving the African countries where they temporarily had a presence. 'This creates ever more desperate attempts by countries to keep the investors, in an industry that has already cost countries too much, by offering better incentives. For instance, a company . . . was able to use the desperation of . . . Uganda for foreign investment to get the government to provide and invest in buildings and infrastructure, secure loans and credit facilities. The company left the country without repaying any of its debts, leaving behind a destitute workforce that did not even have enough money left to pay the bus fare home. And this happened after the company had already abandoned factories in Tanzania and Kenya.' This paper demands to be read.

Martin and Rogers (1995) *J. Int. Econ.* 39 developed the **footloose capital model**, with the basic assumptions that capital is mobile across regions, but moves without owners; the demand linkage is broken, as there is no market increase; the supply linkage is broken as the mobile factor responds to changes in nominal, not real, incomes; and that labour is mobile between sectors, but not across regions. A small selection of the many papers on this model includes Tafenau (2006) *25th SCORUS Conference*; Dupont and Martin (2006) *J. Econ. Geog.* 6; and Toulemonde (2006) *J. Urb. Econ.* 59. R. Baldwin et al. (2003) evaluate this model.

foraminifera Usually marine microorganisms of plankton and *benthic animals with calcite skeletons, found over much of Earth's ocean beds. Foraminifera are very sensitive to temperature, and their fossils may be used to reconstruct past environments; see, for example, the journal *Holocene* (2011) 21, 4 on foraminiferal faunal evidence of twentieth-century Barents Sea warming.

forcing A causal change or perturbation on the climate system. **Radiative forcing** is the net flux of *radiation in or out of a system. The five major greenhouse gases account for about 97% of the direct radiative forcing by long-lived gases. Interannual variations in the growth rate of radiative forcing due to CO_2 are large, and probably related to natural phenomena such as volcanic eruptions and ENSO events, as well as to *anthropogenic activity (Hoffman et al. (2006) *Tellus B* 58, 5). Precisely the same forcing can produce hugely divergent evolutions of the climate system given infinitesimally small differences in initial conditions (Washington (2000) *PPG* 24, 4; E. N. Lorenz 1963). **Orbital forcing** (also known as **astronomical forcing**) is the climatic effect of slow changes in the tilt of the Earth's axis and shape of the orbit.

A **forcing factor** would therefore be a factor that affects forcing (for example, Cuffy and Vimeux (2001) *Nature* 412), but Tooth (2008) *PPG* 32, 1 lists climate, tectonics, and human activity as forcing factors in an arid geomorphology, and Burningham (2008) *Geomorph.* 97, 3–4 writes of sea-level rise as a forcing factor on estuaries. (The literature is littered with references to forcing factors—clearly the buzzwords—but the writers seem to mean major/dominating factors.) Coco and Murray (2007) *Geomorph.* 91, 3–4 go one further, and refer to a **forcing template** of spatial structures and geological constraints in the evolution of sand patterns.

förde An elongated bay in an area of glacial deposition, formed when glacial tongue basins are drowned by rising water levels. The type locations are northern Germany and Denmark.

Fordism Henry Ford (1863–1947) paid workers high wages in return for intensive work. The term Fordism was coined by Antonio Gramsci (1971) to describe a form of production characterized by an assembly line (conveyor belt) and standardized outputs, linked with the stimulation of demand brought about by low prices, advertising, and credit. 'Under Fordism, it was presumed that growing social equity . . . was the sine qua non of economic growth' (Peck (2002) *J. Econ. Geog.* 2). Fordism is associated with the spatial separation of the development of the product, at the centre of research and development (usually in a more developed country), and the actual sites of the production of a standardized product (often in a developing country). See Pietrykowski (1995) *Econ. Geog.* 71, 4. McDowell (2003) *TIBG* 28, 1 writes that in Britain, the old gender order of industrial Fordism is collapsing and the traditional moral certainties of that period, perhaps most dominant in Britain in the 1950s, are being renegotiated.

foredune A ridge of irregular sand dunes, typically found adjacent to beaches on low-lying coasts, and partially covered with vegetation. Hesp (2002) *Geomorph.* 48, 1 reviews the initiation, dynamics, geomorphology, and evolution of foredunes.

foreign direct investment (FDI) The purchase of real assets abroad; through acquiring land, buying existing foreign businesses, or constructing buildings, or mines. **Inward foreign direct investment** is the acquisition of real assets within a country by non-residents.

T. Moran et al. (2005) argue that FDI almost inevitably delivers economic development benefits. 'For host economies, especially developing ones, FDI from other developing countries can add to inflows of other external financial resources, including FDI from developed countries, commercial bank lending, portfolio investment and ODA. For poorer developing countries, it can be significant, accounting for over half of total FDI inflows into several LDCs' (UNCTAD 2006).

However, J. Dunning (1997) characterizes FDI as a bearer of scarce capital and technology that lacks both local linkages and technology transfer. 'Increasing locational "tournaments" to attract FDI . . . may have reduced the benefits to the host countries, as have the increasing skill of the managers of MNEs in making their investments more "footloose"' (Buckley and Ghauri (2004) *J. Int. Bus. Studs* 35). Phelps (2008) *Reg. Studs* 42, 4 argues that FDI is currently skewed towards the interests of TNCs, rather than the interests of local and national communities and governments. Naudé and Krugell (2007) *Applied Econs* 39, 1223 find that neither market-seeking nor re-exporting motives dominate FDI, but good policies made by good institutions in the form of political stability, probably do.

foreign exchange The foreign capital earned by a country's exports. Since the currency of many less developed countries is not accepted by international markets, it often becomes necessary to earn foreign exchange in order to buy imports. See Venables (2004) *World Trade Rev.* 3 on the problems of nations that are small, remote, and poor, and P. Wang (2009) on the **economics of foreign exchange** and global finance.

formal game An exercise in *game theory. Buttyán and Hubaux (2001) outline a formal model of rational exchange. See Padon (1999) *J. Urb. Affairs* 21, 2 on a game theoretic model of local public good provision.

formal sector All those types of employment which offer regular wages and

hours, which carry with them employment rights, and on which income tax is paid—compare with the *informal sector. However, these two sectors are by no means entirely separate; see Arimah (2001) *Afr. Dev. Rev.* 13, 1 on informal and formal sector linkages in Nigeria.

fossil fuel Any fuel found underground, buried within sedimentary rock: coal, oil, and natural gas. Doose (2004) *Geochem. Soc. Pub.* 9 argues that world resources of oil and natural gas, including future discoveries, will only be able to meet demand until roughly 2020. Coal may then be a dominant fuel source, projected to meet demand until roughly the year 2070. Fracking is increasing gas supplies.

foundationalism The use of some secure foundation of certainty (justified belief) in theories of knowledge; 'a methodological claim made for the existence of an underlying structure (a physical system) or an integrated logic (theory) that "explains" individual events' (Blomley (1990) *Urb. Geog.* 11, 5).

fractals A geometric figure, each part of which has the same statistical character as the whole. In a landscape, **self-similar fractals** occur regardless of the scale at which it is viewed; topographic contours are described as self-similar, and a numerical sequence is statistically self-similar if its statistical moments are the same upon resampling. However, in **self-affine fractals**, different geometrical directions are scaled differently to preserve shape or statistical moments; see Sung and Cheng (2004) *Geomorph.* 62, 181; see also Pelletier (2007) *Geomorph.* 91, 3–4. **Fractality** is the state or condition of being fractal.

fractus *See* CLOUD CLASSIFICATION.

fragmentation *See* FARM FRAGMENTATION.

FRAGSTATS A computer software program designed to compute a wide variety of landscape metrics for categorical map patterns. The original FRAGSTATS future patch updates, and releases can be downloaded from the internet.

((⊕)) SEE WEB LINKS
• The Fragstats download page.

Framework Convention on Climatic Change (FCCC) A convention established by the United Nations in 1992 in order to research climatic change, and to implement strategies to slow down, and adapt to, climatic change, replaced by the *Kyoto Protocol.

((⊕)) SEE WEB LINKS
• The UN Framework Convention on Climatic Change.

framing A way of organizing our everyday experience. Sofie Hansen (2011) explains that, for example, we need the frame of play-fighting in order to know not to punch someone really hard. Schmueli (2008) *Geoforum* 39, 2048 describes framing as a 'cognitive process whereby individuals and groups filter their perceptions, interpretations, and understandings of complex situations in ways consistent with their own socio-political, economic, and cultural world views and experiences'.

free atmosphere The portion of the Earth's *atmosphere above the *planetary boundary layer in which the effect of surface friction on the air motion is weak.

free face (fall face) An outcrop of rock which is too steep for the accumulation of soil and rock debris.

free market Essentially, the free market is the price mechanism through which consumers choose goods or services on the basis of their willingness to pay for these. **Free market**

environmentalism is 'an ideology that argues that the free market is the best tool to preserve the health and sustainability of the environment' (Davies in R. Kitchin and N. Thrift 2009).

free trade Trade between countries which takes place completely free of restrictions; there are no *tariffs and *quotas. This allows specialization in member states of free trade areas, and should lower costs because both competition and markets are increased. Beine and Colombe (2006) *J. Econ. Geog.* 7, 1 find that free trade between Canada and the United States clearly leads to more export diversification in the long run, and Keeling (2008) *PHG* 32, 2 finds new regional mobilities driven by the expansion of trade blocs such as the European Union or the North American Free Trade Area. However, Killick (2001) *Dev. Policy Rev.* 19 discredits the neo-liberal argument that free trade and economic liberalization will generate prosperity in the developing world; see also Panagariya (2005) *Foreign Affairs Special edn.* 'Global trade is by no means free trade ... actions on the part of the USA through the US Trade Representative's Office and in relation to trade-related intellectual property rights continually threaten to undermine the legitimacy of trade regimes' (Glasmeier and Conroy in J. Peck and H. Wai-chung Yeung, eds 2003) 'The free trade doctrine is now performed routinely under the auspices of the World Trade Organization. Nevertheless, it remains a local epistemology, whose truth-like status is kept insulated from rigorous challenge by alternative epistemologies and practices' (Sheppard (2005) *TIBG* 30). Feagen (2007) *PHG* 31, 23 calls the free market 'elusive'.

free trade zone A designated area, often within a *less economically developed country, where normal tariffs and quotas do not apply. It is common for the conditions of employment there to be more repressive; such zones are often mandatorily union-free, and working conditions can be harsh (Traub-Werner (2006) *Antipode* 38, 1). See Wei and Leung (2005) *Growth & Change* 36, 1 on the Waigaoqiao free trade zone, Shanghai, and Mullings (2004) *J. Econ. Geog.* 4 on Antigua's free trade zone. See Shen (2004) *Prof. Geogr.* 56, 4 on a shift from cross-border protectionism to cross-border free trade zone in Hong Kong: 'a rearticulation of the state cutting through the geoeconomic and geopolitical time-space envelopes'.

freeze–thaw The weathering of rock said to occur when water, which has penetrated joints and cracks, freezes. In a bright and breezy paper, Murton (2007, *Planet Earth* 4) finds that the long-held explanation that freezing, expanding water shatters rocks is 'probably not very significant in nature because it requires some pretty unusual conditions. The rock must essentially be water-saturated and frozen from all sides, to prevent the piston-like effect of freezing water driving the remaining liquid water into empty spaces or out of the rock through an unfrozen side or crack. So we need to look for another explanation.'

freezing front The edge of frozen or partially frozen ground. In areas of seasonal freezing with no *permafrost, the freezing front moves downward through the earth. In areas of permafrost, the front can also move upwards: 'during frost heaving of soil when the soil is freezing from the top down, water in the soil pores flows upward to the freezing front because of a gradient in the soil moisture pressure (or tension). This occurs even when the soil pores are not saturated. Ice lenses form and grow at or slightly above the freezing front and cause great uplifting forces' (Henry (2000) *US Army Corps of Engineers*).

freezing nucleus A *nucleus on which a water droplet will freeze to form an ice crystal. Ice nuclei thus formed are much less common than *condensation nuclei, but their effectiveness rises as the temperature falls below 0 °C. They grow by *sublimation if the *ambient air is saturated with respect to water. See Auer (1971) *J. Atmos. Scis* 28.

freight rate This cost of transporting goods reflects a number of factors besides basic transport costs, such as the nature of the commodity. Non-breakable, non-perishable items, like coal, are carried most cheaply as they can be carried in bulk on open wagons. The more careful the handling required, the more expensive is the freight rate. Sophisticated manufactured goods can bear high freight rates because of their greater value. Raw materials are carried for less so that they can be moved over greater distances. Hummels (1999) U. Chicago, for example, finds for the United States that the ad valorem freight rate is 7.6% for food and live animals, but only 2.25% for machinery. Falling air-freight rates mean that around 30% of US imports now go by air. However, studies based on this data indicate an implicit willingness to pay for time saved at the rate of around 0.5% of the value of goods shipped for each day saved, indicating the massive premium on proximity (Hummels (2001) mimeo, Purdue U.).

Distance is an important factor. Many freight rates are tapered; that is, the rate per tonne-mile or tonne-kilometre drops as the distance increases, but this change in rates is expressed in a series of distance 'bands' so that, on a graph, the relationship between cost over unit distance and distance would appear as a series of downward steps rather than a smooth diagonal line. Plummer and Sheppard (2006) *J. Econ. Geog.* 6, 5 take to task S. Brackman et al. (2001) for setting transport costs, and thus distance, as an **external quasi-freight rate**.

friction The force which resists the movement of one surface over another. Friction between the surfaces of two mineral grains is related to the hardness of the mineral, the roughness of the surface, and the number and area of the points of contact between the grains. It is of major significance in any study of the movement of sediment since the forces moving the sediment must be greater than the resistance provided by friction. In meteorology, the **frictional force** is the roughness and irregularity of the Earth's surface that reduces wind speeds. The **friction layer**, where this effect is strongest, roughly comprises the lowest 100 m of the atmosphere. See Arya (1985) *J. Appl. Met.* 24, 9.

friction coefficient The ratio of the force that maintains contact between an object and a surface and the frictional force that resists the motion of the object; the ratio of two forces acting, respectively, perpendicular and parallel to an interface, between two bodies under relative motion (Blau (2001) *Tribology Int.* 34, 9). Under the same friction coefficient different normal stresses cause different friction levels (Kame and Yamashita (2003) *Geophys. J. Int.* 155, 3). In a rockfall, the friction coefficient is the strongest factor in determining velocity (Dorren (2003) *PPG* 27, 1). See also Smith et al. (2007) *PPG* 31, 4.

friction of distance As the distance from a point increases, the interactions with that point decrease, usually because the time and costs involved increase with distance. Distance need not be reckoned solely in spatial terms; the frictional effect of distance 'on the ground' is far less in a lowland area with good communications than in an upland area of difficult terrain, but has slackened with improvements in transport and communications. 'Human activities tend to organize with respect to geographic location due to the friction of distance and the consequent competition for advantageous location ... Influencing the relationships between people, place, and activity are the technologies to mitigate the friction of distance' (Miller (2007) *Geog. Compass* 1, 3). Ellegård and Vilhelmson (2004) *Geografiska A* 86, 4 argue that geographical immobility and proximity in everyday life indicate 'the continuing and often neglected importance of the friction of distance'.

friendship, geographies of Geographers of friendship are concerned with: geographies of affect/emotion and the ontological construction of the human; children and young people's geographies and the (re)production of social ordering; and geographies of mobility and transnationalism in a world of increased human spatial movement and social relations at a distance (Bunnell et al. (2012) *PHG* 36, 4, 490). See Jupp in P. Kraftl and J. Horton, eds (2012) on parenting policy and the geographies of friendship: encounters in a children's centre.

fringe belt At the edge of a town or built-up area, a zone of varied land use; 'a zone of spacious, often well-vegetated plots in relatively extensive use, which formed at the urban fringe during a hiatus in the growth of the built-up area and frequently became embedded within that area when it resumed its outward spread' (Whitehand (2007) *Urb. Studs* 43, 11, 2047). Typical features are: Victorian hospitals and cemeteries located beyond the city for reasons of public health; recreation facilities such as playing fields, riding stables, and golf courses; and utilities such as water and sewage works. Many of the functions of the fringe belt have been squeezed out from the town centre due to congestion, high land prices, the need for a special site, or disturbances in the central area.

Sometimes further urban expansion leapfrogs the fringe belt.

front The border zone between two *air masses which contrast, usually in temperature. A **warm front** marks the

(a) ana-fronts

(b) kata-fronts

Fronts

leading edge of a sector of warm air; a **cold front** denotes the influx of cold air. Fronts are intensely *baroclinic zones, about 2000 km long and 2000 km wide, moving at around 14 km per day. In *mid-latitude depressions, fronts develop as part of a horizontal wave of warm air enclosed on two sides by cold air. These **frontal wave forms** move from west to east in groups known as **frontal wave families**.

The basic classification into warm fronts, with a slope of 1 in 100, and steeper cold fronts is further divided by the type of air movement at the front. In **ana-fronts**, the warm sector air is rising, and a succession of cloud types and precipitation results. In **kata-fronts**, the warm sector air is descending, clouds are few, and precipitation is reduced to a drizzle.

frontier That part of a country which lies on the limit of the settled area; the term indicates outward expansion into an area previously unsettled by a particular state. 'The frontier arrives and passes, in a historical sense, replacing one set of social relations with another' (Simmons et al. (2007) *AAAG* 97, 3). Some frontiers have occurred where two nations advance from different directions, leading to boundary disputes. A **settlement frontier** marks the furthest advance of settlement within a state while the **political frontier** is where the limit of the state coincides with the limit of settlement.

Jepson (2006) *Econ. Geog.* 82, 3 links the **agricultural frontier** with land rents, transportation costs, and the institution of private property; the fringe of market-oriented agriculture and ranching, which advances on subsistence farming or uncultivated wilderness, as the case may be. See also Rindfuss et al. (2007) *AAAG* 97, 4 on land change in agricultural frontiers. Pallot (2005) *TIBG* 30, 1 sees the Soviet use of the

peripheries as a place of exile as driven by a powerful state intent on expanding the **resource frontier**; see also Hanlon and Halseth (2005) *Canad. Geogr./ Géogr. canad.* 49, 1 on the greying of resource communities in northern British Columbia. Davidson (2007) *TIBG* 32, 4 and Butler and Lees (2006) *TIBG* 31, 4 write on **gentrification frontiers**.

frontogenesis The development of *fronts and frontal wave forms. Frontogenesis occurs in well-defined areas; for example, off the east coast of North America (see Bell and Bossart, (1989) *QJ Royal Met. Soc.* 115, 498). Fronts are less common in the tropics, where contrasts between *air masses are less marked, but see Chen et al. (2007) *Monthly Weather Rev.* 136, 1.

frost Frozen dew or fog forming at or near ground level. **Black frost** is a thin sheet of frost without the white colour usually associated with frost, occurring when few or no ice crystals are formed, because air in the lower atmosphere is too dry (H. Mavi and G. Tucker 2004). Air below 0 °C is **air frost**. **Radiation frost** (sometimes called **ground frost**) is a ground-level fog formed by nocturnal radiational cooling of a humid air layer so that its relative humidity approaches 100%, characterized by a low vertical ceiling of the cold air layer. Gołaszewski (2004) *Acta Agrophys.* 3, 2 concludes that stratification of cold air temperature on spring nights with radiation frost is much stronger than that on nights with advective-radiation frost. **Hoar frost**, or rime, is a thick coating of white ice crystals on vegetation and other surfaces; Karlsson (2001) *Met. Appls.* 8 finds a correlation between the amount of hoar frost, average wind speed, and difference between dew point and road surface temperature. See also Whiteman et al. (2007) *J. Appl. Met. & Climatol.* 46, 1.

fumigation

frost cracking When the frozen ground reaches very low temperatures, it contracts, splitting up to form a pattern of polygonal cracks. Girard et al. (2013) *Geophys. Res. Letters* 40, 9, 1748 find that: liquid water content has an important impact on freezing-induced rock damage; sustained freezing can yield much stronger damage than repeated freeze-thaw cycling; and that frost cracking occurs over the full range of temperatures measured extending from 0°C down to –15°C. Anderson (1998) *Arct. & Alp. Res.* 30, 4 defines the **frost cracking window**—the optimal thermal range for frost cracking—as between –3º and –10º C. Vegetation and accumulations of organic matter may reduce the likelihood of frost cracking (Murton and Kolstrup (2003) *PPG* 26, 2).

frost creep The net downslope displacement that occurs when a soil, during a *freeze–thaw cycle, expands perpendicular to the ground surface and settles in a nearly vertical direction.

(()) SEE WEB LINKS
• The US Permafrost Association glossary.

frost heaving The upward dislocation of soil and rocks by the freezing and expansion of soil water. See Matsuoka et al. (2003) *Geomorph* 52, 1–2 on frost heaving and sorting.

frost hollow A concentration of cold air in a hollow or valley floor when nighttime *terrestrial radiation is greatest on valley slopes. Air over these slopes becomes colder, and hence denser, and flows downslope. Temperatures in frost hollows can be tens of degrees colder than the surroundings. See Gustavsson et al. (1998) *J. Appl. Met.* 37, 6.

frost pocket *See* FROST HOLLOW.

frost shattering The fracturing of rock by the expansionary pressure

associated with the freezing of water in planes of weakness or in pore spaces; 'the major weathering process in cold areas' (Hall et al. (2002) *PPG* 26, 4). See also Coutard and Francou (1989) *Arct. & Alp. Res.* 21.

frost thrusting The lateral dislocation of soil and rock by the freezing and expansion of water. See Zhu (1996) *Perig. & Permaf. Procs* 7, 1.

frost wedging *See* FREEZE–THAW.

Froude number The ratio of the velocity (v) of a river to its *celerity where celerity is the product of the acceleration due to gravity (g) and the mean depth of flow (d). The Froude number (Fe) is calculated from the equation:

$$F_e = \frac{v}{\sqrt{gd}}$$

Where F_e is less than 1, deeper flow is tranquil. Where F_e exceeds 1, the flow is turbulent. See Lewin and Brewer (2001) *Geomorph.* 40 on Froude numbers and channel patterns.

fugitive species *See* OPPORTUNIST SPECIES.

fumarole A vent in a volcano through which steam and volcanic gases are emitted. See Hildreth et al. (2003) *GSA Bull.* 115, 6 and Walker et al. (2006) *GSA Spec. Paper* 412.

fumigation 1. The trapping of pollutants beneath a stable layer or *inversion, and their transport down towards the ground. See Kim et al. (2005) *J. Atmos. Scis* 62, 6.
 2. Soil fumigation is a chemical or physical process that kills viable weeds, seeds, soil-borne pathogens, and nematodes (Cockx and Simonne (nd), University of Florida, IFAS Extension).

• Cockx and Simonne article on University of Florida website.

functional linkage The link between industries, including information, components, raw materials, finished goods, and transport links. Functional linkages are at the heart of *agglomeration economies. A firm's linkages are classified by their direction of movement: **backward** or **input linkages** are received by the firm—producer services have large backward linkages (Besser (2003) *Rural Sociol.* 68, 4). **Forward linkages** are supplied by a firm to another undertaking: 'consumers/firms like to locate close to their suppliers, as this decreases their living/production costs' (Ottaviano and Robert-Nicoud (2006) *J. Econ. Geog.* 6, 2). See Crozet (2004) *J. Econ. Geog.* 4 on the forward linkages that relate labour migrations to the geography of production.

functional region A type of region characterized by its function, such as a city-region or a drainage basin. See Wheeler and Brunn (2002) *PHG* 26, 4. Coombes et al. in D. Herbert and R. Johnston, eds. (1982) define functional regions for the UK Census.

Gaia hypothesis J. Lovelock (1988) argues that planet Earth—*atmosphere, ecosphere, geosphere, and *hydrosphere—is a single ecosystem/organism, regulating itself by feedback between its *abiotic and *biotic components. Lovelock stresses the overriding importance of Gaia, rather than any individual species. Kleidon (2002) *Climatic Change* 52, 3 concludes that life on Earth tends to enhance carbon uptake.

Gaia tends to equilibrium, but human agency seems to be overriding its regulatory mechanism. 'We have spread thousands of toxic chemicals worldwide, appropriated 40% of the solar energy available for photosynthesis, converted almost all of the easily arable land, dammed most of the rivers, raised the planet sea level, and now . . . are close to running out of fresh water. A collateral effect of all this genetic activity is the continuing extinction of wild ecosystems, along with the species that compose them. This also happened to be the only human impact that is irreversible' (E. O. Wilson 2006).

gaining stream Also known as an effluent stream, this is a stream that is fed, wholly or in part, from groundwater from the underlying *aquifer. As long as the level of the water table is higher than the channel, the flow will be maintained, even during rainless periods. A stream losing water through its bed and banks is, obviously, a **losing stream**.

game theory Game theory models ideas of rational choice in conditions of interdependent decision-making, that is, situations in which the decision of one actor is dependent on the decisions of the other players, and vice versa. The fates of the players are mutually implicated in this way. In geography, game theory is often used to overcome or outwit the environment. A. Dinar et al. (2008) apply game theory to real-life issues in natural resources and the environment. See also Levinson (2005) *Transp. Res. A* 39 on congestion pricing theory.

gamma index (γ) In a network, a measure of connectivity that reflects the relationship between the number of observed links and the number of possible links. The value of gamma is between 0 and 1 where 1 indicates a completely connected network (extremely unlikely in reality). Where e is edges, and v for vertices (nodes),

$$\gamma = \frac{e}{3(v-2)}$$

See Xie and Levinson (2007) *Geog. Analysis* 39, 3.

garden city A planned settlement, as conceived by Ebenezer Howard (1850–1928), with low housing densities, and many parks, open spaces, and allotments; the maximum city size to be about 30000. The first UK garden city was Letchworth, 1903. Howard's ideas were echoed in the construction of *new towns in the UK. A **garden suburb** is a planned suburban development inspired by Howard's ideas. Garden suburbs were built in the late 19th and early 20th centuries, for example London's Hampstead Garden Suburb, 1907. See P. Hall and C. Ward (1999) on Howard's legacy.

gatekeeper An individual—or possibly a group—able to control access to goods and/or services. This may be a body that vets research (see Castree (2006) *PHG* 30, 6); an exercise by the US government to reduce illegal immigration between Mexico and the USA (Herbert (2004) *Antipode*, 36, 1); or men as the gatekeepers of gender equality (Connell (2005) *Signs* 30).

GATT (General Agreement on Tariffs and Trade) An agreement made in 1974 that was intended to reduce tariff barriers, funded by subscriptions provided by the signatory governments. Since voting power reflected the size of the subscription, and the USA was then by far the biggest economy, it initially dominated GATT (Sarre (2007) *Geog. Compass* 1, 5). It was replaced in 1995 by the *World Trade Organization. The **General Agreement on Trade in Services (GATS)** is a regulator of international traded services. Cloke (2008) *Prog. Devt Studs* 8, 1, 7 provides a basic summary of the major components of the GATS agreement.

GaWC A study group and network, with its centre in the Geography Department of Loughborough University, that focuses on the external relations of world cities.

gay geography This is concerned with all the relationships and interactions between human sexuality and space and place. Brown (2009) *Env. & Plan. A* 41, 1496 examines a number of gay spaces and practices in order to consider the different forms of enterprise, transactions, and labour that take place within them.

gaze The way we look at things, for the way we look at things influences what we see; see R. Dubbini (2002). Different people will gaze at a landscape and 'see' different features according to—among others—their gender, race, class, age, and sexuality. Thus, the **male gaze** looks at things from the perspective of a heterosexual man. D. Arnold (2006) outlines the ways in which India's material environment became increasingly subject to the **colonial gaze** at Indian landscapes.

GDP (gross domestic product) The total value of the production of goods and services in a nation over one year. Components for a finished product are not taken into account; only the finished articles (but the definition of 'finished product' varies). GDP is an imperfect measurement of a nation's economy because certain forms of production, especially *subsistence production, are not recorded.

GDP per capita in the most globalized developing countries grew 4.2% annually between 1990 and 2001, 1.9% in the rich countries, and only 0.9% in the least globalized developing countries (D. Dollar and A. Kraay 2001).

G8 A group comprising France, the United States, Britain, Germany, Japan, Italy, Canada, and Russia. The G7/8 Summit deals with macroeconomic management, international trade, and relations with developing countries. Questions of East–West economic relations, energy, and terrorism have also been of recurrent concern. Recently, the summit agenda has broadened considerably to include microeconomic issues such as employment and the information highway, transnational issues such as the environment, crime, and drugs, and a host of political-security issues ranging from human rights through regional security to arms control.

A mechanism for an international exchange of views between the major energy consuming nations was agreed at the G8 Gleneagles Summit in July 2005 (Hammond (2006) *Nat. Resources Forum* 30). See the University of Toronto's G8 information centre; see also D. Della Porta et al. (2006) on the G8 protests in Genoa, 2001.

gelifluction The downslope flow of soil in association with ground ice, occurring in *periglacial environments, where water cannot percolate through the *permafrost. Gengnian et al. (1995) *Permaf. & Perig. Procs* 6, 3 find that abundant groundwater, fine grained sediments, and slopes 10°–30° favour gelifluction. Harris et al. (2003) *ESPL* 28, 12 argue that gelifluction is elasto-plastic in nature.

gelifraction Synonymous with *freeze–thaw, an umbrella term for those physical weathering processes resulting from the expansion of water as it freezes into ice. The main influential factors are altitude, aspect, temperature and precipitation, lithological characters, and structure and relief (Liu and Xiong (1992) *J. Glaciol. & Geocryol.* 14, 4: 332).

geliturbation Any frost-based movement of the *regolith, including *frost heaving and *gelifluction.

gendarme On an *arête, an abrupt rock pinnacle which has resisted *frost shattering.

gender While it is generally accepted that sex is biologically determined, societies construct appropriate behaviour for each gender, thus producing local **gender cultures** (Jones (1998) *Geoforum*, 29, 4). Padmanabhan (2007) *Sing. J. Trop. Geog.* 28, 1 notes that in rural West Africa the **gendered division of labour** extends to labelling certain crops as 'male' or 'female'; with the introduction of new crop varieties, these

gendered plant constructions are renegotiated. Guyat (2005) *NZ Geogr.* 61, 3, writing on bar staff, sees that when 'the dichotomized gender division [was disturbed], it was extremely contradictory and never straightforward because hegemonic performances of masculinities and femininities are so deeply ingrained'.

Gendered power relations often stem from poverty and entrenched structural inequality (McEwan (2000) *Pol. Geog.* 19, 2; see also Budlender (2000) *World Dev.* 28, 70). Bracken and Mawdsley (2004) *Area* 36, 3 argue that fieldwork is one of the key sites of gender discrimination for women in physical geography. Bankey (2001) *Gender Place Cult.* 8, 1 observes that 'the fear of the hysterical woman is also a fear that implies that madness or deviance is rendered visible on and through the body . . . a "logic of visibility" . . . has bifurcated men and women's bodies, spheres and spaces from one another'.

The ordering of space is strongly gendered, and may also reinforce gender stereotypes; when spaces (secluded woodlands, dark alleyways, ill-lit multi-storey car parks) make them feel unsafe, women feel vulnerable, and this will further constrain their movements, so fulfilling the stereotype that women are less adventurous than men. 'While patriarchal structures of inequality often result in the spatial entrapment of women, the spatial boundedness of women's lives can be both enabling and constraining' (J. Lin and C. Mele 2005). Hmm.

Gender may be seen as arising from the norms of (hetero)sex; Robinson (2007) *Gender Place Cult.* 14, 4 finds that the space of international web-brokered marriages 'is one in which women can be seen as active subjects in a transnational space that allows them to act outside, to certain degrees, of kinship-based power'. See Dua (2007) *Gender Place*

Cult. 14, 4 on the importance of gender in constituting the racialized practices of Canada. See also Heynen (2008) *ACME* 7, 1.

The study of **gender and development** considers the importance of gender in development issues: the global diversity of gender roles, relations, and identities such as nation, race, class, age, and sexuality; and the need to integrate gender into research and policy agendas in less advanced economies. See V. Kinnaird and J. Momsen, eds. (2002).

genealogy 1. The study of family history; see Nash (2002) *Env. & Plan. D* 20 (1) 27, who observes that 'genealogy can serve to anchor and protect exclusive national cultures'.

2. The history of systems of thought; the history of those structures within society that 'have produced and shaped the boundaries of knowledge, ideas, truths, representations, and discursive formations in different historical periods' (Crowley in R. Kitchin and N. Thrift 2009). S. Aitken and G. Valentine (2006), p.15 throw light on this concept: in geography 'professional organizations may privilege or reinforce particular fashionable ways of thinking, but there are always dissenting voices. In reality, most ways of knowing are partial or in flux'.

general circulation of the atmosphere The world-scale system of pressure and winds that transports heat from tropical to polar latitudes, thus maintaining the present patterns of world temperatures. This global circulation, influenced by the *Coriolis force, is driven by intense differences in *insolation between the tropics and the poles. Air moves vertically along the *meridians and horizontally with the wind systems, both at ground level and in the upper atmosphere. See

B. Haurwitz (2007). See Yu (2000) *Math. Comput. Simul.* 52, 5–6 for a general circulation model.

general linear models (GLM) A class of statistical models that aims to quantify any correlation between independent variables and dependent variables. R. F. Haase (2011) explains how to use GLM.

generator cell A localized cyclonic cell, caused by fluctuations in the polar front *jet stream, bringing precipitation. See Trapp et al. (2002) *Monthly Weather Rev.* 129, 3.

genetic modification The modification of the characteristics of animals in very precise ways, most recently through the transference of a gene. Animals altered in this way are called *transgenic.* Holloway (Economic and Social Research Council web pages for funding and guidance) provides a useful summary, noting that, to the layperson, the adoption of genetic techniques is portrayed as being 'straightforward and inevitable', which it is not.

(⊕) SEE WEB LINKS

• ESRC web pages for funding and guidance.

genocide A collective, organized attempt systematically to destroy a politically or ethnically defined group. Ideas of 'difference' and '*other' underpin genocide: 'to truly understand how ... contemporary genocides and camps remain legitimate within liberal democratic regimes, as well as to understand how sovereign states discriminate between an authentic life and a life lacking political value, we need to theorize biological constructions of difference and their connections to citizenship' (Mitchell (2006) *PHG* 30, 1). Territoriality is a further component: late 20th-century genocides exhibited a marked spatial

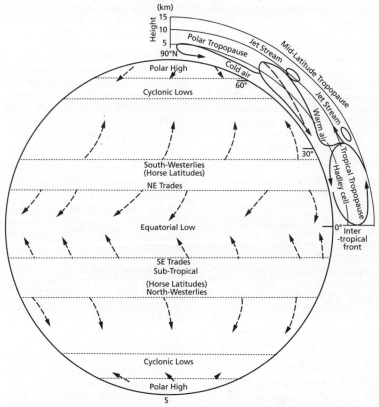

General circulation of the atmosphere

pattern, attacks being more frequent in the peripheral areas of the supposed 'ethnically pure' unit: eastern Croatia in the former Yugoslavia, and north-west Rwanda, for example.

Wood (2001) *TIBG* 26, 1 uses the concepts of Lebensraum, territorial nationalism, forced migration, and ethnic cleansing to explain the production of genocide in Rwanda, and J. Anderson (2010) gives a clear and chilling account of that genocide.

'Ethnic cleansing' (the foul euphemism for genocide) is not achieved solely through mass murder, but through forced expulsions and systematic rape—to 'cleanse' the lineage. 'After a decade of displacement, the legacy of ethnic cleansing endures, forming limits to returns and persistent insecurity for returning communities, thus permanently altering Bosnia's human geography and political future' (Dahlman and Ó Tuathail (2005) *Pol. Geog.*

24, 5). Typically, the overall ploy moves from segregation to isolation and elimination—which explains the vital role that 'safe areas' can play in countering genocide.

The very term genocide has power; see Campbell (2007) *Pol. Geog.* 26, 4. Romeo Dallaire, UN commander in Rwanda, 1994, thought exposure of the term to the Western media was worth a battalion on the ground (S. Power 2002). Do see Joe Sacco's graphic novel *The Fixer* (2003) on Bosnia.

gentrification The rebuilding, renewing, and rehabilitation of depressed areas of the inner city as more affluent families seek to live near to the city centre, trading space and quiet for access to the goods and services of the city centre. (Bourne (2000) *PHG* 24, 153 does not like the term gentrification; 'it is both pompous and now irrelevant. Even "social upgrading" invokes a sense of class-based superiority.')

S. Zukin (1991) describes gentrification as dominated by signature spaces and lifestyles and new middle-class taste formations that get rid of vernacular traditions. Rofe (2004) *J. Austral. Geogr. Studs* 42, 2 writes that gentrification provides a cleaner and more positive identity. Walks and Maaranen (2008) *Urb. Geog.* 29, 4 demonstrate that gentrification is followed by declining, rather than improving, levels of social mix, ethnic diversity, and immigrant concentration within affected neighbourhoods: 'at the same time, gentrification is implicated in the growth of neighbourhood income polarization and inequality'. See Latham (2003) *Urban Studies* 40, 9 on *multiculturalism and gentrification.

Lees (2000) *PHG* 24 and D. Ley (1996) both identify three distinct geographies of gentrification. Lees describes them as international, intra-national, and citywide; while Ley uses international, intrametropolitan, and urban

neighbourhood/intra-urban. Smith (2002) *Antipode* 34, 3 claims that gentrification, 'which initially emerged as a sporadic, quaint and local anomaly in the housing markets of some command-centre cities, is now thoroughly generalized as an urban strategy that takes over from liberal urban policy'; 'the perception of gentrification is no longer about rent increases, landlord harassment, and working-class displacement, but, rather, street-level spectacles, trendy bars and cafes, iPods, social diversity, and funky clothing outlets'. Rérat and Lees (2011), *TIBG* 36, 1, 126 agree: 'gentrification is now sold to us as a sustainable urban form'. L. Lees, T. Slater, and E. K. Wyly (2008) is a fine introduction to the topic. See Phillips (2004) *PHG* 28, 1 on rural gentrification.

geochronology The chronology of the Earth generated from geological information; the science of the absolute and relative dating of geological formations and events.

geocoding (georeferencing) The process used to convert location codes, such as street addresses or postal codes, into geographic (or other) coordinates. Geocoding software is available online, and can offer services such as converting postcode to longitude/latitude, or longitude/latitude to postcode.

geocomputation 'The art and science of solving complex spatial problems with computers. It is focused on the research, development and application of spatial information technologies to address social, economic and environmental problems at human and geographical scales. Geocomputation represents a conscious attempt to move the research agenda back to geographical analysis and modelling, with or without Geographic Information System (GIS) in tow. Its concern is to enrich geography with a toolbox of methods to model and analyze a range of highly complex, often non-deterministic problems' (Xue et al. (2004) *Future Generation Computer Sys.* 20, 7).

(⊕) SEE WEB LINKS
- A survey of geocomputational techniques in spatial data analysis.

geodemographics A form of information technology that enables marketers to predict, through the use of statistical models, the behavioural responses of consumers, based on identity and residential location. Goss (1995) *Econ. Geog.* 71, 2, 171 explains and critiques geodemographics. Singleton (2012) *Geogr. J.* 178, 3, 216 reports that geodemographic analysis shows pupils domiciled within more affluent and less ethnically diverse areas record the highest rates of participation and attainment in the GCSE geography exam (UK).

geodesy The science of the shape and size of the Earth. **Geodetic** is the relevant adjective, and a **geodetic datum** is a set of reference points on the Earth's surface against which position measurements are made, and (often) an associated model of the shape of Earth to define a geographic coordinate system (K.-T. Chang 2014).

geodiversity The variety of rocks, minerals, fossils, landforms, sediments, soils, and associated processes in a place. See Grey et al. (2013) *Procs Geologists' Assoc.* 124, 4, 659.

geodynamics A subfield of geophysics dealing with dynamics of the Earth's lithosphere and mantle and, using data from geodetic GPS, InSAR, seismology, and numerical models. For example, White and Lister (2012) *J. Geodynamics* 7, review the relative motion of India and Asia for the last 100 million years.

geoecology (landscape ecology) An interdisciplinary science which studies the interactions and interrelations in the environment; a view of geo-ecosystems as dynamic entities constantly responding to changes within themselves, their near-surface environments, and external influences, both geographical and cosmic (R. Huggett 1995). See also Shaw and Oldfield (2007) *AAAG* 97, 1.

geoeconomics A branch of *geopolitics concerned with the study of the spatial, temporal, and political aspects of economies and resources. A primary concern and objective of geoeconomics is 'building international partnerships that advance "harmonization", "efficiency", "economic leverage", and "growth" against the supposed threats of political "radicalism", "anachronism", and "anarchy"' (Sparke and Lawson in J. Agnew, K. Mitchell, and G. Cairns, eds (2007), 316).

Geographical Analysis Machine (GAM) A *geocomputation technique developed by S. Openshaw (1997) and co-workers. 'While conceptually simple, the GAM represents one of the first attempts to address, in a geographical manner, a clearly geographical problem' (Atkinson (2001) *PPG* 25, 1).

geographic information systems (GIS) Essentially, the merging of cartography, statistical analysis, and database technology (K.-T. Chang 2014). The critical functionalities of any GIS package are: data capture, storage, management, retrieval, analysis, and display. See Eldrandaly (2007) *Applied GIS* 3, 5, 4 for a really useful guide to the different types of GIS package, and their strengths and weaknesses. See C. P. Lo and A. Yeung (2007) on GIS and spatial problem-solving.

M. Cope and S. Elwood, eds (2009) explain that **qualitative GIS** aims to look at how individuals understand space and what the impacts of these understandings are for the production of socio-spatial relations. For an example, see Verd and Porcel (2011) *Forum Qualitative Sozialforschung/Forum: Qualitative Social Research* 13, 2 on the way qualitative GIS was used to study the social production of urban space. A **geographic resources analysis support system** is a free, open source GIS capable of handling *raster, topological vector, image processing, and graphic data (K.-T. Chang 2014).

geographies A term encompassing the multiple states of the world and the way we understand and are *situated in them. 'Geographies' is not simply the plural of geography—it is argued that the complex world can be seen through many different eyes. There are many geographies and many ways to understand them.

geography 'The science of place and space. Geographers ask where things are located on the surface of the Earth, why they are located where they are, how places differ from one another, and how people interact with the environment' (AAG). 'The subject that bridges the natural and human sciences in understanding societies, places and environments' (Royal Geog. Soc. with IBG). 'A fundamental fascination with, and a crucial method for, understanding the way the world works' (A. Bonnett 2008).

'The nature of geography has meant different things for different people in different places' (P. Hubbard 2002). Thus the Royal Dutch Geographical Society claims that geography 'puts knowing into seeing', while Swedish geography 'manifests itself as a form of scholarly praxis that . . . developed through a process of theoretically informed, hermeneutic reflection

concerned with the endogen principles, and history, of geographical writing' (Olweg (2007) *AAAG* 97, 4). Shaw and Oldfield (2007) *AAAG* 97, 1 identify landscape science as a Russian geographical tradition. Bruneau (2005) *Sing. J. Trop. Geog.* 26, 3 suggests that tropicality has had an epistemologically stronger and more institutionalized relationship with francophone geography than with anglophone geography. Bonnett (2003) *Area* 35, 1 holds that the spirit and purpose of geography should be 'a militant anti-parochialism and a refusal of ethnocentrism'. Abbott (2006) *Sing. J. Trop. Geog.* 27, 3 argues that whiteness 'as an analytical framework' challenges the legitimacy of overseas geographical fieldwork. Barnes and Farish (2006) *AAAG* 96, 4 identify new modes of understanding and representing geography in America that were 'closely wedded to broader geopolitical conditions of war and militarism'. 'The collective work of geographers involves persistently questioning what is left out of the consolidation of any particular geographic account, vision, map, or idea [such as the Iraq War]' (Sparke (2007) *AAAG* 97, 2). 'War is God's way of teaching Americans geography' (Ambrose Bierce).

Geography is one of the structures of how we understand society and space, practised, *inter alia*, by seeing, dwelling, collecting, travelling, mapping, representing, recording, and narrating; an approach which encourages a focus on complexity, multiplicity, and relational thinking. It would take considerable temerity to find a unifying definition throughout the twists and turns that the discipline has taken. Raper and Livingstone (2001) *TIBG* 26, 2 believe that 'there is scope for an ecumenical view of geographical thinking, and the constitutive nature of space–time in the production of entities is central to this view'. Couper (2007) *TIBG* 32, 3, reflecting on the division in geography between physical and human, observes that each discipline has its own system, or language game; parallel ideas may be expressed in different ways in different language games.

geohistory 1. A study of the role of environmental features in history, and the way these features are interpreted and reinterpreted over time. See M. J. S. Rustwick (2007 and 2008).
   2. For F. Braudel (1949), the historical understanding of the spatial and environmental contexts of human activities. See Soja (2010) *Theory & Soc.* 39, 3–4, 361.

geoinformatics The integrated study and examination of hardware, software, tools, techniques, and pedagogy related to the main developments in geographical science, including geographic information systems, remote sensing, photogrammetry, and global positioning systems.

(⊕) SEE WEB LINKS
• Geo Informatics website.

geological column Also known as the **stratigraphical column**, this is the separation of geological time into *eras and periods, as in the table below.

geomagnetism (terrestrial magnetism) The magnetic field of the Earth. The axis runs through the *magnetic poles, whose position varies over time; sometimes the north and south magnetic poles switch places (Maus et al. (2005) *Geophys. J. Int.* 161, 3). The pattern of the magnetic field at any one time will be preserved in contemporary *extrusions of volcanic rock. The study of past magnetic fields, **palaeomagnetism**, can yield information about the creation of new material at the oceanic ridges, and continental drift; see Chauvin et al. (1991) *J. Geophys. Res.* 96, B2.

geomatics The complex topics and techniques that can be assembled under Geospatial Information, namely, geodesy, cartography, photogrammetry, remote sensing, informatics, acquisition systems, global positioning systems, digital image processing, geographic information systems, decision support

Era	Period	Epoch	End date, million years
Quaternary		Holocene (recent)	
		Pleistocene (glacial)	
Tertiary		Pliocene	2
(Cenozoic)		Miocene	
		Oligocene	
		Eocene	
Secondary	Cretaceous		65
(Mesozoic)	Jurassic		135
	Triassic		190
Primary	Permian		225
(Palaeozoic)	Carboniferous		280
	Devonian		345
	Silurian		395
	Ordovician		440
	Cambrian		500
	Precambrian		570

systems, and WebGIS. See Raju in M. V. K. Sivakumar et al., eds (2004).

geomorphic indices A tool for analysing the landforms and evaluating the degree of tectonic activity in a given area; see Keller (1986) *Panel on Active Tectonics*, National Academy Press, Washington, DC; see also Guarnieri and Pirrotta (2008) *Geomorph.* 95, 3–4.

geomorphic systems Open systems where the energy of operation comes from outside, and moves across, the system boundary. Any change in energy or mass causes both the processes and their related landforms to adjust in order to maintain equilibrium in the system. The stability of a geomorphic system will depend on whether the changes are linear, non-linear, or threshold-related, and whether the effects of change are locally damped or propagate through the system (Harvey (2007) *Geomorph.* 84, 3–4).

geomorphic threshold *See* THRESHOLD.

geomorphological sensitivity In a geomorphological context, the degree of response of a landform or system to anthropogeomorpholgical impact. Social factors and natural conditions will affect the degree of sensitivity to anthropological impact; see Rózsa and Novák (2011) *Zeitschrift* 55, 1, 109.

geomorphological threshold A point at which a landform switches, as in the change from an incision stage to an accumulation stage on the slopes of semiarid areas (Gutiérrez-Elorza and Peña-Monné (1998) *Geomorph.* 23, 2–4). An **intrinsic threshold** indicates possibly independent changes within the system; see Hugenholtz and Wolfe (2005) *Geomorph.* 70, 1–2 on the activation of a dune field, while an **extrinsic threshold** indicates a response to change in an external factor; see Brunsden (2001) *Catena* 42, 2–4.

geomorphology The science concerned with understanding the form of the Earth's land surface and the processes by which it is shaped, both at the present day as well as in the past (British Soc. Geomorphology). Since many landforms cannot fully be explained by present-day geomorphic processes, geomorphologists also consider the impacts of past events on the present-day landscape; the landscape is a physical system with a history (S. Schumm 2003). B. Rhoads and C. Thorne (1996) show geomorphology as having the potential to change character as it evolves through time. The **process geomorphology** approach focuses upon the dynamic elements of geomorphological, hydrological, geological, and ecological systems; see D. Ritter et al. (2001).

Harrison and Dunham (1998) *TIBG* 23, 4 stress that landscape change is dominated by uncertainties and probabilities, arguing that an idealist approach to geomorphology (that recognizes the primacy of consciousness) can better appreciate the world's unpredictable and probabilistic nature. Lane (2001) *TIBG* 26 observes geomorphology pays attention to the 'way in which the real becomes contingent to form the actual'; it is place-dependent. Harrison (2001) *TIBG* 26, 3 argues that process geomorphology is essentially reductionist, but that landforms are not amenable to reductionist explanations.

'With humans becoming one of the Earth's major modifiers, human-altered landscape systems are likely to be a prime focus of future research, tackling environmental issues such as deforestation and pollution. In turn, land-surface processes pose a suite of threats to human resources, through the extreme action of soil erosion, slope instability, and river and coastal flooding, and with humans increasingly impinging on the natural environment, hazards research will be of critical importance in most fields of geomorphology' (Stewart in P. Hancock and B. Skinner, eds 2002).

Rhoads (1999) *AAAG* 89, 4 discusses the epistemological nature of knowledge in physical geography, the way in which this knowledge is connected with processes of enquiry and reasoning, the metaphysical assumptions and moral valuations underlying enquiry and reasoning, and the important commonalities and differences between human and physical geography, and between geography and other domains of knowledge. R. H. Grapes, D. R. Oldroyd, and A. Grigelis (2008) *History of Geomorphology and Quaternary Geology* is very well reviewed, but largely confined to Europe. For cultural geomorphology, *see* CULTURAL TURN.

geomorphometry (morphometry, terrain modelling, digital terrain analysis) The quantitative analysis of the morphology of landforms. It uses general geomorphometric measures to capture the basic characteristics of landscapes, like slope, convexity, concavity, aspect, linearity, etc., and **specific geomorphometry**: the use of quantitative measures designed for analysis of specific landscape features. See Pike (2000) *PPG* 24, 1.

Geomorphometry uses input digital elevation models (DEMs, also known as digital land surface models); see Fisher et al. (2004) *TIBG* 29, 1. See Pike et al., eds in T. Hengl and H. Reuter, eds (2007) for a 'brief guide'.

geopolitics 'The study of the relationship between geographical features and international politics. It is a matter of debate whether these features are natural or constructed, what is meant by

international politics and whether the former influences the latter or vice versa. Since the late nineteenth century, "geopolitics" has encompassed several traditions or schools of thought, including one closely associated with the Nazis. This reminds us that geopolitics is also a field of knowledge that developed in the service of state power. Recent scholarship takes a critical approach to geopolitics, seeking to unveil the manner in which politicians discursively construct geopolitical spaces, often by manipulating "geographical facts" for strategic purposes' (Dahlman in C. Gallaher et al. 2009). See Essex *AAAG* 102 1, 191 for a thoughtful paper on the geopolitics of hunger.

Hyndman (2001) *Canad. Geogr./ Géogr. canad.* 45 argues that immigration law produces a '**geopolitics of mobility**'.

Critical geopolitics holds that all truth claims are political: that they are made on behalf of vested political interests and often in the pursuit of political economic imperatives. In this way critical geopolitics manifests a classic concern of poststructuralism to highlight the contingent and political nature of knowledge production. Thus Dalby (1996) *Pol. Geog.* 15 investigates 'the use of geographical reasoning in the service of state power', exploring the way the production of geopolitical knowledge about the relationship between states both exercises political power and affirms identity. Beware: 'after two decades of scholarship on "critical geopolitics", the question of whether it is largely a discursive critique of prevailing knowledge production and geopolitical texts or critique with an implicit, normative politics of its own remains open' (Hyndman (2010) *Pol. Geog.* 26, 247.

geopolitik A view of *geopolitics developed in 1920s Germany. Individuals are subordinate to the state which must expand with population growth, claiming more territory—*Lebensraum—to fulfil its destiny. Bassin (1987) *Pol. Geog. Qly.* 6 argues that the racial doctrines that defined so much of the Nazi variant of fascism were ultimately incompatible with the environmental determinism of geopolitik.

geosciences *See* EARTH SYSTEM SCIENCE.

geosophy The study of geographical knowledge; 'the coalescence during the 1960s of a small number of geographers interested in the social, cultural, and historical foundations of spatial knowledge and behaviour' (Edney (2005) *Cartographica* 40, 1–2); 'the study of the world as people conceive of and imagine it' (John Kirtland Wright, quoted in Keighren (2005) *J. Hist. Geog.* 31).

geospatial The combination of spatial software and analytical methods with terrestrial or geographic datasets. Geospatial object-based image analysis (GEOBIA) is a technique used to analyse remote-sensed images (K.-T. Chang 2014). Hay and Castilla in T. Blaschke, S. Lang, and G. J. Hay (2008) provide a formal definition of GEOBIA, conduct a SWOT analysis of its potential, and discuss its main tenets and plausible future.

geospatial intelligence Earth-referenced information about natural and man-made objects or events with national security implications; the exploitation and analysis of imagery and geospatial information to describe, assess, and visually depict physical features and geo-referenced activities that have national security implications. See Clarke in R. Kitchin and N. Thrift (2009).

geospatial technologies Technologies that enable an ever-expanding range of individuals and social groups to create and disseminate maps and spatial data; Google Earth, Microsoft's Virtual Earth and Wikimapia allow users to add their own geographic information to web-based displays or modify the contributions of others. Other technologies include geotagging, geoblogging, GPS-enabled devices, photographs, narrative text, and video clips that may now be embedded with details about their geographic location.

geostatistics Traditional sampling methods are unsuited to *spatially auto-correlated data; in geostatistics, auto-correlations are not a problem and sampling is less restrictive. For a detailed description of most geostatistical methods, see E. Isaaks and R. M. Srivastava (1989). A web search for geostatistics software will yield any number of programmes. Before choosing a method, note that, in a user-friendly paper, Kent et al. (2006) *PPG* 30, 2 warn that it is vital to specify your scale of interest.

geostrophic wind A theoretical wind, occurring when the *pressure gradient force equals the opposing *Coriolis force (assuming straight/nearly straight isobars; when the isobars are strongly curved, the effect of centrifugal force should be added). The net result is a wind blowing parallel to the isobars. Except in low latitudes, where the Coriolis force is minimal, the actual wind has the same direction as the geostrophic wind. See Danard (1989) *Monthly Weather Rev.* 117, 6 on calculating the surface geostrophic wind in a local area.

Supergeostrophic flow describes winds faster than the expected geostrophic wind, and occurs at a jet entry; see Cunningham and Keyser (1999) *Qly. J. R. Meteorol. Soc.* 125. **Subgeostrophic flow** describes winds slower than the expected geostrophic wind, occurring when air rotates counterclockwise around a low.

geosyncline A thick, rapidly accumulating sediment body, usually parallel to a plate margin, formed within a subsiding sea basin. See Kashfi (1976) *GSA Bull.* 87, 10.

geothermal flux (geothermal heat) Heat from the Earth's interior generated by early gravitational collapse, and later radioactive decay. **Geothermal energy** is a potential energy source in a volcanically active area; see J. Marti and G. Ernst, eds (2005).

geovisualization An abbreviation of geographic visualization, using cartography, cognitive sciences, computer science, data visualization, *GIS, graphical statistics, human–computer interaction design, and information visualization. Geovisualization differs from *cartography in that it affords direct interaction with the content and presentation of what is seen. See Anselin et al. (2006) *Geog. Analys.* 35, 1.

Edsall (2003) *Prof. Geogr.* 55, 2 outlines the advantages of geovisualization systems in exploring complex spatio-temporal information; Kwan (2004) *Geografiska B* 86, 4 discusses geovisualization in time-geographic research; and Gilbert and Masucci (2006) *Trans. GIS* 10, 5 find that the geovisualization of information in ways relevant to people's experiences may be more important to them than 'geographic accuracy'.

geo-web The body of geographically referenced information that is increasingly used to organize and deliver content over the Web; examples include Google Earth, Google Maps, Live Search Maps, Yahoo Maps, and NASA World Wind, which are raising awareness of geography and location as a means to index information. Much of this locational-based data is supplied by users

with no formal background in geography or cartography who 'voluntarily' contribute spatial information over the Web. This is the 'crowdsourcing' of geographical information—that is to say the practice of harnessing the 'power of the crowd' to create collectively represents a pronounced shift in what constitutes 'geographic information' at all levels of data, technologies, and practices. The impetus for this shift is not geographical *per se*. Its emergence and development have been overwhelmingly driven outside academic geography by the Web business and computing communities. Leszczynski (2012) *PHG* 36, 1, 72 explains it all.

gerrymandering Redrawing constituency boundaries for political gain. It involves 'careful drawing of constituency boundaries by a party so that either it wins a particular seat or, more generally, it wins more seats than its opponents' (Johnston (2002) *Pol. Geog.* 21, 1). See Altman (2002) *Pol. Geog.* 21, 1 on evaluating gerrymandering claims, and Chen (2013) *Int. Qly J. Pol. Sci.* 8, 3, 239 on unintentional gerrymandering.

Gestalt theory A theory of perception and social understanding that emphasizes the degree to which individuals understand phenomena as wholes, greater than the sum of their individual parts. 'Gestaltists argue for positive change as a process of mutual and personal discovery' (Hobson (2008) *Geog. Compass* 2, 1). Gestalt is the perception of a pattern or structure as a whole: Barnes (2001) *PHG* 25, 4 defines a **Gestalt shift** as a change marked by abrupt disjunctions rather than inevitable progress, thus resembling, or equating to, a paradigm shift. Swyngedouw (in R. Lee and J. Wills, eds. 1997), refers to the **gestalt of scale**—geographical configurations as a set of interacting and nested scales; 'nineteenth-century middleclass women altered the prevailing *Gestalt* of scale by altering the structures and practices of social reproduction and consumption' (Marston (2000) *PHG* 24, 2).

geyser A jet of hot water and steam, usually as a result of the *geothermal heating of underground water, connected to the surface by a narrow outlet.

ghetto A part of a city, not necessarily a slum area, occupied by a minority group (a **gilded ghetto** is simply a more affluent ghetto). Lifestyles within the ghetto differ distinctly from those of the 'host' population and the prejudices of the host confine the sub-group to particular locations. D. Wilson (2007) describes ghettos in terms of deteriorating levels of poverty, increasing demonization in popular discourse, and their complex relationship with the prison system. 'Any proposition of a particular community in isolation, defined in racial, religious, or regional terms . . . shielded from its powers by fixed boundaries, is destined to end up as a kind of ghetto' (M. Hardt and P. Negri 2000). 'To counter the negative effects of the so-called ghetto life, organizations created alternative spaces that they argued would encourage more appropriate social conduct and would persuade homosexuals to be more supportive of these organizations' political agenda.' (Nash (2006) *Canad. Geogr./Géogr. canad.* 50, 1).

giant nuclei See NUCLEI.

gilgai Undulations of the soil surface, characteristic of many vertisol soils (Fagan and Nanson (2004) *Geomorph.* 60, 1–2). Gilgai result from the shrinking and wetting of soils. See Hallsworth in H. C. T. Stace et al., eds (1968).

Gini coefficient In a *Lorenz curve, a measure of the difference between a

given distribution of some variable, like population or income, and a perfectly even distribution. More simply, it tells us how evenly the variable is spread; the lower the Gini coefficient, the more evenly spread the variable.

GIS *See* GEOGRAPHIC INFORMATION SYSTEMS.

GIS analysis When spatial data are keyed into GIS software, the location of each point is recorded, together with the way in which that point relates to its real position on the surface of the Earth. When this database is loaded and the software turned on, most programs will report on: the number of *points, lines, and *polygons for the theme loaded; the individual polygon areas and the total area for the theme loaded; and the total lengths of linear objects of a particular category.

glacial 1. Of or relating to a glacier.
2. An extended length of time during which Earth's glaciers expanded widely. The most recent glacial epoch, better known as the Pleistocene glacial epoch, previously the Quaternary, covered at least the last 3 000 000 years. A. Penck and E. Brückner (1909) established the classical sequence of four glacial periods in the Alps: the First Glaciation, at its peak some 550 000 years ago, the Second Glaciation, at its peak 450 000 years ago, the Third Glacial Period (Riss) with a climax 185 000 years ago, and the Fourth Glaciation (Wurm) at its height 72 000 years ago. Ice-sheets grew only during parts of these so-called 'glacials', each of which was followed by an *interglacial period.

glacial abrasion The grinding away of bedrock by fragments of rock incorporated in ice. It can be achieved by bodies of subglacial sediment sliding over bedrock, or by individual clasts within ice (Glasser and Bennett (2004)

PPG 28, 3). Abrasion is favoured where effective basal pressures are greater than 1 MPa and where there are low sliding velocities. Landforms of glacial abrasion include striae, grooves, micro-crag and tails, bedrock gouges and cracks. Ice ceases to be an effective agent for abrasion when the weight of the ice is thick enough to bring about plastic flow.

The Boulton model (1996, *J. Glaciol.* 42) sees abrasion as controlled by effective normal pressure, and the Hallet model (1979, *J. Glaciol.* 23) argues that abrasion is highest where basal melting is greatest.

glacial breaching The erosion of a breach between two adjacent valleys when a transfluent glacier flows across the *watershed (divide) between them. See Hall and Jarman in S. Lukas et al. (2004).

glacial debuttressing The exposure of glacially steepened rockwalls, due to the down wastage and retreat of glacier ice. Ballantyne (2002) *Quat. Sci. Revs* 21, 1935, sees glacial debuttressing as 'one of the most important geomorphological consequences of deglaciation in mountain environments'.

glacial deposition The laying down of *sediments which have been removed and transported by a *glacier (see Larson and Piotrowski (2005) *J. Sediment. Res.* 75). The sediments—known as *till, or drift—are deposited when the ice melts; when *ablation is dominant. Glacial deposition is predominant in marginal areas of present and past ice sheets, such as the Eden Valley, and much of the English Midlands, but also occurs in uplands, especially in the form of *moraines. See Escutia et al. in Cooper and Raymond et al. (2007) *Online Procs 10th ISAES* on meltwater production and high sediment discharge in the East Antarctic.

See also DRUMLINS; ERRATICS; KETTLE HOLES.

glacial erosion Glacially eroded landscapes are moulded by abrasion, *debris entrainment, the 'conveyor belt' transport of *moraine on top of a glacier, rock fracturing (Iverson (1991) *J. Glac.* 37), *plucking/quarrying (Hallet (1996) *Annal. Glaciol.* 22), *pressure release (Lewis (1954) *J. Glaciol.* 2), and the action of subglacial water pressure water (Iverson (1991) *J. Glaciol.* 37). Bougamont and Tulaczyk (2002) *Boreas* 32, 1 see sub-ice-stream sediment generation and transport as agents of glacial erosion; Glasser and Bennet (2004) *PPG* 28, 1 argue that 'all landforms of glacial erosion provide evidence for the release of subglacial meltwater'. Glacial erosion can modify the bed of an ice sheet and alter large-scale ice dynamics and mass balance (Oerlemans (1984) *Zeitschrift* 20).

The extent and nature of glacial erosion depend on ice characteristics. *Glen's Flow Law explains glacier behaviour with respect to strain rates and shear stress (Tarasov and Peltier (2000) *Annal. Glaciol.* 30, 1). Alley et al. (2003) *Nature* 424 suggest that, if a glacier digs too deep, strong thermodynamic feedbacks prevent further deepening. For the role of basal velocity in glacial erosion, see Harbor (1992) *GSA Bull.* 104; for the thermal regime, see Sugden (2005) *Geomorph.* 67, 3–4. Other variables are covered by Brook, Kirkbride, and Brock (2004) *Geograf. Annal. A* 86, 3 (rock strength), and Augustinus (1995) *Geomorph.* 14 (rock stress). Paulsen and Wilson (2004) *Glob. & Planet. Change* 42 consider the role of geologic structure, and Olymo and Johansson (2002) *ESPL* 27 provide a useful summary of the roles of rock structure, lithology, and pre-glacial relief. For the response of glaciated landscapes to rapid rock uplift, see Brocklehurst (2007) *J. Geophys. Res.* 112.

In glacially eroded lowlands, relief is confused, weathered rock is stripped away, and *abrasion and *quarrying of the bedrock are active. Drainage is rambling, as earlier patterns are disrupted by erosion and the deposition of *moraines. Lakes and ponds abound, and perched blocks and erratics may be common. See Le Coeur (1988) *Geografiska A* 70, 4 and Munro and Shaw (1997) *Geology* 25. Jansson et al. (2011) *ESPL* 36, 3, 408 give a detailed and helpful account of using a GIS filter to replicate patterns of glacial erosion.

(⊕) SEE WEB LINKS

- Glaciers online's thumbnails of glacially eroded landforms.

glacial–interglacial cycle The most recent third of the Quaternary is most clearly characterized by the alternation between cold glacial periods and warm interglacials. The dominant period of these climate cycles is 100 ka (e.g. J. Imbrie et al. 1993). The terms 'glacial' and 'interglacial' emphasize the changes in ice volume and sea level associated with the appearance and disappearance of large Northern Hemisphere ice sheets (Lang and Wolff (2010) *Clim. Past Discuss.* 6, 2223. About eight major glacial-interglacial cycles have occurred during the past 0.8 million years (Williams in D. J. Dunkerley et al. 1993). Ice core records indicate that greenhouse gases co-varied with Antarctic temperature over glacial–interglacial cycles, suggesting a close link between natural atmospheric greenhouse gas variations and temperature (S. Solomons, ed. 2007).

glacial meltwater The effectiveness of meltwater as an agent of erosion depends on: the susceptibility of the bedrock involved (particularly its structural weaknesses or proclivity to chemical attack); the water velocity and the level of turbulent flow; and the quantity of sediment in transport (Glasser and Bennett (2004) *PPG* 28, 3). A **glacial margin channel** is a stream running between the side of the glacier and the valley sides. Landforms of glacial meltwater erosion include both subglacial and ice-marginal meltwater channels.

glacial surge The swift and dramatic movement of a glacier, associated with the growth of ice up-glacier to unstable proportions and with severe *crevassing. Murray and Porter (2001, *Quat. Int.* 86) link the termination of the surge front on the Svalbard surging glacier Bakaninbreen to the boundary between warm ice and the cold-based snout, and propose that surging takes place by bed deformation and/or sliding in the warm based zone. See Russell et al. (2001) *Glob. & Planet. Change* 28, 1 on glacier surging as a control on the development of proglacial, fluvial landforms and deposits.

glacial trough A deep linear feature, carved into bedrock by glacial erosion that is channelled along a trough or valley. Troughs may be cut beneath ice sheets by a combination of glacial *abrasion and *quarrying. The effects of glacial abrasion are more obvious on the smoothed and polished trough walls whilst the effects of glacial quarrying— such as roches moutonnées and rock basins—are most evident on the valley floors. Glasser and Bennett (2004) *PPG* 28, 3 see troughs as *palimpsest features, developed over successive glaciations.

Hebdon et al. (1997) *ESPL* 22 calculate glacial erosion rates as between 1 and 2 mm a^{-1} in Scotland; Harbor et al. (1988) *Nature* 333 cite rates of 10^{-3} ma^{-1} landslide material in a single glacial cycle.

Although the cross-sectional form of a glacial trough is traditionally described as 'U-shaped', their true cross-sectional morphology is better described by empirical power-law functions (Yingkui et al. (2001) *J. Glaciol.* 47) and by second-order polynomials (James (1996) *ESPL* 21). Furthermore, this cross-sectional morphology varies with lithology (Harbor (1995) *Geomorph.* 14), rock mass strength (Brook and Tippett (2002) *Scot. J. Geol.* 38, 1) and the degree of alluvial modification (James op. cit.).

Modelling studies demonstrate that the cross-sectional morphology of glacial troughs tends towards an equilibrium form (Harbor (1992) *GSA Bull.* 104). Other characteristics of glacial troughs include classic features of glacial erosion such as hanging valleys and divide elimination (Linton in H. E. Wright and W. H. Osburn, eds 1967), whilst the valley patterns associated with these features have been used to quantify the effects of glacial erosion on landscapes (Haynes (1995) *Boreas* 24).

glacier A mass of ice which may be moving, or has moved, overland: when enough ice has accumulated, a glacier will start to move forwards. A glacier may be seen to be the result of a balance between *accumulation and *ablation. Glaciers are classified by their location (*cirque glacier, expanded-foot glacier, valley glacier, *niche glacier, *piedmont glacier), by their function (diffluent glacier, *outlet glacier), or by their basal temperature (*cold glacier, warm glacier, *polythermal glacier). The **glacier balance velocity** is the velocity required to maintain the glacier in equilibrium without change to its geometry. Glaciers where the actual and balance velocities are similar are said to be in balance with the current climate, and are unlikely to experience major changes in flow rate unless the climate changes.

glacier budget The balance in a glacier between the input of snow and *firn, that is, *accumulation, and the loss of ice due to melting, evaporation,

*sublimation, and *calving, that is, *ablation. A glacier grows where the budget is positive and retreats when it is negative.

glacier drainage systems In a warm glacier, meltwater will drain into the ice through a network of veins, *moulins, channels, and conduits, which can connect to an *englacial system, discharging at the snout, or to a *subglacial system of channels, linked cavities, subglacial lakes, and water films. These englacial and subglacial systems may become connected during the melt season; in cold glaciers, however, the superglacial and subglacial drainage systems are always separate. Hydrological features related to the evolution of the glacier drainage system include supraglacial routing, moulin types, and chasms (Yde and Knudsen (2005) *Geografiska A* 87, 3).

glacier hydrology The study of the deposition, distribution, and depletion of snow; the evaluation of hydrologic water balance of basins where appreciable snow occurs; the physics of snow and glacier melt; the storage and transit of liquid water in snow and glaciers, and of possible methods for estimating the rate of stream flow, and the volumes of runoff in basins where snow affects those quantities. See Cooke et al. (2006) *PPG* 30, 5 on glacier hydrology and supercooling; Hock in U. Huber et al. (2005) on melt modelling; and Hock (2005) *PPG* 29, 3 for a review of glacier melt.

glacier mass balance The balance between accumulation and ablation, which is largely dependent on climate. On a valley glacier or ice sheet, mass is added at the top in the accumulation zone and taken away through melting and calving at the terminus in the ablation zone. Consequently the surface profile of the glacier will steepen with increasing accumulation (M. Bennet and N. F. Glasser 2009). Thus, the mass balance gradient is the glacial driving mechanism. See J. Holden, ed. (2012), p. 471 on measuring the mass balance of major ice sheets from space.

glacier response time The duration of the morphometric adjustments following change in net mass balance. This is a vague concept, because the reaction time may depend on the glacier history in an untransparent way. A distinction should be made between response times for glacier *volume* and glacier *length*. When the specific balance changes, the volume will respond immediately. For glacier length it may take a bit more time, in particular when the changes are more apparent at higher parts. 'Numerical experimentation has indeed shown that response times for volume are typically 30% shorter than response times for length' (J. Oerlemans 2001).

glacier retreat (glacier recession) There is no lack of evidence for glacier retreat: for example, Kulkarni et al. (2007) *Current Science* 92, 1 report that nearly 500 glaciers in the Chenab, Parabati, and Baspa basins have shrunk more than 20% over a 40-year period; Joughin et al. (2006) *Nature* 432 report on 'catastrophic retreat' in west Greenland; while Owen and Lehmkuhl (2005) *Boreas* 34 comment on Tibet. Barry (2006) *PPG* 30, 1 provides a useful summary table. See also de Woul et al. (2007) *Geophys. Res. Abs.* 9.

There are reports that some glaciers are extending: Howat et al. (2007) *Climate Dynam.* 28, 1 report that the Mt Shasta, Calif., glaciers have grown by 30% since 1960.

(⊕)) SEE WEB LINKS

• Satellite pictures of the Khatling glacier, Uttarakhand, India, between 1998 and 2013.

glacier sliding See BASAL SLIPPING.

glacio-epoch A multimillion-year period of glaciation. Eyles (2008) *PPP*, 258, 89 reviews the relationship between glacio-epochs over the last 3 Ga of Earth history and phases of supercontinent breakup and assembly.

glaciohydraulic cooling A process that allows water to remain liquid below 0 °C. It is associated with basal water flow through subglacial overdeepenings; as the subglacial melting point changes in response to changing pressure, water may become supercooled, and freeze, but if the adverse bed slope is steep enough, the heat generated will not be enough to match the changed pressure melting point, and it will become supercooled. For more details, see Hooke (1991) *Geol. Soc. Amer. Bull.* 103 and Alley et al. 1998 *J. Glaciol.* 44. Alley et al. (2003) *Nature* 424 view supercooling as a key link between glacier dynamics, basal ice formation, and subglacial landform evolution. For field evidence of glaciohydraulic cooling, see Hooke and Pohjola (1994) *J. Glaciol.* 40 and Lawson et al. (1998) *J. Glaciol.* 44.

Glaciohydraulic cooling is associated with subglacial deformation (Echelmeyer and Zhongxiang (1987) *J. Glaciol.* 33), and ice creep at crystal boundaries (Wolff and Mulvaney (1988) *Ann. Glaciol.* 11). Cooke et al. (2006) *PPG* 30, 5 review recent work on glacial supercooling. Knight (2006) *PPG* 30, 5 suggests that the significance of glaciohydraulic supercooling may have been overestimated.

Glen's Flow Law The relationship between glacier creep and glacier stress can be given by Glen's Flow Law:

$$\varepsilon = A\tau^n$$

where ε is the strain rate, A and n are constants, and τ is the basal shear stress. The constant $n = 3$, and the value of A are dependent on ice temperature, crystal orientation, debris content, and other factors (Davis at Antarctic glaciers.org website). In using this law to model real ice masses, remember that many simplifications may have been assumed (B. Hubbard and N. Glasser 2005).

( ) SEE WEB LINKS
• Website of Antarctic Glaciers organization.

gley soils Soils with mottled grey/yellow patches. **Gleying** occurs in waterlogged soils since oxygen supply is limited there. Under these conditions, *anaerobic micro-organisms extract oxygen from chemical compounds, particularly reducing ferric oxide to ferrous oxide. See Vladychenskii (2011) *Eurasian Soil Sci.* 44, 7, 834.

global Smith (2001) *Transnational Urbanism* notes that each member of the global–local binary has been allocated a distinctive set of attributes; global is a synonym for dynamic, thrusting, open, rational cosmopolitan, dominant, masculinized, and economic, while 'local' suggests authentic, closed, static, nostalgic, defensive but defenceless, feminized, and cultural. However, Ley (2004) *TIBG* 29, 2 argues that 'the global is also the local'.

global care chain A global network of person-to-person contacts between people working as carers, whether paid or unpaid. See Hochschild in W. Hutton, W. and A. Giddens, eds (2000). A common care chain typically entails an older daughter from a poor family who cares for her siblings while her mother works as a nanny caring for the children of a migrating nanny who, in turn, cares for the child of a family in a rich country. Y*eates* (2004) *Feminist Rev.* 77, 79 is excellent on this, as is her state-of-the-art review (Yeates (2012) *Global Networks* 12, 2, 135).

global circulation *See* GENERAL CIRCULATION OF THE ATMOSPHERE.

global city A central location for the leading industries of globalization; specifically financial and specialist business services, with their own forms of production as 'control centres' of the world economy (Knox in P. Knox and P. Taylor, eds 1995; S. Stassen 2000; see also Cochrane and Passmore (2001) *Area* 33, 4). Examples include London, New York, Paris, and Tokyo. Scholz (2000) *Zeitschrift* 1 calls these *acting global cities*, which differ from *globalized places* that house only branches/offshoots of the decision-making structures concentrated in the acting global cities (see also Brade and Rudolph (2004) *Area* 36, 1). Transnational elites have long been associated with the global city (Beaverstock (2002) *Geoforum* 33, 4). Yeung and Olds's 2001 paper for the IRFD World Forum on Habitat labelled Shanghai an 'emerging global city' as opposed to 'hyper global cities' such as New York and London.

Wei and Leung (2005) *Growth & Change* 36, 1 argue that the literature on global cities is too dependent upon a perspective based on major Western cities. Global cities are marked by large-scale in-migration: Wong et al. (2005) *J. Housing & Built Env.* 20 highlight the significance of a temporary migrant labour force in the moulding of Shanghai into a global city, and Beaverstock (2002) *Geoforum* 33, 4 observes that *transnational elites have long been associated with the global city. S. Stassen (1991) has linked global cities with increasing social polarization—a thesis rejected by Nørgaard (2003, *Geografiska B* 85, 2), who prefers 'inequality' to 'social polarization'.

( ) SEE WEB LINKS
• Yeung and Olds on the global city.

global civil society The totality of the multitude of groups operating across borders, beyond the control of governments. See J. Eade and D. O'Byrne, eds (2005).

⊕ SEE WEB LINKS

• *Global Civil Society Yearbook*, full of useful ideas and information.

global commodity chain A network of organizations and production processes resulting in a finished commodity (Ponte (2002) *World Dev.* 30); a chain of nodes from raw material exploitation, primary processing, through different stages of trade, services, and manufacturing processes to final consumption and waste disposal (Brown et al. (2006) *GaWC Res. Bull.* 236). Some have argued that the more inclusive language of value chains should replace the more specific concept of commodity chain.

Smith et al. (2002) *PHG* 26, 1 identify limitations to the global commodity chain framework: a limited treatment of both the state and the labour process in governing the organization of productive systems; problems of linearity and dualism; and the problematic conception of the geographies of linkages in commodity chains. See J. Bair (2009) on frontiers of commodity chain research.

global commons Goods and resources—the oceans, Antarctica, outer space, and the atmosphere—that no person or state may own or control and which are central to life; as such they can be thought of as belonging to the whole of humanity. See J. Vogler (2000).

global energy balance The difference between the total influx of *solar radiation to the Earth's surface and the loss of this energy via *terrestrial radiation, evaporation, and the dissipation of sensible heat into the ground.

globalization The increase in the volume, scale, and velocity of social (and environmental) interactions. Globalization is not new, pre-dating *colonialism. Ash (2004) *TIBG* 29, 2 describes globalization as a politically driven project, led by the US government, the World Bank, the World Trade Organization, and the G8. Jackson (2004) *TIBG* 29, 2 prefers to think of 'globalizing' rather than 'globalized', suggesting that globalization 'might be better thought of as a site of struggle rather than as a foregone conclusion'. The **strong globalization thesis** stresses the primacy of global economic forces over national/domestic political ones, emphasizing the decline of the social democratic politics and the limitations of national governments; see Weiss, in G. Ritzer and Z. Atala, eds (2010), who contrasts it with weak globalization. Tickell and Peck in J. Peck and H. W. Yeung, eds (2003) argue that globalization is not a monolithic phenomenon, but produces its own geography.

Globalization operates on a number of scales; Swyngedouw in G. Clark et al. (2000) refers to the changes in the relationships between geographical scales, which create 'a new scalar migration, especially of the poor'. Taylor (2007) *GaWC Res. Bull.* 238 claims that, generally, globalization is associated with a rescaling argument in which national domination of social practice is dissipating upwards to the global, and downwards to the local. Ash (2004) *TIBG* 29, 2 believes that globalization is creating a 'new, topologically and hierarchically structured economic space which is substantially different from the hitherto dominant world system based on territorially organized and state-regulated economies'. But globalization impinges differently on different places, and in locally specific ways (see OECD (2001) *Devolution and Globalization*); in Dicken's view (2004, *TIBG* 29, 1) bounded

Global energy balance

political places matter—'global changes are manifested most directly at the local level' (P. Dicken 2003). Taylor in W. Dunaway et al., eds (2003) expects globalization to change 'the existing state-centric view' into a city-centric view of the world. 'The global economy thrives on the specialized differences of countries, regions, and cities. But it does need homogenized standards . . . and . . . it also needs standardized built environments' (Sassen (2008) *Urb. Geog.* 29, 2).

Martin (2004, *TIBG* 29, 2) believes that access to key resources and assets are crucial to a nation's (or region's) prosperity under globalizations, while Dicken (op. cit.) underlines the importance of the spatial and territorial range and flexibility of the actors involved. Feminists argue that studies of globalization have tended to privilege the global over the local, and as a result have overlooked how global processes are embodied differently across social categories such as gender and race; see R. Nagar, V. Lawson, L. McDowell, and S. Hanson (2002).

Globalization has many detractors, who claim that global capital privileges profit over local interests and deplore the 'Westernization' of local cultures and what they see as the negation of local identities and autonomies (Amin and Graham (1997) *TIBG* 22, 4). 'There is the argument that the global market has made us all the same anyway because there are the same burger bars, coffee shops and fashion outlets in every town centre and in every continent . . . Maybe this immense crumpling of space—the vastly increased connectedness that undoubtedly *has* happened—has given us an *illusion* of knowledge—a presumption that can be both dangerous, I would suggest, and potentially imperial . . . This particular version of the shrinking world is not just a description of the world as it is, however correct or incorrect. It is

actually part of a project. It is geographical mythmaking, to convince us of globalisation in its current form' (Massey, OU Radio Lecture, November 2006). Others suggest that globalization is a dialectical process; it invades local contexts of action, but new forms of local existence and expression emerge (Bollywood as well as Hollywood; *see* CREOLIZATION). Escobar (2001) *Pol. Geog.* 20, 2 cites *hybridization as proof that culture still 'sits in place'.

In order to rein in excesses of neoliberalism and abuses of state regulation, Falk in J. Dunning, ed. (2003) argues for **globalization with a human face**, based on the consent of citizenry, human rights, participation, and accountability; see also Dicken in J. Peck and H. W. Yeung, eds (2003). 'In the political geographies of globalization, three main themes stand out: dominance, governance, and resistance' (Sparke (2004) *PHG* 28, 6).

global justice movements A global assembly of movements focused on poverty, environmental degradation, racism, sexism, and neo-imperialism. It is often, wrongly, viewed as antiglobalization, but it does criticize the current mode of corporate neo-liberal globalization. It also criticizes inequalities between elites and ordinary citizens within nations of the north and south. But P. Routledge and A. Cumbers (2009) argue that, rather than being indicative of a coherent 'movement', such forms of political agency contain many political and geographical fissures and fault-lines, and are best conceived of as 'global justice networks'.

global pipeline The channels used in distant communications and/or interactions. See Bathelt, et al. (2004) *PHG* 28, 1, 31.

global production network (GPN)

The global network of firms, institutions, and other economic agents which shapes and is shaped by the fundamental processes of knowledge and wealth creation, enhancement, and exploitation; corporate, collective, and institutional elements of organizational power; and, spatial and network embeddedness (Henderson et al. (2002) *Rev. Int. Polit. Econ.* 9, 3). 'As manufactured products are becoming more technologically sophisticated, their production has often involved an increasing number of stages. These production stages can be decoupled and farmed out to a network of geographically dispersed locations to take advantage of lower costs . . . International trade then entails flows of goods belonging to a single industry but at different stages of production. This cross-border multi-staged production process has been facilitated immensely by major improvements in transportation, coordination and communication technologies, and the liberalisation of trade and investment regimes' (G. Nandan (2006) Commonwealth of Australia).

Hesse (2006) *Growth & Change* 37, 4 relates the 'somehow abstract system dynamics of GPNs' to very particular places, while Coe et al. (2004) *TIBG* 29, 4 highlight the dynamic 'strategic coupling' of global production networks and regional assets. See also Coe et al. (2004) *TIBG* 29, 4 on global production systems and intersecting notions of power, value, and embeddedness. The entire issue of (2008) *J. Econ. Geog.* 8, 3 is dedicated to global production networks. See Rutherford and Holmes (2008) *J. Econ. Geog.* 8, 4 on the state's role in *actor networks in GPNs. Coe (2012) *PHG* 36, 3, 388 argues that the global production network is 'perhaps best thought of as a heuristic framework for understanding the interconnectedness and uneven development of the global economy'. That is, it enables students to discover or learn something for themselves.

global space

'Global space . . . is not an isotropic surface of sameness' (Jackson (1996) *TIBG* 21, 3). Doel and Hubbard in A. Mayr et al. (2002) argue that a world city is a 'bounded place plugged into a global space of flows'. 'Neoliberal policies create a global space in which finance capital can range freely in search of ever-increasing profit' (Peet (2007) *5th Int. Conf. Critical Geog.*). Weyland (in A. Oncu and P. Weyland, eds 1997) finds that 'the global (highly paid) corporate and managerial labour force which sustains the "public" multinational business space epitomizing globalism is itself reproduced by the presence of a female "privatized" global space, often shored up by (unpaid) corporate wives as well as (lowly paid) foreign maids'.

G. Ó Tuathail (1996) notes the early modern horizontal organization of space associated with ideas of state sovereignty and the emerging state system; 'such a "chess-board" vision of global space made it possible to authorise and strategize new campaigns of geopolitical pre-eminence' (Hughes (2007) *Geog. Compass* 1, 5).

global value chain

'Sets of inter-organizational networks clustered around one commodity or product, linking households, enterprises and states to one another in the world-economy' (Gereffi et al. in G. Gereffi and M. Korzeniewicz, eds 1994). They can be considered as the specific industrial, geographical, social, and institutional environment within which firm-level production networks operate (Palpacuer and Parisotto (2003) *Glob. Networks* 3, 2), or the entire process of bringing a product from its beginnings to its end use: the extraction of raw materials, design, manufacturing, marketing, distribution, warehousing, delivery,

	Global commodity chains (GCCS)	Global value chains (GVCS)	Global production networks (GPNS)
Disciplinary background	Economic Sociology	Development Economics	Relational Economic Geography
Object of enquiry	Inter-firm networks in global industries	Sectoral logics of global industries	Global network configurations and regional development
Orienting concepts	Industry structure Governance Organisational learning Industrial upgrading	Value-added chains Governance models Transaction costs Industrial upgrading and rents	Value creation, enhancement, and capture Corporate, collective, and institutional power Societal, network, and territorial embeddedness
Intellectual influences	MNC literature Comparative development	International business Trade economics	GCC/GVC analyses Actor-network theory Varieties of capitalism

retailing, and customer support. The value chain perspective stresses the importance of activities other than production itself, notably design, logistics, and marketing; see Kaplinsky et al. (2002) *World Dev.* 30. See also Knorringa and Pegler (2006) *Tijdschrift* 97, 5 and Sturgeon (2005) *Rev. Int. Polit. Econ.* 12, 1 on the governance of global value chains. Gereffi And Lee (2012) *J. Supply Chain Managt* is wonderfully clear.

( ) SEE WEB LINKS
• Global value chains.

global village A term coined by H. Marshall McLuhan (1962 and 1964) who correctly foresaw that continued improvement in electrified communication would eventually lead to instantaneous interaction across the globe. Some use this term to describe the *internet and its effects on human interactions.

global warming The increase in global temperatures brought about by the increased emission of greenhouse gases into the atmosphere. 'Many of the indicators of climate change are suggesting that measurable change is broadly apparent globally, with eleven of the last twelve years the hottest on record across the globe, ice sheets and glaciers melting at rates exceeding scientific expectations, and trends in ecosystems resulting from responses to environmental change apparent, such as coral bleaching and the migration of species poleward and to higher altitudes' (Bardsley and Bardsley (2007) *Geog. Res.* 45, 4).

See Cavan et al. (2001) *Bull. AMS* 82 on western North America, Maheras et al. (2004) *Int. J. Climatol.* 24 and Dünkeloh and Jacobeit (2003) *J. Climatol.* 23 on southern Europe, and Smith et al. (2000) *Int. J. Climatol.* 20 on South Africa, Chile, and south-west Western Australia; see also James J. McCarthy et al., eds (2001).

Nyberg et al. (2007, *Nature* 447) note that the average frequency of major hurricanes in the North Atlantic Ocean decreased gradually from the 1760s until the early 1990s, but has increased significantly since 1995. This trend has been attributed to anthropogenically induced climate change, and natural variability, but the primary cause remains uncertain. Emanuel (*Program in Atmospheres, Oceans, and Climate*, MIT) agrees that there has been a large upswing in the frequency of Atlantic hurricanes since 1995, but observes that there has been no change in the global, annual frequency of tropical cyclones. However, he adds that there is some evidence that the intensity of hurricanes is increasing.

glocal Describing the seamless integration between the local and global; the comprehensive connectedness produced by travel, business, and communications; willingness and ability to think globally and act locally. 'The concept of glocalization captures the dynamic, contingent, and two-way dialectic between the global and the local' (Swyngedouw (2004) *Camb. Rev. Int. Affairs* 17). Hence, Lindell (2009) *Geografiska B* 91, 2, 123 sees organizations across scales as the defining characteristic of the glocal.

Parnreiter et al. (2005) *Mitteilungen der Österreichischen Geographischen Gesellschaft* 147 attribute the preference for 'glocal' financial institutions to using relatively local banks with sufficient global experience to operate in major financial markets, and to so-called 'soft' factors, such as the initiation of business deals, and an in-depth analysis of the company that sets out to issue bonds or shares. 'The hallmark of glocalizing states is the project of *reconcentrating* the capacities for economic development within strategic subnational sites such as cities, city-regions and industrial districts, which are in turn to be positioned strategically within global and European economic flows' (Brenner in J. Peck and H. W. Yeung, eds 2003). See also Cox (2004) *TIBG* 29, 2. Bliss (2005) *NZ Geogr.* 61, 3 argues that responsible **glocal citizenship** makes for a better world.

GNP (gross national product) The *GDP of a nation together with any money earned from investment abroad, less the income earned within the nation by non-nationals. *Compare with* GDP.
GNP per capita is calculated as GNP/population and is usually expressed in US dollars. It may be used as an indicator of *development.

goal function A function that by current optimization can describe the succession of an ecosystem. 'Ecosystems do not, of course, have goals; but their successional development may be described by a goal function' (Jørgensen and Nielsen (2007) *Energy* 32, 5). Phillips (2008) *PPG* 32, 1 reviews theories of evolution and ecosystem development based on goal functions. See F. Müller and M. Leupelt, eds (1998).

Gondwanaland A 'supercontinent': a continuous land surface formed of the now separate units of Africa, Madagascar, Antarctica, Australia, and India. Gondwanaland started in the earliest Paleozoic at the break-up of a Late Proterozoic supercontinent (Veevers (1988) *Geology* 16, 8), and assembled by 650–570 Ma. See Veevers (2003) *Geology* 31, 6.

gorge A deep, narrow cleft in an upland area, usually containing a river. Gorges occur in *karst scenery partly as a result of the collapse of caves, or when the incisive power of a river is greater than the processes of valley-wall erosion.

governance The action, fact, or manner of governing (not solely by governments); the various means used to shape society to a desired end, which include:

- the processes by which governments are selected, scrutinized, and modified
- the capacity of government to generate, establish, and apply policy and legal procedures

- the systems and methods of interaction between the executive, the legislature, and the judiciary
- the relationship between the state, economy, and civil society, and the regulation of economic regimes, and of social authority and practice (Shirlow in C. Gallaher et al. 2009).

Hooghe and Marks (2003) *Am. Polit. Sci. Rev.* 97 distinguish between governance as the dispersion of authority, and governance as task-specific, intersecting, and flexible, while Griffin (2012) *Pol. Studs Rev.* 10, 2, 208 clarifies the terms **multi-level governance**, which is concerned with continuous negotiation among nested governments at several territorial tiers, and **network governance** emphasizing horizontal relations between and across institutions. **Environmental governance** is concerned with: the formal and informal institutions that influence how power is exercised; how public decisions are taken; how citizens become engaged or disaffected; and who gains legitimacy and influence in environmental policy; see Reed and Bruyneel (2011) *PHG* 33, 4, 646.

governmentality A concept of M. Foucault (2003) to encompass the mentalities, rationalities, and techniques used by governments, within a defined territory, actively to create the subjects (the governed), and the social, economic, and political structures, in and through which their policy can best be implemented. In other words, governments try to produce the citizen best suited to fulfil their policies: 'a national form of biopower directed not so much at the management of individual human bodies as at populations and their territories' (M. Hannah 2000). 'Governmentality points at the techniques and rationalities involved in both governing the self and governing others' (Cupers (2008) *Cult. Geogs.* 15). Governmentality also includes ways in which spaces and places are created, and used, in order to pursue policies; see Airriess (2005) *J. Hist. Geog.* 31, 4 on politically contested space in Hong Kong's New Territories. **Colonial governmentality** aimed to construct a wage-earning, property-owning, taxpaying, and literate subject (Legg (2006) *Soc. & Cult. Geog.* 7, 5).

government in exile An exiled administration, such as the government in exile of Tibet that 'claims legitimacy as the official representative of the Tibetan population, performs a number of state-like functions in relation to its diasporic "citizenry" and attempts to make its voice heard within the international community', McConnell (2009) *Pol. Geog.* 28, 343.

grade In geomorphology, a state of *equilibrium in a system. A river is said to be **graded** when it can: transport the 1.5 year water discharge within the bankfull channel; transport all grain sizes supplied by the watershed without net erosion or deposition; and balance any bank erosion with an equal amount of bank sedimentation; that is to say that it has sufficient *discharge to transport its *load, but not enough energy for erosion; if any of the factors controlling discharge changes, the river will react to re-establish grade; see A. S. Goudie and K. A. Gregory (2011) pp. 365–6. Grade over one section of a river course is a **graded reach**. A **graded slope** has an angle at which output, throughput, and input remain in equilibrium.

gradualism A school of thought arguing that, throughout the history of the Earth, geological and biological processes have operated at the rates more or less the same as those observed in the present day (R. Huggett 2009).

granular disintegration (granular disaggregation) A form of weathering where the grains of a rock become loosened and fall out, to leave a pitted, uneven surface. André (2002) *Geografiska A* 84, 3–4 finds that the granular disintegration of crystalline rocks operates at the same rate whatever the environment.

graph theory The term 'graph' can refer to the structure of a network, thus graph theory is the mathematical analysis of the way a network is organized, considering the way all the *nodes in a graph are connected to each other in order to understand its functions. Cantwell and Forman (1993) *Landscape Ecol.* 8, 4, 239 take you through the procedure, step by step.

gravity anomaly The difference between the actual gravitational force and the calculated force. For example, when a plumb line is set up near a mountain range it will be pulled from the vertical towards, and by the gravitational pull of, the mountains—but by far less than expected from calculations. This is a **negative gravity anomaly**. See Nakada et al. (2002) *Technophysics* 351, 263 on the Eastern Kyoto gravity anomaly.

gravity model In human geography, gravity models evaluate, or forecast, the spatial interactions of goods, people, and so on, between origins and destinations. 'The gravity model assumes that the probability of patronizing a facility (facility utility) is proportional to the attractiveness of the facility and inversely proportional to a function of the distance' (Drezner and Drezner (2006) *Geog. Analys.* 38, 4). These interactions are expressed as:

$$I_{ij} = k\,M_i\,M_j\,Dist_{ij}^{-\beta}$$

where I_{ij} denotes the interaction between two locations i and j; M_i and M_j represent the 'masses' measuring the strength of i and j (usually the population numbers of two settlements); $Dist_{ij}$ stands for the distance between i and j; and k and β are constants (G. Robinson 1998). 'While the gravity model is now accepted by economics as a useful tool (it is said to be all over the place at the International Monetary Fund), it is considered only in an incidental manner in economic analysis that employs it' (Isard (2001) *Int. Reg. Sci. Rev.* 24, 3). See Rodríguez-Pose and Zademach (2006) *Tijdschrift* 97, 3 on the gravity model, and mergers and acquisitions.

great circle Any circle on a sphere formed by a plane that passes through the centre of the sphere. The equator and the meridians of longitude are all great circles on the Earth's surface. Any great circle route between two points will represent the shortest line between the two; great circles are therefore relevant to studies of migration and other interactions. Try Zar (1989) *J. Field Ornithol.* 60 for the calculations.

great soil groups The primary classification of global soils into groups; a classification similar to the formations of vegetation. *See* SOIL CLASSIFICATION.

green belt An area of undeveloped land encircling a town, restricting development in the semi-rural areas beyond the built-up zone. The amount of new building is restricted although by no means completely banned. An early green belt was set up in the 1950s around Greater London, in order to limit the spread of suburbs. Other cities have followed this example: in Beijing, all the land in the 'green belt' was forced to keep its status in 2004 without any changes until 2020. 'Under this policy, the trend of central city to form a big pancake would be stopped, while the sub-cities in the region would develop very fast with many small and medium "pancakes" forming outside the 6th Ring Road' (He et al. (2006) *Appl. Geog.* 26, 3–4). Cullingworth calls green belts 'the first article of the British planning creed', but Nadin argues that they 'have become an outmoded and largely irrelevant mechanism for handling the complexity of future change in the city's country-side' (both in J. Cullingworth and V. Nadin 2002).

greenfield site Areas beyond the city where development can take place un-fettered by earlier building and where low-density, high-amenity buildings can be constructed.

greenhouse effect The warming of the atmosphere as some of its gases absorb the heat given out by the Earth. Short-wave radiation from the sun warms the Earth during daylight hours, but this heat is balanced by outgoing long-wave radiation over the entire 24-hour period. Much of this radiation is absorbed by atmospheric gases, most notably water vapour, carbon dioxide, and *ozone, but also by methane and chloro-fluorocarbons. All of these may be called **greenhouse gases**. Without this absorption, which is also known as *counter-radiation, the temperature of the atmosphere would fall by 30–40 °C.

This is the **natural greenhouse effect**. Through human agency, such as the clearance of rain forest, or the in-creased rearing of livestock, the concen-tration of greenhouse gases in the atmosphere is increasing. This is the **enhanced greenhouse effect**. Mea-surements taken at Mauna Loa, Hawaii, show that the concentration of atmo-spheric CO_2, for example, increased by 19.4% between 1959 and 2004. It would

follow, therefore, that increased con-centrations of such greenhouse gases would lead to a rise in global tempera-tures; *see* GLOBAL WARMING.

(⊕) SEE WEB LINKS
• The Carbon Dioxide Information Analysis Centre.

green manure A leguminous crop ploughed into the fields after it matures, since it fixes nitrogen from the air and thus improves soil fertility. See Ajayi et al. (2007) *Nat. Resources Forum* 31, 4.

green politics (greening) A political creed that aspires to the creation of a sustainable society founded on *envir-onmentalism, grassroots democracy, and *social justice. D. Mulvaney (2010) has a really useful A–Z of green politics.

green revolution The development of high-yielding crop varieties (HYVs) for developing countries began in a concerted fashion in the late 1950s. In the mid-1960s, scientists developed HYVs of rice and wheat that were sub-sequently released to farmers in Latin America and Asia. The success of these HYVs was characterized as a 'Green Revolution'.

Early rice and wheat HYVs were rap-idly adopted in tropical and subtropical regions with good irrigation systems or reliable rainfall. Over the following years, the Green Revolution achieved broader and deeper impacts, extending far be-yond the original successes of rice and wheat in Latin America and Asia. 'Pro-ductivity growth associated with HYVs had important consequences. Increased food production has contributed to lower food prices globally. Average cal-oric intake has risen as a result of lower food prices—with corresponding gains in health and life' (Evenson and Gollin (2003) *Science* 300, 5620). 'In order for an African Green Revolution to happen, it is recommended that the agrarian communities and regions reform their land ownership and land tenure from traditional and/or customary laws, in order to enable the ordinary or poor farmers to access and use the land with ease. The farmers need secure and stable tenancy on the land for them to make the necessary investments in modernization of agriculture. Where there are farmers who are tenants on the land, then there should be reform to make the tenancy long term, so that the farmers have operational legitimacy such as land lease or operational leasehold' (Ngambeki (2003) *SMART/AGRI*).

A. Gupta (1998) describes the mixed reception afforded to the green revolution: 'on the one hand, rural people fused environmental and health concerns with a critique of green revolution technology; on the other hand, they welcomed increased wealth and surpluses flowing from the new technology.' 'The post-World War II green revolution in synthetic chemicals transformed New Zealand agriculture and was interpreted in a progressive way. Today, the same chemicals are often seen in anything but this light. This shift in representation can be seen in the ways we interpret texts and how such re-readings often highlight and contest dominant ideas of the past' (Wildblood-Crawford (2006) *NZ Geogr.* 62, 1). Wilson (2001) *TIBG* 26 stresses that the need for initiatives and policies that emphasize environmentally friendly agricultural practices has grown as a result of the green revolution. See also G. Robinson (2004).

'Was there a real alternative to the Green Revolution? Did other development programs contribute to the process of producing higher incomes? Certainly, there were many redistributive programs and investments in health and schooling that contributed to welfare . . . The Green Revolution did raise the productivity of land and water resources. It also raised the productivity of human resources and

did lead to higher wages . . . In short, for the poorest countries, the Green Revolution, late and uneven as it was, was the "only game in town"' (Evenson (2002) *Bureau Econ. Studs Occ. Paper* 19).

grèzes litées Deposits down a hillslope of *imbricated rock fragments bedded parallel to the slope. It may be that the debris is shattered by freeze-thaw, with larger fragments rolling downwards; wind might be the depositional process (Cailleux (1967) *Biuletyn Peryglacjalny* 17; Edlund and Woo (1992) *Geol. Surv. Canad. Part B* 92–1B; Lewkowicz (1998) *Canad. J. Earth Scis* 35).

gross domestic product See GDP.

gross national income (GNI) A broad measure of national prosperity constructed by adding *GDP (Gross Domestic Product) to income accruing from investments outside the country. It is usually expressed in $US.

gross national product See GNP.

gross reproduction rate The number of female babies born per thousand women of reproductive age. The net reproduction rate also takes into account the number of women who cannot or do not wish to have children; in the UK, the cohort net reproduction rate in 2011 was 1.02 children per woman (NISRA, 2011), while in India in 2011 it was 3.0 (Census India, 2011).

gross value added (GVA) The additional value at a particular stage of production, or through image and marketing. It shows the contribution of the factors of production, in increasing the value of a product. GVA measures the contribution to the economy of each individual producer, industry, or sector in a nation. Per capita GVA in the UK for 2012 were:

North East	£16 091
North West	£18 438
Yorkshire and Humberside	£17 556
East Midlands	£17 448
West Midlands	£17 429
East	£19 658
London	£37 232
South-East	£23 221
South-West	£19 023
England	£21 937
Wales	£15 401
Scotland	£20 013

Source: UK Office for National Statistics.

grounded theory This involves the collection, coding, and categorization of qualitative data: 'transcribed text from interviews, focus groups, and researchers' notes, photographs, and other images. The purpose of grounded theory is to build theories from data about the social world such that theories are "grounded" in people's everyday experiences and actions'. Knigge and Cope (2005) *Env. & Plan. A,* 38 (11), 2021 take you through the process step by step. Hypotheses, *models, or theories emerge gradually, through observation, data collection, and systematic reflection.

ground failure *See* AVALANCHE; DEBRIS; LANDSLIDE; ROCK FALL.

ground frost *See* FROST.

ground moraine *See* MORAINE.

ground truth In remote sensing, ground truth refers to the comparison of remotely sensed data with real features and materials on the ground. Search for Art Kalinski, Geospatial Solutions, for a fascinating and wholly relevant account of the use of ground truth websites.

groundwater All water found under the surface of the ground which is not chemically combined with any minerals present, but not including underground streams. The GroundWaterSoftware. com newsletter is really useful.

growth pole A point of economic growth. Poles are usually urban locations, benefiting from *agglomeration economies, and should spread prosperity from the core to the periphery. Planners have tried to create new growth poles; the best known being in the Mezzogiorno of Italy—which 'signally failed' (G.-J. Hospers 2004). Glasmeier (2002) *PHG* 26, 2 blames the rampant enthusiasm for growth poles in the USA for rationalizing and supporting concentrated investment in selected places, 'while downplaying the underlying human needs found in adjacent poor places'. R. Capello and P. Nijkamp (2010) is very useful on this and other development strategies.

growth triangle A cooperative venture among three or more countries; each has different factors of land, labour, capital, and management, which complement the others, generating mutual advantages in external trade and investment. Growth triangles offer economies of scale, widen potential markets, and may act as buffers against recession; see Landingin and Wadley (2005) *J. Int. Dev.* 17 on growth triangles in Asia.

groyne A breakwater running seawards from the land, constructed to stop the flow of beach material moved by *longshore drift.

G-20 (Group of Twenty) A group of twenty of the most important economies on the planet. It includes nineteen independent countries along with the European Union: Argentina, Australia, Brazil, Canada, China, France[†], Germany[†], India, Indonesia, Italy[†], Japan, Mexico, Russia, Saudi Arabia, South Africa, South Korea, Turkey, United Kingdom[†], United States, European Union.

† = also a member of the EU.

guanxi In Chinese society, the clan ties and relationship obligations formed on the basis of kinship, and which are an autonomous unit within the whole of society. Tru Jun and Sin (2010) *Asia Pacif. J. Management* 27, 3, 561.

Guinea current A warm *ocean current off the coast of West Africa, which transports low-saline, warm waters from the Gulf of Biafra south-west along the north Gabon coast.

gully A 'catch all' definition would be 'a water-made cutting, usually steep-sided with a flattened floor', but Bergonese and Reis (2011) *ESPL* 36, 11, 1554 provide six definitions currently in use. An **ephemeral gully** is a small channel eroded by concentrated overland flow that can easily be filled by normal tillage, only to reform in the same location with further runoff (Soil Science Society of America, 1996). Gullying usually occurs in unconsolidated rock and rarely cuts through bedrock. Gullies usually form quickly as a result of destruction of the plant cover; see Malik (2008) *Geomorph.* 93, 3–4. It can be stopped by restoring a vegetation cover, by contour ploughing, and by making terraces and small dams across the hillside; see Xiang-Zhou et al. (2004) *Env. Sci. & Policy* 7, 2; Fu et al. (2000) *Catena* 39, 1.

gust A temporary increase in wind speed, lasting for a few seconds. A typical ratio of gust speed to wind speed in rural areas is 1.6:1, increasing to 2:1 in urban areas due to the effects of high buildings and narrow streets. See Mitsuta and Tsukamoto (1989) *J. Appl. Met.* 28, 11 on the spatial structure of gusts.

guyot A truncated sea-floor volcano occurring as a flat-topped mountain which does not reach the sea surface. Guyots are thought to be associated with *hot spots.

habitat The area in which an organism can live and which affords it relatively favourable conditions for existence. Keagy et al. (2005) *Restor. Ecol.* 13, 3 investigate a scenario whereby some higher-quality habitat is destroyed, in order to determine the quantity of low-quality habitat required to maintain the equilibrium population abundance.
Habitat generality means being able to survive in a multitude of environments. See Brown (2007) *J. Biogeog.* 34, 12.

habitat rehabilitation The management of a degraded habitat aimed at restoring it to 'an ecologically superior state' (Borg et al. (2007) *Geomorph.* 89, 1–2). An understanding of the recovery and reassembly of plant communities is imperative for habitat rehabilitation (Thornton (1984) *Ambio* 28). Borg et al. (op. cit.) argue that rehabilitation projects should provide the appropriate elements for a healthy system; one would then let these elements interact to produce some target state. The problem lies in determining what that target state should be; habitat changes that are beneficial to one species may be detrimental to another; thus, 'in some ways, habitat rehabilitation is based on explicit or implicit species values' (Clough (2003) *Rept. Fish Habitat & Species Recovery Wkshp*).

habitus A socio-cultural milieu in a distinct neighbourhood; the dispositions that shape the actions, bodily movements, tastes, and judgements characteristic of a social class, together with the circumstances that ensure its reproduction. Habitus is improvisatory, operating through human practices rather than through prior conscious thought. See Bridge (2001) *TIBG* 26, 2 on habitus and *gentrification, and Marshall and Foster (2002) *Canad. Geogr./Géog. canad.* 46, 1 on habitus and migration.

hactivism Any operation using computer hacking against an internet site, aiming to disrupt normal operations, but not to cause major damage. T. Jordan (2002) argues that hacktivism is motivated by a libertarian ethos that views the 'free flows of information, securely accessible to all' as 'the highest principle'. Curran and Gibson (2013) *Antipode* 45, 2, 294 are really good on wikileaks, anarchism, and technologies of dissent.

Hadley cell A simple, vertical, thermally direct, tropical, *atmospheric cell. Air rises at the *inter-tropical convergence zone (ITCZ) and drifts polewards, cooling a little through a net loss of radiation, and thereby transferring heat. As the circumference of the Earth decreases with distance from the equator, the moving air has to converge. This combination of cooling and convergence, together with deflection by the *Coriolis force, causes the air to sink around latitude 32°. The air then returns to the equator as surface winds. The basic components of this cell are: weak upper-air easterlies above the ITCZ, subtropical anticyclones at the descending limb, and easterly *trade winds associated with the return of air to the equator, together with *synoptic scale disturbances, such as easterly waves or *tropical cyclones. The westerly subtropical *jet is located at the poleward limit of the Hadley cell.

In the tropics, the rotationally confined, thermally direct Hadley circulations are important for maintaining the observed small meridional temperature gradients in low latitudes (Held and Hou (1980) *J. Atmos. Sci.* 37). The intensification and poleward expansion of the cross-equatorial Hadley cell can lead to westerly acceleration in the winter subtropics, and enhanced vertical shear of the zonal wind in the subtropics and mid-latitudes (Hou (1998) *J. Atmos. Sci.* 55, 14). *See* GENERAL CIRCULATION OF THE ATMOSPHERE.

hail A form of snow made of roughly spherical lumps of ice, 5 mm or more in diameter, formed when a frozen raindrop is caught in the violent updraughts found in warm, wet *cumulo-nimbus clouds. As the drops rise, they attract ice, and as they fall, the outer layer melts, but refreezes when the droplet is again lifted by updraughts, often causing the hailstones to show an onion-like pattern of alternating clear ice (glaze) and opaque ice (rime). A hailstone will descend when its fall speed is enough to overcome the updraughts in the cloud. **Soft hail** is white and low-density as it contains air. See Billett et al. (1997) *Weather & Fcst* 12, 1 on the prediction of hail size and Gavrilov et al. (2010) *Phys. Geog.* 31, 5 on hail suppression.

halo A ring of light around the sun or, more rarely, the moon, caused by the refraction of light by ice crystals. A **cloud halo** is a zone of increased humidity around cumulus clouds; see Lu et al. (2002) *J. Appl. Met.* 41, 8.

halocline Within a sea or ocean, the layer where salinity concentration changes rapidly with depth. See Menviel et al. (2012) *Topical Studs Oceanography* 61–4, 106 on the effects of removing the North Pacific halocline.

halophyte A plant which can grow in saline conditions: salt marshes, estuarine environments, and the lower parts of sea cliffs. Controversy exists over whether halophytes require saline conditions for their existence—or merely tolerate them (Yeo (1998) *J. Exp. Bot.* 49; Glenn et al. (1999) *Crit. Rev. Plant. Sci.* 18).

hamada A desert surface of pebbles/gravel/boulders, formed by the washing or blowing away of the finer material. See Dłużewsk and Krzemień (2008) *Prace Geograficzne*, zeszyt. 118 (it is in English).

hamlet A small settlement without services or shops, and usually without a place of worship.

hanging valley A high-level tributary valley from which the ground falls sharply to the level of the lower, main valley. The depth of the lower valley may be attributed to more severe glaciation: 'the height of a hanging valley reflects the difference in the time-integrated ice discharge in tributary and trunk valleys and therefore increases as the discharge ratio decreases . . . The rate of production of hanging valley relief increases strongly as the ratio of the tributary to trunk ice discharge decreases' (MacGregor et al. (2000) *Geology* 28, 11). The level of a hanging valley may be used as an indicator of glacier ice thickness (Kuhle (2005) *Island Arc* 14, 4).

Brocklehurst et al., *Geomorph.* 97, 1–2, observe that while hanging valleys show glacial modification of a landscape, the overall change in the relief structure of mountain ranges is surprisingly modest.

haptic Relating to the sense of touch, especially when relating to the perception and manipulation of objects. Paterson (2009) *PHG* 3, 6, 766 provides a commendably accessible overview of haptic knowledges in geography that respond to bodily sensations and the reactions of the embodied researcher.

hardpan A cemented layer in the B *horizon of a soil, formed by the *illuviation and precipitation of material from the A horizon, such as clay (forming a **clay pan**), humus (a **moor pan**), or iron (an **iron pan**). See Lee and Gilkes (2005) *Geoderma* 126, 1–2.

Harmattan *See* LOCAL WINDS.

Hawaiian eruption A volcanic *fissure eruption where large quantities of basic lava spill out with very little explosive activity. Hawaiian volcanoes evolve through stages that have been delimited by the compositions of their erupted lavas; see Vazquez et al. (2007) *Geology* 35, 8.

hazard An unexpected threat to humans and/or their property; a hazard includes both the event and its consequences. By this definition, the Indian monsoon is not a hazard, but its failure is. A natural event becomes a hazard through: the social processes which cause some people to be much more at risk from the effects of a hazard; a lack of options to minimize, or escape from, the effects of the hazard; the location of human settlement on potentially dangerous physical sites; the location of human settlement near potentially dangerous economic activity; and activities that aggravate events with natural causes.

'As the range of hazards and vulnerabilities faced by any given community increases, it often becomes possible only to play one kind of risk off against another in search of a "less bad" scenario. Many highly vulnerable communities may deliberately choose to inhabit a hazard prone environment if this reduces other risks, related to income generation for example. Or, should they find themselves in hazard prone zones due to exclusion from formal land markets or for other reasons, they will many times opt to stay in order to maintain those conditions that provide them with the means to reduce daily life risk and vulnerability' (UNDP Expert Group Meeting, Havana, 2002).

Hazard perception is the view which an individual has of a hazard. Robertson (1999) *Geog. Rev.* 89, 4 compares the reality of the tornado hazard with the portrayal in movies, finding the latter seriously misleading. Walker et al. (2002) *Appl. Geog.* 20, 2 report on the failure of regulatory practice to distinguish hazards from industrial accidents. *See* COGNITIVE DISSONANCE.

haze A suspension of particles in the air, slightly obscuring visibility. These particles may be naturally occurring— sea salt or desert dust—or man-made, like smoke.

headcut A nearly instantaneous, radical change in water elevation typically represented by the presence of a waterfall or steep cascade that is eroding upstream. Headcuts often trigger a *gully head; 'headcuts and their . . . gullies, are arguably the most difficult channel disturbances to control or reduce. The most effective method for dealing with headcuts is to prevent them from initiating at all'.

⊕ SEE WEB LINKS

- Website of Plumas Corporation, which focuses on the economics of natural resources.

headward erosion The lengthening of a river's course, by erosion back towards its source. A *knick point on a stream profile suggests that the fluvial system has reacted to a lower base level by headward erosion (Schlunegger and Schneider (2005) *Geomorph.* 69, 1–4). See also Hill et al. (2008) *Geomorph.* 95, 3–4 on the Grand Canyon, USA.

health, geographies of Studies of health from geographical perspectives, including: quantitative and qualitative approaches to map and interpret the geography of health; inequalities in health outcomes and provision/utilization of health services; the impact of migration on health; the impact of health status on migration; health and human modification of the environment; the impact of air/water quality, air/water contamination or pollution, global climate change; and other global environmental changes on health and disease, such as infectious diseases, toxicity, neoplasms, and heat stress. Pearce and Barnett (2012) *PHG* 36, 1, 3 have a very interesting, clear, and nuanced study of the impact of the local economic and social environment upon smoking. Try also Andrews and Evans (2008) *PHG* 32, 6, 759. Issue 5 of *AAAG* (2012), is entirely devoted to geographies of health.

heartland A term suggested by H. Mackinder (1904) to indicate the wealthy interior of Eurasia; whoever controls the heartland would be able to gain a '*hegemony over the entire world'. Bilgin (2007) *Pol. Geog.* 26, 7 shows how this concept has been adopted and adapted to help justify foreign and domestic policy-making in Turkey.

heat island In a city, air temperatures are often as much as 3–4 °C higher than over open country. Causes include: the presence of large vertical faces; a reduced sky-view factor; surface materials of high heat capacities and conductivities; large areas of impermeable surfaces; anthropogenic heat fluxes; and atmospheric pollution (Grimmond

(2007) *Geog. Rev.* 173, 1); for more detail, see Yow (2007) *Geog. Compass* 1, 6.

hedonic pricing A hedonic model of prices is one that decomposes the price of an item into those separate components that determine the price. For example, Stewart and Jones (1998) *Econ. Inquiry* 36, 192 analyse the wages paid to baseball players on the basis of hitting and pitching performance, the player's experience, and so on. A hedonic model would not necessarily separate all the factors that could be separated, only those that affect the usefulness to a buyer of what is being sold.

The UK Office of Fair Trading's *Quantitative Techniques In Competition Analysis*, p. 49, available online, shows exactly how it's done.

hegemony In terms of power and politics, leadership (especially by one state of a federation); more widely, the way in which one social group will represent its interests as being everyone's interests. This ruling group may keep its grip on society either by **social hegemony**—the use of force to maintain order—or, more commonly, by **cultural hegemony**—producing ways of thinking, especially by subtly eliminating alternative views to reinforce the status quo (Laurie and Bennett (2002) *Antipode* 34, 1; Hoelscher (2003) *AAAG* 93, 3). See also S. Zukin (1995) on the cultural hegemony of cities.

Perreault (2007) *PHG* 31, 4 sees the core idea of hegemony as 'ideological leadership and popular consent'. Kohl (2006) *Antipode* 38, 2 discusses neoliberalism as a **hegemonic system**. New approaches to hegemony examine how social identities (of race, gender, etc.) become dominant and how their particular interests are defined and accepted as universal interests; see Gallaher in C. Gallaher et al. (2009).

helicoidal flow A continuous cork-screw motion of water along a river channel.

henge In the UK, a Neolithic circular earthwork with an encircling bank and inside ditch.

heritage Inherited circumstances or benefits; 'a (re)presentation of the past in cultural forms that can hide some voices and confirm others, and . . . formulate certain national identities through which individuals may come to see themselves in their everyday lives' (Bhatti (2008) *Area* 40, 1. Hubbard and Lilley (2000) *Geog.* 85, 3 see the **heritage industry** 'as involving conflicts between the different senses of place, with the distinctive character of a town . . . having resulted from different groups seeking to impose their values on the townscape'. See also While and Short (2011) *Area*, 43, 1, 4 on the production of local heritage and design discourses, and the regulation of conservation and change in the built environment in Manchester, England. Summerby-Murray (2002) *Canad. Geogr./Géog. canad.* 46, 1 argues that the **heritage discourse** is constructed through the creation of memory and the processes of commodification and consumption. See J. Scarpaci (2005) on heritage tourism in Latin America.

hermeneutics The theory and methodology of interpretation. P. Cloke, C. Philo, and D. Sadler (1991) argue that the term is used as the study of the relations between humans and place.

heterogeneous nucleation The freezing of water droplets around a *freezing nucleus. See Khvorostyanov and Curry (2004) *J. Atmos. Scis* 61, 22.

heteronormativity The societal privileging of heterosexuality (Oswin (2008) *PHG* 32, 1); the normalization of man/woman as opposites meant to come together within heterosexual relationships that are based on specific class and race-based relations (Browne (2006) *Antipode* 38, 5). Waitt (2006) *AAAG* 96, 4 writes on resistance to heteronormativity and Sydney's gay games (Oswin (2008) *PHG* 32, 1). Hubbard (2008) *Geog. Compass* 2, 3, 640 explores the theorization of heteronormativity. See the special issue of *Antipode* 34, 5 (2002).

heterotopia 'The coexistence in an impossible space of a large number of fragmentary possible worlds' (M. Foucault 1967). Cyberspace is a good example. See Johnson (2013) *Geog. Compass* 7, 11, 790.

heterotrophe An organism which has to acquire its energy by digesting food that has been manufactured by other organisms. Thus all organisms are heterotrophes except the primary *producers which can manufacture their own food, usually by photosynthesis. **Heterotrophy** is the process whereby existing organic molecules are used by an organism for its energy needs, in maintenance, growth, and reproduction. See Stear (2010) *Limnol. Oceanogr.* 5, 2, 817 on drivers of net heterotrophy in contrasting lakes.

(((⊕))) SEE WEB LINKS
• Website on Fungal biology.

heuristic Describing a teaching technique that promotes discovery. Thus, a **heuristic model** does not answer a question in and of itself, but rather helps create a means for answering a question. Phipps (1988) *Geog. Anal.* 20, 3 explains clearly, using as an example the search for a new house motivated by a housing problem. The searcher can make a *rational* inquiry to reduce this housing stress either by improving the old home, or by finding a new, less stress-inducing home, where cash is not a major issue. Or using a *heuristic* model, the searcher uses the information built up from each house visited to update his/her perceptions of the types of available houses within a price range. See N. Xiao in B. Warf (2010) on heuristic methods in spatial analysis.

hierarchical diffusion *See* DIFFUSION.

hierarchy Ordering of phenomena with grades or classes ranked in sequence; *central place theory, for example, posits a hierarchy of settlements from regional capitals to hamlets. In a **nested hierarchy** each level contains/is composed of the level below, with the individuals in that level being smaller-scale sub-systems of the level above. The metasystem (highest level) does not respond to the lowest sub-system directly, as the dynamics of the sub-system are on too small a scale to be of any consequence; in other words, information across any three hierarchical levels is *buffered*. This suggests that the sensitivity of a system to change takes a different form, depending on where in the hierarchy the drivers of change originate. Whatever your own focus in geography, you will never find hierarchy expressed more clearly than in Couper (2007) *TIBG* 32, 3. Smith and Kurtz (2003) *Geogr. Rev.* 93 suggest a hierarchy of coalitions.

high A region of high atmospheric pressure—in Britain, generally over 1000 mb.

high-order goods and services Goods and services with a high *threshold population and a large *range. Examples include furniture, electrical goods, and financial expertise. These goods are usually *shopping goods.

high-technology (high-tech) Any industry concerned with advanced technology, such as biotechnology,

computers and microprocessors, fibre optics, aerospace, electronics-telecommunication, and pharmaceuticals (Fischer et al. (2005) *Geog. Analys.* 38, 3). In theory such industries are *footloose, since the products have a high value to weight ratio. High-tech hubs in India are Bangalore, Chennai, Delhi, Hyderabad, Mumbai, and Pune (Crescenzi et al. *Econ. Geog.* 12, 5, 1055). Bernstein and Nadiri (1988) *Am. Econ. Rev.* 78 find that new technological knowledge complements R&D in high-technology industries. Florida (2002) *J. Econ. Geog.* 2 finds a significant, positive correlation between his 'bohemian index' and concentrations of high-tech industry. James (2008) *Geog. Compass* 3 has an interesting paper on the gendered geographies of high-tech economies.

hill farming The *extensive animal farming of an upland area. European Parliament's resolution, A6-0327/2008/P6_TA-PROV(2008)0438, highlights the importance of hill and mountain farming in the European Union, particularly stressing the importance of environmental protection, the need to produce quality products, and the need to maintain the cultural landscape as part of the European heritage.

hill fort A fortified site on a hilltop, usually with a ditch and ramparts. The earliest date from the Iron Age but some British examples were created as late as the Dark Ages. Favoured sites for such structures in southern England seem to be chalk uplands.

hillslope Any geometric element of the Earth's solid surface at an angle from the horizontal. Slope profiles are generally surveyed along a straight line (on a map) from the divide, following a line of steepest descent, to the nearest point at their base. The four possible elements of a hillslope profile are: an upper convexity, a free face, a straight slope of almost constant gradient, and a basal concavity. Not all of these features are found in every slope profile, and some may be repeated more than once within a profile.

The amount of the hillside that is convex in profile is usually expressed as a proportion of the total slope length, and most slopes are mainly convex. This proportion of convexity results from the processes acting on the slope, and the effects of the river at the base of the slope; rapid undercutting results in steeper and more convex slopes. With *aggrading streams, more of the slope profile will be concave.

Hillslopes evolve over time in response to the redistribution of their surface sediments, usually with some net removal of material to rivers or the ocean. Where there is a plentiful supply of material, and the removal processes can only move a limited amount for a short distance, the rate of transport is limited by the transporting capacity of the process, which is defined as the maximum amount of material that the process can carry. Hence **transport-limited hillslopes**, where the rate of erosion/deposition is related to the divergence of shear stress, power, or velocity; see Pelletier (2012) *ESPL* 37, 1, 37.

hill wave *See* LEE WAVE.

hinge line A line either side of which *isostatic readjustment is uneven; the line separating uplift from subsidence (Sella et al. (2007) *Geophys. Res. Letts.* 34, L02306). In glaciology, it is the point where a glacier, reaching the coast, starts to float (Debenham (1965) *Geo. J.* 131, 3).

hinterland The area serving, and being served by, a settlement. Kaushik (2013) *Int. J. Sci. Engineering & Res.* (*IJSER*) 1 3 cites the example of Agra (India) and its fertile hinterland. An

information hinterland is the region for which a city provides the best access to information flows (Wang et al. (2007) *Tijdschrift* 98, 1). See Baldacchino (2006) *Asia Pac. Viewpt* 47, 1.

historical geographical materialism Essentially *Marxist geography. See Cox (2013) *Dialogues Hum. Geog.* 3, 1.

historical geography The study of past human geographies; of past landscapes; 'geographic perspectives on the past' (*Historical Geography* website). 'The study of the organization of space in the ... context of time' (X. de Planhol 1994). 'The relationship between humans and their environment in particular socio-cultural, spatio-temporal contexts' (U. Fellmeth et al., eds 2007). 'Historical geography may be viewed as being concerned with the historical dimension in geography, and geographical history with the geographical dimension in history' (A. Baker 2003). Gregory and Healey (2007) *PHG* 31, 5 examine the creation and dissemination of historical GIS databases, and the use of GIS to perform quantitative and qualitative analyses. Livingstone (2000) *Australian Geog. Studs* 38, 1 writes that 'the '"history of geography" might profitably be reconceptualized as "the historical geography of geography"'.

For 'progress reports' on historical geography, see Naylor (2005) *PHG* 29, 5; Naylor (2006) *PHG* 30, 6; and Naylor (2008) *PHG* 32, 2. See Offen (2012) *PHG* 36, 4, 527 for useful summaries of the concerns of historical geography.

historical materialism The general philosophical position of *materialism* begins from the fundamental assertion that matter is the primary substance of all living and non-living things. See R. Williams (1983). *Historical* materialism is differentiated from broader materialism by its insistence on humans'

Hjulström diagram

self-production of reality, and of the worlds humans inhabit; see Mitchell (2004) in J. S. Duncan, N. C. Johnson, and R. H. Schein. It is an analysis of history, most closely associated with Marx, which points out that economic systems underlie the development of history and ideas. See D. Harvey (1996).

historicism A concept that holds that the goal of social and historical inquiry is the formulation and substantiation of general laws of historical development (Bennett in R. Kitchin and N. Thrift, eds 2009). J. M. Blaut (1993), for example, argues that different parts of the world are ranked as being at different stages along a general path of economic development. This view has been vigorously countered; see Lowe and Barnett (2000) *Env. & Plan. D* 18, 1, 53 and E. W. Soja (1989). Rowe (2010) *Western Political Science Association*, has a different view of historicism, seeing it as a form of nostalgia, prominent in postmodernist urbanism.

historiography The study and writing of history; see Trimble (2008) *PPG* 32, 1.

histosol *See* US SOIL CLASSIFICATION.

Hjulström diagram A diagram showing the relationship in a channel between particle size and the mean fluid velocity required for *entrainment. It shows that an entrained particle can be transported in suspension at a lower velocity than that required to lift the particle initially. When the stream velocity slows to a critical speed, the particle is deposited. Note that higher velocities are needed for the entrainment of clay-sized particles because of the electrostatic forces which bind them together.

holism The view that everything is linked to everything else, and should be acted upon as such, and that the whole is more than its parts. In earlier geographies, for example, the *region has been seen as having a distinct identity which does not come entirely from its separate parts. See Antrop (1998) *Landsc. & Urb. Plan.* 41, 3.

hollowing out The reduction of political power; for example, of a member state on joining a supranational body, such as the *European Union; of a local authority by national government; or of a state by global organizations such as the *World Bank or the *International Monetary Fund. The term may also apply to the loss of local government power when national government takes more control (Skelcher (2000) *Pub. Policy & Admin.* 15, 3), but Radcliffe, Laurie, and Andolina (2002) *Transnat. Communities Prog. Seminar* claim that the state is not hollowed out in this process.

Holocene The most recent geological *epoch, stretching from 11 000 years ago to the present day. The Holocene may be divided into:

Early Holocene	Preboreal	$10,300 \pm 200$ years BP
	Boreal	9,000—7,000 years BP
	Atlantic	7,000–5, 000 years BP
Mid-Holocene	Sub-Boreal	5,000–2,5 00 ka) years BP
Late Holocene	Sub-Atlantic	2,500 years BP—present

See K. Alverson (2004) on global change in the Holocene.

During the early Holocene, atmospheric methane (CH_4) concentrations gradually decreased and then reversed direction around 5000 years ago. This latter is the late **Holocene methane anomaly**; see Singarayer et al. (2011) *Nature*, 470, 82 for the orbital explanation, and Zhou (2012) *Holocene* 22, 731, who ascribes it to the expansion of early rice paddy fields.

homelessness, geography of

Homelessness is commonly perceived as an urban problem. Lee and Price-Spratlen (2004) *City & Commy* 3, 1 observe that, in the USA, the homeless, and services for the urban homeless, are mostly located relatively close together, usually in more deprived central city areas, and almost forming a service ghetto. Fagan (2003, *San Francisco Chronicle*) identifies Main Street as the centre of welfare and medical services in San Francisco. For the homeless, such a concentration renders services more accessible, but has a number of disadvantages: 'For many, simply getting to the local night shelter is a frightening experience [and this fear has a] gender dimension, with women especially tending to prefer to use services located away from the main "service hub"' (homeless-research.org. uk). Informal areas—'interstitial niches'—develop in streets, parks, and public car parks, where people find places to set up 'home', sleep, eat, make friends, and look for security, money, and entertainment. These niches are colonized by the homeless at particular times: 'at dusk, the street families lay out boxes, build plastic houses, warm themselves, and cook on small fires. The children rarely have anything. Public space is *their* space only at night' (Morelle and Fournet-Guérin (2006) *Cybergeo, politique, culture, représentations* 342). See also P. Cloke, J. May, and S. Johnsen (2010), which is both informative and well-reviewed.

Differentiation is not merely temporal; it is also social. Homeless Research record a division between 'straightheads', 'smackheads', and 'pissheads': 'what for one homeless person might be experienced as a space of care, might for another become a space of fear because of the dominant presence of a particular homeless sub-group . . . thus embodied and performative geographies can become underpinned by specific spatial logic.' See Cloke, Milbourne, and Widdowfield (2003) *Geoforum* 34, 1 on rural homelessness.

(((●))) SEE WEB LINKS

• Research into homelessness.

homeorhesis While *homeostasis is the maintenance of a steady state, homeorhesis is the preservation of a steady flow. This flow is maintained by positive feedback, thus maintaining the flow by self-adjusting mechanisms.

homeostasis The maintenance of a steady state regulated by self-adjusting mechanisms, under ever-changing conditions. See Lau and Lane (2001) *PPG* 25, 2.

homocline One of a regular series of hills made from a large area of rock strata of uniform thickness and dip.

homogeneous nucleation The spontaneous freezing of water droplets at around −40 °C, as clusters of water molecules within a droplet settle, by chance, into lattice formation of ice, making the entire droplet freeze. See Easter and Peters (1994) *J. Appl. Met.* 33, 7.

homonormativity 'A politics that does not contest dominant *heteronormative assumptions and institutions, but upholds and sustains them, while promising the possibility of a demobilized gay constituency and a privatized, depoliticized gay culture anchored in domesticity and consumption' (L. Duggan 2003). 'In other words, many gays and lesbians want nothing more than to be considered normal so that they might go on with their day-to-day lives as part of the status quo' (Oswin (2008) *PHG* 32, 1). See Brown, Browne, and Lim in V. J. Del Casino Jr et al. (2011), p. 299.

honeycomb weathering The removal of material from sandstone while the cement is preserved, leaving a lacy network; see Rodriguez-Navarro et al. (1999) *GSA Bull.* 111, 8.

honey-pot A location, such as Shakespeare's birthplace, which is particularly appealing to tourists. Planners in National Parks often develop honey-pot sites with car parks, shops, cafés, picnic sites, and toilets, so that other parts of the Park will remain unspoilt; see Pickering and Buckley (2003) *Mt. Res. & Dev.* 23, 3 on Mt Kosciuszko, Australia.

'The tourism hotspots Bondi, Byron Bay, Coogee and Manly are being buried under an avalanche of visitors . . . Councils in these honeypot areas are picking up the tab for everything from dumped cars, and closing illegal backpacker hostels, to increased rubbish collection and vandalism. They complain that although tourism may boost the national accounts, it does nothing for the bottom line of council budgets' (Kennedy (2002), cited by Allon (2004) *Space & Cult.* 7, 1).

horizon (soil horizon) A distinctive layer within a soil which differs chemically or physically from the layers below or above. The **A horizon** or topsoil contains *humus. Often soil minerals are washed downwards from this layer. This material then tends to accumulate in the **B horizon** or subsoil. The **C horizon** is the unconsolidated rock below the soil. These three basic horizons may be further subdivided. Thus, **Ah horizons** are found under uncultivated land, **Ahp horizons** are under cultivated land, and **Apg horizons** are on *gleyed land. The B horizons are also subdivided by means of suffixes: **Bf horizons** have a thin iron pan, **Bg horizons** are gleyed, **Bh horizons** have humic accumulations, **Box horizons** have a residual accumulation of sesquioxides, and **Bs horizons** are areas of sesquioxide accumulation. **Bt horizons** contain clay minerals and **Bw horizons** do not qualify as any of the above. **Bx horizons** or fragipans contain a dense but brittle layer caused by compaction. C horizons are also subdivided: **Cu horizons** show little evidence of gleying, salt accumulation, or fragipan; **Cr horizons** are too dense for root penetration; and **Cg horizons** are gleyed.

Additional suffixes may be used. Some soil scientists use the term **D horizon** for the consolidated parent rock.

In addition to these soil horizons, other layers are distinguished. Thus, the layer of plant material on the soil surface is classified as: the **L horizon** (fresh litter); the **F horizon** (decomposing litter); the **H horizon** (well decomposed litter); and the **O horizon** (peaty). A *leached A horizon is termed an **E horizon** or *eluviated horizon.

horst A block of high ground which stands out because it is flanked by normal faults on each side. See Walsh and Schultz-Ela (2003) *GSA Bull.* 115, 3. It may be that the block has been elevated or that the land on either side of the horst has sunk.

Hortonian overland flow An overland flow of water occurring more or less simultaneously over a *drainage basin when rainfall intensity exceeds the rate of water infiltration in the soil (Horton (1933) *Trans. Am. Geophys. Union* 14; Kirkby (1988) *J. Hydrol.* 100). See also Vigiak et al. (2006) *Geomorph.* 76, 1–2.

hot desert Located on the west coasts of tropical and subtropical climes, these have average temperatures over 20 °C and annual rainfall under 250 mm. Deserts are too dry for most plant species except for *xerophytes. Desert insects, reptiles, mammals, and birds are all adapted to drought. See M. Evenari et al., eds (1985). See also Nash (2000) *PPG* 24, 3 on geomorphological processes in arid areas.

Hotelling model A model of the effect of competition on locational decisions, indicating that locational decisions are not made independently but are influenced by the actions of others (Hotelling (1929) *Econ. J.* 39, 1). Kim (2007) *MPRA Paper* 2742, U. Munich, develops the Hotelling model.

hot spot An area of localized swelling and cracking of the Earth's crust caused by upwelling *magma; deep-seated thermal anomalies fixed with respect to the surrounding mantle (Morgan (1972) *GSA Memoir* 132). Plumes from the lower mantle rise to the surface, forming volcanoes. As plates move over the Earth's surface, the volcanoes delineate and record the history of plate motion, and the radiometric age of seamounts should show a simple progression away from the present-day hot spot locations. See Wessel and Kroenke (1998) *Earth & Planet Sci. Letts* 158, 1–2 on the relationship between a hot spot and its seamounts.

hot towers Immensely tall, tropical *cumulo-nimbus, at the rising branch of a *Hadley cell, that siphons sensible and latent heat upwards, to the equatorial *tropopause. See Montgomery et al. (2006) *J. Atmos. Scis* 63, 1 on vertical hot towers and cyclogenesis.

hub-and-spoke network Hubs are special *nodes that are part of a network linked by spokes, and located in such a way as to facilitate connectivity between interacting places. See O'Kelly (1998) *J. Transp. Geog.* 6, 3, 171. Song and Ma (2006) *Chinese Geogr. Sci.* 16, 3, 211 show how commercial and operational freedoms have led most of the larger airlines to develop hub-and-spoke networks.

hukou A system established in China in 1951 to regulate the movement of people within China, enabling the state to decide where people should live and work, and thus acting as a type of domestic passport system to living and working in a named location. See Ho (2011) *Citizenship Studs* 15, 6-7, 643, and Bosker et al. (2012) *J. Urb. Econs* 72, 2–3, 252, on relaxing hukou.

1.

First pattern of market share

2.

Second pattern of market share

3.

Final pattern of market share

Hotelling model. Two vendors, A and B, are positioned in the centre of their own half of the beach. If A moves closer to the centre, he increases his market share, which leads B to also move to the centre, ensuring that market share is equal again.

human capital Talent (Florida (2002) *AAAG* 94, 2—who finds that the economic geography of talent is strongly associated with high-technology industry location). Lopez Rodriguez et al. (2007) *Reg. Studs* 41, 2 observes that, in the European Union, educational attainment levels are higher in those regions with greater market access. 'The presence and concentration of *bohemians in an area creates an environment or milieu that attracts other types of talented or high human capital individuals (Florida (2002) *J. Econ. Geog.* 2, 1). Liu (2013) *J. Econ. Geog.* (DOI: 10.1093/jeg/lbt024) finds a positive and statistically significant relationship between firm productivity and city-level **human capital externalities** (that is, a workforce whose skills have been provided by bodies external to the firm). **Human capital spillovers** occur when an individual's productivity is increased by interacting with and learning from high-skilled workers (Winters (2013) *J. Econ. Geog.* 13, 5, 799, but see Lui on this.

Human Development Index (HDI)

A compound indicator of *economic development, established by the United Nations, which attempts to get away from purely monetary measurements by combining per capita GNP with *life expectancy and literacy in a weighted average. 'Human Development is a development paradigm that is about much more than the rise or fall of national incomes. It is about creating an environment in which people can develop their full potential and lead productive, creative lives in accord with their needs and interests. People are the real wealth of nations. Development is thus about expanding the choices people have to lead lives that they value' (UNDP).

(🌐) **SEE WEB LINKS**
• The Human Development Index database.

human ecology An approach whereby the human social system and the ecosystem of the planet are seen as a mesh of reinforcing connections, each influencing and being influenced by the others. See G. G. Marten (2001).

human geography The study of the patterns and dynamics of human activity on the landscape. Broadly conceived, human geography includes both those areas of geographic study that are more narrowly focused on human activity, and those that are particularly concerned with human-environment dynamics, or the nature–society tradition (W. Moseley et al., 2007). 'That part of the discipline of geography concerned with the spatial differentiation and organisation of human activity and its interrelationships with the physical environment' (R. J. Johnston et al., 2000). 'More than just analyzing and taking note of where people and their possessions are, however, geography studies the interactions and social habits of humans *within* and *across* all spaces. Basically, geographers critically examine how humans organize and identify themselves in space—how we create "places". Thus, the science of human geography does not look at location per se, as much as it does at the mobility of, access to, and barriers against human processes' (Oas (no date) *Implications* 1, 7). Warf (2004) *Prof. Geogr.* 56, 1 comments that 'human geography has changed so much and so quickly over the last decade that any prediction of its status ten years hence is undoubtedly doomed to failure'. See M. Phillips, ed. (2005).

human influence index An index of human influence on natural environments: based on population density, land transformation (leading to loss and fragmentation of habitats), human access, and power infrastructure. See Sanderson et al. (2002) *Bioscience* 52, 10.

humanism A philosophy that places human welfare and happiness at the centre of ethical decision-making; a 'people-centred philosophy concerned with meanings, values, and understanding our taken-for-granted experiences and "being in the world"' (S. Aitken and G. Valentine, eds 2009). **Humanistic geography** is geography centred on human perception, capability, creativity, experience, and values; 'space and time not only have a form, but . . . they also have a meaning' (Ley (2003) *PHG* 27, 5). It maintains that any investigation will be subjective, reflecting the attitudes and perceptions of the researcher, who may also be an influence on the very field of his study; see Y.-F. Tuan (1974). 'Humanistic geography represents a duality of reason and feeling, science and ethics. As a scientific approach, it is concerned with uncovering the truth regarding people's experiential relationship with place. It does not regard the phenomena under consideration as merely an object of research, rather it bears an ethical message of concern for those objects, be they human beings, nature or place' (Hasson (1984) *Prof. Geogr.* 36, 1).

Human Suffering Index A compound indicator of distress, compiled by 'adding together ten measures of human welfare related to economics, demography, health, and governance: income, inflation, demand for new jobs, urban population pressures, infant mortality, nutrition, access to clean water, energy use, adult literacy, and personal freedom' (Population Crisis Committee, Washington, DC).

humic acid A complex acid formed when water passes slowly through humus. Humic acid is an example of an organic acid in that it is formed from carbon-based compounds. It is significant in chemical weathering and in the formation of soil. See Yadav and Malenson (2007) *PPG* 31, 2. *See also* MODER; MOR; MULL.

Human Suffering Index

- ☐ Data Not Available
- ☐ Minimal
- ☐ Moderate
- ☐ High
- ■ Extreme

Human Suffering Index

humidity The amount of water vapour in the atmosphere. It is more exactly defined as the mass of water vapour per unit volume of air, usually expressed in kg m^{-3}. This is **absolute humidity**. **Relative humidity** is the moisture content of air expressed as the percentage of the maximum possible moisture content of that air at the same temperature and pressure. The **humidity mixing ratio** is the ratio of the mass of water vapour in a sample of air to the mass of dry air associated with that water vapour.

humilis *See* CLOUD CLASSIFICATION.

humus A naturally occurring complex combination of substances resultant of the extensive biochemical breakdown of plant and animal residues (Yadav and Malenson (2007) *PPG* 31, 1). This breakdown is **humification**—the process whereby the simple mineral compounds released by weathering combine with the organic residues to form large, stable organic molecules which bond the soil; see J.-M. Gobat et al. (2004).

Humus can form relatively stable complexes and chelates with polyvalent cations, thus increasing assimilation by plants (D. Alexander and R. W. Fairbridge 1999).

hunger The December 2012 issue of the *Chicago Magazine* shows that hunger is not simply a function of hardship; while some deprived areas ranking at the bottom of the city's community area by almost every economic indicator are 'hungry areas', the wealthiest communities by median income are not the ones with the lowest percentage of food-insecure residents.

(((()))) **SEE WEB LINKS**
• *Chicago Magazine* on hunger.

Hunger Vulnerability Index (HVI) The HVI is a composite of three variables: the percentage by which food availability falls short of 2300 kcals per capita per day in a country; the percentage of the national population that earns less than US $2 per day (measured in 2005 purchasing power

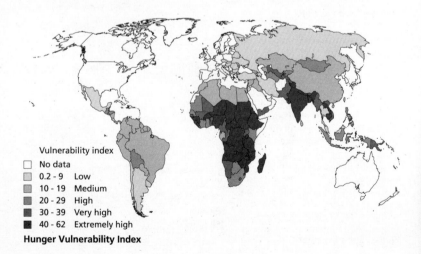

Vulnerability index
☐ No data
0.2 - 9 Low
10 - 19 Medium
20 - 29 High
30 - 39 Very high
40 - 62 Extremely high

Hunger Vulnerability Index

parity [PPP] dollars or 'international dollars'); and the percentage of children below the age of five in a country who are of short stature or suffering from growth failure. These three components reflect food availability within a country, average ability to access food, and a nutritional outcome. The variables are also meant to capture dimensions of vulnerability at three different scales: national (in terms of food supply), household (in terms of income), and individual (in terms of nutritional outcome). The three elements (all percentages below a threshold deemed acceptable) are then averaged to form an HVI score.

hunting and gathering A form of society with no settled agriculture, or domestication of animals, and which has little impact on the environment. The hunting of animals and the collection of edible plants depend on adapting to the environment, rather than changing it. See T. Ingold (2000). Try R. K. Guatam (2011) on the Baigas hunter-gatherers of Central India.

hurricane *See* TROPICAL CYCLONE.

hybrid capitalism A form of capitalism that combines elements of culturally specific imprints—such as employment relations—and globalizing norms—such as international standards of corporate governance (H. W. Yeung (2006) *Econ. Geog.*). The concept of hybrid capitalism offers a way of combining the focus of the global convergence literature on systemic 'drivers' of change with the emphasis on domestic institutional logic found in the 'varieties-of-capitalism' approach (Dyson and Padgett (2005) Anglo-German Foundation). H. W. Yeung (2003) argues that Chinese capitalism will mutate to a form of hybrid capitalism.

hybrid geography Rose (in R. Johnston et al., eds 2000) writes that hybrids 'transgress and displace boundaries between binary divisions and in so doing produce something ontologically new', a notion picked up by Kwan (2004, *AAAG* 94, 4) who recognizes two major divisions within geography: the partition between physical and human geography—of nature and society; and the separation of spatial-analytical geographies which attempt to create 'a mode of disembodied geographical analysis' from social, cultural, and political geographies. Kwan suggests that geographies based on thematic networks would ameliorate 'the polarizing tendency within the discipline'.

S. Whatmore (2002) agrees, arguing that 'it is both more interesting and more pressing to engage in a politics of hybridity . . . in which the stakes are thoroughly and promiscuously distributed through the messy attachments, skills and intensities of differently embodied lives whose everyday conduct exceeds and perverts the designs of parliament, corporations and labour'. Sui and DeLyser *PHG* 36, 1, 111 provide a progress report on this topic.

hybridity The mixture of meanings that emerges when two cultures interact; the new forms that are created when cultures merge. 'Acculturation can be seen as a sort of dialogue between cultures that . . . engenders the creation of new forms of cultural negotiation which themselves materialize as . . . hybridity, creoleness and inbetweeness' (Mitchell (1997) *Antipode* 29, 2). Identities are not stable, but ambivalent. H. Bhabha (1994) claims that subjects at the margins of cultures, or in a *third space, are best placed to resist a domineering culture/narrative, and to create new hybrid forms.

hydration The incorporation of water by minerals. Hydration is believed to be a major force in weathering coarse-grained igneous rocks; see Hughes (1998) *Geoarchaeol.* 3, 4.

hydraulic action (hydraulic force) The force of water, including *cavitation and fluvial *plucking.

hydraulic conductivity The ability of a soil or rock to conduct water. The conductivity of dry soil or rock is low, as water must form a film of water around the soil particles before conduction occurs. Tuli and Hopmans (2004) *Eur. J. Soil Sci.* 55, 1 find that pore geometry and pore-size distribution determine the relations between saturation, hydraulic conductivity, and air conductivity. **Saturated hydraulic conductivity** refers to the maximum rate of water movement in a soil; see Lilly (2000) *Soil Use Manag.* 16.

hydraulic geometry The study of the interrelationships along the course of a river. *Discharge is linked to the mean channel width, the mean channel depth and slope, the suspended *load, and the mean water velocity, and the wavelength of the meander is related to the radius of curvature; see Singh and Zhang (2007) *Hydraul. Procs* 22, 2. See also Smith and Pavelsky (2008) *Water Resource Res.* 44, W03427 on the estimation of hydraulic geometry from space.

hydraulic gradient The rate of change in *hydraulic head with distance. Within a drainage network, topography and valley morphology will accentuate or diffuse the local distribution of flow energy through changes in the hydraulic gradient and constrictions or widening of the river channel (Vandenberghe and Woo (2002) *PPG* 26, 4).

hydraulic head The pressure exerted by the weight of water above a given point.

hydraulic radius Also known as **hydraulic mean**, this is the ratio of the cross-sectional area of a stream to the length of the *wetted perimeter (cross-sectional length of a river bed). The hydraulic radius is a measure of the efficiency of the river in conveying water—if the hydraulic radius is large, there is little friction.

hydraulic resistance A measure of the resistance offered by a material, such as vegetation, to a flow of water. The calculation of flow velocities within stream channels and over hillslopes cannot be made without accounting for the hydraulic resistance experienced by flow over different morphologies (Smith et al. (2007) *PPG* 31, 4). Moore and Burch characterize hydraulic resistance by choosing an appropriate value of the *Manning roughness coefficient, and it can also be characterized using the Darcy–Weisbach friction factor, expressed as a function of the Reynolds number of the flow; see Prosser and Rustomji (2000) *PPG* 24, 2 for the equations. Fisher et al. (2007) *Glob. Change Biol.* 13, 11 recommend the introduction of hydraulic resistance to likely flow paths in the management of forests.

hydraulics The study of the mechanical properties of liquids. Smith et al. (2007) *PPG* 31, 4 summarize the early development of hydraulics. See Boyer et al. (2006) *J. Geophys. Res.* 111, F04007 on flow turbulence and bed morphology, and Wilcox and Wohl (2007) *Geomorph.* 83, 3–4 on three-dimensional hydraulics.

hydraulic sorting The process by which particles of river bed material are sorted into sections of near-uniform particle size due to the change in river *competence throughout its journey from source to mouth; see Attal and Lave (2006) *Spec. Pap. Geol. Soc. Am.* 398, 143. Hydraulic sorting also occurs on beaches; try Hughes et al. (2000) *J. Sed. Res.* 70, 5, 994.

hydric soils Soils that formed under conditions of saturation, flooding, or ponding long enough during the growing season to develop anaerobic conditions in the upper part. Under natural conditions, these soils are either saturated or inundated long enough during the growing season to support the growth and reproduction of hydrophytic vegetation. **Hydric erosion** is the breakdown and transport of soil particles by raindrop erosion and superficial draining. The triggering factor is precipitation—both intensity and duration. However, the relationships between rainfall characteristics, infiltration, drainage, tillage, and soil loss are very complex; see Ballais et al. (2012) *Zeitschr. Geom.*

hydroclimatology The study of the effects of the climate system on the hydrologic cycle over time and space, both globally and locally. Changes in the relationship between the climate system and the hydrologic cycle underlie floods, drought, and possible future influences of global warming on water resources. See M. L. Shelton (2009). Simultaneous variations in weather and climate over widely separated regions are commonly known as **hydroclimatic teleconnections** (Kashida et al. (2010) *J. Hydrol.* 395, 23).

hydro ecology *See* ECOHYDROLOGY.

hydrograph A graph of *discharge, or of the level of water in a river throughout a period of time. The latter, known as a **stage hydrograph**, can be converted into a discharge hydrograph by the use of a stage-discharge rating curve. Hydrographs can be plotted for hours, days, or even months. A storm hydrograph is plotted after a rainstorm to record the effect on the river of the storm event.

hydrography The measurement, description, and depiction of the nature and form of water's topographic features.

hydrological connectivity The transport via watercourses of material, energy, and/or organisms. Hydrological connectivity refers to the passage of water from one part of the landscape to another, and is expected to generate

Hydrograph

some response in the catchment runoff (Bracken and Croke (2007) *Hydrol. Process.* 21, 1749). See Gooseff (2011) *Geog. Compass*, DOI: 10.1111/j.1749-8198.2011.00445.x, for a worked example.

hydrological cycle (water cycle) This is the movement of water and its transformation between the gaseous (vapour), liquid, and solid forms. The major processes are *condensation by which *precipitation is formed, movement and storage of water overland or underground, *evaporation, and the horizontal transport of moisture. The length of time any water stays in the *atmosphere is about eleven days.

hydrological pathways The routes that water takes through the catchment from precipitation inputs to river channel outputs.

hydrology The study of the Earth's water, particularly of water on and under the ground before it reaches the ocean or before it evaporates into the air. This science has many important applications such as flood control, irrigation, domestic and industrial water supply, and the generation of hydroelectric power.

hydrolysis The chemical reaction of a compound with water. Hydrolysis is an important component of soil formation, and of *chemical weathering—for example, as feldspars in granite decompose to make china clay.

hydromorphic Denoting areas with waterlogged soils. *Gley soils form in such conditions.

hydromorphology A subfield of *hydrology that deals with the structure

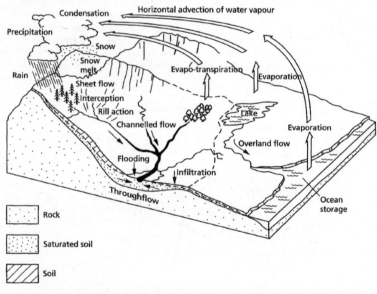

Hydrological cycle

and evolution of Earth's water resources, and the origins and dynamic morphology of those water resources.

hydrosere A successional sequence of plants originating in water.

hydrosphere All the water on, or close to, the surface of the Earth. Some 97% of this water is in the Earth's seas and oceans; of the rest, about 75% is in *ice-caps and sheets, about 25% in surface *drainage and *groundwater, and about 0.03% in the atmosphere.

hydrothermal vent A fissure on the sea floor, usually found in volcanically active areas of the ocean floor. Water that has been heated up to 400°C by underlying magma flows from these vents. This water usually contains dissolved minerals that precipitate out of the vent on contact with the colder ocean water, creating a chimney-like structure. See J. Ishibashi, K. Okino, and M. Sunamura, eds (2015) on the links between sub-sea-floor biosphere and hydrothermal systems.

hygroscopic nuclei *Condensation nuclei which are hygroscopic, i.e. which tend to attract and condense ambient water vapour.

hyperconcentrated flow An intermediate condition between debris flow and normal streamflow (Katoka and Natajo (2002) *Sedimentol.* 49, 2); a liquid–solid two-phase flow, in which the liquid phase is a syrup made of water and fine sediment less than 0.01 mm in diameter, and the solid-phase consists of coarse particles larger than 0.05 mm suspended in the syrup (Chien, ed. (1989), cited by Xu and Yan (2006) *Geomorph.* 73, 1–2).

hypermobility Highly mobile travelling. Khisty and Zeitler (2001) *Systemic Practice & Action Research* 14, 597 define hypermobility as the maximization of physical movement, while Gao et al. (2012) *Canad. Geogr./Géog. canad.* 56, 4, 439 are more specific, defining hypermobility as 'a rapid switching of economic links among manufacturers, importers, and retailers'. Adams (1999) OECD argues that in mall-scale pedestrian societies, everyone knows everyone. In hypermobile societies old-fashioned geographical communities are replaced by aspatial communities of interest 'we spend more of our time, physically, in the midst of strangers'.

hyperreality A state supposedly more real—more sensation-packed—than reality, which can seem flat and dull. A **hyperreal procedure** would not use observations taken from real things; rather, it would use photographic representations to construct a world. Baudrillard shows that immaterial representations of reality begin to blur with material reality especially in highly commodified societies that depend heavily on advertising and image manipulation, such as the USA. These images, simulations, and signifiers he calls hyperreality. D. DeLyser, ed. (2010), p. 233 is excellent on this.

hyporheic zone A spatially and temporally dynamic mixing zone at the interface between groundwater and surface water. Hydrologic exchange between streams and hyporheic zones is important in keeping surface water in close contact with chemically reactive mineral coatings and microbial colonies in the subsurface. For an exhaustive treatment of the topic, see Harvey and Wagner in J. Jones and P. Mulholland, eds (2000).

hypsography The study of landforms and processes in physical geography.

hysteresis The impact of its history on a physical system. The future development of the system depends on its history, for its progress is determined by the direction in which the reaction is occurring. It is normally described by a bivariate plot in which the value of one variable is dependent on whether the other variable is increasing or decreasing. For example, when plotting the relationships between solute concentrations and river discharge

during storm events, very different solute concentrations are measured at the same discharge on the rising and falling limbs of the storm hydrograph.

hythergraph A plot of monthly rainfall against monthly temperature over a year. *See* CLIMOGRAPH.

Hysteresis

ice When the water content of snow falls below about 8% ice grain growth occurs by vapour flux of the air in the pores. The resulting growth of ice grains is very slow. In general, growth in wet snow speeds up with increasing water content (D. McClung and P. Schaerer 1993).

ice, internal deformation The internal deformation of ice is achieved by *creep and large-scale faulting and folding. See Arenson et al. (2002) *Permafr. Periglac. Procs* 13, 117.

ice age A time when there are ice sheets on the continents, within which there may be *interglacial periods of milder climate. Ice ages last for some tens of millions of years, separated by intervals of *c*.150 million years. See Raymo and Huybers (2008) *Nature* 451, 284–5, DOI:10.1038/nature06589, on glaciation and Earth's orbit; Easterbrook (1963) *GSA Bull.* 75, 12 on Pleistocene glacial events and sea-level change; Gore et al. (2001) *Geol.* 29 on East Antarctica (ice free at the last glacial maximum); Bond et al. (1997) *Science* 278 on the Little Ice Age; and Barry (2006) *PPG* 30, 3 on glacier retreat since then. A. Goudie (2000) is strong on the stratigraphic information from ocean and ice cores, and the causes of climate change.

ice-albedo feedback *See* ALBEDO.

ice-cap A flattened, dome-shaped mass of ice, similar to an *ice sheet, but under 50 000 km^2 in area. See Colgan and Sharp (2008) *J. Glaciol.* 54, 184, 28 on oceanic and atmospheric influences on net accumulation on the Devon Ice Cap, Nunavut, Canada.

ice contact feature A landform formed in contact with a glacier. For **ice contact terrace**, see *kame terrace. See Hickson (2000) *Geomorph.* 32, 3–4 on **ice-contact volcanism**.

ice fall An area of *extending flow marked by crevasses, where the gradient of a glacier steepens. The rapid flow at this point causes the ice to thin. See P. Knight (1999).

ice floe A flat section of floating ice which is not attached to ground ice. See Meylan et al. (1996) *J. Geophys. Res.* 101.

ice flow A glacier flows because the ice within it deforms in response to gravity. The major controls on ice flow are temperature and *shear stress, which is dependent on the ice thickness and the surface slope of the glacier. This can be summarized as:

$$\tau = \rho g (s - z) \sin \alpha$$

where τ is the shear stress at a point within the glacier, ρ is the density of ice, g is the acceleration due to gravity, α is the surface slope of the glacier, s is the surface elevation, and z is the elevation of a point within the glacier.

The basal shear stress is given by:

$$\tau = \rho g h \sin \alpha$$

where τ is shear stress, ρ is the density of ice, g is the acceleration due to gravity,

h is the thickness of the glacier, and α is the surface slope of the glacier. See M. Bennet and N. F. Glasser (2009).

ice front The floating ice cliff at the seaward end of an *ice shelf. See Frezzotti and Polizzi (2002) *Annal. Glaciol.* 34, 1 on ice fronts as indicators of climate change. An ice front responds rapidly to temperature rise: in the Jakobshavn glacier in Greenland a floating **ice tongue** collapsed and caused the calving ice front to retreat several kilometres (http://www.geog.cam.ac.uk/research/projects/greenlandicesheet/).

ice lens An underground ice segment, often with convex upper and lower surfaces. *See* PALSA.

ice-marginal channel A channel cut by meltwater, running parallel to the ice margin, and found most often in a glacier's *ablation area, where rates of surface ablation and meltwater production are highest. Parallel sets of ice-marginal channels may form when the ice margin retreats. See Wadham et al. (2001) *Hydr. Procs* 15, 12.

ice mound In a *periglacial landscape, a swelling in the ground due to the expansion of an underground *ice lens. *Solifluction displaces the material at the top of the mound, so that when the ice lens melts a depression forms, often surrounded by a rampart. *See* PINGO.

ice patch, A small stretch of ice in the mountains that may have been in place for thousands of years; a 'residual ice mass without motion but with melting mechanisms which imply adjustments and small movements'. González-Trueba et al. (2011) *Geografiska Series A* 93, 57.

ice polygon (ice wedge polygon) A three- to six-sided polygon of *ice wedges, formed by *ice segregation and

the drying and shrinking of sediments, and becoming more regular with age. See Matsuoka (1999) *Polar Geosci.* 12.

ice scar A 'healed' cut in a tree that has been damaged, for example, by the impact of an avalanche or a glacier. Using *dendrochronology, ice scars can be dated; see Lemay and Bégin (2012) *ESPL* 37, 13, 1353.

ice segregation The creation of discrete bodies of ground ice in *periglacial conditions. Segregation develops well in silts (which seem to have the optimal pore size), depends on a moderate freezing rate, and often occurs under stones, which have high thermal conductivity. See Hall (2002) *S. Afr. J. Sci.* 98 and Murton et al. (2006) *Science* 314.

ice sheet A continuous mass of glacier ice with an area over 50000 km^2; Typically an ice sheet is 1–3 km thick. Payne and Sugden (2006), *ESPL* 15, 7 find that both the vertical amplitude and the spatial distribution of bedrock basins and ridges are important in determining the pattern, rate, and extent of ice-sheet growth. Alley et al. (2005) *Science* 310, 5747 report that the Greenland and Antarctic ice sheets, if completely melted, could raise sea levels by about 70 m. The resulting freshwater fluxes may affect oceanic circulation. An **ice core** is a column of ice extracted by drilling into an ice sheet.

ice shelf A sheet of ice extending over the sea from a land base, fed by snow falling on it, or from glaciers on the land. Smith et al. (2007) *Antarc. Sci.* 19, 1 record the constant, but minor, retreat of George VI ice shelf, Antarctica, since the mid-1980s, concluding that while atmospheric warming has been critical in driving this retreat, other mechanisms and feedbacks, such as changes in sea ice extent or oceanography, may be involved.

ice stream Within a glacier or ice sheet, a stream of ice moving faster than, and not always in the same direction as, most of the ice. Ice streams respond to the weight of the ice, and to gravity. They are slowed by side drag from the slower-moving ice that flanks them, and by back pressure from the ice sheet itself (Stokes et al. (2007) *Earth Sci. Revs* 81, 3–4).

ice wedge In a *periglacial landscape, a near-vertical ice sheet, tens to hundreds of metres long, tapering into the ground to depths of 12 m. Individual wedges usually meet another wedge orthogonally. Matsuoka (1999) *Polar Geosci.* 12 records an expansion of 16 mm across an ice wedge at –17 °C. **Ice wedge casts** are sediment-filled, fossil ice wedges; see Harris et al. (2005) *Geomorph.* 71, 3–4 for the ice-wedge casting processes.

Ice-wedge networks form as ice fills the tension fractures resulting from cooling. They form in stable, aggrading or sloped surfaces, with winter temperatures from –15 °C to –35 °C. Modelled polygonal networks self-organize through interactions between fractures, stress, and re-fracture (Plug and Werner (2008) *Permaf. & Periglac. Procs* DOI: 10.1002/).

iconography 'A set of symbols in which people believe, encompassing elements of national feeling, from the state flag to the culture transmitted through the state's schools' (P. Taylor 1993). O'Sullivan (2001), *TIBG* 29, 3 illustrates the Christian symbolism expressed in late medieval European urban forms. See also Leib et al. (2000) *GeoJournal* 52. Ponder on this: 'if Geography is prose, maps are iconography' (Lennart Meri 2013).

identity The individuality of a being or entity. National identity may be based on myths; 'like cultural identities, diasporic national identities are actively constructed' (Mavroudi (2007) *Glob. Networks* 7, 4), but Nairn (2003) *New Left Rev.* 23 observes that 'for growing masses of people . . . issues of identity are not metaphorical but [are] treasured, if deplorable, bits of cheap plastic: everyday life and death'. Identity may emerge from the production and reproduction of space; see Saarinen (2004) *Tourist Studs* 4 on tourist destinations and their identity, and try M. Tewdwr-Jones and P. Allmendinger (2006). Dittmer and Larsen (2007) *Soc. & Cult. Geog.* 8, 5 write that national identity is constructed by the interpellation of subjects from competing discourses that all emphasize different aspects of the lived experience of the inhabitants. The use of the term 'interpellate' is questionable.

In physical geography, identity is established using 'methods of ontology based on meaningful collections of attributes' (Raper and Livingstone (2001) *TIBG* 26).

identity, geographies of The study of identities and space: 'we are in the midst of a redefinition of space. In the very moment that national and ethnic boundaries are breaking down we encounter paradoxical reinvestments in homeland, territorial integrity, localism, regionalism, and race- and ethnocentrism' (P. Yaeger 1996). Thus, much of the work revolves around the relationship between cosmopolitanism and transnationalism, with an emphasis on migratory movements and the formation of new kinds of transnational social spaces, identities, and relations. 'Consciousness is always in motion and forms of identity such as a subaltern cosmopolitanism reflect particular relationships in particular times and places. Groups and individuals perform their own sense of cosmopolitanism dependent on context—their own particular locations in various axes of power as well as the broader structuring forces of global capitalism and geopolitics' (Cheah in P. Cheah and B. Robbins, eds 1998). For many, forms of citizenship are 'flexible, multicultural and polycultural, and dissenting' (Mitchell (2007) *PHG* 31, 5).

'Self-perceived liberals often position themselves as fatigued defenders of an identity politics and a form of multiculturalism that has now gone too far. Identity politics is projected (on the left) as either undermining efforts for redistribution or impeding the struggle against a neolib' (Mitchell (2004) *PHG* 28, 5); see S. Huntington (2004). **Identity politics** are based around cultural identity, such as *disability *ethnicity, *gender, *race, or *sexuality.

ideology At its simplest, any ideology could be defined as a general view that informs people's actions. See D. Gregory (1978).

idiographic Of, or relating to, the study of anything unique. The idiographic approach has been the cornerstone of *regional geography. In biogeography, **idiographic analyses** are detailed analyses of single species range; see Fattorini (2007) *Divers. & Distrib.* 13, 6.

igneous rock A rock formed from molten *magma, deep beneath the Earth's surface, subsequently coming to the surface as an *extrusion, or remaining below ground as an *intrusion. The plutonic mode of igneous rock formation is characterized by 'intrusive emplacement under compressional tectonic stress regimes. The volcanic mode is dominated by the eruption of large volumes of magma, through fault controlled dyke structures in extensional tectonic stress regimes' (Neuberger and Reuther (1987) *Int. J. Earth. Scis* 76, 1). The nature of the rock also depends on

the rate at which it cooled; intrusions of magma cool slowly, allowing enough time for large crystals to form, while extrusions cool quickly, leaving little time for crystal growth; see R. Le Maitre (1989).

illuviation The *downward translocation from the A *horizon and subsequent precipitation in the B horizon of clay-sized particles in a soil. The **illuvial horizon** is the B horizon, in which there is redeposit ion or entrapment of matter brought down from above.

image A picture built up by an individual from the social and physical *milieux experienced from birth; see A. Scott's work (2002) on the production of images, symbols, and messages in manufacturing and services. 'Geography is a visual discipline, working with the comparative power of vision' (Matless (2003) *Antipode* 35, 222) and with a past reliant on visual aids such as maps, globes, models, slides and photographic illustrations. While geography has 'always entailed making and interpreting images' (D. Cosgrove, 2008), 'images continue to be thought as either reliable or deceptive'. See A. Buttimer et al., eds (1999) on texts and images as mediators in the production of spaces.

imagined community This term was first used (B. Anderson 2006) to describe a nation, for even in the smallest nation, people will never know, meet, or hear about all their fellow-countrymen, but most people will have in their minds an image of their nation as a community. Thus, A. Latham et al. (2009) describe an imagined community as 'a group of people, united in the sense of their community, this unity being aided by newspapers, magazines, poster hoardings, cinema, radio, and television'.

imagined geographies (imaginative geographies) The ways that other places, people, and landscapes are represented; how these imaginings reflect the preconceptions and desires of their inventors, and the relations of power between subjects and objects thereof (E. Said 1979). Thus, with independence, Singapore's ruling party sought to transform 'a group of people with divergent orientations ... into a populace of loyal citizens' (L. Kong and B. Yeoh 2003). D. Gregory (2004) argues that, since 9/11, the United States has re-imaged the world into 'us' and 'them', and 'light' against 'darkness': '*orientalism [is] abroad again'. Jeffrey (2007) *Area* 38, 2 argues that viewing the 1992–5 conflict in Bosnia-Herzegovina as a humanitarian, rather than a political, crisis, justified military and NGO intervention. See Schwartz (1996) *J. Hist. Geog.* 22, 1 on imaginative geographies of the 19th century.

imagined places 'Imagined places are not fairytale places; they are not just fantasy. In one way or another, imagined places have a connection with a place that exists geographically' (E. van Alphen 2005). 'Imagined' is used not to mean 'false' or 'made-up', but 'perceived'. 'It refers to the perception of space created through certain images, texts, or discourse' (Kassim Al-Mahfedi, (2011) *The Criterion*). P. Shurmer-Smith and K. Hannam (1994) argue that images are not the reality but are our only access to reality. Imagined places are constructed by their representations in art, film, literature, maps, and media; see Gorman-Murray (2007, *MC Magazine* 9, 3) on Sydney's King Street, imagined as a gay/lesbian place by the gay/lesbian media.

imbricated Of deposits, laid down in overlapping sheets (like roof tiles), as on steep slopes in *periglacial environments (Millar (2006) *Catena* 67, 1).

IMF *See* INTERNATIONAL MONETARY FUND.

immigration The movement into another country of someone as a permanent resident. Governments may try to restrict immigration by using quotas, discounting family ties, or being reluctant to accept asylum seekers or refugees (Kofman (2002) *Pol. Geog.* 21, 8). Skenderovic (2007) *Patterns Prej.* 41, 2 identifies the definition and exclusion of the Other as characteristic of radical-right politics, while Abu-Laban and Garner (2005) *Urb. Aff. Rev.* 40, 4 compare the Canadian view of immigration (as a national policy) with that of the USA (as individual choice). See Liu (2001) *Soc. & Cult. Geo.* 1, 2 on space, place, and scale, and the construction of racial-ethnic and immigrant identities.

imperialism 'Unequal economic, cultural, and territorial relationships based on domination and subordination' (Jones and Phillips (2005) *AAAG* 95, 1). Imperialism dominates the external trade of the subordinate nation (E. Arghiri 1972), taking raw materials from the 'colony' and in return selling it finished goods, thus depressing manufacturing industry in the 'colony'; see Zanias (2005) *J. Dev. Econ.* 78, 1. It can exist without the creation of formal colonies; Domosh (1999) *TIBG* 24, 4 suggests a multiplicity of imperialisms. The **imperialist gaze** saw subsistence landscapes as empty, despite the presence of indigenous people. Colonized spaces were 'stripped of preceding significations and then re-territorialized according to the convenience of the ... imperial administration' (D. Harvey (1989 and 2003), who sees the projection of US power on the world as a **'new imperialism'**).

import penetration The extent to which the country is dependent upon its imports, to reduce which governments may resort to *protection-ism (M. Kamiya 2002).

import substitution A strategy that encourages domestic manufacturing in order to reduce imports of manufac-tures, save foreign exchange, provide jobs, and lessen dependency. Puga and Venables (1999) *Econ. J.* 1109, 455 find that, while import substitution maintains diversity, trade liberalization generates higher levels of welfare. See Panagariya (2004) on problems with import substitu-tion in India, and Yu Zhou (2008) *World Devt* 36, 11, 2353 on export orientation and import substitution in China.

inceptisol *See* US SOIL CLASSIFICATION.

incidence matrix A matrix that shows the relationship between two classes of objects. Horner-Devine et al. (2007) *Ecology* 88, 6, 1345, who provide crucial help for the mathematically challenged, use presence–absence matrix in which each row represents a species or taxon, and each column represents a sampling site. Within each incidence matrix the presence of a species, taxon, or opera-tional taxonomic unit (OTU) at a specific site is denoted by a 1, and its absence is denoted by a 0. Williamson (1978) *J. Ecol.* 66, 911 is recommended.

incised meander A *meander formed when a *rejuvenated river cuts deeper into the original meander; see O. Catu-neanu (2006). An **intrenched meander** is an incised meander with a symmetri-cal cross-valley profile, an **ingrown meander** one with an asymmetrical profile (Rich (1914) *J. Geol.* 22).

inclusion 1. The action or state of in-cluding or of being included within a group or structure. The principle that binds together the present UK govern-ment's (and indeed Europe's) social policy is the reduction in the number of people and places experiencing '*social

exclusion' from mainstream societal activities and the promotion of 'social inclusion' into these activities (Social Exclusion Unit 2001), and the Sustainable Development Commission promotes social inclusion and environ-mental justice (SDC 2002). Stewart in P. Askonas and A. Stewart, eds (2000) thinks that an understanding of social inclusion involves 'the recognition and acceptance of the historically situated and therefore contingent character of prevailing institutional realizations of justice and concomitantly privileged conceptions of the good life obtaining at any point in time'. Gray in Askonas and Stewart (op. cit.) argues that the disrup-tion of solidaristic social formations (particularly states) as a consequence of economic globalization renders any meaningful form of social inclusion im-possible. Nonetheless, Mitchell (2011) *AAAG* 101, 2 shows the way Marseille's city politicians have worked directly with both the representatives of major ethnic and religious groups and with the major capitalist players in the city, including elite entrepreneurs and members of the local mafia, to promote inclusion, based on ethnic ties, trust, and reciprocity, in marked contrast to Paris. I. M. Young (2000) insists that demographic prac-tices depend on inclusion. See also Holt (2008) *PHG* 32, 2.

2. A solid, liquid, or gaseous foreign body enclosed in a mineral or rock. **Melt inclusions** are small parcels of melt(s)—most are less than 0.0001016 cm across—that are entrapped by crystals growing in the magmas and eventually forming igneous rocks.

indeterminacy The state of being vague and poorly defined. Indeterminism is a key idea in *chaos, catastrophe, and complexity theories. S. Lash, B. Szers-zynski, and B. Wynne (1996), writing on environmental issues, note that 'a recog-nition of the intrinsic indeterminacy of

Indifference curve

the future, and of all our roles in shaping that future, imposes on us all the duty to take responsibility for the future we are making'.

index of centrality Many centrality indices are based on shortest paths, measuring, for example, the average distance from other vertices (Brandes and Pich (2006) *Int. J. Bifur. & Chaos* 17, 7). Simmons and Jones (2003) *Prog. Plan.* 60, 1 use it as a measure of the external market served by a city, and Channell and Lomolino (2000) *Nature* 403 use it to distinguish between 'central species' and 'peripheral species'. See Beauchamp (1965) *Behav. Sci.* 10 for an 'improved index of centrality'.

index of circulation *See* ROSSBY WAVES.

index of segregation A measurement of the degree of residential segregation between two sub-groups inside a larger population. See Johnston et al. (2005) *Urb. Studs* 42, 7.

index of vitality (I_v) An index to indicate the growth potential of a population:

$$I_v = \frac{(\text{fertility rate} \times \% \text{ aged } 20-40)}{(\text{crude death rate} \times \text{old-age index})}$$

indifference curve In economics, a graph of the various levels of *utility achieved at different prices through buying two commodities or services, for example, a mortgage and commuting costs. At a given price, various combinations of the two would yield the same total utility, hence the term 'indifference'. The concept is used in the *Alonso model, and the monocentric city model; see Bertaud and Brückner (2005) *Reg. Sci. & Urb. Econ.* 35, 2.

indigenous Originating in and characteristic of a particular region or country. **Indigenous peoples** are peoples defined in international or national legislation as having a set of specific rights based on their historical ties to a particular territory, and their cultural or historical distinctiveness from other populations that are often politically dominant; K. S. Coates (2004). **Indigenous knowledge** describes those distinctive ideas, skills, and information held by indigenous peoples, usually contrasted with Western scientific knowledge. Briggs (2013) *Prog. Devt Studies* 13, 3, 231 notes that the promise of using indigenous knowledge in development strategies has not been fulfilled. Read his *Final Thoughts*, p. 39–40; well worth it.

indivisibility The minimum scale at which any technology can be used; almost all manufactured capital goods are indivisible. The indivisibility of a tractor makes it uneconomic for many small farmers in the developing world (V. Balasubramanian et al. 1998). It is crucial in finding optimum factory size (Stykolt and Eastman (1960) *Econ. J.* 70, 278), and in economies of scale (Elliethy (2006) *J. Int. Dev.* 4, 4). Edwards and Starr (1987) *Amer. Econ. Rev.* 77, 1 hold that **indivisible labour specialization** brings scale economies.

induced demand In *transport geography, the additional traffic resulting from additions to road capacity (Levinson (2008) *J. Econ. Geog.* 8, 1). Goodwin and Noland (2003) *Appl. Econ.* 35 find that building new roads really does create extra traffic. See also St John and Pouder (2006) *Growth & Change* 37, 2.

induration The hardening or cementing process that consolidates sediments to form sedimentary rocks. Sedimentary rocks usually become more and more indurated over time. See Hermann (2008) *J. Coastal Res.* 24, 6, 1357 on induration and sand dunes.

industrial cluster *See* CLUSTER.

industrial district *See* MARSHALLIAN INDUSTRIAL DISTRICT.

industrial diversification The deployment of investment and labour over a wider range of activities. Wei et al. (2007) *Econ. Geog.* 83, 4 show that diversification in China is led by local governments, and entrepreneurs embedded in thick local networks. R. Boschma and G. Capone (2014) analyse the process of industrial diversification in the *EU-27 and European Neighbourhood Policy countries in the period 1995–2010.

industrial geography The study of the spatial arrangement of industry; 'the study of production and production systems across all sectors' (Harriman in G. Gaile 2002). 'We are currently in another time of paradigm conflict, with those who focus on the sociology and consumption of production and favour different styles of "theorizing" even while there is a revival of interest in analytical, regional science type research' (Stafford (2003) *Industr. Geogr.*). 'Industrial geography has always been a bellwether subdiscipline reflecting methodological changes within economic geography' (T. Barnes and M. S. Gertler 1999). The 'central question of industrial geography as to why industrial activities grow/decline in particular places' (Watts cited in Stafford, op. cit.). 'Social networks . . . play an important role in determining who can successfully start a new venture in an industry, and where they can start it. Potential entrepreneurs need access to a variety of resources—including knowledge of the key technologies in the industry—to begin operations and to compete successfully; [and] social networks facilitate access to these resources' (Sorenson in D. Fornahl et al. 2003). McCann and Sheppard (2003) *Reg. Studs* 37, 6–7 argue for the redevelopment of analytical industrial location theory.

industrial inertia The survival of an industry in an area after its original locational factors are gone. It is not unusual in industry, as sunk costs are high, and, over time, enterprises tend to reproduce *external economies (Rota and Vanolo (2007) *EURA Conf. Paper*).

industrialization The development of manufacturing industry from a predominantly agrarian society. Characteristic features of industrialization include the application of scientific methods to solving problems; mechanization and a factory system; the division of labour; the increased geographical/social mobility of the labour force; and capital deepening (Atack et al. (2005) *J. Econ. Hist.* 58, 3). These are also features of *capitalism, and capitalism is not the same thing as industrialization.

'Industrialization is widely seen as the most important social and economic change of the last 200 years . . . with the power to modify or destroy any pre-existing social arrangement' (A. Janssens 1993). Industrialization encourages urbanization; for example, Chen (2002) *China Econ. Rev.* 13, 4 finds that those Chinese provinces and municipalities with higher per capita GDP are more urbanized. As industrialization continues, there is a sectoral shift in employment: the sequence is illustrated by the data for Shanghai.

Year	1950	1995	1999	2002
Primary industry	3%	4.3%	1.8%	1.6%
Secondary industry	76%	63.8%	48%	47.4%
Tertiary industry	21%	31.9%	50.2%	51%

Source: Zhang (2006) *Urban Affairs Review* 9, 42.

This change from manufacturing to commerce and services 'to a considerable extent, means [from] more [to] less stable jobs' (Aguilar (1997) *GeoJournal* 43, 4).

Industrialization is often seen as a solution to poverty; see Fukunishi et al. (2006) *UNIDO*. Yamagata (2006) *Cambodian Econ. Rev.* finds that the Cambodian garment industry offers scope for the poor to increase their wages substantially. But Blumer (1969/70) *Studs Compar. Int. Dev.* 5, 3 holds that early industrialization brings 'a series of new demands on life' (such as a search for

superior economic status) that lead to 'disorganization'.

industrial location policy The neoclassical view of the role of the state in economic activities is 'inherently suspicious of government interference in private sector decision making, and its justification for industrial location policy relies primarily on the promotion of mobility, both with respect to capital and labour. Behaviouralists are more particularly concerned with the effects of industrial incentives on capital mobility, and specifically whether or not industrial location incentives change the location preferences of entrepreneurs in favour of designated regions' (R. Hayter 2004). 'Industrial location policy is . . . the result of corporate policies to minimize union strength, to avoid taxes, and to exploit new markets' (T. J. Sugrue 2005). See Siaka (2011) *Int. Research J. Manag. & Bus. Studs* 1, 7, 197 on the pattern of industrialization under economic liberalization.

industrial location theory 'The general critical factors of industrial location are transportation, labour, raw materials, markets, industrial sites, utilities, government attitude, tax structure, climate, and community. In addition, for international location considerations, four general factors are identified: [the] political situation of foreign countries, global competition and survival, [and] government regulations [are critical]' (Badri (2007) *J. Bus. & Public Affairs* 1, 2—who also reviews industrial location theory). **Industrial organization**—the structure of an industrial unit—is a component of industrial location theory; see McCann and Sheppard (2003) *Reg. Studs* 37, 6–7.

industrial revolution Although there is some discussion about its timing, the industrial revolution is generally accepted as occurring in Britain in the late 18th and early 19th centuries. K. Pomeranz (2000) explains the industrial revolution though 'Europe's armed exploitation of other resource-rich regions, particularly in the Americas'.

industrial specialization The domination of a limited range of industries in a region. Sölvell et al. (2006) *DRUID Summer Conf.* analyse industrial specialization in terms of absolute number of employees, specialization, and dominance, and identify nineteen 'three-star' regional clusters in the ten new EU states. Svaleryd and Vlachos (2002) *J. Int. Econ.* 57 find that the financial sector has greater impact on industrial specialization among OECD countries than do differences in human and physical capital.

industry A sector of the economy in which firms use similar factor inputs to create a group of related products and/ or services. The extraction of natural resources, including agriculture, is **primary industry**; manufacturing is **secondary industry**; **tertiary industry** comprises services, including transportation and communication; **quaternary industry** deals with a range of producer services from banking to retailing to real estate; and **quinary industry** with consumer services such as education, health care, and government information. With economic development, industry moves from primary industry to progressively less material-oriented activities such as textiles and clothing, until, over time, tertiary industries dominate; the sequence is illustrated in the figure.

The relative dominance of the latter three categories over the former two in many industrialized countries has informed discussions of a *post-industrial society, in which Miozzo and Soete (2001) *Tech. Forecast. & Soc.* 67, 2 identify research and development, design, marketing, distribution, and after-sales maintenance as essential.

Phelps and Ozawa (2003) *PHG* 27, 5 propose four industrial phases.

Industry. Industrial change in Japan 1990–2002
Source: Ström, 2005, *Soc. Sci. Japan* J 8, 2

	Protoindustrial	Industrial	Late industrial	Post-industrial
Sectorial basis	Agriculture-manufacturing	Manufacturing	Manufacturing	Services-manufacturing
Division of labour	Intra-firm, intra- and inter-sectorial	Inter-firm and inter-sectorial	Intra- and inter-firm, intra-sectorial	Inter- and inter-firm, inter- and intra-sectorial
Source of accumulation	Exports →domestic	Domestic → exports	Domestic → exports	Exports → domestic

inequality The fact of **global inequality** is not disputed; the world is routinely divided into developed/developing, North/South, more/less/least economically developed, high/medium/low development, and so on. The richest 2% of adults in the world own more than half of global household wealth (J. B. Davies, S. Sandstrom, and A. Shorrocks 2006). There is no consensus about the causes of global inequality, not least because of the difficulties in measuring inequality: if we scale income by some index of health, there is more inequality in the world than if we consider income alone (Deaton (2006) *NBER W. Paper* W12735).

Many writers blame globalization: Ash (2004, *TIBG* 29, 2) argues that 'TNCs and international banks are deeply embedded in the production of inequality, through their influence over governments, their tax evasion schemes, their poor environmental record, their labour market practices, their locational mobility, their market power, their dumping practices.' Cornia and Court (2001, *Policy Brief 4* World Inst. Dev. Econ. Res.) contend that the liberalization of domestic banking associated with globalization has caused rises in income inequality; Rowson (2000) *healthmatters* 41 agrees.

There is also debate over the impact of *structural adjustment measures on inequality. In 2004, *SAPRIN claimed that structural adjustment polices are the largest single cause of inequality; at the other end of the spectrum, Ferreira (2004) *World Bank Policy Working Paper* 1641 asserts that, under a programme of structural adjustment, the population of Tanzania in poverty fell from 65% in 1983 to 51% in 1991.

Whether global inequality is increasing or decreasing is by no means clear. The UNDP (2005) reports that the ratio of the assets of the world's richest 20% to the poorest 20% increased from 30:1 in 1960 to 78:1 in 1994, yet Firebaugh and Goesling (2004) *Amer. Journal Soc.* 110, 2 report that global income inequality declined in the last decades of the 20th century.

There are those who hold that there is a relationship between **regional inequality** (as measured by the *Gini coefficient) and level of development (as measured by per capita GDP), but the data in the table on p. 255 do not offer robust support to this view.

National inequality has grown markedly; since 1980, virtually every country in the world has become more unequal (Gilbert (2007) *Geog. Compass* 1, 3). During the 1980s and 1990s, income inequality rose in the OECD countries, particularly escalated in China, Vietnam, and South Asia, shot up in Latin America, and went ballistic in Russia in the 1990s (2005, *World Bank* 45). See Alderson and Nielsen (2002) *Amer. Journal Soc.* 107 for an exhaustive study of inequality in sixteen OECD countries.

Region	Gini coefficient for per capita GDP (ppp US$) 2003	GDP per capita (ppp US$) 2003
Sub-Saharan Africa	72.2	1856
South Asia	33.4	2897
East Asia and Pacific	52.0	5100
Latin America and Caribbean	57.1	7404
High Income OECD	67.0	30181

Source: UNDP (2005) 55 and 222.

In Britain, inequality rose in the 1980s and 1990s. Changes since 2000 are less clear, but the proportions of 'exclusive wealthy'—the people with so much wealth that they can exclude themselves from the norms of society—are at their highest since 1990 (Joseph Rowntree Foundation 2007). The rich and poor now live further apart; while in some cities over half of all households are now breadline poor, wealthy households have concentrated in the outskirts and surrounds of major cities, and exclusive wealthy households have increasingly been gathering around London. Since 2000, there seems to have been little progress in reducing geographical polarization.

Does inequality matter? Gilbert (2007) *Geog. Comp.* 1, 3 argues that inequality creates social ghettos and increased violence, and blocks progress for the poor. 'Inequality is not inherently wrong—as long as three conditions are met: society as a whole is getting richer; there is a safety net for the very poor; everybody, regardless of class, race, creed, or sex, has an opportunity to climb up through the system' (*The Economist*, 15 June 2006). Korpi (2008) *J. Econ. Geog.* 8, 2 finds a positive link between wage inequality and labour market size, due to increasing upper income levels as labour market size increases. See D. Dorling (2011).

((()) SEE WEB LINKS

- The Joseph Rowntree Foundation study of poverty and wealth across Britain.

infant mortality The number of deaths in the first year of life per 1000 children born. Poor sanitation is a major factor: this is the 'urban penalty'/'urban graveyard effect' (Mercier and Boone (2002) *Hist. Geog.* 28, 4). Gyimah (2006) *Ethnic. & Health* 11, 2 finds that mortality is more of a reflection of socio-economic inequalities than ethnic-specific socio-cultural practices, but notes that wider birth spacing and longer breastfeeding also reduce infant mortality. T. Boley (1999) sees infant mortality as an index of social tension and cultural values.

infiltration Water entering soil. Lipiec et al. (2005) *Geophys. Res. Abs.* 7, 02621 find that tillage increases soil porosity, and thus infiltration. Reduced tillage and stubble retention practices can greatly reduce surface sealing and improve infiltration rates. Soil type is significant—sands take in water faster than loams and clays; see Koelliker and Duell (1990) ASAE Paper 902568 on the effects of burning on infiltration. Infiltration may not occur if the speed of the water is too great.

Infiltration capacity is the maximum rate at which water can infiltrate the soil, and a key determinant of the volume of surface run-off, which occurs when rainfall exceeds a soil's infiltration capacity. **Relative infiltration capacity** is the ratio of mean precipitation intensity to saturated hydraulic conductivity (Berger and Entekhabi (2001) *J. Hydrol.* 247, 3).

informal sector (informal economy) 'A group of household enterprises or unincorporated enterprises owned by households that includes informal own-account enterprises, which may employ contributing family workers and employees on an occasional basis, and enterprises of informal employers, which employ one or more employees on a continuous basis. The enterprise of informal employers must fulfil one or both of the following criteria: size of unit below a specified level of employment, and non-registration of the enterprise or its employees' (ILO Report of the Fifteenth International Conference of Labour Statisticians, Geneva, 1993).

'The informal sector plays an important and controversial role. It provides jobs and reduces unemployment and underemployment, but in many cases the jobs are low-paid and the job security is poor. It bolsters entrepreneurial activity, but at the detriment of state regulations compliance, particularly regarding tax and labor regulations. It helps alleviate poverty, but in many cases informal sector jobs are low-paid and the job security is poor' (GDRC). See also B. Guha-Khasnobis et al. (2000).

( ) SEE WEB LINKS
• The informal sector.

information city A city whose residents work mostly in education, consulting, banking, services, computer programming, computer manufacturing, and medical equipment. Those in the *creative industries (R. Florida 2002) could be considered as elements of both in some cases. Is Birmingham, UK, an information city? Webster (2007) *City* 5, 1 thinks not. Is Bangalore? Liang (2005) *Wld Info. City Conf.* says yes, and writes of 'the other information city': 'the proliferation of non-legal media practices ranging from pirated VCD's, DVD, MP3's to grey market mobile phones and pirate book sellers on the streets.'

information society A society built on information storage, retrieval, and transmission, *time–space compression, *post-Fordism, *flexible accumulation, and the advance of finance capital, which is characterized by networking, *globalization, and the flexibility, individuality, and instability of work. 'Information technologies, both as a process and product, contain embedded social knowledge, and thus represent the construction of new social norms and institutions' (Eischen (2001) U. Calif., Santa Cruz, *W. Paper Series* 1009). 'The information society can be usefully characterized as a universe at the intersection of three distinct but interdependent spaces: the geographical space, the social space, and the informational space' (Ekbia and Shurman, *Information Society*, special issue). See F. Webster (2007) on the work of key theorists of the information society in light of social and technological changes, such as the rapid growth of the Internet and accelerated globalization.

information system An integrated assemblage of elements that collect, store, and process data and deliver information, knowledge, and digital products. K.-T. Chang (2014) adds 'the interaction between people, processes, data, support operations, management, and decision-making'.

infragravity wave (IGW) A sea wave that is low in amplitude (frequencies of 0.05 to 0.003 Hz), has long periods (> 30 seconds), and can form standing waves that periodically oscillate the still water level of a surf zone. See Beetham and Kench (2011) *ESPL* 36, 14, 1872 for several explanations for the generation of IGWs.

infrastructure The framework of communication networks, administration, and power supply; 'the fact that a direct link between investment in transport infrastructure and national or regional economic development has not been shown to exist has not stopped politicians, lobbyists and promoters from asserting that great benefits will accrue from such investments' (Scrafton (2002) *J. Transp. Geog.* 10, 1). 'In developed countries, where there is already a well-connected transport infrastructure network of a high quality, further investment in that infrastructure will not on its own result in economic growth' (Banister and Berechman (2001) *J. Transp. Geog.* 9, 3). Nonetheless there is some empirical analysis that supports the idea that infrastructure growing ahead of productive activities would stimulate growth (Reardon et al. (2001) *World Dev.* 29, 3). Kirkpatrick (2004) *Publish. Res. Quart.* 20, 3 thinks that African governments' telecommunications monopolies are 'obstructive'.

ingrown meander *See* INCISED MEANDER.

inner city An area at or near the city centre with dilapidated housing, derelict land, and declining industry (also called the **twilight zone**, or **zone of downward transition**). Langlois and Kitchen (2001) *Urb. Studs* 38, 1 use a general deprivation index to locate deprivation in central Montreal (but note that it isn't confined to the inner city). W. J. Wilson (1996) asserts that the decline of US inner cities reflects *deindustrialization, combined with the suburbanization of a small number of middle-class African Americans. 'Following the restructuring process which swept away the traditional manufacturing economy of the inner city twenty-five years ago, new industries are transforming these former postindustrial landscapes. These creative,

technology-intensive industries include Internet services, computer graphics and imaging, and video game production, and are integral to the production of the "new inner city" of the 21st century' (T. A. Hutton 2008).

innovation The introduction and adoption of something new. An **innovation wave** is the diffusion of an innovation from its point of origin. Beginning at the innovation epicentre, it is adopted by a very few, close to the focus. Through time, the peak of the wave moves away from the source as more distant locations adopt the innovation. Consolidation follows across the whole of the region concerned; finally, as the innovation saturates the region, diffusion slows down until the stage of maximum acceptance.

An innovation wave may occur when larger cities take up the innovation, which then filters down the urban hierarchy. The largest cities have benefited from many innovation cycles, and, in consequence, have developed a broader diversity of activities, and attained higher levels of social and organizational complexity—thus making them more likely to adopt further innovations (Pumain et al. (2006) *Cybergeo:systèmes, modèl., géostat.* 343).

input–output analysis A view of the economy that stresses sectorial interdependence—the output of one sector is often the input of another. See R. K. Miller et al. (1989) and Fleischer and Freeman (1997) *Annals. Tour. Res.* 24, 4 on input–output analyses of tourism.

InSAR (Interferometric Synthetic Aperture Radar) A remote sensing technique which uses radar satellite images. The US Geological Service has an excellent site on InSAR. See Baltzer (2001) *Prog. Phys. Geo.* 25, 2 for an account of the use of InSAR in forest mapping and monitoring.

inselberg A steep-sided hill composed mainly of hard rock, which rises sharply above a plain in tropical and subtropical areas; see Römer (2007) *Geomorph.* 86, 3–4. *See also* BORNHARDT.

insolation The *solar radiation hitting the Earth's surface. Global variations in insolation are critical in the *general circulation of the atmosphere. 'Climatic changes result from variables in planetary orbits which modulate solar energy emission and change seasonal and latitudinal distribution of heat received by the Earth. Small insolation changes are multiplied by the *albedo effect of the winter snow fields of the Northern Hemisphere, by ocean-atmosphere feedbacks, and, probably, by the stratospheric ozone layer' (Kukla (1972) *Boreas* 1, 1). Insolation *forcing functions include the eccentricity of the Earth's orbit, the angle from the vernal equinox to perihelion, obliquity, and time (Huybers (2011) *Nature* 480, 229).

instability 1. The condition of a slope just before the initiation of mass movement. See (2008) *Geomorph.* 94, 3–4, special issue.
    2. The condition of a parcel of air which has positive *buoyancy, and thus a tendency to rise through the *atmosphere. It is the temperature of the parcel relative to the *ambient air which is critical, since this affects densities. A parcel remains unstable if it cools more slowly than the surrounding, stationary air.
    Absolute instability occurs when an air parcel, displaced vertically, is hastened in the direction of the displacement; such a move will end when the temperature of the parcel is at one with its surroundings. The **potential instability** of a parcel of air can be calculated

using a *tephigram. **Conditional instability** occurs when a parcel of air would become unstable if lifted by some other force, such as a mountain, the moistening of lower-level air, or when a lower, stable parcel of air is overlain by an unstable layer. **Convective instability** is a tendency of an air parcel towards instability when it has been lifted bodily until completely saturated, which often occurs via the inevitable stirrings in the atmosphere. The atmosphere is said to be unstable when the fall of temperature with height of the environmental air is more rapid than that experienced by a rising air parcel, so that the parcel continues to be less dense than the ambient air. See E. Fedorovich et al. (2004).

institutional thickness 'The localised conditions that are crucial to the particular articulation of localities and regions within wider scale processes of economic transformation' (Henry (2001) *Env. & Plan. A* 33). Institutional thickness includes trade associations, voluntary agencies, sectoral coalitions, concrete institutions, and local elites— their effects on local policy, and their consensus institutions: common agreements, shared views and interpretations, and unwritten laws.
    E. Sheppard (2002) is sceptical: 'how, exactly, does institutional thickness promote local economic development and prosperity? Further, how do we explain the fact that some very successful regions do not appear to be underpinned by a highly concentrated and co-ordinated matrix of institutions?' Does institutional thickness stimulate growth, or vice-versa? (A. Wood and D. Valler 2004).

integrated assessment (IA) An interdisciplinary process which combines, explains, and communicates knowledge from a range of disciplines in order wholly to weigh up an entire chain of causes and effects. Van Amstel (2005)

Env. Sci. 2, 2–3 uses an integrated assessment model to evaluate six methane emission reduction strategies on future air temperature. For a second example, see Mendonça et al. in M. van Asselt et al., eds (2001).

integrated land-use and transportation models

Models which attempt to represent human behaviour in a realistic way, measuring and representing the decision-making that leads to the organization of human activities in time and space. These range from daily household decisions—what to do, where and when to do it, and by which mode of transport—to long-term decisions relating to land use and transport networks. See T. Nisbet, H. Orr, and S. Broadmeadow (2004) for using integrated land use planning to resolve problems of sedimentation.

Despite examples such as these, Couclelis (2005, *Env. & Plan. A* 37, 8) argues that 'despite more than four decades of effort to integrate the two fields, the place of land-use models in planning remains problematic'.

((⊕)) SEE WEB LINKS
- The Heartlands Initiative.

integrated water management

'A process that promotes coordinated development and management of water, land, and related resources, in order to not only maximize economic and social welfare, but also ensure equity and sustainability' (Chao (2005) *Env. Informatics Archs.* 3). Radif (1999, *Desalination* 124) urges that integrated water resources management be based 'on the perception of water as an integral part of the ecosystem, a natural resource and a social and economic good'.

The general objective is the environmentally sound, equitable, and sustainable utilization and development of water resources. This may include supply management, demand management (water conservation, transfer of water to uses with higher economic returns, etc.), water quality management, recycling and reuse of water, economics, conflict resolution, public involvement, public health, environmental and ecological aspects, socio-cultural aspects, water storage (including long-term storage or water banking), conjunctive use of surface water and groundwater, water pollution control, flexibility, regional approaches, weather modification, sustainability, etc. (Bouwer (2000) *Ag. Water Manage.* 45).

In a study of the *Loess Plateau, China, Chen et al. (2007) *PPG* 31, 4 list the factors to be considered: 'basic farmland construction, plantation of cash trees, firewood and conservation wood, roads, water banks, agricultural structure modifications, and local people's needs (such as daily fuel, daily drinking water, education, medical care and other factors regarding improving living conditions).' Remedial measures include the use of *check dams; afforestation and the planting of shrubs and grasses on slopes (*see* REHABILITATION ECOLOGY); and land closure, in order for ecosystems to rehabilitate naturally, without further human intervention (see Zhang (2005) *Acta U. Agric. Suecia* 126). 'Natural recovery by land closure is strongly recommended on the steep slopes' (Chen et al., op. cit.).

In natural rehabilitation, the degraded soil may be improved (Jia et al. (2005) *Forest Ecol. & Manage.* 217), and the biodiversity may be increased steadily over time. However, revegetation usually takes a long time, especially from an extremely degraded state (Bellot et al. (2001) *Land. & Urb. Plan.* 55). Zhang (op. cit.) estimates that several decades to hundreds of years may be needed in the Loess Plateau. R. Quentin Grafton (2011) is the stand-out text.

integration 1. **Social integration** The process whereby a minority group, particularly an ethnic minority, adapts to the host society and where it is accorded equal rights with the rest of the community. Lemanski (2006) *GeoForum* 37, 3 reports on social integration in post-apartheid urban South Africa: 'different races are not only living peacefully in shared physical spaces but also actively mixing in social, economic and, to a lesser extent, political and cultural spaces.' But Nagel and Staeheli (2008) *Soc. & Cult. Geog.* 9, 4 call integration 'a socio-political process by which dominant and subordinate groups negotiate the terms of social membership'.

Lemanski notes a social continuum of integration (greeting in the street, visiting homes, inter-marriage) and a continuum of **spaces of integration**, for example physical space (shared neighbourhood), economic space (common employment type), social space (cross-race friendship), political space (common involvement in civic organizations), and cultural space (shared sense of belonging).

2. **Economic integration** can be the breaking down of trade barriers between nations. 'Economic integration has a direct effect on internationalisation by reducing transaction costs, and information costs are a part of transaction costs. Being in a currency union has been shown . . . to have a significant effect on bilateral trade' (Rose (1999) *CEPR Discussion Paper* 329, London). The effect of sharing a common currency is significant even when exchange rate volatility is taken into account (Guerin (2006) *World Econ* 29, 2). **Horizontal integration** occurs across different sectors; **vertical integration** occurs within a hierarchy; see Farrington (2007) *J. Transp. Geog.* 15, 5.

intellectual property A creative work, or a device or invention that can be patented or copyrighted. Different knowledges have been variously adopted, reworked, and contested at different scales, including attempts to defend and (re)construct more collective notions of knowledge and property (S. Wright, U. Newcastle, UK). 'Endemic healing floras and native intellectual property diffused back and forth between Europe and the colonies, transforming forever the geography of tropical plant pharmacopoeias and folk healing traditions' (Voeks (2004) *AAAG* 99, 4). Intellectual property rights laws are best understood as particular, culturally defined systems for codifying knowledge, employed to discipline objects, phenomena, and social relations (Parry (2002) *Antipode* 34, 4).

intensity The size of an earthquake, based on the impact on people, buildings, and the ground surface. Isoseismal lines represent equal earthquake intensity. The Modified Mercalli intensity scale is the world standard; the Medvedev–Sponheuer–Kárnik (MSK) is popular in Europe. Although differing in detail, the MM and MSK scales are much alike. As they are descriptive, and based on recorded damage, **intensity values** can be calculated and isoseismal lines drawn for present day and historic earthquakes (Chester (2001) *PPG* 25, 3).

intensive agriculture Agriculture with a high level of capital and labour inputs, and high yields. 'Intensive cropping systems, in which large areas of land [are] cultivated permanently (or with relatively short fallow periods) through the use of labor, irrigation water, and its dissolved nutrients, knowledge, and other inputs, develops synergistically with socially and culturally complex human societies' (Vitousek and Chaplin (2013) *Elementa: Science of the Anthropocene*). Phillips (1998) *Appl. Geog.* 18, 3 provides a gloomy list of the effects of agricultural intensification in Italy: declining frequency of crop rotations; increased use of chemical fertilizers and pesticides; more up- and downslope tillage; and enlarged fields. Mechanization has led to a decline in draught cattle numbers, causing loss of agricultural employment, and a fall in soil fertility. See Kasente et al. (2000) *UNRISD Occ. Paper* 12, on the intensification of non-traditional agricultural exports.

interaction The action between points, each upon the others; 'business concentration clearly facilitates more intensive social interaction, and interaction intensity is argued to be a key element differentiating between types of client-supplier relations' (Bennet et al. (1999) *TIBG* 24, 4). An **interaction model** describes the reactions of two or more processes or systems as they affect each other; see Eradus et al. (2002) *J. Transp. Geog.* 10, 2.

interception The retaining of raindrops by plant leaves, stems, and branches. Xiao et al. (2000) *Hydrol. Procs* 14 find that rainfall frequency is more significant than rainfall rate and duration in determining interception. Guevara-Escobar et al. (2007) *J. Hydrol.* 333, 2–4 evaluate interception around the canopy of a single tree.

interdependence In a heartening review of the term, Smith et al. (2007) *Geog. Compass* 1, 3 declare that interdependence conveys a sense of relying on and being responsible for others. 'Today we all gather here to declare our interdependence. Today we hold this truth to be self-evident: we are all in this together' (Will Smith, Live8, 2005). With reference to **economic interdependence**, and in a study of German and Chinese interdependence, Dunford et al., *Eur. Urb. & Regional Studs* (2013) 20, 1, 14

find that economies are driven not mainly by supply-side but by a set of demand-side mechanisms, as recognized in Keynesian and Marxist as opposed to neoclassical models.

interdisciplinarity The combination of two or more academic disciplines into one investigation; using skills and knowledge that cross traditional subject boundaries. See the special issue of *Area* (2009) 41, 4.

interface The zone of interaction between two systems or processes. Estuaries are interfaces between *fluvial and marine systems (Bate et al. (2002) *Water SA* 28, 3).

interflow Water that moves downslope through soil pores (as opposed to *throughflow, which is soil water moving downslope along impermeable soil horizons—it's difficult to separate the two processes in the field). Interflow is of major importance for run-off generation and groundwater recharge (Flugel and Smith (1998) *Env. Model. & Sofw.* 14, 2).

interglacial A long, distinct period of warmer conditions between *glacials when the Earth's glaciers have shrunk to a smaller area—the present, Holocene, period is an interglacial. Soon (2007) *Phys. Geog.* 28, 2 argues that the persistence of insolation forcing at key seasons and geographical locations, taken with closely related thermal, hydrological, and cryospheric changes, suffice to explain transitions in paleoclimates.

Intergovernmental Panel on Climate Change (IPPC) The leading international body for the assessment of climate change, established by the United Nations Environment Programme (UNEP) and the World Meteorological Organization (WMO) in 1988 to provide the world with a clear scientific view on the current state of knowledge in climate change and its potential environmental and socio-economic impacts. The IPCC is a scientific body under the auspices of the United Nations (UN). It reviews and assesses the most recent scientific, technical, and socio-economic information produced worldwide relevant to the understanding of climate change. It does not conduct any research nor does it monitor climate-related data or parameters.

intermediate beach A beach form somewhere between the high-energy *dissipative beach and the low-energy *reflective beach, and containing elements of both these two basic types of beach. For example, the upper part of an intermediate beach can be steep, and show characteristics of a reflective beach. Intermediate beaches tend to form near shore bars, which will dissipate some of the wave energy.

intermediate technology Mid-level technology, often the most suitable level for economic development. Chinese aid projects in Africa, for example, consist largely of intermediate technology: Chinese workers are paid at local rates, train their local equivalents, and aim to leave projects in a state where they can be locally managed, operated, and maintained so that even the spare parts could be locally produced (Mawdsley (2007) *Geog. Compass* 1, 3). The *Intermediate Technology Group* is now called *Practical Action*.

internal colonialism The practice of exploitation by a nation-state within its own boundaries, In many states, indigenous communities suffer from poverty and isolation. *Dependency theorists argue that this situation arises when the state treats its indigenous people in the same way as the European powers treated their colonies. 'The population of an internal colony faces severe discrimination in civil society, violence (as broadly defined), and restricted access to the state's resources' (Das and Chilvers in R. Kitchin and N. Thrift 2009).

internal deformation A distortion within a material, such as glacier ice: individual ice crystals react to stress by elongating, or by melting and recrystallizing, and ice may also shear along separate shear planes; see Hambrey et al. (2005) *J. Geophys. Res.* 110, F01006 and Mair et al. (2001) *ESPL* 23. See Lewkovicz and Harries (2005) *Geomorph.* 69, 1–4 on internal deformation in permafrost.

internal migration The temporary or permanent relocation of its population inside a *nation-state. China has the largest internal migration; in 2003, the so-called floating population reached 140 million, most of them rural labourers moving from the countryside to cities and coastal areas (Ping and Shaohua (2005) *Chinese Acad. Soc. Scis*). The rapid growth of rural–urban migration in China results from the privatization of agriculture, and the transition to a market economy (Ping and Pieke (2003) *Inst. Sociol., Chinese Acad. Soc. Scis*).

international division of labour The allocation of various parts of the production process to different places in the world. 'The world economy is organized through horizontal and vertical linkages of an international division of labour, in which the modes of integration and geographical scopes vary over time' (Appelbaum et al. in G. Gereffi and M. Korzeniewicz, eds 1994). Grinberg (2007) *Glob. Pov. Re. Grp* observes that whilst some countries have tended to concentrate within their boundaries the great bulk of the skilled labour force and therefore of the most complex labour processes (the USA, western Europe, and Japan), others have been mainly transformed into sources of a relatively cheap and disciplined, though less skilful, labour force (mainly East and South-East Asia). A third group of countries have been turned into reservoirs of consolidated surplus population relative to the needs of the accumulation process (for instance, most of Sub-Saharan Africa and parts of Asia, Northern Africa, and Latin America) and, in some cases, have eventually become new sources of cheap and disciplined labour (parts of South Asia, Vietnam, and the Caribbean Basin). See Hardy (2007) *Eur-Asia Studs* 59, 5 on the new international division of labour and Poland.

International divisions of labour involve transfers of reproductive labour beyond territorial boundaries. Thus, Filipina workers take on the household tasks of middle-class women abroad, while they themselves may hire poorer workers to do their housework in the Philippines (Boyle (2002) *PHG* 26, 4). *Service offshoring* is an extension of outsourcing as firms apply an international division of labour to service tasks (Bryson (2007) *Geografiska B* 89). 'To many Marxist scholars, the rise of East Asia as a global factory can often be "read off" as an empirical proof of the working of . . . spatial fixes, particularly in relation to the rise of the new international division of labour' (Cheung (2007) *Econ. Geog.* 83, 4).

international financial system (global financial system)

The collective name for the various official and legal arrangements that govern international financial flows in the form of loan investment, payments for goods and services, interest and profit remittances. The main elements are the surveillance and monitoring of economic and financial stability, and provision of multilateral finance to countries with balance of payments difficulties (WHO). With the decline of *Bretton Woods, much of the power over the international financial system, and especially power over how

credit is created, bought, and sold, has transferred back to money capitalists. With the growth of electronic telecommunications, these money capitalists are able to operate on a global scale and with a speed of reaction which states find very difficult to emulate. In this new international financial system, 'the balance has shifted from a financial structure which was predominantly state-based with some transnational links, to a predominantly global system in which some residual local differences in markets, institutions, and regulations persist'. (Thrift and Leyshon (1994) *Pol. Geog.* 3, 4, 299). See B. Christophers (2013).

International Fund for Agricultural Development (IFAD)

A UN agency for ending rural poverty in developing countries.

SEE WEB LINKS
• The IFAD website.

International Geosphere–Biosphere Programme (IGBP)

A programme focused on acquiring basic scientific knowledge about the interactive processes of biology and chemistry of the Earth, including human modifications, as they relate to global change.

SEE WEB LINKS
• Website of IGBP.

International Labour Organization (ILO)

An agency of the United Nations whose aims are: 'to promote and realize standards and fundamental principles and rights at work; to create greater opportunities for women and men to decent employment and income; to enhance the coverage and effectiveness of social protection for all; and to strengthen tripartism and social dialogue' (ILO).

SEE WEB LINKS
• Website of the ILO.

International Monetary Fund

'The work of the IMF is of three main types. Surveillance involves the monitoring of economic and financial developments, and the provision of policy advice, aimed especially at crisis-prevention. The IMF also lends to countries with balance of payments difficulties, to provide temporary financing and to support policies aimed at correcting the underlying problems; loans to low-income countries are also aimed especially at poverty reduction. Third, the IMF provides countries with technical assistance and training in its areas of expertise. Supporting all three of these activities is IMF work in economic research and statistics. In recent years, as part of its efforts to strengthen the international financial system, and to enhance its effectiveness at preventing and resolving crises, the IMF has applied both its surveillance and technical assistance work to the development of standards and codes of good practice in its areas of responsibility, and to the strengthening of financial sectors. The IMF also plays an important role in the fight against money-laundering and terrorism' (IMF website).

The IMF doesn't get a good press from geographers, largely through its role in promoting *structural adjustment policies: as a result of austerity programmes sponsored by the World Bank and the IMF, 'the 1980s and 1990s were the "lost decades" for development in much of Africa' (Carmody (2008) *Geog. Compass* 2, 1). Watts et al. (2005) *PHG* 29, 1 describe the IMF as one of the ungoverned bodies that regulate global capitalism, and D. Harvey (2004) argues that localized financial crises are in some sense intentionally managed by various US-sponsored agencies such as the International Monetary Fund under the guise of neoliberal economics, 'thereby maintaining the United States as the global hegemon'.

(⊕) SEE WEB LINKS
• The IMF website.

International Organization for Migration (IOM)

The leading inter-governmental organization in the field of migration working closely with governmental, inter-governmental, and non-governmental partners, the IOM works 'to help ensure the orderly and humane management of migration, to promote international cooperation on migration issues, to assist in the search for practical solutions to migration problems, and to provide humanitarian assistance to migrants in need, including refugees and internally displaced people. The IOM Constitution recognizes the link between migration and economic, social, and cultural development, as well as to the right of freedom of movement'. For the role of the IOM in resolving border disputes, see Andrijasevic and Walters (2010) *Env. & Plan. D* 28, 6, 977.

(⊕) SEE WEB LINKS
• Website of IOM.

international trade theory

The theory that countries will produce and export goods that require resources (factors) which are relatively abundant, and import goods that require resources which are in relative short supply. B. Ohlin's (1967) model of international trade states that those products a country or region produces relatively better than the competitors have a comparative advantage, and that comparative advantage dictates who trades what, and with whom. Andreson (2010) *Geog. Compass* 94–105 explains this very clearly.

internet

A global network of connected computers (Kitchin (1998) *PHG* 22). Warf (2007) *Sing. J. Trop. Geog.* 28, 2 identifies costs, wealth, literacy, and telephone penetration rates as determinants of internet uptake. Zook (2002) *Jour. Econ. Geog.* 2, 2 observes that 'although both the internet and venture capital have been viewed as independent of geography, the development of this industry continues to highlight the continuing relevance of regions and place-based relations'. Thus, Gorman (2002) *Tijdschrift* 93, 5 finds that website production is dominated by New York and San Francisco.

In a North Carolina study, Palm and Danis (2002) *Tijdschrift* 93, 5 find that the internet has little impact on housing search patterns. Adams and Ghose (2003) *PHG* 27, 4 describe the vast, variegated space of international and multicultural communication used by the 'Indian diaspora' and residents of India. See Wharf and Vincent (2007) *Area* 39, 1 on Arab geographies of the internet, J. Rodgers (2003) on the internet and activism, and Grubesic and O'Kelly (2002) *Prof. Geogr.* 54, 2 on measuring access to commercial internet sites.

interpellation This is 'call and response'; and it explains how one class allows the other class to exert influence over them. This concept may be used in mutual interactions, as with managers and workers. Workers end up seeing themselves in the ways that the managers describe them; 'this naming of others in the workplace is, in turn, internalized by workers themselves, so they come to conform to or recognize themselves in the managerial naming' (McDowell et al. (2007) *Econ. Geog.* 83, 1, 1).

intersectionality The interconnections and interdependence of race with other categories. The originator of this term, Crenshaw (1991) *Stanford Law Rev.* 43, 6, considers the various ways in which race and gender interact to shape the multiple dimensions of Black women's employment experiences.

Wills (2008) *Antipode* 40, 1 believes that geography has real strength as a discipline for the study of intersectionality: 'to understand the ways in which class intersects with other social cleavages, with very different e/affects.' Understanding such intersectionality at the local level can help in mapping larger-scale trends in the geography of class and employment; see A. Ong (2006). See also Brown (2012) *PHG* 36, 4, 541.

interstadial A warmer phase within a *glacial which is too short and insufficiently distinct to be classed as an *interglacial. See Phillips et al. (1994) *Geology* 22 on interstadial climatic changes, and Ringberg et al. (2003) *Boreas* 32, 2 on interstadial events and glacial varves.

intertextuality The way that any texts acquire meaning through the way they are described or evoked by other texts; 'it is generally accepted that every text is related to other texts and that these relations are constitutive for the generation of the meanings of the text' (Alkier at *Open Journals* online). Aitken in R. Kitchin and N. Thrift (2009) is very good on this concept.

inter-tropical convergence zone (ITCZ) That part of the tropics where the opposing north-east and south-east *trade winds converge. It is not a continuous belt, more like a necklace with groups of clouds as the 'beads'; in places there may be two or more 'strings'. The zone is narrower over the oceans, and broader over the continents, where other wind systems may be involved; in West Africa the ITCZ is the convergence of the Guinea *monsoon and the *Harmattan (Miskolczi et al. (1997) *J. Appl. Met.* 36, 5). See Yancheva et al. (2007) *Nature* 445 on the influence of the ITCZ on the East Asian monsoon.

The ITCZ moves north and south; moving more over land, and arriving in the summer in each hemisphere. Its position is affected by the apparent movement of the overhead sun, the relative strengths of the trade winds, and the changing locations of maximum sea-surface temperatures (R. D. Thompson 1998). This means that the movements of the ITCZ are highly unpredictable. In May 2005, the African portion of the ITCZ was 1° south of the average; a period of drought near the Sahel. In July 2003, it was around 0.9° north of the average; heavy rains fell in Ethiopia, with flooding in Khartoum. The movements of the ITCZ also affect bush/forest fires in South-East Asia (P. Kershaw et al. 2001). Over the oceans, the ITCZ is broad, and often loses its identity.

intervening opportunities theory

Stouffer (1940) *Am. Soc. Rev.* 5, 6 argues that the spatial distribution of competing (alternative) intervening destinations influences decision-makers' choices (Cascetta et al. (2007) *Papers Reg. Sci.* 86, 4). See Alix et al. (1999) *J. Transp. Geog.* 7, 3 on Montreal as an intervening transhipment opportunity between Quebec and Europe.

intrazonal soil A soil affected more by local factors than by climate, unlike a *zonal soil. For example, waterlogging creates *gley soils, and a limestone bedrock will produce a *rendzina.

intrenched meander *See* INCISED MEANDER.

intrusion A mass of igneous rock which has forced its way, as *magma, through pre-existing rocks, and then solidified below the surface of the ground; hence **intrusive rock**. See Jousset et al. (2003) *J. Volc. & Geoth. Res.* 125, 1–2.

Igneous intrusions are classified according to their shape and size: *dykes are **discordant**, since they cut across the bedding or fabric of pre-existing rocks, *sills are **concordant**, since they intrude between layers. *Plutons are the largest intrusions, typically spheroidal, and ranging from *stocks to *batholiths.

Intrusion

invasion The establishment of species in ecosystems they are not native to; see, for example, Russell-Smith et al. (2004) *J. Biogeog.* 31, 8 on the invasion of rain forest into eucalyptus woodland in Australia. Campbell et al. (2002) *J. Biogeog.* 29, 4 outline a methodology which might be able to be used to predict invasions by alien plant species. 'Humans cause invasions, humans perceive invasions, and humans must decide whether, when, where and how to manage invasions' (Richardson et al. (2008) *PHG* 32, 2).

invasion and succession 1. A model of change used in *urban ecology to represent the effects of immigration on the social structure of an urban area. Invasion and succession involve a chain reaction, with each preceding immigrant wave moving outwards and being succeeded by more recent, poorer immigrants (R. E. Park, E. W. Burgess, and D. McKenzie 1925/1974). This model saw immigrant enclaves as transitional stages on the road to eventual acceptance and integration in the larger (American) society (Clark in R. Waldinger and M. Bozorgmehr, eds 1996). Van Kempen (2007) *J. Housing & Built Env.* 22, 1 argues that divided cities cannot be explained without using concepts like invasion, succession, and filtering.

2. In plant ecology, the introduction and subsequent spread of a species. Diamond (1974) *Science* 184 provides a dynamic model of successional processes, and R. H. Groves and F. Di Castri (1991) distinguish between invasion and succession.

invasive species A species is invasive if it: is not native to the biogeographical region in question; has adapted to its new location; is becoming more profuse/more widely distributed; is an irritant; and results from anthropogenic activity. Henderson et al. (2006) *PPG* 30, 1 find that intact natural areas are becoming more invaded and degraded.

inversion The increase of air temperatures with height. (This is the reverse of the more common situation in which air cools with height.) Inversions occur: when strong, nocturnal, *terrestrial radiation cools the Earth's surface and therefore chills the air which is in contact with the ground (Lapworth (2003) *Qly. J. Royal Met. Soc.* 129); when cold air flows into valley floors, displacing warmer air (LeMone et al. (2003) *J. Atmos. Scis* 60); where a stream of warm air crosses the cool air over a cold *ocean current (Vihma et al. (2003) *Boundary-Layer Met.* 107); where warm air rises over a cold front (Ross and Orlanski (1978) *J. Atmos. Scis* 35, 3); when air from the upper *troposphere, subsiding in a warm *anticyclone, is compressed and *adiabatically warmed. The boundary between the top of the cold air and the beginning of the inversion is an **inversion lid**. *See also* TRADE WIND INVERSION.

inverted U A particular shape of a line on a graph, such as on income equality with per capita income; see Kuznets (1955) *Am. Econ. Rev.* 45. Grote (2008) *J. Econ. Geog.* 8, 2 finds an inverted U in the establishment of foreign banks in Germany.

invisibles Imported and/or exported items that are not goods. Invisibles include government grants to overseas countries and subscriptions to international organizations, payments for shipping services, travel, royalties, commissions for banking and other services, transfers to or from overseas residents, interest, profits, and dividends received by or from overseas residents.

involution 1. The refolding of two *nappes differing in age so that parts of the younger nappe lie below older rocks. **2.** The convolution of layers of ground under *periglacial conditions. To form thermokarst involutions, ice-rich permafrost must thaw, drainage conditions must be poor, and sediments must vary in texture or composition. In addition, the sediments should be susceptible to fluidization, liquefaction, or hydroplastic deformation (Murton and French (1993) *Perm. & Periglac.* 4, 3). The repetition of differential heave and soft-loam settlement promotes decimetre-scale involutions in near-saturated soils subject to deep seasonal frost penetration (Ogino and Matsuoka (2007) *Perm. & Periglac.* 18, 4).

ionosphere A layer of the *atmosphere containing ions and free electrons. The ionosphere is warmed as it absorbs solar radiation, and it will reflect radio waves. Three bands are recognized: E, F, and F2, at about 110, 160, and 300 km above the Earth. The ionosphere is in a state of constant motion, and is affected by tidal forces and by the Earth's magnetic field.

irradiance The rate of flow of radiant energy through unit area perpendicular to a solar beam. **Total solar irradiance** is the dominant driver of global climate (Mendoza (2005) *Advances Space Res.* 35, 5). See also Mishchenko et al. (2007) *Bull. Am. Met. Soc.* 88, 5.

irrigation The supply of water to the land by means of channels, streams, and sprinklers in order to permit the growth of crops. Without irrigation arable farming is not possible where annual rainfall is 250 mm or less, and irrigation is advisable in areas of up to 500 mm annual rainfall. To some extent, irrigation can free farmers from the vagaries of rainfall and, to that end, may be used in areas of seemingly sufficient rainfall because irrigation can supply the right amount of water at the right time. 'The contrasts in adjustments and attitudes over the use of water for agriculture stem largely from the inconsistent provision of information

from the scientific community, and differing ideologies, namely environmental protection versus economic development' (Yang et al. (2006) *Geogr. Res.* 44, 2). 'Essential complements to water pricing are water distribution rules and technological choices at critical nodes in the delivery system that allow farmers flexibility in conserving water in response to higher prices. Among supporting institutions, water users' associations seem a higher priority than water markets' (Dinar and Mody (2004) *Nat. Resources Forum*, 28, 1).

Writing on Ecuador, Cremers et al. (2005) *Nat. Resources Forum*, 29, 1 judge that bottom-up processes of awareness, capacity-building grassroots' claims, collective action, and mobilization are critical in establishing water rights, together with governmental political will and an enabling political environment. In a Malaysian study, Johnson (2000) *Geog. J.* 166, 3 show that tertiary intervention has increased the capacity of the farmers to unofficially control the distribution and supply of the water resource, resulting in a significant over-supply of water; inefficient water use; and a reduction in yields without a reduction in incomes. Lam (1994), cited by Kurian and Dietz (2004) *Nat. Resources Forum* 28, 1, finds a negative correlation between land-holding inequality and irrigation management performance. See Reinfelds et al. (2006) *Geogr. Res.* 44, 4 on sustainable diversion limits in river systems.

isentropic (isentropic surface) In meteorology a surface of constant potential temperature. Isentropic surfaces slope very gently upwards towards the cold air in the presence of a horizontal temperature gradient, and winds flow along these surfaces. See Ambaum (1997) *J. Atmos. Scis* 54, 4 on the isentropic formation of the tropopause. In **isentropic analysis**, the forecaster looks at the atmosphere in three dimensions, considering constant potential temperature surfaces. Air parcels move up and down these surfaces; therefore, the forecaster can see where the moisture is located and how much moisture is available. Try the UW Isentropic Analysis and Modelling Group home page.

island arc An arcuate island chain, mostly of volcanic origin, formed when *oceanic crust plunges into the *mantle where it undergoes *subduction. The *magma thus formed creates a chain of submarine volcanoes—eventual islands; see Jicha et al. (2006) *Geology* 34, 8 on magma flow in the Aleutian Island arc. See also Takahashi et al. (2007) *Geology* 35, 3.

island biogeography The relationship between area and species number on islands, as an equilibrium between immigration and extinction. See R. H. MacArthur and E. O. Wilson (1967) and R. Whittaker (1998).

Island biogeography is also relevant to continents, when plant and animal communities are effectively reduced to islands in a sea of cultivation or urbanization; Meadows (2001) *PPG* 25, 1 writes on *habitat islands*—distinctive communities in a sea of other types of community.

isodemographic map A cartogram which represents nations, states, counties, and so on, in proportion to the size of their population. See A. Cliff and P. Haggett (1988).

isogloss A line on a map which represents the geographical boundary of regional linguistic variants. By extension, the term also refers to the dialect features themselves. For example, in the USA north of the Mason–Dixon line, the word greasy is pronounce with an 's'; south of the line, it is sounded with a 'z' (L. Campbell 2004). Isoglosses are usually highly simplified representations and do not depict an abrupt transition. Sometimes, a **bundle of isoglosses** may occur, where a number of isoglosses lie close enough together to indicate a true *dialect boundary; see W. Wolfram and N. Schilling-Estes (1998).

isoline/isopleths Any line on a map joining places where equal values are recorded: an **isohyet** connects points of equal rainfall; an **isophene** connects places with the same timing of similar biological events; an **isotherm** joins places of equal temperature; and an **isotim** is a line drawn about a source of raw materials or a market where transport costs are equal.

isostasy The continental crust of the Earth has a visible part above the surface and a lower, invisible one. The balance between these two is isostasy. **Erosional isostasy** is a 'plausible mechanism' (Maddy et al. (2000) *Geomorph.* 33, 3–4) for differential crustal movement, in which the crust realigns in response to erosional unloading and depositional loading (M. A. Summerfield 1991). The lack of evidence for significant surface denudation may suggest that erosional isostasy may be no more than a positive feedback response to uplift initiated by other influences (Maddy (1997) *J. Quat. Sci.* 12). **Glacio-isostasy** results from the cessation of pressure from glacier ice on glacier melting, with unloading leading to uplift (Lambeck (1995) *J. Geol. Soc. London* 152). **Hydro-isostasy** is similar to glacio-isostasy; here, the loading/unloading of the continental shelves due to changes in sea level may have been significant in driving continental uplift.

isosteric Of that part of the atmosphere having uniform density.

isotropic Having the same physical properties in all directions.

jet stream 'A sinuous, relatively narrow ribbon of air that encircles the globe at an altitude of around 11 km, sometimes reaching speeds of 400 kph. They form where there is a substantial variation of air temperature within a relatively short geographical distance' (Meteogroup, 26 Nov. 2006). The westerly jet streams (J) in the Northern Hemisphere lie where significant breaks occur in the tropopause.

The **polar-front jet stream** is a frontal wind, not necessarily continuous, blowing west to east just below the *tropopause, parallel to the surface fronts and moving with them, and draining the air rising from the fronts. It is strongest at the 200–300 mb level, and swings between latitudes 40 and 60 °N;

speed and location vary from day to day with the Rossby waves. This jet marks the polar *front, and has important effects on *convergence and *divergence in the upper air. Divergence is more frequent in the downstream, poleward sector of the jet core, and is associated with extratropical cyclone development. At the 'jet entrance', the pressure gradient steepens, and the wind becomes super-*geostrophic, leading to high-level convergence. The polar-front jets often 'steer' the movement of major low-level air masses.

A strong polar-front jet is associated with rapidly moving *depressions (Galvin (2006) *Weather* 15, 11), but the jet stream and its accompanying depressions in the surface can be diverted

Jet stream. Identified by J on the chart

by *anticyclones. This jet splits into the polar front jet and subtropical jet, which develop each year in March through to early Southern Hemispheric spring (Bertler et al. (2006) *Advances Geoscis* 6, 83386).

The **westerly subtropical jet** is at the poleward limit of the *Hadley cell, around 30°N and S; the northern subtropical jet is strongest at the 200 mb level, and above the Indian subcontinent. This is one of the most powerful wind systems on Earth, at times reaching speeds of 135 m s^{-1}, and it follows a more fixed pattern than the polar-front jet. It results from the poleward drift of air in the Hadley circulation and the *conservation of angular momentum. Chen and Yen (1986) *Mon. Wea. Rev.* 114, 255032570, are among many to link this jet with the South Asian summer *monsoon.

The **easterly tropical jet** develops during the summer months at 15°N, and is strongest at the time of the summer monsoon. The **stratospheric subpolar jet stream** blows at a height of 30000 m, being westerly in winter and easterly in summer.

Jevons Paradox This paradox, also known as the **rebound effect**, holds that as technology improves environmental efficiency, the gains will be eaten up by increased exploitation of the environment. See Polimeni (nd), *Int. J. Interdisciplinary Soc. Scis* 2, 2, 383.

joint A crack in a rock without any clear sign of movement either side of the joint.

jökulhlaup 'A "catastrophic" flood, often generated by an ice-dammed lake outburst'; for example, the volcanic eruption and concomitant catastrophic flooding from Grímsvötn, Iceland, in November 1996 (Tweed and Russell (1999) *PPG* 23, 1). Alho (2003) *Geograf. Annal. A* 85, 3–4 illustrates the severe

and extensive jökulhlaup-induced process of land degradation.

Jurassic The middle period of *Mesozoic time stretching approximately from 190 to 136 million years BP.

(((∰))) SEE WEB LINKS
• An overview of the Jurassic period.

just city Democracy, equity, diversity, growth, and sustainability are indispensable to the creation of just cities (Fainstein (2005) *Urb. Affairs Rev.* 41): 'I did not attempt a justification for choosing these values but simply assumed agreement on them.' These are problematic: 'illiberal majorities can make democracy indifferent to minority rights; the high cost of achieving equity [can create] resentment . . . diversity can lead to social breakdown; and growth . . . benefits most those who already have the most. Sustainability may diminish growth' (Fainstein (2006) *Lecture on the Just City*). She later writes, 'if the city is composed of diverse communities, conflict arises over whose definition of justice should prevail' ('The Resurgent City'). P. Marcuse et al. (2009) provide an outstanding review of the different debates and theories that have emerged over time with respect to the 'just city'.

(((∰))) SEE WEB LINKS
• MIT's 'Just Jerusalem'.

justice 'Geographers have long been concerned with questions of social justice. In recent years there has been a particular emphasis on the processes through which certain social groups such as women, ethnic minorities, children, lesbians and gay men and disabled people are excluded from, or marginalized within, everyday spaces Research on these topics inevitably has led to questions about how academics can address these issues in ways that are sensitive to, and empower those participating in, such studies and consequently to debates about professional ethics' (Valentine (2003) *PHG* 27; see also D. Sibley 1995).

Geographical investigations of equity, fairness, and justice have centred on spatial equality, territorial justice, and minimum standards (Hay (1995) *TIBG* 20, 4): 'what is just at one geographical scale for one set of workers might be unjust for other workers elsewhere.' While aeronautical engineers working in Seattle might feel they are justly treated by Boeing, Mexican workers stitching seat-covers for Boeing aircraft may not enjoy equal pay and conditions. However, 'the logic of capital is ... not about ... social equity, human happiness, or environmental justice' (N. Castree et al. 2004).

Raustiala (2005) *Fordham Law Review* rejects the US Federal Court's notion that 'American law is tethered to territory—that simply by moving an individual around in space, the rights that individual enjoys wax and wane ... spatial location ought not woodenly foreclose the existence of constitutional rights for noncitizens subject to American power outside the boundaries of the United States'. Osofsky (2008) *Villanova Law School* 53 explores ' "wormholes" in the US legal system that transport people ... into another timespace in which basic protections are absent'. He is referring to rendition flights. Of equal interest is Jeffrey's paper (2011) *TIBG* 36, 3, 344 on the Dayton Peace Accords over Bosnia and Herzegovina, where legal processes of justice—in the USA—remained remote from the locations where the crimes took place, and whereby spaces of 'transitional' justice were marginalized.

The Brundtland Report establishes the pursuit of intra- and intergenerational equity as key dimensions of future progress towards sustainability. 'If the ability to live in an environment which meets reasonable standards of quality and tolerable levels of environmental risk is included as a basic need ... pursuing environmental justice is ... central to the broader framing of sustainable development' (Walker and Bulkeley (2006) *Geoforum* 37, 5). See A. Dobson, ed. (1999) on *sustainable development and intergenerational justice.

just-in-time system (JIT) A system of production which aims to deliver all the necessary inputs, such as raw materials, components, and labour, just in time for the appropriate stage of production. This is essentially a flexible system of manufacturing, linked with small production runs and flexible methods of production. It is largely credited with being a Japanese innovation. 'Dell [computers] has created a distinctive geography based upon a small number of geographically dispersed, strategically located assembly sites connected into close concentrations of suppliers needed for its just-in-time system ... "we are not experts in the technology we buy. We are experts in the technology of supply chain integration"' (G. Fields 2004).

Kamchatka current A cold *ocean current off the peninsula of the same name in Siberian Russia. See Cokelet et al. (1996) *J. Phys. Oceanogr.* 26, 7.

kame An isolated hill or mound of stratified sands and gravels which have been deposited by glacial meltwater. Lundqvist (2003) *Geografiska A* 85, 1 is recommended; see also Bennett et al. (2000) *Quat. Sci. Revs* 19, 7. **Kame terraces** are flat-topped, steep-sided ridgs of similar *fluvio-glacial origin, running along the valley side. They are *ice contact features, formed between the side of a decaying glacier and a valley wall; see Lundqvist (1995) *Geomorph.* 11, 3 and Bennett et al. (2000) *Quat. Sci. Revs* 19, 7 on lake-bed kame terraces. **Moulin kames** form below *moulins (Carlson et al. (2005) *Geomorph.* 67, 3–4).

karre A collective name (pl. *karren*) for the shallow channels formed by solution on exposed limestone. **Kluftkarren** are enlarged joints; **rillenkarren** are very closely spaced small runnels; small radiating rillenkarren are the overflow of surface solution pans. **Rinnenkarren** are both longer and deeper, as much as 10 m in length and 0.5 m in depth. They may develop as a result of coalescence of small channels. Their walls are sharp, in contrast with the large, rounded hollows known as **rundkarren**. The types of karren can be explained as the results of the solution process under different hydro-dynamical behaviour (Ferrarese et al. (2003) *Geomorph.* 49, 1–2).

karst (labyrinth karst) The deep canyons of limestone formed by *carbonation. Initially the limestone shows the wide, deep tissues known as bogaz; these widen and deepen into long gorges known as karst streets with other, cross-cutting, lines of erosion. The remnant of this carbonation is **tower karst**. See Zhu and Chen (2006) *Speleogen. & Evolution Karst Aquifers* 4, 1 on **cone karst**, tower karst, and tiankengs (giant dolines), and Harmon et al. (2006) *GSA Spec. Paper* 404 for perspectives on karst geomorphology, hydrology, and geochemistry. Karstic features do occur in non-carbonate rocks; see Uagodo et al. (2011) *Zeitschr.* 55, 1, 1.

kata- From the Greek *kata-*, down-sinking, as in a *katabatic wind. The term is also used to describe the sinking of air in the warm sector of a *depression at a cold or a warm **kata-front**, bringing about a large-scale inversion of temperature at the fronts, which are fairly inactive. At a kata-warm front, cloud development is limited to cirrus and high stratus, and precipitation is restricted to light rain; at a kata-cold front strato-cumulus is common, and precipitation is similarly moderate.

katabatic wind A gravity-driven atmospheric current, which is forced by cooling air adjacent to a sloped surface (Gallée et al. (1996) *J. Appl. Met.* 35, 7). Descending, *adiabatically warmed katabatic winds are *föhn winds. Cold katabatic winds result from the slumping down of very cold, and hence dry, air.

Coastal Antarctica is dominated by katabatic gales; the gentler katabatic flows of hill slopes produce *frost hollows.

kettle hole Large masses of ice can become incorporated in glacial *till and may be preserved after the glacier has retreated. When one of these bodies of ice finally melts, it leaves a depression in the landscape; a kettle hole. **Kame and kettle topography** is hummocky terrain evolved by melt-out, comprising pitted or kettled outwash (*sandur).

Khamsin *See* LOCAL WINDS.

kinematic wave theory A theory developed by Lighthill and Whitham (1995) *Procs Royal Soc., London, Series A* 229 which has applications in geomorphology. See Howes and Abrahams (2003) *Geomorph.* 53, 1–2 for a kinematic wave equation for overland flow, and Van de Wal and Oerlemans (1995) *J. Glaciol.* 41 on high-velocity kinematic waves in glaciers, generated by a major accumulation of snow above the *firn line, and causing a *glacial surge when the wave reaches the snout.

kinetic energy The energy of motion; in geomorphology, the energy used by wind, water, waves, and ice. Kinetic energy for channelled flow is defined as:

$$\frac{MV^2}{2}$$

where M is the mass of water, and V is the mean velocity.

kinship Family relationships: connection by blood, marriage, or adoption. Megoran (2006) *Pol. Geog.* 25, 622 shows that kinship networks in Uzbekistan/Kyrgyzstan form an integral part of the individual's meaning and orientation in the world.

knick point (nick point) A point at which there is a sudden break of slope in the *long profile of a river. In areas of uniform geology, the presence of a knick point may be evidence of *rejuvenation; the river is forming a new, lower profile cutting first from the mouth of the river and working upstream as *headward erosion takes place. See Crosby and Whipple (2006) *Geomorph.* 82, 1–2 and Schlunegger and Schneider (2005) *Geomorph.* 69, 1–4.

A **knickzone** is a steep reach caused by more resistant lithology, by an increase in shear stress downstream of a confluence, or by surface uplift (Bishop et al. (2005) *ESPL* 30, 6).

Knickzones form in response to base level fall, which is ultimately driven by *eustasy, *drainage capture, and *tectonics (Pederson and Tressler (2012), *Earth & Plan. Sci. Letts.* 345–48, 171). Abbhül et al. (2011) *ESPL* 36, 1464 analyse distinct knickzones in river profiles on the Western Escarpment of the Andes.

knowledge-based organizations

Those organizations that successfully create new knowledge as the basis for new products and services, usually associated with large urban agglomerations. See Henry and Pinch (2000) *Geoforum* 31, 2, 191 on the **knowledge community** of Motor Sport Valley. Today, knowledge is considered among the most important resources, and an essential driving force of economic growth; see Y. Aoyoma, J. T. Murphy, and S. Hanson (2011), section 6.1. The **knowledge-based economy (KBE)** generates information rather than goods and services. M. Castells (1996) sees the production of information and of knowledge-intensive goods and services as characteristics of post-industrial societies; others see it as a product of globalization (see J. Dunning, ed. (2004) and Schamp et al. (2004) *Eur. Plan. Studs* 12, 50). Larner (2007) *TIBG* 32, 3 reports on the use of long-standing informal networks and opportunistic links with expatriate experts in the KBE of New Zealand.

Not everyone sees the rise of the KBE as beneficial. Von Osten in P. Spillmann, ed. (2004) points out that 'controlled access to knowledge goods and information ... creates new global differences in power, new forms of resistance and subversive practices', and Chi-ang Lin (2006, *Eco. Econ.* 60, 1) is troubled by the persistent emphasis on knowledge and economic growth at the expense of poverty reduction and environmental conservation. See Hudson (1999) *Eur. Urb. & Reg. Studs* 6, 1.

knowledge flows The mobility and migration of the highly skilled in knowledge-based economies is interwoven in complex ways with flows of knowledge. Hughes (2007) *PHG* 31, 4 summarizes the literature. Park (2003) *Papers Reg. Sci.* 82, 2 demonstrates that the internet has greatly contributed to the diffusion of codified knowledge and technology to China and the Asia Pacific Rim, but the distribution of knowledge is spatially uneven. Thus, he argues (2004, *PHG* 28, 3) 'we should not abandon policies to reduce the disparity in the knowledge-based economy'. Egger et al. (2007) *Ann. Reg. Sci.* 41 introduce factor mobility in the **knowledge-capital model** to explain why European nations are less specialized than US regions.

knowledge in geography Harrison (2002) *Geoforum* 33, 4 considers the way certain presuppositions and idealizations over the nature of understanding and meaning are or have been built into our (social scientific) modes and methods of explanation. For Brown (2004) *TIBG* 29, 3 the key questions are: what are the motivations behind chosen research topics and questions? What are the reasons for favouring one particular methodology in tackling a given research question?

Lane (2001) *TIBG* 26 stresses the need to achieve the closure required to make things amenable to study; B. L. Rhoads and C. E. Thorn (1966) wonder how place shapes our understanding. Couper (2007) *TIBG* 32, 3, drawing on Wittgenstein, observes that parallel ideas may be expressed in different ways in disciplines in geography: for example *see* HYSTERESIS; CONTINGENCY. See also Rose (1997) *PHG* 21, 3 on situating knowledges. Vallance (2007) *Geog. Compass*, 4 detects the emergence of multiple economic geographies of knowledge, the diversity of learning processes in different types of organizational and spatial setting, and the circulation of mobile forms of knowledge across geographical space.

Kondratieff cycle An economic theory that states that Western capitalist economies are susceptible to extreme performance volatility as they expand and contract over the years. In contrast with what is referred to as the business cycle, the Kondratieff Wave (1935, *Rev. Econ. Stats* 17) holds that these fluctuations are in fact part of much longer cycles known as 'super cycles' that last between 50 or 60 years or longer depending on factors such as technology, life expectancy, etc. and thus must be examined over their entirety to be best understood (2012 *Platform Economic Geography* is strongly recommended).

Each cycle lasts 50–60 years and goes through development and boom to recession. The first cycle was based on steam power, the second on railways, the third on electricity and the motor car, and the fourth on electronics and synthetic materials. I think there is now a fifth cycle, based on *knowledge. Alas, Smith (2003) *PHG* 27, 1 calls the Kondratieff cycle one of the 'big and spectacular accounts of the world', exaggerated in order to simplify, pretending that the partial is the whole.

Kuro Shio A warm *ocean current which, fed by the North Equatorial current, runs from the Philippines to Japan, thence feeding into the North Pacific current.

Kyoto Protocol An international agreement linked to the United Nations Framework Convention on Climate Change, which commits signatories to internationally binding emission reduction targets; a protocol 'recognized as the most important global agreement of the late twentieth century, not only for fixing greenhouse gases (GHG) emission limits to be achieved by industrialized nations by 2012, but also for providing three flexible mechanisms through which industrialized countries can achieve their emission reduction objectives. These mechanisms: Joint Implementation (JI), Emission Trading (ET), and the Clean Development Mechanism (CDM) allow industrialized countries to meet their emission reduction targets through joint efforts or through projects abroad rather than through domestic actions alone. They give industrialized countries the opportunity to earn emission reduction credits anywhere in the world, at the lowest cost possible' (Olawuyi (2010) *Law, Environment & Devt J.*, p. 21).

Kerr (2007) *Area* 39, 4 examines the efficacy of national climate programmes to meet emissions reduction targets: only four of 21 countries with defined programmes demonstrate improved emissions trends following their inception and in only one is the change statistically significant. See also Bailey (2007) *AAAG* 97, 3.

labour The manual or intellectual work which is one of the *factors of production. **Labour intensive companies** seek to find more profitable locations for their activities across the globe. The European Union project (CIT2–CT2004–001695) reports that this delocalization trend has its greatest impact on 'intermediate' countries (such as Greece) which face competition from advanced economies on quality, and from developing countries, on price. S. S. Barney (2001) suggests that China may only be able to maintain its comparative advantages in labour-intensive manufacturing by moving production from eastern coastal regions, where labour costs have risen, to the cheaper western regions.

Sanyal and Menon (2005) *Econ. Dev. & Cult. Change* 53, 4 find that **labour disputes** have had detrimental effects on the location of new investment in India. Herod (2002, ESRC Seminar on Geographies of the New Economy) argues that the 'threat of capital flight has allowed corporations to play workers against each other, [undermining their] bargaining power', and Lambert (2005) *J. Indus. Relations* 47, 3 documents the demise of **organized labour** in Australia.

The **labour market** is the mechanism whereby labour is exchanged for material reward. R. Johnston et al. (2000) argue that all labour markets are territorially constituted, the results of the relationships between home, residential setting and workplace, and between production and reproduction, and that the geographical scales over which they operate have become increasingly intertwined and interactive. R. Martin (with P. Morrison, ed. 2003), argues that 'there is no such thing as the national labour market, instead a mosaic of local and regional markets'; thus, **segmented labour market theory** sees the labour market as composed of self-contained sub-markets. Sunley and Martin (2002, *ESRC*) show that the UK government's schemes for tackling youth unemployment have varied significantly across different parts of the country, depending on local labour markets, and Gordon (1995) *TIBG* 20, 2 focuses on the significance of labour market segmentation for migration conditions.

Monastinotis (2007) *Area* 39, 3 presents a set of **labour market flexibility** indicators for the UK. Bande and Karanassou (2006) *ersa06* show that the degree of labour market flexibility differs between high and low unemployment regions.

labour, geographies of The study of the spatial contexts and scales in which workers live, and organize, and work to address particular economic and political problems. Mitchell (2011) *Antipode* 43, 2, 563 believes that 'putting workers and their practices and interests right at the heart of our analyses and making these ontologically prior in our theorizing' are at the heart of labour geography. Smith and Butler (2000) *Env. & Plan. A* 32, 9 see it as the ways in which labour, largely political labour, has influenced the geographical landscape; 'understanding work requires knowledge of the wider economic, political, and cultural context in which employment takes place'. Herod (2003) *Int. Labor & Work.-Class Hist.* 64 argues that 'workers make their own geographies, though not under the conditions of their own choosing'.

Major themes are: the role of work and employment in contemporary capitalist society (M. Pacione (2005) and Cumbers et al. (2003) *Env. & Plan. A* 35, 9); workers' socio-geographical positionality in explaining employment within and beyond the workplace (Smith (2000) *Env. & Plan. A* 32, 10 on economic practices in eastern Europe, and Herod (2000) *Env. & Plan.* A 32, 10 on workers, neoliberalism, and globalization); workplace structures and identity (Dowell in T. Barnes et al., eds (2002) and Wills et al., guest editorial (2000) *Env. & Plan. A* 32, 9).

Gazier (2006) Background Paper, 11th EU–Japan Symposium, discusses the differentiated impact of globalization on work, and Gritsai (2005) *GaWC Res. Bull.* 162 and Winther (2001) *Urban Studs* 38 write on the evolution of a changing spatial division of labour. See Yildirim (1999) *Work, Empl. & Soc.* 13 and Whitelegg (2003) *Antipode* 35, 2 on spatial changes in employment relations. MacKinnon et al. (2008) *Env. & Plan. A* 40, 6 detect a major shift in industrial relations 'away from national collective bargaining to a system of localized company bargaining'. See Jones in Valler and Wood, eds (2004) on geographies of local labour regulation, James (2007) *Geog. Compass* 2, 1 on gendered geographies of labour, Wills (2001) *Antipode* 33, 3 on geographies of union organization; Wills and Lincoln (1999) *Env. & Plan. A* 31, 8 on cultures of work; and Jackson and Palmer-Jones (1999) *Dev. & Change* 30, 3 on the embodiment

of employment. See also Lier (2007) *Geog. Compass* 1, 4 for a review of labour geography.

labour theory of value The Marxist contention that the value of a product reflects the amount of labour-time needed to make it. If the capitalist pays low wages that do not reflect the labour expended, he will obtain surplus capital. This may be seen as exploitation, which can lead to class conflict. Brown (2008) *Camb. J. Econ.* 32, 1, 1 25 is really useful.

labyrinth karst Deep canyons of limestone formed by *carbonation. The wide, deep fissures known as bogaz widen and deepen into long gorges (*karst streets*) with other, cross-cutting, lines of erosion. The remnant of this carbonation is *tower karst. In the late stages of evolution, labyrinth karst is replaced by limestone towers (Brook and Ford (1978) *Nature* 275).

laccolith An intrusion of igneous rock which spreads along bedding planes, forcing the overlying strata into a dome. Rocchi et al. (2002) *Geology* 30, 11 delineate the two-stage growth of laccoliths at Elba Island, Italy.

lacustrine Of lakes, especially in connection with sedimentary deposition. **Lacustrine plains** result from the infilling of a lake. Soil parent materials are usually fine grained, well sorted, and often *varved. Ground surfaces are level to gently inclined and slightly concave. Landform elements include lakes, playas, some ox-bows, and some lagoons. For **lacustrine terraces**, see Korotkii et al. (2007) *Russ. J. Pacific Geol.* 1, 4.

lagoon A bay totally or partially enclosed by a spit or reef running across the entrance, known in the Baltic as a *haff*; see Miotk-Szpiganowicz et al. (2007) *2nd MELA Conf. Abstr.*

Peak rainfall

Peak discharge

Lag time
(time between peak rainfall and peak discharge)

Increased rainfall and discharge →

Falling limb

Rising limb

Time →

Lag time

lag time (lagged time) The interval between an event and the time when its effects are apparent; for example, in the storm *hydrograph.

lahar A rapidly flowing, high-concentration, poorly sorted sediment-laden mixture of rock debris and water (other than normal streamflow) from a volcano (Lavigne et al. (2007) *Forum Geografi* 21, 1); 'transient sediment-water flows whose properties are unsteady so that the sediment load fluctuates during the flow' (Lavigne and Thouret (2003) *Geomorph.* 49, 1). Lahars most commonly occur when a crater lake or an ice-dammed lake suddenly overflows; perhaps because of an eruption, the collapse of a dam, heavy rain, snow melt, or the mixing of a *nuée ardente with lake water. See (28 Sept. 2007) *GeoNet* on the lahar from Mt Ruapehu, New Zealand. Berti and Simoni (2007) *Geomorph.* 90, 1–2 propose a method for delineating lahar-hazard zones in volcanic valleys.

laissez-faire economics The view that a *market economy will perform most efficiently if it is free from government intervention, and is subject only to market forces. This view takes no account of environmental degradation (see Dean and Lovely (2008) *Economist's View* on China) or social justice (see Antle (1993) *Amer. J. Agric. Econ.* 75, 3). 'The free-market is not . . . a natural state of affairs which comes about when political interference with market exchange has been removed. In any long and broad historical perspective the free market is a rare, short-lived aberration. Regulated markets are the norm, arising spontaneously in the life of every society. The free market is a construction of state power' (J. Gray, 1998). Gray shows convincingly that laissez-faire capitalism has a very ambivalent relationship to democracy as it demands serious social control (see Wills (2000) *PHG* 24, 4). In the UK, L. Elliot and D. Atkinson (1998) argue that this control has been achieved partly through the extension of insecurity into the nooks and crannies of everyday life for the majority of the population. Vojnovic (2003) *Geograf. Annal. B* 85, 1, in a study of Houston, Texas—'the archetype laissez-faire city'—finds that, despite the local laissez-faire rhetoric, government intervention in Houston's growth has been vital. Additionally 'there is a strong contradiction between neoliberal laissez-faire in the East and growing regulative intervention in the West' (M. Redclift and G. Woodgate, eds 1997).

laminar flow A non-*turbulent flow where the movement of each part of the fluid (gaseous, liquid, or plastic) has the same velocity, with no mixing between adjacent 'layers' of the fluid. It may be seen at low velocities in a smooth, straight river channel (see Smith et al. (2007) *PPG* 31, 4).

land breeze A wind blowing from land to sea (an offshore wind) which develops in coastal districts towards nightfall. Pressure is higher above the land than above the sea, as the land cools more rapidly in the evening, and air therefore moves seawards along the *pressure gradient. See Holmer and Haeger-Eugensson (1999) *Phys. Geog.* 20, 2.

land capability The potential of land for agriculture and forestry depending on its physical and environmental qualities. The main factor investigated is soil type, but climate, gradient, and aspect are also considered. Present land use is not taken into account.

🌐 **SEE WEB LINKS**
• A model of changes in land capability for Tasmania.

land classification Land classification is essential for geographers, planners, and, increasingly, for environmental scientists (Owen et al. (2006) *Landsc. & Urb. Plan.* 78, 4). There is 'no single correct way to describe reality and solve practical questions' regarding the classification of land cover, and all classifications are subjective. Thus, the quality of the classification depends on the skill of the interpreter (Lofvenhaft et al. (2002) *Landsc. & Urb. Plan.* 58, 2–4). The Food and Agriculture Organization of the United Nations Land Cover Classification System uses an initial dichotomous phase, where the user derives the main land cover type, and a subsequent modular-hierarchical phase, where a land cover class is defined by determining one classifier at a time.

🌐 **SEE WEB LINKS**
• The FAO Land Cover Classification System.

land consolidation A type of *land reform which aims to give each farmer one relatively large plot of land rather than scattered, small parcels of land. Issues in land consolidation include: the consent of the peasants; an assessment of the value of current farmland holdings; consolidation as the promoter of rural development; the assignment of new farmland for each household; the application of modern technologies; the control of corruption; an appeal system; the reimbursement of expenses by local or national government; and, controversially, population control (Zhou (1999) *EUI W. Paper* 99, 1). Kirit and Nagarajan (2004) *Occ. Paper, Nat. Bank Agric. & Rural Dev.* 31 record huge benefits from land consolidation in the Indian state of Tamil Nadu.

land cover The observed biophysical cover of the Earth's surface. Strictly speaking, it describes vegetation and man-made features, and omits bare rock and water, although, in practice, these elements are often included under this term.

land degradation A noted decrease in, or any adverse influence on, the *carrying capacity of land, revealed by: decreased soil fertility; increased soil erosion; removal of vegetation cover; and general negative changes in land use which compromise the supply and quality of ground and surface water, biodiversity, the carbon-oxygen cycle, and general air quality (anon. (2000) *Mine Policy & Mine Monitoring Unit*). Land degradation is seen as 'an unavoidable by-product of mining'; but see Singh (2004) *EnviroNews* 10, 1.

In Australia, land clearing and the use of European farming systems have been major causes of severe land degradation, including *salinization and erosion (Pannell and Ewing (2006) *Agric. Water Manage.* 80, 1–3). Douglas (2006) *Geog. Res.* 44, 2 links globalization in South-East Asia with land degradation, but finds that local changes may have a greater impact. Davis (2006) *Geog. J.* 172, 2 finds that land degradation in dryland Morocco is commonly blamed on overgrazing, despite existing evidence that ploughing marginal lands and over-irrigation are the primary drivers: 'all too often the outcomes of neoliberal reforms on the environment . . . include increased pollution of air, earth and water, and land degradation in the form of deforestation, soil exhaustion, salinization and erosion.' Thornes (2007) *Geog. Res.* 45, 1 emphasizes the need to take forage ecology into account when managing grazing in relation to land degradation.

Mambo and Archer (2007) *Area* 39, 3 use remote sensing to detect and map susceptibility to land degradation, and Kiunsi and Meadows (2006) *Land Deg. & Dev.* 17 present a land degradation model. See also A. Conacher (2001).

land economics The study of land use and of the factors which influence and shape it. Land values are central to this study as land use and land values are interlinked; see the journal *Land Economics*.

landform segmentation Minár and Evans (2008) *Geomorph.* 95, 3–4 introduce the concept of elementary forms (segments, units) defined by constant values of fundamental morphometric properties and limited by discontinuities of the properties.

Land Information System A system for the capture, storage, manipulation, analysis, and display of land-use data. See Tabi et al. in A. Bationo et al., eds (2007) on Nigeria, and Karikari et al. (2003) *Prog. Dev. Studs* 3 on Ghana.

landless workers' movement (landless movement) In Brazil, a social movement aiming to fight for access to the land for poor workers in general, land reform, and to counter social issues hindering access to land, such as racism, sexism, media monopolies, and unequal income distribution. Ondetti (2006) *Latin American Pols & Soc.* 48, 2, 61 is very helpful.

land-locked state A nation with no access to the sea. Globally, there are twenty-six land-locked states ranging in size from the tiny Vatican City to Mongolia. Geography has put land-locked countries in a position of disadvantage. To be land-locked is more detrimental for a country's development than at any time in the past; see A. Chowdhury and S. Erdenebileg (2006).

land management The way land is used is driven by the interplay of economic, social, and environmental factors. Land management is about finding the right balance of these, often competing, factors that allows sustainable land use. Whenever land is put to use, it is subject to stresses that can be quite high, as in the case of intensive agriculture. These stresses may not only affect the biodiversity of the land, such as through the loss of hedgerows, but can also have widespread social impacts—for example, by detracting from the land's natural beauty. Research on multifunctional land uses looks at land from a holistic perspective to understand how different patterns of use cause different stresses, what the consequences of these stresses are, and how they can be managed or corrected. For example, Amundson (2001) *Ann. Rev. Earth & Planet. Scis* 29 argues that the store of terrestrial carbon can be increased by proper land management practices. Pressing land management issues make sound ecological information and conservation thinking a major imperative (Foster (2002) *J. Biogeog.* 29). Bowman (2001) *J. Biogeog.* 28, 5 argues that ecologists must learn to harness the power of environmental history narratives to bolster land management practices designed to conserve biological heritage. See Eriksen (2007) *Geog. J.* 173, 3 on fire as a savanna land management tool, and Lin and Ho (2005) *AAAG* 95, 2 on the role of the Chinese socialist state in land management.

((⚙)) SEE WEB LINKS

• EC Environment Research website.

land reform A sweeping change in land tenure. It usually involves the breaking-up of large estates and the widespread redistribution of the land into smallholdings, but may also be *land consolidation. 'Where issues of justice and equity, [and the] sustainable and productive use of land are addressed in conjunction with one another, land reform can make a major contribution to the future of the Republic of South Africa, to sustainable livelihoods, and to the broader issue of ecologically sustainable development' (Attfield et al. (2004) *Third World Q.* 25, 2). The redistribution of

land is a complex, and slow, business; in 1994 whites owned about 87% of South Africa's farmland. The new government set a target for at least 30% of it to be transferred to blacks by 2014. More than a decade on, only 4% had changed hands; see Oudtshoorn (3 May 2007) *The Economist.*

land rent theory Any theory that explains rental value of parcels of land, for land rents are not exclusively determined by the intrinsic qualities of any parcel of land. Sivam (2002) *Habitat International* 26, 523 is very good on the urban residential land market in India. *See also* BID-RENT THEORY; VON THÜNEN MODEL.

land rights The legal entitlement to land. See Sjaastad and Cousins (2009) *Land Use Policy* 26, 1, 1 on the formalization of land rights in the South, and Patel et al. (2011) *Environ. & Urbanization ASIA* 2, 1, 45 on Mumbai.

landscape 1. All the visible features of an area of land, the appearance of an area, or the gathering of objects which produce that appearance: 'the expression of interaction between humans and their environment' (C. Sauer 1925). In a rather grumpy paper, Minca (2007) *PHG* 31, 2 complains that ideas of the geographical landscape have remained essentially immune to the extraordinary body of critical reflection in the writings of Denis Cosgrove, Stephen Daniels, James Duncan, and other British cultural geographers. 'We may think of individual "landscapes" as being compromised, partial, contested and only provisionally stable as modes of ordering the world and our engagement with it. If so, this suggests that we should not think of individual landscapes as discrete pieces of territory because they are supported by, and help to sustain, the interests of mere sections of any given society.

Alternatively, we might think of landscapes as being formed in relation to other landscapes and conceptions of landscape. In that case, perhaps also we should base our analysis in terms of the interconnectedness of landscape, its links with other landscapes, other geographies' (J. Gold and G. Revill 2000). 'Landscape is far from being a neutral concept—[it is] is deeply rooted in a performative and ideological mode of engagement with site' (M. Shanks, site 2006). Revil (2007) *TIBG* 32, 2 argues for 'landscape as a mode of governance'. Henderson in C. Wilson and P. Groth, eds (2003) argues that landscape in geography can be broken down into landscape as *Landschaft; landscape as social space; the epistemological landscape ('the material revelation of human practice and thought'); and the apocryphal landscape ('landscape as a way of seeing').

B. Rhoads and C. Thorn (1996) define the geomorphological landscape as 'the totality of surface landforms', referred to by Thwaites and Bell (2004) *SuperSoil*, as the *geoscape*.

2. An overview of a specific phenomenon, and its structures, which might not be material. Parto (2005, *MERIT-Infonomics Res. Memo* 2005–001) writes that 'the new **landscape of governance** . . . is marked by a departure from territorially defined spaces of the post war nation states . . . governance in this new landscape increasingly has to rely on formal and informal, supra- and sub-national institutions of the nation states if it is to attain a similar degree of legitimacy as governance at the national territorial scale'. See Hashim (OECD 2002) on Malaysia, and Robinson and Shaw (2001) *Reg. Studs* 35, 5 on north-east England. Parto (op. cit.) contends that 'each subsystem is the product of its "own" institutional landscape'. Make of that what you will.

The term is similarly used in **landscapes of care**; see Moon et al. (2006) *TIBG* 31, 2; Gleeson and Kearns (2001) *Env. & Plan. D* 19, 1; and Williams (2002) *Sol. Sci. & Medi.* 55, 1. See the Health, Place and Society Research Group at Queen Mary College, U. London.

landscape conservation The planning and management of the scenic and wildlife resources in geographical and environmental systems. See Pinto-Correia (2000) *Landsc. & Urb. Plan.* 50, 1–3 on the possible integration of different interests and policies for nature conservation, and Watson et al. (2005) *J. Biogeog.* 32, 8 on matrix management in the conservation of fragmented landscapes.

landscape ecology Within this study, a landscape is a 'heterogeneous land area composed of a cluster of interacting ecosystems that is repeated in similar form throughout' (R. T. T. Forman and M. Godron 1986). Landscape ecology specifically addresses the importance of spatial configuration for ecological processes (M. Turner et al. 2001). The central themes are: detecting a pattern, and expressing it in quantitative terms; identifying and describing the agents of pattern formation; understanding the ecological implications of the pattern; spatially and temporally characterizing the changes in pattern and process; and managing landscapes to achieve human objectives. For ten differing definitions of landscape ecology, see the United States Regional Association for Landscape Ecology site.

landscape evaluation An attempt to assess the landscape in objective terms. Sometimes a consensus of views on the landscape is sought so that particular landscapes may be chosen as being outstandingly beautiful. Landscape description studies try to identify important items such as topography or buildings. Some kind of ranking method may be attempted to compare one landscape with another. Using this idea, Leopold attempted to calculate how close a landscape was to being unique. Personal preferences may also be used.

landscape evolution The processes through which the landscape emerges. Earth's landscapes reflect the interaction of climate, tectonics, and denudational processes operating over a wide range of spatial and temporal scales; these processes can be considered catastrophic or continuous, depending on the timescale of observation or interest.

Perron and Fagherazzi (2012) *ESPL* 37, 1, 52 suggest a sequence of landscape evolution. Firstly, variability in initial conditions can give rise to steady-state landscapes in which the characteristics of individual landforms vary about the mean, even in a homogeneous system with constant forcing. Secondly, landforms can have a range of mean equilibrium dimensions, either because different initial conditions evolve to different 'attractor' states, or because landscapes evolving from different initial conditions towards a single state converge too slowly to reach equilibrium under natural conditions. Thirdly, a landscape that experiences a perturbation or a change in process may begin to evolve towards a different equilibrium state, potentially leading to *hysteresis or rapid changes in topography. See Himsah and Korup (2012) *ESPL* 37, 2, 249 for an overview of key themes in landscape evolution.

A **landscape evolution model** is a mathematical theory describing how the actions of various geomorphic processes drive and are driven by the evolution of topography over time. Tucker and Hancock (2010) *ESPL* 35, 1, 28 provide a useful summary. Dymond and Rose (2011) *Geomorph.* 132, 1–2, 29 offer an alternative phenomenological approach to modelling landscape evolution, and compare their model to the highly erodible Waipaoa catchment in New Zealand. There is good agreement between the predicted and observed topography.

landscape matrix The most extensive and connected landscape type, playing the dominant role in landscape functioning. The characteristics of matrix structure are the density of the patches (porosity), boundary shape, networks, and heterogeneity. The type of landscape matrix surrounding a *patch can mitigate the negative effect of habitat isolation for a given species, according to its degree of matrix habitat use (Barbaro et al. (2007) *J. Biogeog.* 34, 4). V. Ingegnoli (2002) describes three matrix types: continuous, with a single, dominant element type; discontinuous, with a few co-dominant element types; and web-shaped, with connected corridors of prevailing functions.

landscape morphology In *geomorphology, the form and spatial structure of the *landscape. Landscape morphology has a direct influence on water movements, soil, physical and chemical properties, and on the productivity of the vegetation cover. In *human geography it is the material formation of the landscape, its shaping and reshaping, in which social structures and cultural worlds are enfolded.

landscape pattern The existing pattern in the landscape whose characteristics are filtered by constraints of scale and resolution (in space and time) of the reception system. Technically, this system is defined by physical sensors; the human version is called visual perception (Bartel (2000) *Ecol. Model.* 130, 1–3). *Landscape ecology seeks to associate ecological processes with landscape pattern (Griffith et al. (2002) *Sing. J. Trop. Geog.* 23, 1). See Lü et al. (2007) *PPG* 31, 6 on the interactions between landscape pattern and nutrient transport processes. Misclassification can cause large errors in landscape pattern indices (Foody (2007) *PPG* 31, 4). See Li and Wu, in J. Wu and R. Hobbs, eds (2007).

landscape preference It is argued that most cultures have a preferred landscape: for example, Eleftheriadis et al. (1990) *Env. Manage.* 14, 4 find that, out of six European groups, Greeks preferred sea landscapes most, and Italians, Austrians and Yugoslavs (*sic*) least. Wherret (1999) *Landsc. & Urb. Plan.* 45, 4 uses the internet to detect preferred Scottish landscapes (but doesn't tell us what they are). D. M. Williams (2002) theorizes that landscape preference and sense of place must be understood as existing in juxtaposition to other places and scales. See T. Daniel and R. Booster (1976) on predictive models for estimating landscape preference and scenic beauty. Herzog et al. (2000) *En. & Behavior* 32, 3 observe that young children display higher landscape preference, since they see landscapes as 'playscapes', while teenagers are more interested in social concerns.

landscape sensitivity The potential for landscape dysfunction as a result of human impacts; the likelihood of change (Miles et al. (2001) *Catena* 42, 2 landscape health). In reality, landscape sensitivity is difficult to determine, as land uses tend to be confounded with climate type, and assessment of dysfunction is highly related to land use (Hobbs and MacIntyre (2005) *Glob. Ecol. & Biol.* 14, 1).

Brunsden (2001) *Catena* 42, 2 states that the landscape sensitivity concept concerns the likelihood that a given change in the controls of a system or the forces applied to the system will produce a sensible, recognizable, and persistent response. Landscape sensitivity (potential instability) may develop as a function of environmental factors acting over a long time period. Inherent landscape sensitivity becomes manifest when change reaches critical thresholds triggered either by extreme climatic events or by cumulative human impact or when both act in tandem (Marker (2003) *Geog. J.* 169, 1).

landscape studies The examination of the structure and organization of landscapes; of interrelationships between place-making, personal and social memory; of landscape as the location of cultural and ecological patterns, processes and histories; of the social life of landscapes—their reception by different societies; of 'how and why places come to be understood and experienced as thresholds of time-space in different societies, including how places constitute geographies of belonging through and beyond urban and national space' (Till (1999) *Ecumene* 6).

Landscapes may be viewed not simply as 'scenes' into which humans are inserted, but rather as the products of human activity, shaped through and shaping cultures (C. Sauer 1925). An important development in the study of the cultural landscape is the incorporation of social theory: 'the thrust of the new landscape studies [was] to consider landscapes as part *of a process* of cultural politics, rather than as the outcome of that process' (B. Bartley et al. 2002). 'Within landscapes are particular sites—monuments or markers—which facilitate and direct the process of "collective memory" and through which social groups situate their identities in time and place' (Inwood et al. (2008) *Soc. & Cult. Geog.* 9, 4). For the social life of landscapes, see *Site/Lines*, the journal and online forum of the Foundation for Landscape Studies.

Landschaft A concept of landscape which attempted to classify landscapes, usually distinguishing between the natural and the cultural landscape. The term became part of modern geographical currency in Germany towards the end of the 19th century. 'It did not do so unambiguously, however, and this has had unfortunate consequences for the translation of the term into landscape in the English-speaking literature' (A. Baker 2003).

landslide A form of *mass movement where the displaced material retains its form as it moves. Cruden and Varnes (*USGS Fact Sheet* 2004–3072) divide landslides into slides, falls, flows, topples, and lateral spreads. Probably the world's largest landslide occurred in south-west Iran in 1937, when a segment of the Kabir Kuh ridge, about 15 km long, 5 km wide, and 300 m thick, slid off the mountain, with enough momentum to travel 20 km.

Landslides are prompted by an increase in *pore pressure through snow melt, through precipitation, and through spring action. These all reduce the friction which binds the mass to the slope. See Hungr. et al. (2008) *Geomorph.* 96, 3–4 on the technique of constructing a regional magnitude-cumulative frequency curve for landslide prediction.

land surface process models These describe the energy, water, carbon, and nutrient fluxes on a local to regional scale using a set of environmental land surface parameters and variables. In principle many of these inputs can be derived through remote sensing (Bach and Mauser (2003) *Trans. Geosci. & Rem. Sens* 41, 7).

land tenure The nature of access to land use. Common forms of land tenure are owner-occupied farms, and tenancies which basically involve payment (in the form of labour, cash, or *share-cropping) to the landlord from the tenant. A *plantation is owned by an institution and uses paid labour. Collectives may own land together and work together, sharing any profits. Under traditional communal land tenure systems in Ghana, the direct descendants of a landholder inherit their rights without losing them to a larger group (Asbere (1994) *J. Black Stud.* 24, 3). Formal land tenure laws and regulations can be conceptualized as 'one of many state simplifications which have the character of maps' (J. Scott 1998). 'In the hands of powerful lawgivers, such legal "maps" not only summarize "reality"; they also have (intended and unintended) transformative effects' (Larsson (2007) *Pol. Geog.* 26, 7).

land use zoning The segregation of land use into different areas for each type of use: agricultural, industrial, recreational, and residential. Planners have sought to pursue *sustainable development using more mixed land use zoning in order to reduce the demand for travel, and greater coordination between transport and land use planning, including support for public transport (Haughton and Counsell (2004) *Geog. J.* 170, 2). See Greed (2006) *Tijdschrift* 97, 3 on the problems for women in cities divided by traditional land use zoning, and Degg and Chester (2005) *Geog. J.* 171, 2 on land use zoning in the mitigation of seismic/volcanic hazards.

La Niña A state of unusually cool sea surface temperatures in the western Pacific, which are associated with lowered precipitation in the southern USA and western South America, but an increase in the frequency of *tropical cyclones in the Atlantic, and heavier rainfall in Indonesia and Australia. La Niña can be seen as the other extreme from *El Niño.

'There is strong statistical and modelled evidence that persistent, La Niña-like, cooler sea surface temperatures . . . produce multiyear droughts not only in the Great Plains and the Southwest, but also in the Mediterranean region of Europe, the Pampas region of South America, the steppes of Central Asia, and the outback of Western Australia' (Goodrich (2007) *Geog. Compass* 1).

lapili *See* PYROCLAST.

lapse rate In meteorology, the rate at which stationary or moving air changes temperature with a change in height. *See* ADIABAT; ENVIRONMENTAL LAPSE RATE.

latent heat The quantity of heat absorbed or released when a substance changes its physical state at constant temperature. The release of **latent heat of condensation** in the rising air of a *hurricane is the chief force fuelling that meteorological phenomenon. See Grossman and Rodenhuis (1975) *Monthly Weather Rev.* 103, 6. The **latent heat flux** is the flux of heat from the Earth's surface to the atmosphere that is associated with evaporation or transpiration of water at the surface and the subsequent condensation of water vapour in the troposphere. It is an important component of Earth's surface energy budget.

lateral accretion The build-up of sediments, as in a recurved spit (Kumar and Sanders (1974) *Sedimentol.* 21, 4), or a meander (Brooks (2003) *Geomorph.* 54, 3–4).

lateral erosion Usually of rivers; erosion of the banks rather than the bed. In a stream or river, it results in undercutting of the banks or terrace formation (Hanson et al. (2006) *Geomorph.* 76, 1–2).

lateral moraine *See* MORAINE.

laterite Thick, red, and greatly weathered and altered strata of tropical ground. *Horizons are unclear and the nutrient status of the soil is low. See Bourman (2007) *Geog. Res.* 45, 3. Laterite is soft but gets brick-hard when exposed to the air. **Lateritic soils** (*latosols) are tropical/equatorial soils characterized by a deep weathered layer from which silica has been *leached, a lack of *humus, and an accumulation or layer of aluminium and iron sesquioxides. **Laterization** is the formation of lateritic soils, taking place in warm climates where bacterial activity takes place throughout the year. D. Nahon (1991) argues that ferric *duricrusts result from long-term laterization processes and as a consequence are extremely widespread in tropical environments.

latifundium A large farm or an estate (pl. *latifundia*), particularly in Latin America. The estate is farmed with the use of labourers who sometimes lease very small holdings from the landowner. See Simmons et al. (2007) *AAAG* 97, 3 on the Amazon Land War.

latitude Parallels of latitude are imaginary circles drawn round the Earth parallel to the equator. The parallels are numbered according to the angle formed between a line from the line of latitude to the centre of the Earth and a line from the centre of the Earth to the equator. Those regions lying within the Arctic and Antarctic circles, having values of 66.5° to 90°, are termed **high latitudes**. **Low latitudes** lie between 23.5° north and south of the equator, i.e. within the tropics. **Mid-latitudes**, also known as temperate latitudes, lie between the two.

latosol A major soil type of the humid tropics with a shallow A *horizon but a thick B horizon of clay, sand, and iron and aluminium sesquioxides which, respectively, give it a red or yellow colour. Much of the silica has been leached, and latosols tend to be of low fertility.

Laurasia One of the two original continents which broke from the supercontinent, Pangaea, by *continental drift. As a supercontinent, Gondwanaland lasted much longer than Laurasia, and was hotter (Veevers (1995) *Geology* 23, 12).

lava *Magma which has flowed over the Earth's surface. The *viscosity of lava depends on its silica content, pressure, and temperature. Temperature is the most important factor. **Basic lavas** have a low silica content and flow freely (Arculus (1976) *GSA Bull.* 87, 4); **acid lavas** are more viscous (Wilson and Grant (2006) *GSA Spec. Paper* 405). *See* EXTRUSION.

law A theory or hypothesis which has been confirmed by empirical evidence. The **geography of law** explores the relationship between law and geography, 'particularly in relation to globalisation—of law, commerce, environmental change, and society—which renders relations between the local and the global more significant' (J. Holder and C. Harrison 2002). Braverman et al. (2013) *Buffalo Legal Studies Research Paper Series 2013-032* is excellent.

law of retail trade gravitation *See* REILLY'S LAW.

law of the sea A framework, agreed to by the majority of maritime nations, for administering the seas. It recognizes seven administrative zones: internal waters, the *territorial sea, the *contiguous zone, the *continental shelf, exclusive fishing zones of up to 200 miles from a nation's coastline, exclusive economic zones of the same extent as the fishing zones, and the high seas. See M. Freeman (1991).

layer 'The visual representation of a geographic dataset in any digital map environment. Conceptually, a layer is a slice or stratum of the geographic reality in a particular area, and is more or less equivalent to a legend item on a paper map. On a road map, for example, roads, national parks, political boundaries, and rivers might be considered different layers' (*ESRI GIS dictionary*).

Census Tracts
Roads
Bus Routes
Shopping Centres
Industrial Sites

layer

leaching The movement of water down the *soil profile. This results in the movement of cations, sesquioxides, *clay colloids, and *humus to the lower soil *horizons. Specific types of leaching include: lixiviation (the removal of the soluble salts containing metallic cations); the removal of *chelates; *lessivage; and, in tropical soils, *desilication.

lean production A method of production focused on getting the right things to the right place at the right time in the right quantity in order to achieve perfect work flow, while minimizing waste, being flexible, and able to change.

Rutherford (2004) *PHG* 28, 4, 425 is good on this.

least developed country (LDC) At the time of the 2003 review of the list, the following three criteria were used by the UN: low income, in the light of a three-year average estimate of the gross national income per capita (under $750 for cases of addition to the list, above $900 for cases of graduation); weak human assets, as measured through a composite Human Assets Index; and economic vulnerability, as measured through a composite Economic Vulnerability Index.

((()) SEE WEB LINKS
- UNCTAD's Least Developed Countries Report.

Lebensraum Literally, 'living space', the room needed for a nation's expansion. The term was introduced by Friedrich Ratsel (1896) and became a central concept in Nazi propaganda.

lee The side (lee side) sheltered from the wind. Somewhat confusingly, a **lee shore** is the shore towards which the wind is blowing. **Lee depressions** occur when *lee troughs develop into *lows. They are frequent in winter where mountains block low-level air streams. See R. Barry and R. Chorley (1992).

lee trough As a stable air column rises to cross over a ridge, it shrinks in the vertical, therefore *diverging in the horizontal. This gives it a negative relative *vorticity. Accordingly, in the Northern Hemisphere, the flow is deflected anticyclonically: to the right. Since low pressure always lies to the left of an airstream in the Northern Hemisphere, the pressure is lower on the lee of the mountain. See Weisman (1990) *Monthly Weather Rev.* 118, 4.

lee wave A wave motion in a current of air as it descends below an upper layer of stable air, after its forced rise over a mountain barrier, which sets up vertical oscillations (also called a **standing, rotor, hill,** or **mountain wave**). The wavelength is *c*.5–15 km with an amplitude of *c*.500 m. Lee waves are often indicated by higher wind speeds and *lenticular clouds.

legal geography Broadly, the study of the relations between law and space. This may include: the regulation of land use through planning and property; the regulation of personal movement and migration; theorizing the relationship between formal and informal means of regulation and people and place; markets; governance and social justice issues; and the movement and political expression of minority groups and interests. All of these are regulated in some way through the mechanism and institution of the law. See N. Blomley, D. Delaney, R. T. Ford (2001) for a reader on legal geography. Martin et al. (2011) *PPG* 34, 2 present a conceptual framework that integrates lawyers into existing legal geography scholarship.

leisure, geography of Leisure and consumption have become major forces in contemporary society, with social, economic, and geographical implications (Dodson (2000) *Tijdschrift* 91, 4). Geographies of leisure study the relationships between leisure and other social practices and behaviours related to human movement; the spaces produced by leisure activities (gardens, heritage sites, parks, theme parks, and so on) and the meanings given to these spaces, together with the spatial patterns of people's behaviour in their free time. See Hubbard et al. (2008) *PHG* 32, 3 on the urban geography of 'adult entertainment' (yes, really) and Jayne et al. (2008) *PHG* 32, 2 on geographies of alcohol and drunkenness. D. Crouch (1999) sees leisure, tourism, travel, recreation, time off work, holidaying (however the activity is labelled) as an 'encounter'; between people and places, people and people, people and nature, and people and their past.

lenticular cloud A lens-shaped cloud formed above a peak as air rises over mountain barriers and condensation occurs. *See* LEE WAVE.

lenticularis *See* CLOUD CLASSIFICATION.

lesbian geographies The geographies of the intangible and physical space that is defined by lesbians; the 'ways in which lesbians organize themselves in both urban and rural landscapes' (Browne and Nash in R. Kitchin and N. Thrift (2009). See *Soc. & Cult. Geog.* (2007) 8, 1 for a collection of useful papers on this topic.

less economically developed country (LEDC) A country with low levels of economic development. Indicators include high birth, death, and infant mortality rates (typically over 20; over 30; and over 50 per thousand, respectively); more than 50% of the workforce in agriculture; and with low levels of nutrition, secondary schooling, literacy, electricity consumption per head, and GDP per capita—generally below $US1 000 per capita. The word 'economically' was included in this term in order to redress perceptions of the people of financially poorer nations as uncivilized, irrational, ignorant, and unskilled.

lessivage The translocation of *clay colloids in a soil, with no change in their chemical composition. Characteristic features of lessivage are often observed in the soils without hydrological barriers hampering, or preventing, the vertical migration of soil water and mass transfer processes (Zaidel'man (2007) *Eur. J. Soil Sci.* 40, 2).

levée A raised bank of alluvium flanking a river. The bank is built up when the river dumps much of its *load during flooding. New Orleans was founded on a natural levée; elevations as high as 19 feet above sea level provided the driest land for settlement. French colonists created their own levées, and ever since permanent settlement began, residents have been expanding the levées. See Turner (2007) *Tech. in Soc.* 29, 2 on the repetitive disasters resulting from levée failure around, and within, New Orleans.

lexical diffusion In geolinguistics, phonological changes proceeding through the lexicon, word by word. See Britain (1997) *Paper Presented at NWAV-26.*

life course geography The way ideas of time and the life course are represented spatially. One technique is the use of life maps to enable people 'to be the tellers of their own stories' (Worth (2011) *Area* 43, I 4, 405). This method reinforces the participatory focus of the research. An alternative or closely matching term is **life transition**, and the December 2011 issue of *Area* devotes much of its space to this latter term.

life expectancy The average number of years which an individual can expect to live in a given society, normally derived from a national *life table. Life expectancy is usually given from birth but may apply at any age, and because, in all societies, mortality rates tend to be rather high in the first year of life, life expectancy at birth is usually significantly lower than at 1 year old. Women consistently have a longer life expectancy than men, especially in *more economically developed countries where the risks of childbirth are less than those in *less developed countries.

The lower life expectancies for less economically developed countries generally reflect high *infant mortality rates, but by the age of 70, the years of life remaining to an individual are, globally, very similar. Thus, the strong correlation between *GDP per capita and life expectancy becomes weaker as the age of an individual increases.

life space The limited time and space which an individual has in which to pursue a necessarily limited range of opportunities. Life space is the interaction of the individual with her or his behaviour setting. L. Lofland (1998) notes that the allocation of differing amounts of life space to different realms will vary by the social class of a city's inhabitants.

life table A summary of the likelihood of living from one age to any other. In a life table, a hypothetical *cohort of 100 000 births is set up and then the loss by deaths is shown for each year of life. Averages of losses are calculated for a given year, and from this the actual diminution of the cohort is shown.

life world (lifeworld) The taken-for-granted dynamic of everyday experience that largely happens automatically, without conscious attention or deliberate plan (Seamon in T. Mels, ed. 2004).

lift force The upward force produced when fluid rises over a particle. In watercourses, the particle moves up from the bed into the flow when the lift force exceeds the gravitational force provided by the mass of the particle. See Cornelis (2004) *Geomorph.* 59, 1–4.

light industry The manufacture of relatively small articles (toasters as opposed to girders), using small amounts of raw materials. In consequence, the *material index is low, and such industries are more *footloose than heavy industries.

lightning An emission of electricity from cloud to cloud, cloud to ground, or ground to cloud, accompanied by a flash of light. It results from variations of electrical charge on droplets within clouds and on the Earth's surface. This variation may be caused by the break-up of raindrops, the splintering of ice crystals, or differences between splintered ice crystals and soft hail. As a *cumulus cloud develops, the frozen upper layer becomes positively charged, and most of the cloud base negatively charged, with positively charged patches. The negative charges are attracted to the earth, which has a positive charge. When the electrical field strength gets to about 1 MVm^{-1}, the electrical insulation of the air breaks down. The result is a *leading stroke*, or *stepped leader*, from cloud to ground, which creates a conductive path between them. The return stroke, from earth to cloud, follows the same pathway (only millimetres across), with a charge up to 10 000 amps. The intense heat of the stroke engenders light, and a violent expansion of the air, making waves, heard as *thunder. Not all the negative charge may be released; there may be several return strokes, each prefaced by a downward dart leader which reactivates the channel. Where the path between ground and cloud is clearly visible, **forked lightning** is seen. The illumination of other clouds by a concealed fork is **sheet lightning**.

De Pablo and Rivas Soriano (2007) *Monthly Weather Rev.* 135, 7 report on an inverse correlation between lightning flash rates and index values of the *North Atlantic Oscillation, and Gołkowski (2011) *Acta Geophysica*, 59, 1, 183 investigate variations in the global atmospheric electric circuit.

Ball lightning has been described as a sphere of glowing light meandering through the lower air. Little is known about it, but there are a few fascinating

19th-century observations accessible from the American Meteorological Society website.

(⊕) SEE WEB LINKS
• American Meteorological Society.

limestone pavement A more or less horizontal, bare limestone surface, cut into by grikes (fissures) running at right angles to each other, leaving clints (upstanding areas of limestone) between them.

liminal In between. J. Anderson (2012) uses this term to describe adolescents, who are in between childhood and adulthood.

limnion The lower layers in a body of water that are marked by low temperatures and insufficient light for photosynthesis. Levels of dissolved oxygen are low.

limnology The study of all the biological, chemical, meteorological, and physical aspects of freshwater ponds and lakes. See P. O'Sullivan and C. Reynolds, eds (2004, 2005) and Lau and Lane (2001) *PPG* 25, 2 on shallow lake ecosystems.

line In a *GIS, a one-dimensional feature or a feature that has length. Many line features also have width in real life, but as with points, GIS software assumes that they don't (M. N. DeMers 2009).

linear wave theory Linear wave theory was first presented by Airy (1845) to provide methods to calculate wave characteristics such as height, period, and speed using a sinusoidal model of the waveform. Based on the ratio of water depth, h, to deep-water wavelength L_o, three different wave regions can be identified: deep water ($h/L_o > 0.5$), intermediate water ($0.5 > h/L_o > 0.05$), and shallow water ($h/L_o < 0.05$). This is exceptionally well explained by Masselink in J. Holden, ed. (2012). Linear wave theory relies on the assumption of closed circular particle orbits, but this assumption breaks down in the transition zone approaching shore, where waveform changes occur.

line-haul costs The costs of transporting goods over a route, but not loading or unloading. The selection of any shipment route depends on the total sum of its network access, line haul, interlining, terminal transfer, and network egress costs (Southworth and Peterson (2000) *Transp. Res. C* 8). Furuichi and Koppelman (1994, *Transp. Res. A* 28) report that business travellers value time to the airport twice as much as time on the aeroplane (**line-haul time**). The 200-year-long collapse in transport time has now been replaced by a plateau in general **line-haul speeds**, but time/space continues to collapse locally, for example through new fixed links or high speed trains (Knowles (2006) *J. Transp. Geog.* 14, 6).

linguistic distance The degree of contrast between two *dialects, measured by the percentage of items differing from a fixed set. See Gooskens (2005) *DiG* 13 on the geographical and demographic determinants of linguistic distance, and Scapoli et al. (2005) *J. Theor. Biol.* 237, 1 on the link between linguistic similarity and linguistic distance.

link The route or line joining two *nodes.

linkages Flows of inputs and outputs to and from a manufacturing plant in association with other plants. Movements of matter are **material linkages** as opposed to **machinery** and **service linkages** (such as information, advice, and maintenance). Individual plants are also tied together by **forward linkages**—supplying customers, and **backward linkages**—with their suppliers.

Horizontal linkages occur between plants which are engaged in similar stages of a manufacturing process.

literary geography The combination of the study of literature and a geographical perspective (Ridanpää (2007) *Soc. & Cult. Geog.* 8, 6); how and why the geographical setting is important in the plot or mood of a text or texts, and how the setting itself advances the plot, as in E. M. Forster's *A Passage to India*; the study of textual meaning as it is differentiated across space (Saunders (2011) *PHG* 33, 4, 436). See Sharp (2000) *Area* 32 on fictive geographies, and Kong and Tay (1998) *Area* 30.

lithification Processes by which sediments are converted into hard rock. These include the expulsion of air, or the suffusion into the rock of *cementing agents in solution, like quartz.

lithology The physical character of a rock or rock formation; its composition, structure, texture, and hardness. A **lithogenous** sediment consists of particles derived from the physical and chemical breakdown of rocks and minerals.

lithosphere The Earth's *crust, and that upper layer of the *mantle which lies above the *asthenosphere. Defant and Drummond (1993) *Geology* 21, 6 believe Mount St. Helens to be an example of the partial melting of the subducted lithosphere. The strength of the continental lithosphere is likely to be contained within the seismogenic layer; variations in the thickness of this strong layer determine the heights of the mountain ranges it can support (Maggi et al. (2000) *Geology* 28, 6). Constantin et al. (1996) *Geology* 24, 8 find that magmatic intrusion is a fundamental process in the construction of the oceanic lithosphere. See Scoppola (2006) *GSA Bull.* 118, 1 on the westward rotation of the lithosphere relative to the underlying mantle.

litter The layer of dead and dying vegetation found on the surface of the soil (some soil classifications assign the litter layer to the A *horizon, shown as A00). Cellulose, hemicellulose, and lignin make up 20–30, 30–40, and 15–40% of the total litter mass, respectively; the exact values are governed by plant type and age, onsite conditions, and yearly environmental variations (Yadav and Malenson (2007) *PPG* 31, 2).

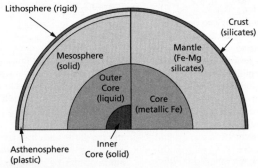

The Earth's constituent layers
Source: http://faculty.weber.edu/bdattilo/shknbk/notes/cmpstrct.htm

Little Climatic Optimum The time, *c.* AD 750–1200, when Europe and North America were much warmer; see Peterson (1994) *Clim. Change* 26. See Nunn (2000) *NZ Geogr.* 56, 1 on the transition between the Little Climatic Optimum and the *Little Ice Age.

Little Ice Age The phase between AD 1550 and 1850 when temperatures were globally lower than they are at present (Thompson et al. (1986) *Science* 234, 4774). **Little Ice Age glacierization** occurred over about 650 years and can be defined most precisely in the European Alps (*c.* AD 1300–1950) when extended glaciers were larger than before or since. **Little Ice Age climate** is defined as a shorter time interval of about 330 years (*c.* AD 1570–1900) when Northern Hemisphere summer temperatures (land areas north of 20° N) fell significantly below the AD 1961–1990 mean (Matthews et al. (2005) *Geografiska A* 87, 1).

littoral drift The amount of sediment transported by *longshore currents.

lixiviation *See* LEACHING.

load The matter transported by a river or stream. **Solution load** is dissolved in the water. **Suspension load** refers to undissolved particles which are held in the stream. On the river bed, the material of the *bed load jumps by *saltation, or rolls along the bed. The deposits forming a channel bed are known as **bed-material load**. Conventional theory is that meanders develop to an equilibrium form which is related to discharge and sediment load (Hooke (2007) *Geomorph.* 91, 3–4). See Singh et al. (2007) *Geomorph.* 86, 1–2 for a three-dimensional model for sediment transportation dynamics.

loam An easily worked, fertile soil, composed of *clay, *silt, and *sand, roughly in a ratio of 20 : 40 : 40.

A **clay loam** has a clay content of 25–40%, a **silt loam** has more than 70% silt, and a **sand loam** has between 50 and 70% sand. Loams heat up rapidly, drain neither too slowly nor too easily, and are well aerated.

local For a term so frequently employed in geography, 'local' is difficult to define. If we consider 'local' as relating or restricted to a particular area, or one's *neighbourhood, 'particular area' and/or neighbourhood need to be defined, so the problem is compounded. More interesting are the overtones attached to this term. Evans (2004) *Area* 36, 3 notes 'the suite of values that have become attached to the "local", such as inclusivity, collaborative action, self-determination in terms of policy and action, and responsiveness to real-world needs'. In terms of food production, Goodman (2003) *J. Rural Studs* 19 shows how the 'local' is associated with ecology, differentiation, and quality, with the latter being tied to relational ideas of tradition, trust and 'place' (see also Feagan (2007) *PHG* 31, 1). Allen's supposed truism needs to be evoked here: 'the local is not everywhere the same' (2003, *J. Rural Studs* 19).

Most geographers acknowledge a continuum between local and global; thus 'whatever lure the local may hold for landscape geographers it must become one of understanding how particular places, particular landscapes fit into a larger, scalarly complex, mosaic' (Mitchell (2001) *PHG* 25, 2). And rather than positing the global and the local as a dichotomy, Ç. Keyder and his collaborators (1999) advocate a dialectic in which the global and the local are mutually constitutive.

local climate The climate of a small area such as a moorland or city—a mesoclimate—falling between a *microclimate and a *macroclimate. At this scale, such variables as local winds, *albedo, relief, slope, and *aspect are of considerable significance. D. Wheeler and J. Mayes (1997) fit local climates into a larger picture for the British Isles. *See* URBAN CLIMATE.

local economic development policy Enterprises promoted by the public sector, or by public–private partnerships, to attract new enterprises and investment. Local economic development is conditional in its geography on land use plans and on the permissions made within the context of those plans (Cox (2004) *TIBG* 29, 2. See also Muir et al. (2000) *Area* 32, 4). Polèse and Shearmur (2006) *Papers Reg. Sci.* 85, 1 argue against using local economic development strategies as a means of arresting population and employment decline. Nonetheless, diversification into new markets is a key objective of local economic development policy in north-east Scotland (Chapman et al. (2004) *TIBG* 29, 3). See Bartik (2005) *Growth & Change* 36, 2 on local economic development incentives. Schragger (2010) *Chicago Law Rev.* 77, 1 disagrees with all the above by concluding that we 'do not know enough to be able to predict how one [local economic development] policy or another will affect city growth and decline'.

local food systems (LFS; local food networks) Local food systems include community gardens, community-supported agriculture, farmers' markets, food cooperatives, and community gardens, all of which may be described as direct agricultural markets (Hinrichs (2000) *J. Rural Studs* 16), or **local food networks** (Jarosz (2000) *Agric. & Hum. Values* 17). Local food systems will reduce food miles, and should make for fresher food, and support both localized production practices and local crops or livestock.

Hendrickson and Heffernan (2002) *Sociol. Ruralis* 42 describe food in a local system as 'rooted in a space that enables and constrains production and consumption through its own unique characteristics'. Anderson and Cook in J. M. Harris, ed. (2000) argue that local food systems can supplement and complement larger-scale food systems. Bellows and Hamm (2001) *Agric. & Hum. Values*, 18, 3 concur: 'the realities of a "local food system" necessitate an integration of local and non-local, and conventional and sustainable in local food systems.' Feagan (2007) *PHG* 31, 1 argues that the local in LFS will have to be contingent on the place—its social, ecological, and political circumstances—going on to warn against xenophobia, 'place purity', and anti-democratic orientations in local food systems.

local governance The institutions around local governments engaged in the design and implementation of economic and social policy: business elites, community leaders, development corporations, training and enterprise councils, and voluntary groups. Local governance, regime theory, and new regionalism literature argue that horizontal networks, consisting of public, private, and non-profit agents, diminish the significance of formal hierarchical political-territorial structures (Kubler and Heinelt in H. Heinelt and D. Kubler, eds 2005). However, Rosen and Razin (2007) *Tijdschrift* 98, 1 show that they only reshape the intervention of the central state in local development. See Tewdwr-Jones (2003) *Int. Plan. Studs* 8 on place-making as the tool of effective local governance.

localism A range of political philosophies that prioritize the local over regionalism and centralized government. Localism generally supports the local production and consumption of goods, local control of government, and the promotion of local history, local culture, and local identity. Featherstone et al. (2011) *TIBG* 37, 2, 177 argue that localism is being mobilized as part of an 'anti-state', 'anti-public' discourse to build support for an aggressive round of 'roll-back' neoliberalism.

localization economies Advantages arising from the localization together of a number of firms in the same type of industry. Desmet and Fafchamps (2005) *J. Econ. Geog.* 5 distinguish between localization economies—benefits which derive from being located close to other firms in the same industry—and urbanization economies—associated with closeness to overall economic activity. Boschma and Weterings (2005) *J. Econ. Geog.* 5, 5 list knowledge externalities and a skilled labour market. Viladecans-Marsal (2004) *J. Econ. Geog.* 4 refers to localization economies as associated with a city's specialization. However, Malmberg and Maskell (2002) *Env. & Plan.* A 34, 3, 129 call this 'an elusive concept'.

local knowledge A familiarity with and understanding of our day-to-day life; 'the tacit and explicit knowledge possessed and used by people who share the same culture' (A. Rogers, N. Castree, and R. Kitchin 2013). See Ley (2003) *PHG* 27, 5, 537 for a social history of local knowledge.

local winds Local winds blow over a much smaller area than global winds and have a much shorter time span. **Hot winds** originate in vast *anticyclones over hot deserts and include the Santa Ana (California), the Brickfielder (south-east Australia), the Sirocco (Mediterranean), the Haboob (Sudan), the Khamsin (Egypt), and the Harmattan (West Africa).

Cold winds originate over mountains or other snow-covered areas and include the Mistral, funnelled down the Rhone Valley, and the *Bora. Some local winds, such as the Southerly Burster of Australia, are associated with cold fronts. Other local winds include *land breezes and sea breezes. *See also* MOUNTAIN WIND; FÖHN.

location Absolute location is expressed with reference to an arbitrary grid system as it appears on a map. **Relative location** is concerned with a feature as it relates to other features.

Transnational companies can move their production plants, and their call centres, with ease from place to place—from continent to continent indeed—at the slightest whiff of lower costs. 'So, rather than erasing the importance of material location, it could well be argued that current trends in the world economy have in some senses reinforced its significance because as companies can move about more easily they can respond to even the smallest differences between places.' Massey (2006) BBC/OU Open2.net.

locational advantage Although it is true that transport costs have become less significant, and that new computer and telecommunication technologies have reduced the importance of location for production, locational advantage is still important in industrial location; for example, the overwhelming majority of logistics providers possess a geographic speciality (and even the worldwide integrated firms either have branches in, or have acquired firms with, strategic geographic locational advantage). 'This is because transportation is primarily a local industry that requires rich, geographically-specific knowledge, even if the industry deals with international and widely geographically dispersed transactions' (Aoyama et al. (2006) *Prof. Geogr.* 58, 3). For a second example, see

Lee and Rodrigue (2006) *Growth & Change* 37, 4 on locational advantage and differential growth within the Korean port system. See Murphy (2012) *Transportation Res. A* 46, 1, 91.

location-allocation model A mathematical model used to establish optimal locations. The model takes account of the location and demand of the customers, the capacity of the facilities, and operational and transport costs. Rushton (1988) *Econ. Geo.* 64, 2 provides an easily comprehensible discussion of the use of a location-allocation model to provide 'an explicit framework for diagnosing accessibility problems, measuring the efficiency of location decisions and the current levels of settlement efficiency [and the] generation of viable alternatives for action by decision makers'. A more technical, but still admirably clear, report on the use of location-allocation models in planning health care in Ghana is provided by Møller-Jensen and Kofie (2001) *Geografisk Tidsskrift* 101. See Hodgson (1988) *Soc. Sci Med.* 26, 1, 153 on an hierarchical location-allocation model for primary health care delivery in Goa.

location coefficient Also known as the location quotient, this expresses the relationship between an area's share of a particular industry and the national share. Thus, the locational coefficient for a given region equals:

$$\frac{\% \text{ employed in a field in a given region}}{\% \text{ employed nationally in that field}}$$

See Guimarães et al. (2007) *J. Reg. Sci.* 47, 4.

location-specific Locational factors bestow *competitive advantages if they offer resources that are not easily transferred to other locations. Such resources are either tangible (like natural resources or labour) or intangible (like expertise, or specialized services).

Location-specific resources influence the locations of transnational corporations (TNCs), especially if they complement the expertise of the TNC; the more easily the TNC can influence the location-specific resources, the more likely it is to invest in the professional development of its human capital. The TNC is then less likely to substitute one location for another one. In a study of the location-specific resources of Vienna, Windsperger (2006 *South East Eur. J. Econ. & Bus.* 20) identifies geographic distance; transport infrastructure; specific human capital; management know-how; historic ties; a multicultural environment; and a high quality of life, as major factors in location specificity.

location theory A group of theories which seek to explain the siting of economic activities. Various factors which affect location are considered such as localized materials and *amenity, but most weight is placed on transport costs. Witlox (2000) *Tijdschrift* 91, 2 advocates a relational view on industrial location theory.

Loch Lomond Stadial A stadial characterized by small ice caps and *cirque glaciers in the Scottish Highlands. See Sissons (1979) *Nature* 280.

locked zone Along a tectonic *rift, an area where plates are still attached to each other. The Cascadia subduction zone stretches from California to British Columbia. All along this zone, the subducting plates are forced beneath the North American Plate, but around 30 km down the plates have become locked. Below this zone the plates are more pliable, allowing them to move more readily past each other. This freer movement deeper down causes strain to accumulate along the locked zone.

Once that strain is great enough to overcome the friction that keeps the plates locked, the fault will rupture, causing earthquakes.

lodgement The release and consolidation of debris from a glacier. Lodgement of particles occurs when the frictional drag between a particle and the glacier bed exceeds the shear stress resulting from the moving ice. The latter aligns fragments of this debris, known as **lodgement *till**, in the same direction as the flow of the glacier. Ruszczyńska-Szenajch (2001) *Quat. Sci. Revs* 20, 4 distinguishes between **hard lodgement till** for deposits released from a glacier sole mainly due to friction, and **soft lodgement till** for sediments released from the base of a glacier due to the melting of ice in a more water-saturated subglacial environment.

loess (löss) Any unconsolidated, non-stratified soil composed primarily of silt-sized particles. The origin of loess is in dispute. Some writers believe the deposit to be wind-borne (Mason et al. (1999) *Geomorph.* 28, 3–4); others note the occurrence of the soil in *periglacial environments (Pawelec (2006) *Geomorph.* 74, 1–4) or stress the importance of glacial grinding in the production of silt-sized particles (Sculier et al. (1998) *Geomorph.* 23, 1). Sweeney (2005) *Quat. Res.* 63, 3 points to the frequency of dust storms in deserts; Smith et al. (1991) *ESPL* 16 stress the importance of *salt weathering. The loess sequences of north-central China preserve the longest and most detailed record of *Quaternary climate change found on land. See Haberlah (2007) *Area* 39, 2 for a clear and concise summary on loess. Loess is very fertile but very difficult to conserve; soil erosion in most of the Loess Plateau is 5 000–10 000 Mg/km^2 per year, and 20 000 Mg/km^2 per year in some places (Chen et al. (2007) *PPG* 31, 4). See also Cheng (2007) *Tillage Res.* 94, 1.

logit model Model used to represent choice between two mutually exclusive options; for example, a commuter may decide to drive to work or to use public transport. The **multinomial logit model** is mathematically simple (Andrew and Meen (2006) *Papers Reg. Sci.* 85, 3), and is widely used, but imposes the restriction that the distribution of the random error terms is independent and identical over the alternatives, causing the cross-elasticities between all pairs of alternatives to be identical, and this can produce biased estimates. The **nested logit model** allows the error terms of pairs or groups of alternatives to be correlated, but the remaining restrictions on the equality of cross-elasticities may be unrealistic (Hunt (2000) *Papers Reg. Sci.* 40, 1). Logit models which allow different cross-elasticity between pairs of alternatives include: the **paired combinatorial logit** (Koppelman and Wen (2000) *Transp. Res.* B4); the **cross-nested logit** (Vovsha (1997) *Ann. Transp. Res. Board, Washington, D.C.*), and the **product differentiation model** (Filippini (1999) *Int. J. Econ. Bus.* 2). The major weakness of logit models is the implication that the choice between any two groups of alternatives depends solely on the characteristics of the alternatives being compared, and not upon the characteristics of any other alternatives in the choice set.

longitude The position of a point on the globe in terms of its *meridian east or west of the prime meridian, expressed in degrees. These degrees may be subdivided into minutes and seconds, although decimal parts of the degree are increasingly used.

longitudinal data Information on one or more areas over periods of time; for example, on changes in human reproduction (Reher et al. (2007) *Pop. & Dev. Rev.* 33, 4).

long profile A section of the longitudinal course of a river from head to mouth, showing only vertical changes (Harmar and Clifford (2007) *Geomorph.* 84, 3–4). See Jain et al. (2006) *Geomorph.* 74, 1–4 on stream power plots along long profiles. The theoretically smooth curve shown by such a profile may be interrupted by breaks of slope which can result from bands of resistant rock or from *rejuvenation (Larue (2008) *Geomorph.* 93, 3–4).

longshore drift The movement of sand and shingle along the coast. Waves usually surge onto a beach at an oblique angle and their *swash takes sediment up and along the beach. The *backwash usually drains back down the beach at an angle more nearly perpendicular to the coast, taking sediment with it. Thus there is a zigzag movement of sediment along the coast. **Longshore currents**, initiated by waves, also move beach material along the coast. The term **littoral drift** is synonymous.

long wave *See* EARTHQUAKE.

lopolith A large *intrusion which sags downwards in the centre, forming a saucer-shaped mass. Wilson (1956) *GSA Bull.* 67, 3 interprets lopoliths as funnel-shaped in cross-section with the most basic rocks at the bottom. See O'Driscoll et al. (2006) *Geology* 34, 3 on the Great Eucrite intrusion of Ardnamurchan.

Lorenz curve A cumulative frequency curve showing the distribution of a variable such as population against an independent variable such as income or area settled. If the distribution of the dependent variable is equal, the plot will show as a straight, 45° line. Unequal distributions will yield a curve. The gap between this curve and the 45° line is the inequality gap. Such a gap exists everywhere, although the degree of inequality varies. *See* GINI COEFFICIENT.

Lösch model A model of central places developed by A. Lösch (1954)

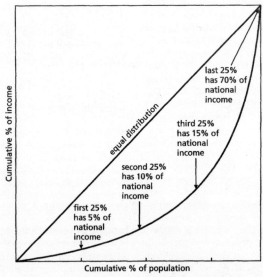

Lorenz curve

which is less narrow than that of Christaller, in that it treats the range, threshold, and hexagonal hinterland of each function separately. The resulting pattern of central places is much more complex than Christaller's, and yields a continuous, rather than a stepped, distribution of population sizes. See Esselbichler in A. Leyshon et al., eds (2011) pp. 33–4.

löss *See* LOESS.

low A region of low atmospheric pressure. In Britain, the term low is generally applied to pressures of below 1 000 millibars.

low-order goods and services
Goods and services with a low *range and a low *threshold population like daily newspapers, and hairdressing. The goods are often convenience goods.

Luminescence Dating (Optically Stimulated Luminescence Dating; OSL) All sediments include low levels of uranium, thorium, and potassium which produce ionizing radiation, which accumulates over time, and is absorbed and stored within the sediment. If this store is expelled from the sediment, it will generate luminescence, the amount produced being proportional to the accumulated dose. If the annual radiation dose is known, the age of the sediment can be calculated:

$$\text{Age} = \frac{\text{total accumulated radiation dose (Gy)}}{\text{annual radiation dose}}$$

Sediments rich in quartz and feldspar, which cannot be dated by conventional radiocarbon methods, can be absolutely dated ($\pm \sim 10\%$) within a range of 100 000 to 200 000 years. Aeolian sediments are ideal for luminescence dating, but samples from other sedimentary environments have been successfully dated. Li et al. (2012) *Holocene* 22, 397 describe the methodology well.

lunette *See* SAND DUNE.

lynchet In earlier, possibly prehistoric, times, small belts of uncultivated land were sometimes left between ploughed areas. These now can be seen in the landscape as part of a system of terraces. See Klingelhofer (1991) *Trans. Sm. Philos. Soc.* 81, 3.

lysimeter A device for the direct estimation of evapotranspiration, composed of a vegetated block of soil 0.5–1 m^3 to which the amount of water added is known, and from which the amount lost as run-off or percolation may be measured. Richards (2004) *PPG* 28, 1 explains further.

macro- Large-scale. Thus, a **macroclimate** is the general climate of a region extending across several hundred kilometres; and **macrometeorology** is the study of large-scale meteorological phenomena—from *monsoons to the *general circulation of the atmosphere.

macroecology The integration of perspectives over broad spatial and long temporal scales into the study of ecology (Brown and Maurer (1989) *Science*, 243). Macroecology has moved from the absolute advocacy of small-scale over large-scale studies to an advocacy of some large scales in preference to others; Blackburn and Gaston (2002) *Glob. Change & Biogeog.* 11, 3 argue that there is no general sense in which one scale of study is better than any other. Although macroecology arose from geographical ecology, it has diverted from a geographical perspective. Ruggiero and Hawkins (2006) *Glob. Change & Biogeog.* 15, 5 argue that present trends in macroecology adopt a predominantly statistical approach to looking for rules and regularities 'that increasingly ignore the spatial distributional aspects linked to mapping. 'In a vigorous reply, Blackburn and Gaston, 2006 [*J. Animal Ecol.* 75], stressed that there is much more to macroecology than mapping, and emphasized that 'macroecological patterns will only be elucidated effectively through the formulation of relevant hypotheses and related statistical approaches' (Kent (2007) *PPG* 31, 6).

macrotidal Having a large tidal range; J. L. Davies (1980) defines *microtidal* as having a range less than 2 m, *mesotidal* 2–4 m, and *macrotidal* over 4 m. Areas with a tidal range over 10 m are sometimes distinguished as *hypertidal* (Allen (2000) *Quat. Sci. Revs* 19). **Macrotidal estuaries** have their dynamics dominated by tidal forcing, as compared to the forcing imposed by the river (Regnier and Steefel (1999) *Geol. & Cosm. Acta* 63, 9). Reichmüth and Anthony (2007) *Geomorph.* 90, 3–4 explore the relationship between tidal influence and intertidal bar–trough systems in a macrotidal setting. Macrotidal sandy beaches are thought to represent optimal conditions for coastal dune development (Ruz and Meur-Ferec (2004) *Geomorph.* 60, 1–2).

magma The molten rock found below the Earth's crust which can give rise to *igneous rocks. Magma occurs in *subduction zones, at continental rift zones, *hot spots and *mid-oceanic ridges, and may rise as *mantle plumes—see Foulger et al. (2005) *GSA Spec. Vol.* 388, on plates, plumes, and paradigms. Magma composition is an important control on the geomorphology of lava flows and volcanoes (Carrasco-Núñez (1997) *Geol. Mag.* 134, 2); see Morgan and Ghen (1993) *Nature* 364 on magma supply. **Magmatic stoping** is the detaching and engulfing of pieces of the country rock, but its role in the location of plutons is disputed: see Žák et al. (2006) *Int J. Earth Sci.* 95 versus Glazner and Bartley (2006) *GSA Bull.* 118.

magnetic pole reversal The Earth's magnetic field resembles a bar magnet located at the Earth's centre, with its axis emerging at the magnetic poles. The north and south magnetic poles have repeatedly changed places, at irregular intervals, while the axis has stayed in place. When *igneous rocks form, they take up the prevailing pattern of the Earth's magnetic poles. As a result, areas of the ocean bed where *sea-floor spreading has taken place are characterized by **magnetic stripes**—parallel bands of igneous rock with differing magnetic polarity. See Nicolosi et al. (2006) *Geol.* 34 on the Marsili Basin, Italy.

magnitude Of an *earthquake, an expression of the total energy released.

malapportionment The existence of electoral districts with wildly unequal populations (Norman et al. (2007) *Pol. Geog.* 26, 1). Unintentional malapportionment happens when population growth varies in electoral districts. Redistricting is then needed to even out voter numbers; see Johnson et al. (2006) *Pol. Geog.* 25, 5.

Malthusianism In 1798, Thomas Robert Malthus (1766–1834) published his *Essay on Population* in which he put forward the theory that the power of a population to increase is greater than that of the Earth to provide food. He asserted that population would grow geometrically, while food supply would grow arithmetically. When population outstrips resources, **Malthusian checks** to population occur: misery, vice, and moral restraint.

Malthus' predictions were not borne out in 18th-century Britain, perhaps because of the increases in food output, and emigration to the colonies; A. Macfarlane (1997) refers to the 'great escape' from the Malthusian trap by large chunks of humanity. Malthusian theory is deployed to underpin the theory of European historical superiority by arguing that Europeans, uniquely, have generally (and rationally) avoided the Malthusian disasters of overpopulation while non-Europeans (irrationally) have not done so and therefore not developed as Europe has.

'While old-style Malthusians simply saw an inevitable tendency of human population to outgrow food production, neo-Malthusian scholars offered a more complicated, political argument: overpopulation would cause resource depletion and hunger which in turn would lead to political instability threatening Western interests and world peace' (Flitner and Heins (2002) *Pol. Geog.* 21, 3). Brander (2007) *Can. J. Econ./Rev. can d'écon.* 40, 1 sees lower fertility as the central element in achieving sustainable development, 'which is very much a modern translation of what Malthus wrote in 1798'. Furlong et al. (2006) *Int. Interact.* 32, 1 find that the neo-Malthusian factors are significant, but not dramatic, factors in boundary disputes.

mammatus Breast-shaped lobes of cloud, hanging from the undersurface of a *cumulo-nimbus anvil, signifying negative *buoyancy. **Mammilated** means smooth and rounded.

mandate A statement of empowerment, for example, through an election: 'representing humanity ultimately requires legitimation through some sort of people's mandate' (Dicken (2004) *TIBG* 29, 1). The term also describes a territory, once part of the German or Ottoman Empires, governed by a member of the League of Nations, 1919–39.

mangrove swamp A number of types of low trees and shrubs, growing on mud flats in tropical coastal areas where the tidal range is slight. Mangroves are significant agents of *progradation along tropical coasts, and are especially well developed in South and East Asia. See Saintilan and Williams (1999) *Glob. Ecol. & Biogeog.* 8, 2. Concedda et al. (2011) *AAAG* 101, 6, 1259 report an increase in the area of mangrove swamps in Senegal, but a decline in the capacity of these mangrove forests to provide key ecosystem goods and services.

Manning equation Also known as the **Mark–Manning formula**, the **Gauckler–Manning formula**, or the **Gauckler–Manning–Strickler formula**, this is an empirical formula estimating open channel flow, or free-surface flow driven by gravity:

$$V = \frac{k}{n} R_h^{2/3} S^{1/2}$$

where V is the cross-sectional average velocity (L/T; ft/s, m/s); k is a conversion factor of (Length$^{1/3}$/Time), 1 m$^{1/3}$/s for SI, or 1.4859 ft$^{1/3}$/s U.S. customary units, if required. (Note: (1 m)$^{1/3}$/s = (3.2808399 ft)$^{1/3}$/s = 1.4859 ft$^{1/3}$/s); n is the **Manning coefficient**, and is unitless; R_h is the hydraulic radius (L; ft, m); and S is the slope of the water surface or the linear *hydraulic head loss (L/L) ($S = h_f/L$).

manor The smallest area of land held in the Middle Ages by a feudal lord, with its own court for minor offences. It usually consisted of a village, the lord's holding (*demesne*), and *open fields farmed on the three-field system. See Z. Razi and R. Smith (1996).

mantle The mantle, with a density of up to 3.3 g cm^{-3}, and a thickness of some 2800 km, lies between the *crust and the Earth's *core. The upper layer, immediately below the *Moho discontinuity, is rigid, forming the lower *lithosphere; the lower layer is the *asthenosphere. Zhou

et al. (2002) *Lithos.* 62, 3 see the interaction between continental crust and subcontinental mantle as expressing geodynamic processes deep in the subcontinental lithospheric mantle. See Husson (2006) *Geology* 34, 9 on upper mantle flow.

manufacturing industry Secondary *industry.

map A cartographic representation of selected spatial information; the cartographer decides what to include and what to leave out. 'Maps are active; they actively construct knowledge, they exercise power, and they can be a powerful means of promoting change' (J. Crampton 2010). We tend to restrict the term to visual maps, but spatial information can be shown on a computer screen, through braille, or a spoken description, and these may also be described as maps. See Pontius et al. (2007) *AAAG* 97, 4 on the Behavioral Landscape Model and spatial patterns. **Map generalization** is the decreasing of the detail on a map when reducing its scale, for the map scale determines the size and number of graphic objects that can be put on a map.

Map reading is the interpretation and analysis of map images; an understanding of the physical and psychological processes in map reading helps in map-making. See Lobben (2007) *AAAG* 97, 1 and Griffin et al. (2006) *AAAG* 96, 4. *See also* MENTAL MAP.

map projection Both a method of mapping a large area and the result of doing so. The Earth is a sphere; a map is flat, so that it is impossible to produce a map which combines true shape, true bearing, and true distance. 'The usefulness of a particular **map projection** and the justification behind its transformation from a three-dimensional spherical surface to a two-dimensional plane surface is all too often judged from the aesthetic or suspicious eye of a naïve map reader rather than by the science behind its creation' (M. Monmonier 2004). See also Monmonier (2005) *PHG* 29, 2.

Mercator's projection exaggerates the size of the northern continents; it has been criticized as overemphasizing the importance of Europe and North America, although such was not Mercator's intention. No projection is perfect: for example, Mollweide's and Peters' are **equal area projections** (correct in area), but distort shapes. **Azimuthal projections** show true direction; **gnomic projections** show the shortest straight-line distance between two points; **orthographic projections** convey the effect of a globe. **Interrupted projections** show the Earth as a series of segments joined only along the equator. Details of the projection used are given below each map in a good atlas.

marginalization The process by which individuals and groups are prevented from fully participating in society. Marginalized populations can experience barriers to accessing meaningful employment, adequate housing, education, recreation, clean water, health services, and other social determinants of health. Both community and individual health are deeply affected by marginalization. The Ontario **marginalization index** is based on four criteria: residential instability, material deprivation, dependency, and ethnic concentration. **Geographies of marginalization** consider the way space

Map projection. Mercator's projection is drawn from the shapes and shadows of landmasses that are projected onto a cylinder of paper encircling the earth, when the earth is lit from the centre.

is fashioned to privilege some groups and marginalize others (Trudeau and McMorran in V. Del Casino et al. 2011).

🌐 **SEE WEB LINKS**
• Ontario marginalization index.

marginal land Land which is difficult to cultivate, and which yields little profit.

margin of cultivation The distance from a market where the revenue from a product exactly equals total cost. If production costs are the same whatever the distance from the market, transport costs determine the margin of cultivation. See Imai (1985) *Annals Tohoku Geog. Assoc.* 37, 4, 279.

mariculture This is marine aquaculture; the cultivation of marine organisms in the open ocean, or in an enclosed section of the ocean, or in seawater-filled ponds or tanks. Suryanata and Umemoto (2003) *Env. & Plan. A* 35, 2, 199 point out the problems associated with marine aquaculture in Hawaii.

maritime Of climates near the sea coast. Such climates have less extremes of temperature, both diurnally and seasonally, than their continental counterparts.

market An arena in which buyers and sellers exchange goods and services, usually for money; it doesn't have to be a physical location. **Market area analysis** is the investigation of how a firm's market area is established. Barker (2000) *Papers Reg. Sci.* 79 presents a shopping model and its application to market area analysis. See also Atsuyuki et al. (2008) *Geogr. Analy.* 38, 1 and Bruinsma et al. (2002) *Tijdschrift* 91, 3 on airports.

 Geographies of markets are concerned with: markets as heterogeneous arrangements of people, things, and socio-technical devices; the insight that multiple frames of reference are mobilized in everyday market activities; and the performance of diversity and difference in concrete market contexts with an attention to mobility in network capitalism (Berndt and Boekler (2011) *PPG* 35, 4, 559).

market cycle A series of periodic—usually one-day—markets such that a trader moves from one location to the next in a weekly cycle.

market economy An economy, characteristic of *capitalism, in which the major parts of production, distribution, and exchange are carried out by private individuals or companies rather than by the government, whose intervention is minimal. Market economies are characterized by specialized production, the freedom to exchange commodities between individuals, and the use of the *market mechanism to determine prices. See Peck (2008) *PHG* 32, 1 for definitions of the market economy, and social market economy.

market mechanism The interaction of supply, demand, and prices. Page (2005) *TIBG* 30, 3 compares two approaches to water supply: 'there are those who claim that water is scarce and, like all scarce goods, it is best allocated through a market mechanism. Water must be turned into a commodity to be sold at the full price of its production, including environmental externalities, if it is to be efficiently and equitably distributed and conserved. . . . On the other side are those who argue that water, though scarce, is in its essence not commercial and that such a process of commodification . . . is little more than theft of a common good.'

market potential The intensity of possible contacts with markets. The so-called Harris market potential uses bilateral distances to weight surrounding locations' GDP:

$$MP_{\text{Harris},\,i} = \sum_j \frac{GDP_j}{\text{dist}_{ij}}$$

A more sophisticated version of Harris's market potential uses road travel times for lorries as weights, instead of great circle distances:

$$MP_{\text{Traveltime},\,i} = \sum_j \frac{GDP_j}{\text{traveltime}_{ij}}$$

See Holl (2011) *J. Econ. Geog.* DOI 0.1093/jeg/ibr030 on the productivity impacts of market potential and transport infrastructure investment in the manufacturing sector in Spain.

Markov chain analysis A probability process that can be used to analyse a chain of events through time, such as the housing rental market. One of the earliest geographers to use this probabilistic process is Clark (1965) *AAAG* 55, 2, whose instructions in this method reward careful study.

marl Clay, usually alluvial, rich in soft calcium carbonate. See Cerdà (1999) *Soil Sci. Soc. Am. J.* 63 on comparative erosion rates in marl, clay, limestone, and sandstone.

Marshallian industrial district A spatial concentration of specialized small and medium-sized firms, supported by a variety of regional institutions for promoting coordination, learning, and innovation (Lee and Saxenian (2008) *J. Econ. Geog.* 8, 2). Dense interactions among a large number of competing and cooperating firms create an external economy favouring technological innovation and learning (Wang (2007) *J. Dev. Stud* 43, 6).

Marxist geography A theory and methodology for understanding the political economy of the world—and

changing it. It highlights the dialectical processes between the natural environment, spatial relationships, and *social formations* ('layers of relations— economic, social, cultural, political— arranged in space in patterns generally similar, but particularly different, under a given mode of production', Peet (1979) *AAAG* 69). At all scales, the capitalist mode of production assumes a centre-periphery pattern: 'uneven development is integral to capitalism. Capitalism produces and remakes landscapes and natures of differentiation and is sustained through their continual reworking . . . . Rent, profit (and loss), urbanization (and counter-urbanization), devaluation (and speculation), gentrification (and revalorization), imperialism (and revolution), and enclosure (and globalization) are forceful spatial-temporal projects' (Castree and Gregory (2007) *AAAG* 97, 1).

Perkins (2006) *Antipode* 38, 1 observes an emphasis within Marxist political thought on the social production of nature, and Castree (1999) *TIBG* 24, 2 seeks to show how a certain kind of Marxism, and a certain reading of the nature of capitalism and class, can offer the basis on which questions of identity, difference, and otherness can be articulated. While Sage (2006) *Prof. Geogr.* 58, 1 comments that constant contextualizing through Marxism often seems to limit, rather than widen, the scope of locally derived concepts, Cambridge (2005) *PHG* 29 thinks that the Marxian critique of capitalism has been absorbed and normalized in geography, such that 'we are all political economists now'. See A. Merrifield (2011) on magical Marxism.

masculinity 'A highly differentiated identity category with its own internal relations of alliance, subordination and dominance' (R. Connell 1995). McDowell (2002) *Env. & Plan. D* 20, 1 explores the social construction of masculinities among white working-class youth in contemporary Britain; see Milam (2008) *Soc. & Cult. Geog.* 9, 2 on Thai masculinities. B. van Haven and K. Hörschelmann, eds (2004) is very well reviewed.

mashup Also known as **map hacking**, this is a website or web application that combines content from more than one source; a tool overwhelmingly suited to geographers. Google Maps Mania is free. You might find Ting (2009) *Access to Knowledge* 1, 2, helpful.

mass balance The relationship between the masses of the components of a landform, particularly in glaciers, where zero mass balance describes a situation where accumulation balances ablation, a positive mass balance describes an increase in glacier mass, and a negative mass balance describes a reduction in glacier mass. See M. M. Bennett and N. Glasser (2009), p. 43, for an excellent summary of mass balance and global sea-level rise. Alexander et al. (2011) *Geograf.* 93, 1, 41 describe a simple steady-state mass-balance model for the Franz Josef glacier, New Zealand.

mass budget The balance between the accumulation and ablation of ice in a glacier. Its components are accumulation, ablation, and iceberg calving. A glacier that is not in balance with the ambient climate loses/gains mass and thereby raises/lowers global sea level (IAMU, 2006–7). The flow of solid ice governs the dynamical response of the glacier to a change in its mass budget. Machguth et al. (2007) *Geophys. Res. Abs.* 9, 06249 compare the output of three distributed mass-balance models of differing complexity.

mass conservation, principle of For a river, sediment continuity. Rivers adjust their width, depth, and slope to keep a balance between the water and sediment supplied from upstream and exported at the outlet; see Dietrich et al. (1999) *Hydrol. Procs* 13.

mass movement (mass wasting) The movement downslope of rock fragments and soil under the influence of gravity. The material concerned is not incorporated into water or ice, and moves of its own accord, but slides are often triggered by increase in water pressure on rocks and soil. **Granular mass movements** entail transport by large (>0.06 mm) solid grains, mixed with less-dense intergranular liquid or gas, and include rock avalanches, debris flows, pyroclastic flows (Iverson and Vallance (2001) *Geology* 29, 2). See also Deplazes et al. (2007) *Terra Nova* 19, 4. A widely used classification of mass movement uses the combination of types of movement (*falls, *topples, *slumps, *slides, and *flows) with the nature of the material (bedrock, *debris, and fine soil); see R. Chorley et al. (1985). Many cases of mass movement include more than one type of movement.

mass strength The strength of a rock's resistance to erosion. Mass strength will vary according to the innate strength of the rock, the jointing/bedding of the rock, and its state of weathering. See Selby (1980) *Geomorph.* 24 on rock mass strength classification.

mass transfer In geomorphology, the movement of materials, especially of sediment. Sedimentary mass transfers move eroded sediments from their source area to an area of temporal storage or long-term deposition in sinks. Rates of sediment transfer are not only conditioned by competence of geomorphic processes but also by the availability of sediment for transport. Studies on mass transfers and sedimentary source-to-sink fluxes generally refer to the development of sedimentary budgets. See Beylich (2011) *Zeitschr. Geom.* 55, 2, 145 for a very clear account.

materiality (materialization) In the early 2000s, many scholars of human geography looked to 'rematerialize' the discipline, that is to say, to ground their studies in material culture, or to be concerned with socially significant differences of gender, class, race, sexuality, or (dis)ability; to ground them in 'reality' (Jackson (2000) *Soc. & Cult. Geog.* 1, 9). However, Anderson and Tolia-Kelly (2004) *Geoforum* 35, 669 observe that 'social and cultural geography is increasingly marked by a profusion of new materialisms that afford very different styles of "materialist" theoretical-empirical work'. Try Anderson and Wylie (2009) *Env. & Plan. A* 41, 318.

mathematical geography The study of the interactions of geography and mathematics, such as the Earth's size and shape, time zones, and motion. See Withers (2006) *PHG* 30, 6.

matrix *See* LANDSCAPE MATRIX.

Maunder minimum Between 1645 and 1715, a period of reduced solar activity (sunspots) that coincided with the coldest part of the *Little Ice Age. This may be a causal connection; Inesol et al. (2011) *Nature Geoscience* 4, 753 tie variability in solar activity to terrestrial climate impacts in the form of warmer winters in some places and colder winters in others. See also Diodato and Bellocchi (2012) *Holocene* 22, 589.

maximum sustainable yield The maximum yield of a renewable resource consistent with maintaining its stock. Arlinghaus et al. (2002) *Fish & Fisheries* 3, 4 review the concept.

meander A winding curve in the course of a river. A *sinuosity of above 1.5 is regarded as distinguishing a meandering channel from a straight one. The dimensions of a meander are related to the square root of water discharge, Q:

$$\lambda = k_1 Q^{0.5}$$
$$A_m = k_2 Q^{0.5}$$
$$w_c = k_3 Q^{0.5}$$

where λ is meander wavelength, A_m is meander amplitude, w_c is the channel width, and k_1, k_2, and k_3 are coefficients whose value varies with location. Bank resistance controls meander wavelength, and is positively correlated with meander wavelength (Abrahams (1985) *ESPL* 10, 6). Meander bends with a low radius of curvature have deeper pools and more lateral migration; the ensuing bank erosion reduces pool depth via higher sediment inputs. Meander bends with smaller pools have less lateral migration, but use energy vertically, thereby deepening pools and fostering pool-riffle sequences (Hudson (2002) *Phys. Geog.* 23, 2).

Eaton et al. (2006) *ESPL* 31 argue that slight initial variations in the shear stress distribution result in local net scour along the banks, increasing local transport capacity. In a positive feedback, this leads to further local net scour, which is mostly vertical until the channel banks begin to fail, at which point bank erosion triggers negative feedback that increases channel sinuosity. Once one initial bend develops, others inevitably follow; see also Parker et al. (2011) *ESPL* 36, 1, 70 for a new model of river meandering. See Millar (2005) *Geomorph.* 64 on equations for the meandering–braiding transition, and Pittaluga and Seminara (2011) *ESPL* 36, 1, 20 on nonlinearity and unsteadiness in river meandering. The **meander belt** is the total width across which the river meanders.

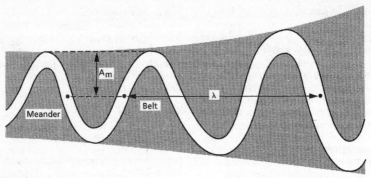

Meander

mean information field (MIF) In
*diffusion, the field in which contacts
can occur. It generally takes the form of a
square grid of 25 cells, with each cell
being assigned a probability of being
contacted. The possibility of contact is
very high in the central cells from which
the diffusion takes place, becoming
markedly less so with distance from the
centre, that is, there is a *distance decay
effect. See Bivand (2008) *J. Reg. Sci.* 48, 1.

means of production The elements
needed to produce goods and services:
land, labour, and *capital. Any production process depends upon a particular
material configuration of the means of
production (Hudson (2008) *J. Econ.
Geog.* 8, 3). Marx believed *capitalism to

be characterized by a split between the
capitalist owners of the means of production, and the proletarians, with only
their labour to sell. Akram-Lodhi (2005)
J. Agrar. Change 5, 1 notes that the 1980s
decollectivization of Vietnamese agriculture has divorced the landless from
the means of production.

mechanical erosion Erosion by
physical means, with no chemical
change. The main agents of physical
erosion are wind, water, and ice.
Mechanical weathering is the splitting
up *in situ* of rock, with no chemical
change.

media geographies Media geographers study the ways geographies are

0.0096	0.0140	0.0168	0.0140	0.0096
0.0140	0.0301	0.0547	0.0301	0.0140
0.0168	0.0547	0.4431	0.0547	0.0168
0.0140	0.0301	0.0547	0.0301	0.0140
0.0096	0.0140	0.0168	0.0140	0.0096

Mean information field

communicated through social media; 'the way this communicated knowledge affects the audience's life-world; and the close connections between geographers and journalists who share an interest in important societal questions, whilst using altered approaches' (Zimmermann (2007) *Aether*, 1, 59.

medial moraine *See* MORAINE.

medicine, geography of (medical geography) A field of geography that considers the distribution of specific diseases and human characteristics in relation to the geographical and topographical features of regions, countries, or the world. In its initial incarnation, medical geography was an enquiry into the environmental correlations of diseases; see Brown and Moon (2004) *J. Hist. Geog.* 30. Medical geographers also became interested in the spatial distributions of those medical facilities, from hospitals to clinics to GP surgeries, that society 'invents' to treat diseases (Philo (2004) *NZ Geogr.* 63, 2). Parr (2002) *PHG* 26, 2 lists geographical variations in incidences of disease, and the complex intersections between health, illness, place, and space, as key issues, while calling for more attention to the body. Medical geographers map/represent disease incidence (new cases) or prevalence spatially, to identify social or environmental factors associated with disease and locate areas where risk is particularly high. 'Most medical geographers are well aware of the statistical challenges associated with the analysis and representation of aggregate geographic data—the small numbers problem, the *modifiable areal units problem and the *ecological fallacy are perhaps most often of concern' (Yiannakoulias et al. (2005) *Canad. Geogr./Géogr. canad.* 49, 3). Philo (op. cit.) argues that 'the human being—living, experiencing, thinking, feeling, often anxious, perhaps

suffering, maybe dying—risks obliteration when expressed as "patients", disease entities, figures in graphs, or dots on maps'. See Sui (2007) *Geog. Compass* 1, 3 on GIS and medical geography, and Smyth (2008) *PHG* 32, 1 on the geographies of health inequalities.

Mediterranean climate A climate of hot, dry summers and warm, wet winters, characteristic not only of Mediterranean lands, but also in California, central Chile, and the extreme south of Africa. In summer, the climate is dominated by *subtropical anticyclones, and *trade winds prevail. In winter, mid-latitude depressions bring rain. *Local winds are of great significance.

Mediterranean soils Formed in *Mediterranean climates, these soils owe their red colour to the *leaching of clays in the wet winters, which releases iron. Leaching is rare in summer so that a carbonate *horizon often builds up.

megalith Any free-standing large stone sited by humans, mostly dating from *c.*3000 to 2000 BC. C. Holtorf (2001) points out that all interpretations of megaliths, whether as tombs or 'mnemonic markers', can support the idea that megaliths were meant for a future time.

megalopolis A many-centred, multi-city, urban area of more than 10 million inhabitants, characterized by complex networks of economic specialization. Vicino et al. (2007) *Int. J. Urb. Reg. Res.* 31, 2, in a study of the US Bowash Megalopolis, reveal a complex socioeconomic pattern structured by class, education, housing tenure, housing age, and race and ethnicity. They identify distinct clusters: 'affluent places', 'places of poverty', 'Black middle class places', 'immigrant gateway places', and 'middle America places'. The *Metropolitan*

Institute Census Report Series 05:01 (2005) is an excellent resource on this.

melting pot A concept of a number of ethnic groups, cultures, and religions in a society fusing together to produce new cultural and social forms. Brueckner and Smirnov (2007) *J. Reg. Sci.* 47, 2 suggest that under mild conditions of interconnectedness, a society proceeds inexorably toward homogenization, with waves of foreign immigration, introducing unconnected groups into the population, providing short-term infusions of heterogeneity. As links form between these new groups and the rest of the population, the homogenization process typically reasserts.

melt-out till Subglacial melt-out *till reflects the amount of sediment in transport at the time of melting; supraglacial melt-out till is located at the margins of a glacier, and includes lateral and terminal moraines. Larson et al. (2003) *GSA Abstr. with Progs.* 35, 299 suggest that distinctive silt pellets in melt-out tills indicate *glaciohydraulic supercooling.

meltwater Glacial meltwater is produced by the melting of glacier ice at the surface, or by pressure and *geothermal heat at the base, forming *subglacial streams. Some surface meltwater may percolate through the ice to emerge at the base. Both subglacial and proglacial **meltwater erosion** are extremely effective agents of landscape change. Water often flows under pressure within and below decaying ice, pressure that enables meltwater to flow upslope, and carry large quantities of debris, which promotes *abrasion.

Glacial meltwater landforms form subglacially; at or near the ice margin; and in proglacial settings. **Ice margin meltwater** follows the edge of the glacier when drainage routes have been cut off by ice. **Tunnel valleys** occur below the ice and can cut steep-sided, flat-floored valleys; see Cutler et al. (2002) *Quat. Int.* 90, 1. **Spillways** are channels cut by streams overflowing from *proglacial lakes; see Thieler et al. (2007) *Pal. Pal. & Pal.* 246, 1. **Coulees** are canyons which result from the sudden and violent release of water from ice-dammed lakes when the barriers which impound them are breached. See Shaw (2002) *Quat. Int.* 90, 1 on the meltwater hypothesis for subglacial bedforms, and Hannah and Gurnell (2001) *J. Hydrol.* 246, 1 and Pohl et al. (2005) *Atmos-Ocean* 43, 3 for models of meltwater action. For meltwater and the reconstruction of former ice sheets, see Clark and Mix (2002) *Quat. Sci. Revs* 21, 1–3. See also Glasser et al. (2004) *Boreas* 33, 3 on meltwater processes as evidence for multiple glaciations.

memory The recollection of the past. We construct meaning about our lives and worlds through the places we create, inhabit, or visit. Till (1999, *Ecumene* 6) describes places as 'fluid mosaics of memory, metaphor, matter and experience, [which] create and mediate social spaces and *temporalities'. Hoelscher and Alderman (2004, *Soc. & Cult. Geog.* 5, 3, special issue), observe that 'memory and place conjoin to produce much of the context for modern identities'. Look for O. Jones and J. Garde-Hansen (2012).

Tolia-Kelly (2004, *TIBG* 29, 3) writes on **re-memory**: 'a conceptualization of encounters with memories, stimulated through scents, sounds and textures in the everyday . . . souvenirs . . . [and] signifiers of "other" narrations of the past, not directly experienced'.

mental health, geographies of Essentially, how and why the social and physical environments matter for mental health and psychological well-being in human populations. See S. Curtis (2011).

Parr (2000) has a useful paper in *Health & Place*, which is the most important journal on this topic.

mental map A map of the environment within an individual's mind that reflects the knowledge and prejudices of that individual. A mental map indicates the way an individual acquires, classifies, stores, retrieves, and decodes information about locations. Ojanen and Vuohula (2007) *J. Common Mkt Studs* 45 describe Finland as lacking a prominent presence on the mental map of other EU governments. Walker (2000) *Int. Affairs* 76, 3 notes that the overall American mental map of Europe is becoming clearer: it includes Turkey, and 'may include Ukraine and even Russia'. Sbragia (2007) *Politics* 27, 1 describes the EU as a complete puzzle for most Americans. It is not a major actor in international relations, is not represented in the UN, and yet can force Microsoft and other iconic American corporations to change their behaviour if they want access to the European market. 'The American mental map does not have a place for such an entity.'

Mercalli scale A measurement of the intensity of an *earthquake.

Modified Mercalli scale of earthquake intensity:

    I. Felt by very few, except under special circumstances.

   II. Felt by a few persons at rest, especially on the upper floors of buildings.

  III. Felt noticeably indoors, although not always recognized as an earthquake. Vibration like passing lorry.

  IV. Felt by many indoors during daytime, but by few outdoors. Some awakened at night. Vibration like lorry striking building.

   V. Felt by nearly everyone; many awakened. Some breakages, disturbances of trees, telegraph poles.

  VI. Felt by all; many run outside. Some heavy furniture moved.

 VII. Everyone runs outside. No damage in well-built buildings; moderate damage in ordinary structures; considerable damage in poorly constructed buildings.

VIII. Considerable damage except in specially constructed buildings. Disturbs people driving cars.

  IX. Damage even in specially designed structures. Buildings shifted from foundations; ground cracked; underground pipes broken.

   X. Ground badly cracked. Railway lines bent. Landslides considerable.

  XI. Few brick-built structures remain standing, if any. Bridges destroyed. Broad fissures in the ground.

 XII. Total damage. Waves observed on ground surface. Objects thrown upward into the air.

Adapted from US Geological Survey.

mercantilism The view, current in early modern Europe, that one nation's gain depends on another nation's loss; a trading nation prospers only if it encourages the export, but discourages the import, of manufactures. I. Wallerstein (1980) is worth reading. 'Compared to the US, Japan is thought to actively engage in mercantilism, accumulating substantial foreign reserves through exclusionary trade and industrial practices' (Pantulu and Poon (2003) *J. Econ. Geog.* 3).

meridian An imaginary circle along the world's surface from geographic pole to geographic pole; a line of longitude. Meridians are described by the angle they form west or east of the **prime**

meridian, which has a value of 0° and runs through Greenwich, England.

meridional circulation In meteorology, air flowing longitudinally, across the parallels of latitude. **Meridional flow** occurs in *atmospheric cells, resulting in part from longitudinal temperature variations. See Song et al. (2008) *J. Climate* 21, 19. The **meridional temperature gradient** is the change in temperature along a *meridian. Von Storch (2008) *J. Climate* 21, 18 finds that an increase in the meridional gradient of sea surface temperatures strengthens the atmospheric circulation.

mesa A steep-sided plateau or upland, formed by the erosion of nearly horizontal *strata. A mesa resembles a *butte, but is very much wider; see Brookes (2001) *Geomorph.* 39, 1–4.

mesoclimate *See* LOCAL CLIMATE.

mesometeorology The study of middle-scale meteorological phenomena; between small features, like *cumulus clouds, and large features, like anticyclones. **Meso-scale** describes systems, or patterns of systems between small and *synoptic scale *c.*10–100 km across.

Mesozoic The middle *era of Earth's history, from around 225 to 190 million years BP.

(()) SEE WEB LINKS
• .An overview of the Mesozoic era.

metadata This describes data, specifically the nature of a dataset: who wrote it and when; how the data were created and why; the computer network where the data were created; and the standards used.

metageography The spatial framework on which we hang our knowledge; a framework we take for granted. We speak, for example, of Europe, but rarely ponder what that means. Do we define Europe by membership of the European Union? Or by entries to the Eurovision song contest? A **metageographic community**, is 'a multiscaled network of like-minded individuals, institutions, and organizations where physical nearness is not essential for the expression or maintenance of collective identity. An MGC is metageographic because it exists only as the result of mediated interaction over distance; that is, it exists "above the map" in transmission that is not bound by the proximity of its members in localized physical space' (Nashleanas (2011) *AAAG* 101, 3).

metagovernance A counter-process to *governance whereby economic and political coordination is achieved despite the limitations of governments, states, firms, governance, and clans (Jessop (2004) *Rev. Int. Pol. Econ.* 4, 3).

metamorphic aureole An area of rock altered in composition, structure, or texture by contact with an igneous *intrusion. See Richards and Collins (2002) *J. Metamorph. Geol.* 20, 1. **Metamorphic rocks** have been changed from their original form by heat or by pressure beneath Earth's surface; for example, limestone to marble, shale to slate, slate to schist. **Dislocation metamorphism** (a widespread but non-universally recognized concept—see Sobolev et al. (2002) *Bull. Voronzeh State Univ.* 2: *Geology*) occurs through friction along fault planes or *thrust planes. **Regional metamorphism** (also called **dynamic metamorphism**) occurs during an *orogeny; see Patison et al. (2001) *Australian J. Earth Scis* 48, 2.

metaphor Geographers are interested in the metaphors people use to explain their perception of a place or a process. For example, Hulme (2012) *TIBG* 37, 3,

346 looks at some of the ways in which writers, geographers, and artists have imagined the island of Cuba. He quotes A. Gaztambide-Geigel (1996) in imagining Cuba as a key; 'if the American continent lies behind the door of its eastern coastline, then Cuba—the largest island in the Caribbean—looks as if it's being inserted into the lock'. This metaphor partly explains the US thinking behind the Cuban Missile Crisis of 1962.

metapopulation In ecology, a place-specific assemblage of species that are connected. A **metapopulation construct** consists of several spatially discrete populations, along with areas of suitable habitat that are either occupied or unoccupied. In classical metapopulation theory, each population progresses relatively decoupled from the other populations (Hogan (2011) *Encyclopedia of Earth* website).

metasomatism The change in *country rock brought about by the invasion of fluid; see Storey et al. (2004) *Geology* 32, 2. Granites are formed at great depths by invading granulizing fluids; see Langer (1966) *Annals NY Acad. Scis* 136, 1.

meteoric water Groundwater originating from *precipitation.

meteorology The study and science of all aspects of the atmosphere—weather and climate—aiming to understand the physical and chemical nature of the atmosphere, its dynamical behaviour, and its complex interactions with the surface. It includes both short- and long-term weather forecasting, and the determination of past and future climatic change. Meteorology now includes the study of the atmospheres of other planets (S. Dunlop 2001).

methodological individualism (MI) One of the many views of the role of individuals in the definition and explanation of social phenomena. Since social wholes are apparently made up of human beings and are caused by their actions, they can and should be explained in terms of the actions of individual human beings. In this view, for example, culture is thought to be represented by the thoughts, attitudes, and behaviours of individuals, rather than by group dynamics. L. Udehn (2001) attempts to cover everything of importance that has been written on methodological individualism.

metrology The scientific study of measurement, especially the standardization and definition of the units of measurement in science. See Pike (2001) *Prof. Geogr.* 53, 2.

metropolis A very large urban settlement usually with accompanying suburbs. No precise parameters of size or population density have been established. 'Metropolises, are spaces of mobility. The growth of cities during the 20th century represented a constant increase in the mobility and in interconnections among places, regions, cities and countries around the world. In fact, it was precisely the development of communication and transportation technologies that made this growth possible' (L. F. de Mello and E. Marandola 2005).

As with other metropolises, Berlin's 'wealth of facilities, as well as their scatter across the metropolis, can be understood only in the light of the city's history and, paradoxically, its troubles' (Grésillon (1999) *Cult. Geog.* 6). Soja, in G. Bridge and S. Watson (2012) believes the metropolis era to be at an end.

micelle A sub-microscopic aggregate of molecules within a *clay colloid.

microclimates The climates of those parts of the lower *atmosphere directly affected by the features of the Earth's surface. The height of this part of the lower atmosphere varies according to the size of the influencing feature, but could be four times the height of the feature (Barry (1970) *TIBG* 49). Local factors commonly override the influence of larger scale macroclimatic factors (Vitt et al. (1994) *Arct. & Alpine Res.* 26). See Lindberg (2006) *6th Int. Conf. Urb. Climate* on modelling the urban microclimate.

micro-credit Lending small amounts of money to very poor households at commercial rates, rather than at the 'usurious rates of loan sharks' (S. Buckingham-Hatfield 2000). The classic model is the Grameen Bank, initiated in Bangladesh by Mohammed Yunis. Similar micro-credit schemes have been set up throughout the developing world, and have been enormously effective, not only in alleviating *poverty and improving child nutrition, but also in increasing the voluntary use of contraception. N. Burra et al., eds (2005) assess the impact of micro-credit. Poverty trends would probably not be reversed, 'but booming micro-credit would at least speak for a world where justice would have a greater place' (Santiso (2005) *Int. J. Soc. Sci.* 57, 185).

(⊕) SEE WEB LINKS
• Grameen Bank website.

microsimulation A computer analysis of individual behaviour, be it, for example, people or vehicles: 'a methodology that is concerned with the creation of large-scale population microdatasets and with the analysis of policy impacts at the microlevel. In particular, microsimulation methods aim to examine changes in the life of individuals within households and to analyse the impact of government policy changes for each individual and each household' (D. Ballas et al. 2005).

middle-income trap The phenomenon of hitherto rapidly growing national economies stagnating at middle-income levels and failing to graduate into the ranks of high-income countries. See Aiyar et al. (2013) *IMF Working Paper No.* 13/71.

mid-latitude depression An area of low atmospheric pressure occurring between 30° and 60°. This low is some 1500–3000 km in diameter and is associated with the removal of air at height and the meeting of cold and warm *air masses in the lower atmosphere. At the fronts between the air masses, a horizontal wave of warm air is enclosed on either side by cold air. The approach of the warm front is indicated by high cirrus cloud. The cloud then thickens and lowers, and rain falls. As the warm sector passes over, skies clear and the temperature rises. The cold front is marked by heavier rain and a fall in temperature. The warm front advances at 20 to 30 miles per hour, whilst the cold front can move forward more quickly at 40 to 50 miles per hour. In consequence, the cold front eventually pinches out the warm air, lifting it bodily from the ground to form an *occlusion. See R. Barry and R. Chorley (2003).

mid-oceanic ridge *See* OCEANIC RIDGE.

migration The movement of people from one place to another. The terms **in-migration** and **out-migration** are used for **internal migration**; **voluntary migration** refers to unforced movements, **compulsory migration** describes the expulsion of minorities from their country of birth by governments, or by warring factions. In the case of *commuting, migration is a daily act, but, because there is no change of residence, a purist would not call commuting a migration, preferring the term *mobility. Bell and Muhidin (2009) *UNDP, Hum. Devt Repts. Res. Paper 2009*/30, claim that 'internal migration is the most significant process driving changes in the pattern of human settlement across much of the world' and compare internal migration between countries.

In **innovative migration** people move to achieve something new (see Waxman in D. Mittelberg 1999). **Economic migration** is the movement of people from a poor to a richer area; Gerber (2006) *Int. Mig. Rev.* 40, 3 argues that market conditions are key determinants of contemporary migration patterns within Russia, for example. Boswell (2003) *Int. Affs.* 79, 3 finds that, in the EU, domestic political-electoral pressures and the institutional context have damaged longer-term strategies of migration management, refugee protection, and

relations with third countries. The distinction between economic and betterment migration is unclear. Hoey (2005) *J. Contemp. Ethnog.* 34, 5 writes on **non-economic migration**. Findlay and Li (1997) *Area* 29, 1 see the migration act as a manifestation of an individual's identity.

Some political geographers, such as J. Hyndman (2005) refer to **migration wars**—conflicts, usually by nation-states—over who to let in and who to keep out. As globalization proceeds, countries have invested heavily in border enforcement, increasing patrols along borders, building fences, and using infrared technology to keep people from crossing borders illegally. Mitra and Murayama, *Inst. Developing Econs Res. Paper #137*, provide a wealth of detail on the inter-state migration for India. Collins and Huang (2012) *Area* 44, 3,

India—the number of migrants by state

270 present a useful review on the relative value of different research methods, and the challenges of designing, executing, and analysing different methodological approaches in migration research.

migration-development nexus

The connections between *migration and *development, particularly arising from outmigration from the Global South. Taylor and Singh (2013) *Econ. & Pol. Weekly* 48, 24 discuss the migration-development nexus within the Doaba region of east Punjab.

migration systems theory

This suggests that migration flows acquire a measure of stability and structure over space and time, allowing for the identification of stable international migration systems, which are characterized by relatively intense exchanges of goods, capital, and people between certain countries and less intense exchanges between others. An international migration system generally includes a *core receiving region, which may be a country or group of countries, and a set of specific sending countries linked to it by unusually large flows of immigrants (Massey et al. (1993) *Pop. & Dev. Rev.* 19, 3, 431).

Milankovitch cycles

There are three interacting, astronomical cycles in the Earth's orbit around the sun: in the shape of the elliptical orbit ($c.$95000 years); in the axis of rotation ($c.$42000 years); and in the date of perihelion (the time of year when the Earth is closest to the sun, about 21000 years) (M. Milankovitch 1930). There is also a 100 ka cycle; see Berger (1978) *J. Atmos. Scis* 35.

The medium-scale cycles are poorly known: Campbell et al. (1998) *Geology* 26 recognize up to eight cycles, but cannot explain their origins. Ware (1995) *Fisheries Oceanog.* 4 demonstrates four short-term cycles: 2–3 years, 5–7 years (El Niño–Southern Oscillation), 20–25 years (bidecadal oscillation), and a poorly resolved, very low-frequency oscillation with a 50–75-year frequency. See Harris (2005) *PPG* 29, 2 for an excellent summary. Imbrie and Imbrie (1980) *Science* 207 show that Milankovitch cycles satisfactorily explain the major fluctuations in ocean temperature.

milieu

The sphere in which individuals live and which they are affected by, including tangible objects and people, social and cultural phenomena, and the *images which influence human behaviour. This is a huge theme in Canadian geography; see *Canad. Geog./Géog. canad.* (2000) 44, 1; (2002) 46, 1 and 46, 4; (2004) 48, 3; and (2006) 50, 4. Buciek et al. (2006) *Geografiska B* 88, 2 consider the ability of the Danish *kulturmiljø* (cultural milieu) to capture and incorporate the 'multiple and often contradictory cultural practices of different groups of actors and, not the least to transgress the often rather static and confined view on local history, which often results from the *heritage perspective'.

In economic geography, Waxell (2005) *Geografiska regionstudie* 62 finds the local milieu to be important for knowledge transfer, and Maskell (2004) *PHG* 28, 1 questions the view that tacit knowledge transfer is confined to local milieux whereas codified knowledge 'may roam the globe almost frictionlessly'. See also Gareau (2008) *Antipode* 40, 1 on the *neoliberal milieu.

military geography

The spatial expressions and impacts of military power; a field of study 'intimately associated with state military discourses of military power; a broad political geography, focused on the spatiality of armed conflict; and research from across the social sciences on the political economies and sociocultural geographies of militarism, particularly in nonconflict situations' (Woodward (2005) *PHG* 29, 6, 718).

military-industrial complex In a nation, those industries that provide *materiel* for the military, together with the military establishment itself. As defined by M. L. Stackhouse (1971) with reference to the USA, 'a specific set of impersonal, interlocking institutional structures, rooted in the history of urban industrial society and in the specific conflicts in America's emergence as a world power. It is governed by generalized value patterns that transcend ordinary political ideology; it is undergirded by a matrix of personal, contractual, and fiscal linkages, and it is centred in the relationships of the Department of Defense to major corporations'. Barnes (2008) *Geoforum* 31, 9, 3 refers to it as a 'cyborg entity'.

mimesis 'The belief that we should strive to produce as accurate a reflection of the world as possible' (J. Duncan and D. Ley 1993); 'Less-than-conscious embodied practices; the subconscious inculcation of the specific "rules of the game" of a particular field' (Holt (2008) *PHG* 32, 2). See Malmberg and Maskell (2002) *Env. & Plan. A* 34 on spatial clusters, spin-offs, and mimesis.

mineralization The breakdown of organic residues in soils; an essential process in humus formation. See Andriesh (2006) *18th World Congr. Soil Sci.*

mineralized zone An enriched zone of mineral deposits around an *igneous intrusion. Rousell and Trevisiol (1988) *Mineralium Deposita* 23, 2 identify the sequence of events in the mineralized zone of the Wanapitei complex, Ontario.

minimum efficient scale of production (mes. of production) The smallest possible size of a factory compatible with profitable production, below which *economies of scale don't apply. Mes. can be defined in terms of production, or of employment. The turnover of a single product is ultimately bounded by the minimum efficient scale and consumer demand for a specific product (Frenken and Boschma (2007) *J. Econ. Geog.* 7). A firm may lower its mes. by buying in components, or through technological change (M. Jovanović 2006).

misfit stream A stream that appears to be too small to have eroded the valley in which it flows; a stream whose volume is greatly reduced or whose meanders show a pronounced shrinkage in radius. It is a common result of drainage changes effected by capture, glaciers, or climatic variations (Beaty (1990) *Canad. Geogr./Géog. canad.* 34, 2).

mist A suspension of water droplets which restricts visibility to between 1 and 2 km.

mixed farming The farming of crops and livestock under unified management.

mixed methods research Combining qualitative and quantitative methods in a single study. De Lisle (2011) *Caribbean Curriculum* 18, 87 explains further.

mobility A research field that spans 'studies of corporeal movement, transportation and communications infrastructures, capitalist spatial restructuring, migration and immigration, citizenship and transnationalism, and tourism and travel' (Hannam et al. (2006) *Mobilities* 1, 1). T. Cresswell (2006) addresses 'how the fact of movement becomes mobility. How, in other words, movement is made meaningful, and how the resulting ideologies of mobility become implicated in the production of mobile practices.' T. Cresswell and P. Merriman 'provide a foundational understanding of work that is ongoing across mobilities scholarship' (Shaw (2012) *Area* 44, 1, 128).

The **mobilities paradigm** emphasizes that all places are tied into at least thin networks of connections that stretch beyond each such place, and mean that nowhere can be an 'island'. It goes beyond the imagery of 'terrains' as spatially fixed geographical containers for social processes, and calls into question 'scalar logics such as local/global as descriptors of regional extent', and suggests a set of questions, theories, and methodologies 'rather than a totalizing or reductive description of the contemporary world' (Sheller and Urry (2006) *Env. & Plan. A* 38, 2). Thus, McNeill (2008) *PHG* 32, 3 considers the hotel as a space 'operating between fixity and flow, locating and refreshing mobile bodies, [and] embedding them in relatively fixed networks within particular cities'.

modal split The varying proportions of different transport modes which may be used at any one time. The choices of modes may be determined by the costs, destinations, capacities, and frequencies of the modes together with the nature of the goods carried and their destinations; Buchanan et al. (2006) *J. Transp. Geog.* 14, 5 find that, in commuter behaviour, the key variable determining modal split and trip length was the distance the residence was located from the *central business district. Titheridge and Hall (2006) *J. Transp. Geog.* 14, 1 find that location, urban form, levels of car ownership, and the mixture of different social classes in the ward of residence are important determinants of modal split.

model A representation of some phenomenon of the real world made in order to facilitate an understanding of its workings. A model is a simplified and generalized version of real events, from which the incidental detail, or 'noise', has been removed; an 'abstraction of an object, system or process that permits knowledge to be gained about reality by conducting experiments on the model' (Clarke in P. Atkinson et al., eds 2004). Malmberg (1992, *Geografiska B* 74, 2) defines idealist models as 'designations of theories that lack a specification of how this model is related to processes in time and space'.

In the 1960s, the quantitative revolution in geography saw an emphasis on models: basic mathematical equations and models, such as gravity models, deterministic models, such as von Thünens and Weber's location models, and stochastic models that use probability. The 'cultural' turn—the counter-positivist response from human geography—stressed the weaknesses of models. O'Sullivan (2004) *TIBG* 29, 3 writes of geocomputational modelling, that 'the compositional model represents a theory about the world, rather than the world itself. The end result is a model . . . whose behaviour may be almost as intractably difficult to account for as the world it represents'. Connecting the model back to the world it represents is difficult for a number of reasons, principally the *equifinality problem, which makes it impossible to judge the relative merits of alternative models on purely technical grounds.

In physical geography, the basic requirement of a model is that it includes the important phenomena controlling a system, yet is restricted in complexity. One solution is to use a lumped model, which is constructed using average or 'lumped' representation of the relevant phenomena. Thus a lumped model of a catchment area would assume that catchment is homogeneous, by using representative soil properties and precipitation inputs. All this is reasonably clearly set out in Maurer et al. (2010) *J. Amer. Water Res. Ass.* 46, 5, 1024. Or try V. P. Singh, M. Fiorentino, eds (1996), section 10.2.1.

mode of production The way in which society organizes production. K. Marx and F. Engels (1974) claimed that under the **capitalist mode of production**, bourgeois democratic society is naturalized, so that the way people act and feel is universalized as 'human nature', in order to preserve the status quo. M. Sahlins (1972) identifies a peasant **family mode of production**, which he describes as pre-capitalist, while F. Ellis (1988) stresses the flexibility of the family in agricultural and non-agricultural activities. This rather deterministic view of the critical influence of the mode of production has been challenged, for example, because of the Eurocentric nature of the argument. Nonetheless, geographers recognize that each mode of production creates its own geography: for example, capitalism seems to be inextricably wedded to *uneven development (see, among many, Walker (1978) *Rev. Radical Pol. Econs* 10, 3).

moder A humus which is less acid than *mor but more acid than *mull. See Fons and Klinka (1998) *Geoderma* 86, 1 on mormoders and leptomoders.

modernity The knowledge, power, and social practices which emerged in 17th-century Europe. Modernity was not associated solely with 'newness', but also with beliefs in rationality and 'progress', and came to be seen as a central attribute of Europe, which the rest of the world was expected (or compelled) to adopt. 'The substitution of new designs for old will be a progressive move, a new step up the ascending line of human development' (Bauman in J. Krieger, ed. 2001; read the whole entry). Larsen (2006) *Geografiska B* 88, 3 sees Western modernity as characterized by the binary oppositions of self–other, nature–culture, and future–past. I. Wallerstein (2004) argues that modernity is concretely expressed as the modern world system. Claval (2007) *Tijdschrift* 98, 2 sees the large city as an expression of 19th-/20th-century modernity, while Taylor (2007) *TIBG* 32, 2 sees the subordination of cities to territorial states as the most important geographical attribute of modernity. Tomic and Trumper (2006) *Antipode* 38, 3, writing on Chile, find that the concept of cleanliness as an expression of the modernity claimed by neoliberalism for a small part of Chilean society comes 'at a high human cost': precarious labour under harsh conditions and low wages. Murphy (2006) *AAAG* 96, 1 argues that a core tenet of modernity is an abrupt break with the past, but D. Harvey (2003) argues against the 'modernity as break' thesis.

modernization theory All modernization theories are based on the assumption that less-developed areas are merely at an earlier stage along a single development path already experienced by the most-developed capitalist economies; Rostow's model is an example, and the *flying geese theory is an East Asian variety of modernization theory.

modifiable areal units Areas studied in geography, such as states, counties, census areas, or *enumeration districts, which may be arbitrary units unconnected with any existing spatial patterns developed—the boundaries of many African states, for example, cut across cultural and ethnic regions. In many cases the investigator has to use the units for which data are available rather than the units which are more suited to the investigation. Guo and Bhat (2004) *Transportation Res. Record* 1898 contend that the fundamental issue is the inconsistency between the analyst's definition of areal units, and the decision-maker's perception of residential neighbourhoods, and propose a way out of the problem.

Moho discontinuity (Mohorovičić discontinuity) The boundary in the Earth's interior between the *crust and the upper *mantle, marking a change in rock density from 3.3 (crust) to 4.7 (mantle). Clague and Straley (1977) *Geology* 5, 3 see the oceanic Moho as the boundary separating partially serpentinized ultramafic rocks, above, and fresh tectonized ultramafic rocks below.

The Moho occurs at about 35 km below the continents (but can be depressed to nearly 50 km depth; see Fuis et al. (2008) *Geology* 36, 3), and at around 10 km beneath the oceans.

Moh's scale A relative scale of hardness, based on the ability of a harder mineral to scratch a soft one. The ten minerals in the scale are: talc, gypsum, calcite, fluorite, apatite, feldspar, quartz, topaz, carborundum, and diamond. Use Moh's scale by testing an unknown mineral against one of these standard minerals. Whichever one scratches the other is harder, and if both scratch each other they are both of the same hardness.

moist coniferous forest Forest found only on the west coast of Canada and the USA where the climate is cool and moist throughout the year. Very large evergreens such as redwoods and Douglas fir grow, and undergrowth is sparse.

moisture budget (moisture balance) The balance of water, as represented broadly by the equation:

balance = precipitation − (runoff
+ potential evapotranspiration
+ the change in soil-moisture)

Moisture balance values of less than 1.0 indicate an annual excess of PET over P, a moisture deficit. Areas where values fall below 0.5 are regarded as drylands.

moisture index A measure of the water balance of an area in terms of gains from precipitation (P) and losses from *potential evapotranspiration (PE). The moisture index (MI) is calculated thus:

$$MI = \frac{100(P - PE)}{PE}$$

See Smakhtin and Hughes (2007) *Env. Model. & Software* 22, 6 on drought indices.

mollisols In *US soil classification, soils with a rich humus content, developed under grassland. *Rendzinas and *chestnut soils fall into this category. Kravchenko et al. (2011) *Chin. Geogra. Sci.* 21, 3, 257 provide a clear and useful account.

monadnock An isolated peak, the relict of long-term subaerial denudation, now standing above the *peneplain. The term comes from Mount Monadnock, New England; see Fowler-Billings (1949) *GSA Bull.* 60, 8.

monetary policy, geography of The aim of monetary policy is to control the supply of money, often targeting a rate of interest for the purpose of promoting economic growth and stability. The official goals usually include relatively stable prices and low unemployment, but monetary policy and aims vary spatially. See Chandra Mekha (2011) *Int. Conf. on Humanities, Geog. & Econs.* on Indian monetary policy.

money, geography of The interactions between money, *place, and *space, ranging from the global scale (*uneven development, *capitalism, *circuits of capital), to regional and local, such as variations in access to *financial services. Of considerable interest is the continued predominance, despite *time–space compression, of certain financial centres, such as Tokyo, New York, Frankfurt, London, and Amsterdam; see Faulconbridge (2007) *Growth & Change* 38, 2 on the changing landscape of European financial centres.

'Money is not just an economic entity, a store of value, a means of exchange, or even a "commodity" traded or speculated on for its own sake; it is also a social relation' (Martin (1999) in Stuart Corbridge). 'Value can be thought of as generated through relations and things which, via the material and social practices of the economy (production, exchange and consumption), come to be regarded as socially useful, helpful, uplifting or, more narrowly but generally, as fundamental to everyday life going on "as normal". These flows encompass the exchange of value embodied in products and may involve the exchange of money for work or the capacity to work, which could lead to an augmentation of future production and/or consumption' (Hudson (2008) *J. Econ. Geog.* 8, 3).

B. Cohen (1998) argues that the globalization of money has undermined the power of sovereign nation-states: 'a system predicated upon the ability to draw a line between the "inside" of the nation state and the "outside" of the international system has been shattered by the growing frequency and volume of monetary flows.' Lee (2002) *PHG* 26, 3 argues that the decision-makers in specialized investment houses and banks, all too ready to shift hot money around the globe, construct geographies through the intersection of their assessments of internal conditions in emerging markets and their power to engage with them. Lee did not forecast the resultant crash into global depression. Unfortunately.

There is surprisingly little on fraud in the geographical literature. *ID Analytics* examine how US identity fraud rates vary by geography, finding that the highest rates of identity fraud occur in New York state, followed by California, Nevada, Arizona, and Illinois, while the least risky states tend to be in the upper Midwest and upper New England.

monoculture A farming system given over exclusively to a single product. Writing on globalization and agrarian change in the Pacific Islands, Murray (2001) *J. Rural Studs* 17 lists some disadvantages of monoculture: 'increasing pollution, soil degradation and ground water depletion; further concentrating economic power, property ownership and social inequalities; and contributing to urbanization as displaced small growers migrate to towns and cities. Moreover, the Tongan economy has been left more vulnerable to global economic fluctuations.' However, Castree (2003) *PHG* 27, 3 writes that 'it is assumed to be self-evident that monocultures are undesirable. Yet they are linked to high yields, while their implied opposite—a world of mixed, non-chemical agriculture—might not so readily feed current population numbers.'

monopoly The provision of a good or service by a single supplier, who then has the power to set prices, since there is no competition. See Holmes and Stevens (2004) *J. Econ. Geog.* 4 on monopoly, geographic concentration, and business size.

monsoon Colloquially, a sudden wet season within the tropics, but, more explicitly, a seasonal shift of air flows, cloud, and precipitation systems.

Monsoons have been described in West and East Africa, northern Australia, Chile, Spain, and Texas, but the largest is the south-west monsoon. This Asian monsoon is the atmospheric response, complicated by the presence of water vapour, to the shift of the overhead sun, and therefore zone of maximum heating, from the Tropic of Capricorn in late December to the Tropic of Cancer in late June. Associated with this response are major changes in *jet stream movements and a meridional shift of the rain-bringing *inter-tropical convergence zone.

In winter, pressure is high over central Asia; winds blow outwards, and some depressions, guided by upper-air westerlies, move from west to east, bringing rain. During the spring, a *thermal low develops over northern India. Rains, related to a trough in the upper air, enter Burma in April/May, and India in late May/early June. In early summer, the direction of the upper air changes, and the tropical easterly jet stream is semi-permanent about 15° N for the rest of the summer. With this change, the monsoon 'bursts', giving heavy rain across the southern half of the subcontinent.

By late June there is a continuous southerly flow of warm, moist air into the monsoon trough lying across northern India. This flow is a continuation of the south-east *trades, altered to south-westerlies by the *Coriolis force as they move north, across the equator, bringing huge quantities of water vapour, and reaching the west coast of India as the south-west monsoon. Here, the rainfall is *orographically enhanced. *Monsoon depressions are formed in association with these air flows, steered by the now easterly jet. Subsiding upper air prevents rainfall in the Thal and Thar deserts of the north-west of the subcontinent. By autumn, the easterly jet stream is replaced by a narrow band of the westerly subtropical jet stream, which follows the southern Himalayas. The south-west monsoon begins to retreat in September. Thereafter, the north-east trade winds dominate, and *hurricanes are common in the Bay of Bengal. See O'Hare (1997) *Geo* 82, 3 and 4 for an exegesis on the Indian monsoon, and Bhaskara et al. (2001) *Meteor. Applications* 8 on forecasting of the Indian south-west monsoon.

monsoon depression A summer low pressure system affecting South Asia, 1000–2500 km across, with a *cyclone circulation about 8 km. Lasting two to five days, it comes roughly twice monthly, bringing 100–200 mm of rain per day. See Daggupaty and Sikka (1977) *J. Atmos. Scis* 34, 5.

monsoon forest A tropical forest found in *monsoonal regions. Trees are semi-deciduous or evergreen, in order to withstand drought. Monsoon forest is more open, and has more undergrowth, than *tropical rain forest. Kuznetsov and Kuznetsova (2013) *Biology Bull.* 40, 2, 187 provide a comprehensive and detailed survey of tropical monsoon forest in Vietnam.

Monte Carlo model A simulation model using computational algorithms, and relying on repeated random sampling to obtain numerical results; it lets the user see all the possible outcomes of any decisions made and assess the impact of risk.

mor A forest humus characterized by an organic matter on the soil surface in matted horizons, reflecting mycogenous decomposers, and with an abrupt boundary between the organic horizon and the underlying mineral soil. See Olsson (2005) *Glob. Change* 11, 10.

moraine The rocks, boulders, and debris that are carried and deposited by a glacier or ice sheet. Moraine is often partly stratified, since some may have been formed under water. **Ablation moraine** is common on retreating glaciers, and coarse, because meltwater has washed out finer particles. As the glacier shrinks, lateral moraines are deposited at the side of the glacier. Where two lateral moraines combine, a central, **medial moraine** may be formed. **Ground moraine**, also called a *till sheet, is a blanket covering the ground. Other moraines have been moulded by ice parallel to the direction of ice movement. These include **fluted moraines** which are long ridges, possibly formed in the shelter of an obstruction; see Benn (1994) *Sedimentol.* 41. *Drumlins are streamlined moraines. **Dump moraines** are ridges formed approximately transverse to flow from material delivered to the margin of a glacier by ice flow. They mark the stationary position of an ice margin. The size of dump moraines depends on the rate of ice flow and the debris content of the ice (Murray in J. Holden 2012). A **hummocky moraine** is a strongly undulating surface of ground moraine, with a relative relief of up to 10 m, and showing steep slopes, deep, enclosed depressions, and meltwater channels, which have formed from the meltout of supraglacial or englacial material. Boone and Eyles (2001) *Geomorph.* 38, 1 suggest that the growth and decay of ponds in repeated cycles of subglacial till failure and landform development can produce hummocks, but Lukas (2006) *PPG* 30, 6 notes that the term hummocky moraine can be problematic.

End moraines/terminal moraines are till ridges, usually less than 60 m high, marking the end of a glacier. In plan, they are crescentic, corresponding with the lobes of the glacier; a well-developed end moraine shows that the ice front was there for some time. Krzyszkowski (2002) *Sed. Geol.* 149, 3–4 argues that end moraines often constitute alluvial fans, formed at the ice margin by the redeposition of supraglacial material. See Dyke and Lavelle (2000) *Canadian J. Earth Scis* 37 on major end moraines in Canada.

Recessional moraines mark stages of stillstand during the retreat of the ice; a transverse moraine is a recessional if the up-glacier surface shows streamlining. **Rogen moraines** are fields of transverse moraines, 10–30 m in height, up to 1 km long, 100–300 m apart, and often linked by cross-ribs. Their origin is uncertain. **De Geer moraines** form where a glacier meets its *proglacial lake, and consist of till, layered sand, and lake deposits (Evans et al. (2002) *J. Quat. Sci.* 17, 3). **Push moraines** occur when a glacier is retreating in the melt period but advancing in the cold season; see Krüger et al. (2002) *Norwegian J. Geog.* 56. Glaciohydraulic supercooling may aid the development of moraines created by melt-out from basal ice (Lawson et al. (1998) *J. Glaciol.* 44).

moral geography Studies of the interactions and variations within morality, place, and space, such as the analysis of global variations in moral beliefs and practices; 'perspectives on human geography with a normative emphasis' (D. Smith 2000). Smith stresses the need for geographers to engage with moral issues, asking: does distance diminish responsibility? Should we interfere with the lives of those we do not know? Is there a distinction between private and public space? Which values and morals, if any, are absolute, and which cultural, communal, or personal? And are universal rights consistent with respect for difference?

Castree (2007) *Int. J. Reg. Res.* 31, 4 argues that all workers, knowingly or not, operate with moral geographies: 'moral geographies matter because they are the ethical basis for all worker solidarity and division, at whatever geographical scale happens to interest us.' Whitehead (2003) *Ethics, Place & Env.* 6, 3 argues that, in addition to geographical space, geographical scale also plays a crucial role in the construction and maintenance of moral frameworks. See also R. Lee and D. Smith, eds (2004).

morbidity Ill health. **Morbidity rates** are of two types: the *prevalence rate* (the numbers suffering from a specific condition at any one time), and the *incidence rate* (the numbers suffering from a particular condition within a given time period, usually a year).

more economically developed country (MEDC) A country with: low birth, death, and infant mortality rates (characteristically around 10; around 10; and under 12 per thousand, respectively); less than 10% of the workforce in agriculture; high levels of nutrition, secondary schooling, literacy, electricity consumption per head; and per capita GDP generally above $US2000. 'Economically' was included in this term some 30 years ago, to redress the perceptions of financially richer nations as more civilized, and more rational.

more-than-human world (more-than-human geographies) The non-human world, and its geographies: 'worlds, characterized by lively processes and impure forms, co-existing in inhabited landscapes' (Lorimer (2012) *PHG* 36, 5, 593). The more-than-human approach emphasizes the ways in which humanity totally depends on non-human things, which are often beyond its control or understanding.

morphodynamics The study of the evolution of landscapes and seascapes in response to the erosion and deposition of sediment (Parker (2006) *U. Illinois*); 'the interaction between the form and the processes' (I. Livingstone and A. Warren 1996). See Stephenson (2006) *PPG* 30, 1 for a review of coastal morphodynamics, and Swart and Zimmerman (2009), *Ann. Rev. Fluid Mechs.* 41, 203 on the morphodynamics of tidal inlet systems. For morphodynamic modelling, see Abad and García (2004) *Wld Water & Env. Resources Congr.* and Federici and Seminara (2003) *J. Fluid Mechs* 487.

morphogenesis The process or processes that initiate the development of a landform; hence **morphogenetic factors**. See Kompani-Zare et al. (2011) *Catena* 86, 3, 150.

morphogenetic region Any region whose geomorphology and geomorphological processes are recognizably distinct, 'wherein regional landforms reflect regional climates' (R. Chorley, S. A. Schumm, and D. E. Sugden 1985).

morphological impact assessment system (MImAS) A tool for assessing the impact of modifications to a river on river morphology and associated ecological quality. The inputs to the tool are:

- length and type of engineering modification
- length of the river or reach to be assessed
- channel type.

Each type of modification has a particular impact rating for each channel type, which is multiplied by the length of channel affected, and scaled to the reach assessment length:

$$\text{impact} = \left(M \times (R/L) \times 100 \right)$$

where M is the modification impact rating, L is the length of the modification, and R is the length of the reach being assessed. See Gilvear and Jefferies in J. Holden (2012) p. 360.

morphological mapping Mapping landform segments and facets using breaks of slope and inflections (Savigear (1965) *AAAG* 55). 'Landforms are discrete units that can readily be defined and verified at different scales by proven techniques, are therefore acceptable integrated classifiers of the landscape, and can be used to divide it into discrete segments' (Bocco et al. (2001) *Geomorph.* 39, 3–4).

morphology 1. The study of form; the term is now used as a synonym for the form itself, as in the morphology of the landscape or of the city.

2. In geolinguistics, the phonological shapes of words that adapt to special grammatical functions. See U. Ammon (2005).

morphometry The process of measuring the external shape and dimensions of landforms (for example, Federici and Spagnolo (2004) *Geografiska A* 86, 3 on cirques), or living organisms (for example, Wilfert et al. (2006) *J. Biogeog.* 33, 11 on termites).

morphotectonics 'Morphotectonics is a rather poorly defined term, but most geomorphologists would probably agree that it concerns causal linkages between topography and underlying tectonic or structural features within the crust' (French (2006) *PPG* 30). Morphotectonic features may reveal recent seismic activity (Caputo et al. (2006) *Terra Nova* 18, 3); can provide an insight into past sediment flux to depositional basins (Jones et al. (2004) *Terra Nova* 16, 3); or illuminate drainage basin evolution (Ghassemi (2005) *Basin Res.* 17, 3). A. Scheidegger (2004) claims that morphotectonics is fundamental to the understanding of landscape evolutions.

Morphological mapping

mother cloud A cloud from which another cloud develops; for example strato-cumulus cumulo-genitus (from cumulus). See Curic et al. (2002) *EGS XXVII Abstract* #3084.

moulin (glacier mill) A rounded, often vertical hole within stagnating glacier ice. Meltwater, heavily charged with debris, swirls into the hole. 'Moulins are ephemeral features in such an environment' (James (2003) *Geomorph.* 55, 1–4). Some of this debris settles out at the base of the moulin, forming a **moulin kame** after the ice retreats.

mountain building The creation of uplands by movements of the Earth's crust. This is often, but not necessarily, associated with an *orogeny. Possible causes of the uplift of mountains are: a decrease of the density of the crust causing it to rise, possibly above a *hot spot; thickening of the crust at *collision margins; subduction below continents, causing them to rise; overthrusting of sedimentary rocks during collisions; and compression at *converging or at *conservative plate margins.

mountain ecology A change in altitude up a mountain equals a latitudinal shift away from the equator, and vegetation types change accordingly. The sequence in the San Francisco mountains is hot desert cactus at 200 m; oak scrub at 800 m; coniferous forest at 2000 m; and alpine tundra at 4000 m (Woodward (1993) *Ecol. Applications* 3).

mountain meteorology Upland areas are cooler than lowlands of the same latitude, since temperatures fall by 1 °C/150 m. The heating of a valley floor may cause *anabatic winds, and colder and heavier air at the peaks may spill down as a *katabatic wind. When warm, moist air rises over mountain barriers, *orographic rain falls; when this air has passed over the barrier, it descends and

is adiabatically warmed, which may bring *föhn winds. The warmed air can now hold the remaining moisture, so a 'rain shadow' of drier air develops in the lee of mountains. Lee depressions form downdraught of mountains, and major barriers such as the Himalayas affect the upper atmosphere. See C. D. Whiteman (2000).

mountain wind A wind formed when dense, cold air flows downslope from mountain peaks, or from a glacier. *See* KATABATIC WIND; BORA.

mud flat An accumulation of mud in very sheltered waters. Mud is carried into *estuaries and sheltered bays and settles at low water. See Proffitt et al. (2005) *J. Ecol.* 93, 2.

mud flow The *flow of liquefied clay. Debris of all sizes, including large blocks of rock, may be transported by mud flows. In a study of northern Italy, Luino (2005) *Geomorph.* 66, 1–4 finds that mud flows in small drainage basins are usually triggered when the rainfall has reached a value equal to 10–20% of the local mean annual rainfall (MAR). Mud flows in larger basins are triggered when rainfall is equivalent to 15–30% of the local MAR.

mudslide A mass movement of *debris lubricated by melting ice and snow, or by heavy rain. Meiton and Razafimahefa (2011) *Zeitschr*, 55, 1 find that extreme *precipitation, together with anthropological features, such as felling and deforestation, and periodic bush fires are the major factors initiating mudslides. *See also* JÖKULHLAUP; LAHAR.

mull In soil science, a mild humus produced by the decomposition of grass or deciduous-forest litter. Earthworms and other soil fauna mix this humus thoroughly with the mineral content of the soil. See Ampe and Langohr (2003) *Catena* 54, 3.

multicell storm A severe local storm comprising a succession of convective cells. The first cell is formed by a powerful updraught of moist air. As the ensuing downdraught stifles this updraught, a new updraught develops to the right of the first. This sequence continues until the surface air cools. See Fovell and Tan (2000) *QJ Royal Met. Soc.* 126, 562.

multiculturalism 'The political accommodation by the state and/or a dominant group of all minority cultures, defined first and foremost by reference to race or ethnicity; and more controversially, by reference to nationality, aboriginality, or religion' (Modood in I. McLean and A. McMillan, eds 2003). Nash (2005) *Antipode* 37, 2 describes multiculturalism as 'the diverse policies and ways of thinking about societies characterised by cultural plurality'.

   Liberal multiculturalism focuses on cultural diversity, celebrating ethnic variety, and teaching tolerance. Howard-Hassman (1999) *Canad. Pub. Policy/Analyse de pol.* 523 observes that the public policy of liberal multiculturalism encourages private, individual choices of identity; 'paradoxically, this liberal policy also encourages identification with Canada and Canadian citizenship'. Nash (2003) *PHG* 27, 5 contends that neoliberal multiculturalism 'delegitimates the politics of class and redistributive justice and provides opportunities for more overtly reactionary versions of nationhood to emerge'.

   Critical multiculturalism is concerned with 'majorities' as much as 'minorities', and considers the institutions and practices forming the whole society. It sees inequalities of power, and racism, as central, and emphasizes recognition and rights. 'The continuous invocation of "oppression" in critical multiculturalism, and related calls for "counter-hegemonic" action, look incongruous: oppression becomes a thoroughly discursive, contestable and pluralized modality, and "hegemony" itself is an impossible attempt to constitute imaginary unity out of palpable diversity' (McClennan (2001) *Ethnicities* 1, 3).

multi-model inference (MMI) A method of simultaneously comparing the evidence for multiple hypotheses, each represented as a model. Millington et al. (2011) *Geog. Compass*, DOI:10.1111/j.1749-8198.2011.00433.x review two complementary MMI techniques—model selection and model averaging—and highlight examples of their use by biogeographers.

multinational corporation *See* TRANSNATIONAL CORPORATION.

multinational state A state which contains one or more ethnic groups, as identified by religion, language, or colour. Bell (1999) *Cult. Geogs.* 6, 183 encounters iconic references to the pre-Revolutionary cultural traditions of Muslim central Asia in Tashkent, with the aesthetic of a modernized, multinational (yet Russianized) society, arguing that political elites have a difficult task in rallying citizens around symbols of national autonomy, while not undermining their own legitimacy, which still rests in large part on Soviet-era institutions.

multiple nuclei model A representation of urban structure based on the idea that the functional areas of cities develop around various points rather than just one in the central business district. Some of these nuclei are pre-existing settlements, others arise from urbanization and *external economies. Distinctive land use zones develop because some activities repel each other; not all land users can afford the high

Multiple nuclei

costs of the most desirable locations, and some require easy access.

multiplier The economic consequence of an action, intended or otherwise. Once a location becomes established as a production centre, economies of agglomeration (which increase with increasing regional size) tend to give it permanent cost advantages over other locations. These scale and cost advantages are reinforced by the relatively higher wages paid to workers in the scale-intensive industries, which are thought to act, by means of a Keynesian income multiplier, as a stimulus to local markets, resulting in additional scale economies, which, in turn, lead to further growth of regional exports (Leichenko (2000) *Econ. Geog.* 76, 4).

Holloway and Kneafsey (2000) *Sociologia Ruralis* 40 argue that farmers' markets provide a local multiplier effect. In services, Glückler (2005) *Env. & Plan.*

A 37, 10 observes that continuous collaboration with a client may produce multiplier effects through client referrals and provide cheap access to new potential clients and future business. See Peck (2002) *J. Econ. Geog.* 2 on the multiplier effects of major factory closures: reduced demand, spending power, savings, and investment.

multi-scale landscape morphometry Digital images of landscape features may be captured in a sequence of successively coarsened resolutions. Hierarchy theory (see O'Neill et al. in M. G. Turner and R. H. Gardner 1991) suggests that patterns observed at any given resolution will, to some extent, weaken the patterns observed at finer resolutions and will themselves be constrained by coarser resolution patterns. Establishing the range of scales over which landscapes show hierarchically nested spatial patterns makes it possible to identify the scales at which fine-scale processes affect global scale patterns and vice versa. The selection of appropriate scales of measurement is fundamental to this process, and is generally guided by principles of hierarchy theory, since hierarchy incorporates relational links between nested levels. Bailey (1987) *Landscape and Urban Planning* 14 suggests a hierarchy of criteria for multi-scale ecosystem mapping and Tay et al. (2005) *Geosci. & Rem. Sens. Letts, IEEE* 2, 4 provide a simple framework to generate multi-scale digital elevation models and extract topologically significant multi-scale geophysical networks.

mythical geography 'Symbolic places that celebrate history and invest locations with mythical meaning provide a sense of identity in place and time; they fuse history and geography in terms of myth and memory' (Azaryahu and Kellerman (1999) *TIBG* 24, 1). 'The Greeks used the mythical geography of their land to re-create a physical view of the past that their poets, priests and politicians used as a paradigm for contemporary behaviour, and they drew upon the world around them, not just to illustrate that past, but also in many ways to create it' (J. Boardman 2002). 'Mythical geography is inherently holographic, in the sense that its interpretation is not tied to the identification of specific physical locations; each human being has the universe within himself and myths are meant to be discovered and understood by everyone as if they had been created for him alone' (C. Carpentier de Gourdon 2002).

naive geography The incorporation of the knowledge that people have about their surroundings including *GIS and cartography, in order to make spatial information and decision support more accessible (here, naive means instinctive, not stupid, geography). 'The naive geography world is based on the human mind . . . It is the set of informal ideas concerning geographical relationships' (M. A. Nelson 2003). For a comprehensive review of naive geography, see Egenhofer et al. in A. Frank and W. Kuhn, eds (1995).

nappe Part of a broken recumbent fold which has been moved forward over the rock formations beneath and in front of it, finally covering them; see van Hinsbergen et al. (2005) *Geology* 33, 4 on nappe stacking in the Aegean region.

narratives of place *See* PLACE.

nation 'Governable space' (Watts (2004) *TIBG* 29). The ethnic or cultural nation has ethnicity as the basis of membership, with culture as a key cohesive element. The cultural interpretation regards 'nation' as something organic, evolving on the basis of an aesthetic, primordial community, whereby ethnic group solidarity is based on ties passed down from the past, and hence the modern nation may be related to historical ethnicity. Culture and language are crucial cohesive elements of this community. In contrast, the territorial or civic nation has territory as the basis of membership and citizenship

as the cohesive force (Robinson et al. (2001) *Pol. Geog.* 20, 8). Mavroudi (2007) *Glob. Netwks* 7, 4 shows that the nation is not a 'given' entity; see Uimonen (2003) *Glob. Netwks* 3, 3. Although many *multinational states exist, E. Gellner (1983) argues that a mismatch between national and political units engenders nationalist movements; see Sekulic (1997) *Nats. & National.* 3, 2. L. Kong and B. Yeoh (2003) hold that, in postindependence Singapore, the creation of a fixed national identity has denied the population's innate hybridity; see also Chua in J. S. Kahn, ed. (1998).

nationalism This refers to the territorial expression of identity: a sense of belonging to a group or community associated with a particular territory. Often national identities do not correspond with the territory of the nation-state; many French-Canadians, for example, view Quebec as a distinct nation and believe it should become its own nation-state; see M. Henderson (2007). Most states in the world are multinational states, meaning that they contain a variety of nationalities. However, Singh (2012) *Social Sciences* explains how, 'despite demographic and cultural diversities, geography, history and creative political construction have combined to produce the Indian variant of nationalism'.

National Park An area less exploited by humans, containing sites of particular scenic or scientific interest, and protected by a national authority.

'Geography, and our identification with it, give us a sense of place. Geography also affects our national identity, and for many National Parks is the fundamental reason for their establishment as parks' (United States National Parks Authority).

There may be conflicts between the conservation of the natural environment and public access; 'where this happens, priority must be given to the conservation of natural beauty' (Lord Sandford 1974). Imrie and Edwards (2007) *Geog. Compass* 2 record protests by locals over the Maya Biosphere Reserve, Guatemala. See also Milian (2007) *Cybergeo* on the politics of 'sites naturels' in the Pyrenees.

Attempting to resolve post-colonial tensions, the management of many national parks in the developing world is shared between indigenous communities and the non-indigenous; see Waitt et al. (2007) *TIBG* 32, 2 on the Uluru National Park. For more on management, see T. Prato (2005) and for the economic benefits of a national park, see S. L. Patiño's study of the Madidi National Park, Bolivia.

nation-state The term 'nation-state' is a mix of two linked, though different concepts, 'nation' and 'state'. Nations are groups of people who believe themselves to be linked together in some way, based on a shared history, language, religion, other cultural practices, or links to a particular place. States are legal and political entities, with power over the people living inside their borders. In this way, states are associated with territorial sovereignty. The concept of a nation-state fuses together the nation—the community, and the state—the territory (Gilmartin in C. Gallaher et al. 2009), yet the nation-state rarely occurs in practice. Indeed, Walby (2003) *Sociol.* 37, 3 argues that the nation-state is more mythical than real: there are more nations than states; several key examples of presumed nation-states are actually empires; there are diverse and significant polities in addition to states, including the EU and some organized religions; and polities overlap, and rarely politically saturate the territory where they are located. Nonetheless, D. Slater (2004) argues that as the nation-state is an important crossroads between the global and the local; it is the critical site for geopolitical study, and Rodrik (2013) *Econ. Geog.* 89, 1, 1 argues that the nation-state is the main determinant of the global distribution of income, the primary locus of market-supporting institutions, and the chief repository of personal attachments and affiliations. See Mountz (2003) *Antipode* 35, 3 on a poststructuralist approach to geographies of the nation-state. **Nation-building** has as its aim the unification of its people, via state action, to promote stability and viability. See Schatz (2003) *Kellogg Institute Working Paper #303* on nation-building in Kazakhstan, which, at final independence comprised 39% ethnic Kazakhs. See also A. Kapur (2002) p. 129 on India and nation-building.

natural area 'A geographical area characterized both by a physical individuality and by the cultural characteristics of the people who live in it.' (Zorbaugh in J. Lin and C. Mele 2005). Most cities contain numbers of natural areas which are delimited by informal boundaries, such as canals, parks, railways, rivers, and main roads. 'Culture is the agent, the natural area the medium, the cultural landscape the result' (C. Sauer 1925).

natural increase The surplus or deficit (negative natural increase) of births over deaths.

natural region A region unified by its physical attributes, especially its latitude, relief and structure, and location. Lane et al. (2004) *Australian Geog. Studs* 42, 3 argue that the 'natural' region is a highly contested concept. Even so, in Canada, federal and provincial park and protected area system plans adopt natural region representation approaches (Lemieux and Scott (2005) *Canad. Geog./ Géog. canad.* 49, 4).

natural resource Any property of the physical environment, such as minerals, or natural vegetation, which is exploited by humans. Natural resources may be classified as *renewable and *non-renewable. The journal *Natural Resources Forum* covers technical issues of extraction, development, and efficient exploitation of resources, and international, multidisciplinary issues related to sustainable development and management of natural resources, particularly in developing countries and countries in transition.

natural resource evaluation Assessing the economic value of the natural environment. Key considerations include: an assessment of the importance of each resource to human life in a particular context, how far one resource can be replaced by another, whether the resource is a stock or a flow, and whether it is exhaustible or renewable. For the analytic tools used in natural resource evaluation, see S. Schueller et al. (2006).

natural resource management (NRM) The management of natural resources to promote development that is economically viable, socially beneficial, and ecologically *sustainable; S. Moore and S. Rockloff (2006) analyse the Australian government's attempts to effect nationwide changes in the management of agricultural areas and rangelands; see also Robins and Dovers (2007) *Geogr. Res.* 45, 3 on NRM in Australia, and Perreault (2006) *Antipode* 38, 1 on the

exclusion of Bolivia's poor resulting from the restructuring of water and gas resources. For the methodologies used in natural resource management, see Viergever et al. (2006) *Geomatics World* 4 and G. Herath and T. Prato (2006).

natural vegetation In theory, the grouping of plants which has developed in an area without human interference; since most landscapes have been changed by humans, it is argued that there are relatively few areas of truly natural vegetation left; many biogeographers use, instead, the concept of **potential natural vegetation**—the vegetation that would exist today if humans were removed from the scene and if the plant succession after their removal were telescoped into a single moment. The time compression eliminates the effects of future climatic fluctuations, while the effects of humans' earlier activities are permitted to stand. The potential natural vegetation reveals the biological potential of all sites (US Env. Protect. Agency).

nature 'Nature, in the common sense, refers to essences unchanged by man; space, the air, the river, the leaf' (R. W. Emerson 1836); 'a leisure-time concept . . . something to be consumed' (D. Harvey 1985). Bakker and Bridge (2006) *PHG* 30, 1 observe that resource and environmental geography have conceptualized nature in predominantly physical terms: as an assemblage of things independent of (although related to) society, whose properties (responsiveness to technology, geographical location) and social utility are revealed by science. 'In a world with multiple realities of "nature" that are constructed by diverse groups with differing motivations and access to power, decision-making in environmental management can become a contest over whose knowledge is "right"' (Pedynowski (2003) *PHG* 27, 6).

'Contemporary privatization is re-making nature–society relations as property. Privatization innovates and proliferates new forms of property such as patents for genetic information, markets for water, and tradable credits for polluting. In so doing, privatization transforms people's relationships to themselves, each other, and the natural world' (Mansfield (2007) Lecture, Ohio State University). See Castree (2003) *PHG* 27, 3; see Mrozowski (1999) *Int. J. Hist. Archaeol.* 3, 3 on the commodification of nature. See Sheppard (2011) *Dialogues Hum. Geog.* 1 46: it's clear and thought provoking.

nearest neighbour analysis The study of a spatial phenomenon in order to discern any regularity in spacing by comparing the actual pattern of that phenomenon with a theoretical random pattern. For the methodology, see Mak (2003) *Geog.* 516, 4. Alternatively, most *GIS software can do it.

nearshore In a coastline, the area comprising the swash, surf, and breaker zone; the area in which waves are forced to break owing to the shallowing of water closer to the shoreline (J. Holden, ed. 2012).

nearshoring The relocation of work in services from the country of origin of that work to a lower-wage nation nearby. Within Europe, there are major nearshoring centres in Bulgaria, the Czech Republic, Hungary, Poland, Slovakia, and Romania. Meyer (2006) mpra.ub.uni-muenchen.de/5785/, is strong on relevant data for eastern and central Europe.

nebkha See SAND DUNE.

nehrung A synonym for *barrier beach.

neighbourhood A district forming a community within a town or city, where inhabitants recognize each other by sight. See C. Moughtin (2005). Neighbourhoods evolve their own distinctive characteristics, or subculture; see Shaw et al. (2004) *Urb. Studs* 41, 10. Schnur (2005) *Tijdschrift* 96, 5 shows that local social capital can make the crucial difference in neighbourhood development.

neighbourhood effect The power of the environment to structure social contacts, plus the empirical fact that contact across class boundaries makes a consensual impact on partisan choice (Miller (1978) *Brit. J. Pol. Sci.* 8). McAllister et al. (2003) *Brit. J. Pol. Sci.* 31 provide 'very strong' evidence that members of each social class in the UK were much more likely to vote Labour than Conservative in the low-status than in the high-status areas. 'This is entirely consistent with the concept of the neighbourhood effect, but alternative explanations are feasible.' Johnston et al. (2004) *Polit. Geog.* 23, 4 find that neighbourhood effects are among the strongest influences on voting patterns. This term is not to be confused with neighbourhood effects, such as 'the hypothesized negative effects of living in poverty concentration neighbourhoods on various individual outcomes such as employment, earnings, school performances, and "deviant' behaviour,"' (Hedman et al. (2013) *J Econ. Geogr.* DOI: 10.1093/jeg/lbt042.

neoclassical economics Neoclassical economics focuses on the behaviour of individual consumers and firms. Consumers and firms interact in markets, which set the prices of goods and services. Taking into account their limited incomes and the prices set in the market, consumers buy the goods and services that maximize their well-being. Taking into account the prices set in the market, firms make decisions, including location decisions that maximize their profits. Firms compete with one another, and this competition ensures that they will seek out and take advantage of profitable opportunities (Cortright (2006) Brookings Inst. Met. Policy Prog.).

Neoclassical economics suggests that surplus labour and *capital will move to areas where labour and capital are in short supply, so that uneven development will be eradicated (Taylor (2007) *Globalizations* 4, 4). Other important ideas include *cluster development and trickle-down economics—but see Basu and Mallick (2007) *Camb. J. Econ.* 32, 3.

neo-colonialism The control of the economic and political systems of one state by a more powerful state, usually the control of a developing country by a developed one. It is marked by the export of *capital from the least economically developed countries (LEDCs) on the periphery to the controlling more economically developed countries (MEDCs) at the core, a reliance by the LEDCs on imported manufactures from the MEDCs, and adverse *terms of trade for the periphery or satellite nations (Blattman et al. (2004) *Harvard Inst. Econ. Res. Disc. Paper* 2040). The means of control are usually economic, including trade agreements, investment, and the operations of *transnational corporations, which are often seen as the primary instruments of neo-colonialism; but see Law (2000) *east asia cult. crit.* 8, 1. Forstorp (2008) *Studs Philos. & Educ.* 27 argues that 'globalization' and 'knowledge society' are expressions of neo-colonialism.

neo-conservatism A particular version of conservatism that holds that security is best attained by using state power to spread freedom and democracy, if necessary by force and without international cooperation; a view

strengthened in the USA after 9/11. Elden (2007), *Int. Politics* 44, 1 writes that the war on Iraq violated territorial sovereignty, even though neo-conservatives see territorial sovereignty as an absolute. As a result, in the USA, funding programmes addressing 'natural' hazards have been cut to fund counter-terrorist strategies (Graham (2006) *Int. J. Urb. & Reg. Research* 30, 2); government spending on public services has decreased (McGary (2004) *Philosophy & Geog.* 7, 1); and public–private sector partnerships replace traditionally public-sector activities (O'Neill et al. (2002) *Health Prom. Int.* 15, 2). See also Yamazaki (2002) *Geopolitics* 7, 1 on neo-conservatism in Japan; Peunova (no date) *Grad. Stud. Rev.* 2, 2 on a Russian retro-empire; and Mahs (2001) *Urb. Geog.* 22 on neo-conservatism and urban homelessness.

Neogene The current geological period of the *Cenozoic era, consisting of the Miocene, Pliocene, *Pleistocene, and *Holocene epochs. The Neogene began 23 million years ago (E. Martin and R. Hyde 2008).

neogeography New forms of geographical knowledge, using the World Wide Web and created by the collective actions of many individuals and processed and displayed online, rather in the way that bird counts have used volunteers to conduct a census of birds in the Western Hemisphere since 1900. These neogeographical data range from place tags on virtual globes to GPS data that can be mapped and combined with other data to create large, dynamic, open datasets. Connors et al (2012) *AAAG* 102, 6, 1267 have a long and detailed survey, with many case studies.

neoglacial A time of increased glacial activity during the *Holocene. Lamoureux and Cockburn (2005) *Holocene* 15 suggest that increased moisture, perhaps driven by shifts in oceanographic conditions, were important in North American neoglacials.

neoliberalism It seems generally to be agreed that this term refers to the liberalizing of global markets associated with the reduction of state power: state interventions in the economy are minimized; privatization, finance, and market processes are emphasized; capital controls and trade restrictions are eased; free markets, free trade, and free enterprise are the buzzwords. Beyond that, definitions become more partial; D. Harvey (2005), for example, speaks of 'the doctrine that market exchange is an ethic in itself, capable of acting as a guide for all human action... [the reduction of] the obligations of the state to provide for the welfare of its citizens... unfettered individual rights'. Harvey further argues that privatization writ large, financialization (*sic*), the skilful manipulation of crises practically and discursively to spread 'market rule', and state redistributions to the private sector and the rich have been neoliberalism's modus operandi. Thus, mainstream scholars depict neoliberalism as an innovation of capitalism, while *political economists view the rise of neoliberalism as a response to a capitalist crisis of accumulation. Additionally, Birch and Mykhnenko (2009) *J. Econ. Geog.* 9, 3, 355 argue that whereas **neoliberal ideology**—based on abstract economic concepts such as free market efficiency— represents a global discourse that developed at diverse sites around the world, neoliberalism as a state-led project produced national varieties of neoliberalism in which deregulation, privatization, and trade liberalization were pursued for different political reasons, in different ways, and to different extents.

Case studies of the impacts of neoliberalism abound; see, for example: Budds (2004) *Sing. J. Trop. Geog.* 25 on Chile;

McCarthy (2006) *AAAG* 96, 1 on British Columbia; Ryan and Herod (2006) *Antipode* 38, 3 on Australia and Aotearoa; Gökarkisel and Mitchell (2005) *Global Networks* 5, 2 on Turkey; Fisher (2006) *Soc. Justice* 33 on race, neoliberalism, and 'welfare reform' in Britain; and Emery (2006) *Soc. Justice* 33, 3 on contesting neoliberalism in South Africa. See Liverman and Vilas (2006) *Ann. Rev. Env. & Resources* 31 on neoliberalism and the environment.

Attention has also been paid to **governance/governmentalities of neoliberalism**: see Larner and Butler (2005) *Studs Polit. Econ.* 75 on New Zealand; McCarthy (2004) *Geoforum* 35, 3. **Spaces of neoliberalism** are addressed by N. Brenner and N. Theodore, eds (2003), R. Basu (2004), W. Larner and W. Walters (2004); and **scales of neoliberalism** by Kohl and Warner (2004) *Int. J. Urb. & Reg. Research* 28, 4.

neopatrimonialism A mode of government based on informal political ties and exchanges; the head of state and his agents mostly exert their power through personal whim and material incentives. Erdmann and Engel (GIGA-WP-16/ 2006) argue that neopatrimonialism is 'not a cut-and-dried concept'. In one view, the use of state resources by elites to buy loyalty among the public is corruption, or bad *governance. In another, it is an effective way of the poor to be directly linked to the state and obtain the support and goods they need' (N. Castree, R. Kitchin, and A. Rogers, 2013).

neotectonics The study of the causes and effects of the movement of the Earth's crust in the Neogene, i.e. the late *Cenozoic era. See Székely et al. (2002) EGU *Stephan Mueller Spec. Pub. Sers.* 3 on neotectonic movements and their geomorphic response.

nesting The way in which one network fits into a larger one. For example, Inkpen et al. (2007) *Area* 39, 4, see scale as a **nested hierarchy**. Species communities often exhibit **nested assemblages**, the species found in species-poor sites representing subsets of richer ones (Schouten et al. (2007) *J. Biogeog.* 34, 11).

net primary productivity In ecology, the amount of energy which primary producers can pass on to the second *trophic level; see Zhao et al. (2012) *PLoS ONE* 7, 11 DOI:10.1371/journal.pone.0048131.

net radiation The balance of incoming solar radiation and outgoing terrestrial radiation (also called **net radiative balance**).

(((•))) SEE WEB LINKS
- Michael Pidwirny on net radiation and the planetary energy balance.

network A system of interconnecting routes which allows movement from one centre to the others. Most networks are made up of nodes (*vertices*), which are the junctions and terminals, and links (*edges*), which are the routes or services which connect them. 'Networks constitute the new social morphology of our societies, and the diffusion of networking logic substantially modifies the operation and outcomes in the processes of production, experience, power and culture' (M. Castells 1996). Dicken (2004) *TIBG* 29, 1 describes **situated networks** as a generic form of social organization. Brown et al. (2008) *GaWC Res. Bul.* 236 see the world economy as based upon economic nodes connected as chains/networks. See Glückler (2007) *J. Econ. Geog.* 7, 5 on economic geography and the evolution of networks.

Network connectivity is the extent to which movement is possible between points on a network—cities' housing firms with high service values will record large measures of network connectivity. Taylor et al. (2002) ESRC attempt to measure the global **network connectivity of cities**. See also Taylor and Catalano (2001) *GaWC Research Paper* 61 for global, and banking, network connectivity.

network society Although Castells is the authority on this, his definition of the network society seems wilfully maladroit; 'the global web of horizontal communication networks that include the multimodal exchange of interactive messages from many to many both synchronous and asynchronous' (Castells (2007) *Int. J. Communication*). Hampton and Wellman (2001) *Amer. Behavioral Scientist* 45, 476 have an interesting paper on Netville, Ontario: see especially pp. 491–2.

neural network An artificial neural network (ANN) is a way of processing information that is inspired by the way biological nervous systems such as the brain process information. It is made up of a large number of highly interconnected processing elements (neurones) working together to solve specific problems, such as pattern recognition or data classification. Like people, ANNs learn by example, adjusting the connections that exist between the neurones. 'Neural networks, with their remarkable ability to derive meaning from complicated or imprecise data, can be used to extract patterns and detect trends that are too complex to be noticed by either humans or other computer techniques.' This quote is from Christos Stergiou and Dimitrios Siganos, 'What is a Neural Network', Imperial College.

(((•))) SEE WEB LINKS
- Imperial College computing website.

névé An alternative term for *firn.

new economic geography A modelling approach to economic

geography that stresses the importance of the larger picture; for example, concentrations of local successful economic activity drive up land prices, which then discourages new entrepreneurs from locating there. Nobel prize winner Krugman (1998) *Oxf. Rev. Econ. Policy* 14, 2, 7–17 calls this the combination of centrifugal and centripetal forces. Fujita and Krugman (2004) *Papers Reg. Sci.* 83, 139 define this as 'a body of research initially stemming from international trade theory which fundamentally attempts to explain the formation of a large variety of economic agglomeration (or concentration) in geographical space'.

new economy 'The economic frenzy of the 1990s . . . [typified by] new styles of production, novel forms of consumption and organization in everyday life, new horizons of planning and logistics of mobility, [and] new forms of materiality and sensibilities' (Löfgren (2003) *Glob. Netwks* 3, 3). It is being built around smaller enterprises distributed in a much more dispersed geography, relies on the old economy for physical capital and a portion of demand, and the location of business activity in the new economy is far less tied to the cost-based logic that has shaped and reshaped the distribution of industry in the old economy. It relies to a growing extent on telephone-based networks for production and delivery, but also has leading-edge layers that require face-to-face contact (Beyers (2002) *J. Econ. Geog.* 2, 1). For J. Bryson et al. (2004) the key players in the new economy are service activities, although Chiang (2008) *Geog. Compass* 2, 1 focuses on 'the critical role of manufacturing', which spurs innovation spillover and generates multiplier effects. However, for Wood (2002) *J. Econ. Geog.* 2, 2 the term signifies that things are not what they used to be 'even if we are not certain what they are'. Ettlinger (2007) *PHG* 32, 1 argues that

'the popular characterization of the new economy in terms of "time–space compression" requires revision to consider the complex content of *what* is being compressed over time and across space'.

new industrial spaces Agglomerated production systems that coordinate inter-firm transactions and the dynamics of entrepreneurial activity (Moulaert and Sekia (2003) *Reg. Studs* 37, 3). See Hansen and Winther (2007) *Danish J. Geog.* 107, 2 on Copenhagen.

new international division of labour (NIDL) A global division of labour associated with the growth of *transnational corporations and the *deindustrialization of the advanced economies. The most common pattern is for research and development in *more economically developed countries, and cheap, less skilled labour in *less economically developed countries. See Marin (2006) *J. Eur. Econ. Ass.* 4, 2–3 on outsourcing. The impacts of the NIDL have been uneven: between nations, where some benefit more than others, and within nations, in locally specific ways; Martin (2004) *TIBG* 29, 2 refers to the latter as a 'new sub-national development paradigm'. See OECD (2001) *evolution and globalization*.

newly industrializing country (NIC) A country which has seen rapid industrialization since the 1970s. The economic success of newly industrializing countries has been attributed to shared values and high levels of mutual trust within cultural groups (Batley in V. Desai and R. Potter 2002). See Sum in H. Beynon et al., eds (2001) on the 'embedded exportism' of East Asian newly industrializing countries. See also N. Forbes and D. Wield (2002).

new political economy The methodology of new political economy rejects the split between agency and structure characteristic of classical political economy. Instead, it builds on structuration theory, institutional economics, and *public choice theory, to develop an integrated analysis which combines theories which analyse agency in terms of instrumental rationality with contextual theories which analyse structures institutionally and historically.

new regionalism A territorially fluid, process-based, diverse, collaborative, and open mode of governance: see Smith (2007) Regional Studies Association, Lisbon; Wallis (2002) *Munimet*; I. Sagan and H. Halkier, eds (2005); and Rossi (2004) *Int. J. Urb. Reg. Res.* 28, 2).

E. LeSage and L. Stefanick (2004) provide a useful summary table:

New regionalism	Old regionalism
*Governance	Government
Process	Structure
Open	Closed
Collaboration	Coordination
Trust	Accountability
Empowerment	Power

Lovering (1999) *Int. J. Urb. & Reg. Res.* 23, 3 differentiates between the 'new regionalism in thought'—'that sub-set of ideas in policy related economic geography which converge on the claim that "the region" is displacing the nation-state as the "crucible" of economic development'—and the 'new regionalism in practice'—an economic analysis 'overwhelmingly dominated by a "productivist" and micro-economic focus. That is, it treats regional economic activity as if the driving influences on the regional economy *as a whole* can be understood by looking at only one type of regional economic actor.' For other criticisms of new regionalism, see Bull (1999) *3rd World Qly.* 20; Rossi (2004) *New Int. J. Urb. & Reg. Res.* 28, 2; and Sandstrom (2001) *J. of Urban Affairs* 23, 5 (2001).

new town A newly created town, planned to relieve overcrowding and congestion in a major conurbation. In the UK, new town housing was arranged in *neighbourhood units of around 5 000 people with their own facilities such as shops, schools, and medical centres, and housing densities were low—about five houses per hectare. All British new towns were planned to give a balance of social groups. See S. Ward (2002) on planning the 20th-century city; D. Frantz and C. Collins (1999) on Celebration, a master-planned city near Disney World; and Popke (2000) *Growth & Change* 31, 2 on a new town in South Africa.

new trade theory The theory that in a world of imperfect competition, international trade is driven as much by increasing returns and external economies as by comparative advantage. 'What matters, instead, is the overall pattern of trade: the broad pattern of what countries produce is determined by things like resources and climate, but there's a lot of additional specialization due to economies of scale, and there's much more trade, especially between similar countries, than you would expect from a purely resource-based theory' (Krugman, U. of Calgary, Dept. of Economics). Martin and Sunley (1996) *Econ. Geog.* 72, 8, 259 explain more fully.

(((•))) SEE WEB LINKS

- Website of U. of Calgary, Dept. of Economics, 'Krugman on Krugman'.

new urbanism A way of fostering a community by basing its plan on an old European village, with homes and businesses clustered together. Residents can walk to shops, businesses, theatres, schools, parks, and other important services. It is a movement 'intent on improving quality of life in cities by changing urban form'; McCann in R. Kitchin and N. Thrift (2009). This movement, according to Talen (1999) *Urb. Studs* 36, 8, 1361 'lives by an unswerving belief in the ability of the built environment to create a "sense of community"'. Read her review of this belief.

niche A set of ecological conditions which provides a species with the energy and habitat which enable it to reproduce and colonize. A niche is usually identified by the needs of the organism. See Chave et al. (2002) *Am. Nat.* 159 on niche differentiation and the maintenance of species diversity. Etienne and Olff (2004) *Am. Nat.* 163 develop a simple model uniting a neutral community model with niche-based theory.

nimbo-, nimbus Referring to clouds bringing rain, as in nimbo-stratus clouds.

nitrogen cycle The cycling of nitrogen and its compounds through the *ecosystem. Most plants obtain the nitrogen they need as inorganic nitrate from the soil solution. Animals receive the required nitrogen they need for metabolism, growth, and reproduction by the consumption of living or dead organic matter containing nitrogen in molecules.

(⊕) SEE WEB LINKS
• Michael Pidwirny on the nitrogen cycle.

nitrogen fixation The alteration of atmospheric, molecular nitrogen to nitrogen compounds. The fixation mechanisms are: biological micro-organisms (such as those in the root nodules of leguminous plants), *lightning and other natural ionizing processes, and industrial processes. See Vitousek et al. (2002) *Biogeochem.* 57–8 on modelling nitrogen fixation.

nivation The effects of snow on a landscape. The main processes are backwall failure, sliding and flow, niveo-

aeolian sediment transport, supra- and ennival-sediment flows, niveo-fluvial erosion, development of pronival stone pavements, accumulation of alluvial fans and basins, and pronival solifluction (Christiansen (1998) *ESPL* 23, 8). For the factors influencing the effects of nivation, see Raczkowska (1995) *Geograf. Annal. A* 77, 4. Nivation may produce the shallow pits known as **nivation hollows**. In time, these hollows may trap more snow and may deepen further with more nivation so that *cirques or *thermocirques are formed; see Kariya (2005) *Catena* 62, 1.

node Accessibility of a node is given by the travel cost between a node and another node, or a weighted average of travel cost to a number of nodes in the network (Holl (2007) *J. Transp. Geog.* 15, 4). Transport nodes show a distinct hierarchy, and the importance of a transport node appears to correlate strongly with its effects on the surroundings (van den Berg and Pol (1998) *Env. & Plan. C* 16, 4). Factors that determine the importance of a transport node are: the situation of the transport node with respect to major metropolitan regions; the number and type of the converging transport modes; the number of destinations that can be reached from the transport node, and the frequency of the transport moves. The more important a transport node, the greater the potential regional economic development (van den Berg and Pol (1997) *Eur. Reg. Sci. Assoc.* 37th Eur. Congr.). See also de Langen and Visser (2005) *J. Transp. Geog.* 13, 2. See Devlin et al. (2008) *J. Transp. Geog.* 16, 1 on **arc-node topology**—the topological data structure ArcInfo uses to represent connectivity between arcs and nodes.

nomadism A form of social organization where people move from place to place. Nomadism incorporates the advantage of mobility; traditionally no-madic groups were able to exploit natu-ral resources at dispersed locations. 'In the discourse of modernization and so-cial change nomadism's place is usurped by agriculture' (Kreutzman (2003) *Geog. J.* 169, 3). As international boundaries are increasingly well defined, with bor-der guards, nomadism declines: N. M. Shahrani (1979) uses the term **closed frontier nomadism**, with sedentariza-tion and confined migration cycles.

Governments try to immobilize no-mads for the purposes of taxation as well as to improve their health and literacy; the mobility of capital and labour in the age of empire produced a spatial order where nomadism had to be seen as a cultural problem (Noyes (2000) *Cult. Geogs* 7). Conversely, Drakakis-Smith (2007) *Mobilities* 2, 3 shows how local authorities keep some families mobile and excluded. Mouffe (quoted in Pugh (2007) *Area* 39, 1) writes: 'against those who ... advocate "nomadism" I am con-vinced that radical politics cannot avoid "territorialization" and that all forms of territorialization should not be perceived as machines of capture.' B. Jordan and F. Düvell (2003) argue that **global eco-nomic nomadism** requires a redefini-tion of citizenship beyond national borders which involves shared duties for those who have access, and rights for those who remain on the outside.

nominal On a *GIS map, a geographic feature that has a name only.

nomothetic A scholarly approach that stresses the value of general principles that apply everywhere in space and time.

nonconformity A series of sedimen-tary strata overlying an igneous or metamorphic rock.

non-ecumene The uninhabited or very sparsely populated regions of the world. It is not easy to draw boundaries

between the *ecumene and the non-ecumene as regions of dense occupation merge into sparsely populated regions. If there is a boundary, it is not static.

non-linear (nonlinear) Describing a system in which the output is not uni-formly related to the input (not a straight-line graph). All the component parts of the Earth system interact on each other, which is an expression of non-linearity.

See Phillips (2003) *PPG* 27, 1 on non-linearity in geomorphic systems.

non-place The ubiquitous place, an indistinct location or route: shopping malls, airports, supermarkets; anywhere and nowhere.

non-renewable resource A finite mass of material which cannot be restored after use, such as natural gas. (S. Barr and P. Chaplin (2004) argue that cultural heritage is a non-renewable resource.) Each non-renewable resource has a theoretical depletion time—a ratio of its stock and rate of use. Non-renewable resources may be sustained by *recycling, which Jenkins (2000) *Econ. Ecol.* 34, 3 claims tends to be built into traditional cultures.

non-representational theory **Representational** describes the ways that words, images, or other media 'capture' the characteristics of things existing or occurring outside these media. However, M. Shapiro (1988) argues that representations do not 'imi-tate' reality but are the practices—visual, aural, oral, quantitative, or qualitative—through which things take on meaning and value. To put this simply, 'nature' is not a real, material actor but a socially constructed object. This means that ac-ademics have to move beyond mere representation. M. Doel (1999) provides a helpful parallel. The task of a painter, he says, is not to paint an object, or even to represent it—but to be that object's effects. This does not mean that descriptions are no longer admissible. 'Close descriptions can still be offered of particular encounters, attending in the process to the situated, embodied sense-making work being (unavoidably) undertaken by the peoples involved, [the work] that makes those encounters what they are' (Laurier and Philo (2006) *Area* 38, 4).

Non-representational theorists argue, that 'representation' generates an unwavering, deadened picture of the world. They emphasize knowing through connection and participation; the spotlight is on the process, rather than the outcome—'it ain't what you get but the way that you got it'.

That this theory has been taken up by human geographers owes a very great deal to the work of Nigel Thrift, whose major work on the topic was published in 2007. This 'folded mix of the witnessed and witnessing world' (Dewsbury (2003) *Env. & Plan. A* 35, 11) chimes with the concept of *hybrid geographies. Lorimer (2008) *PHG* 32 waxes lyrical about non-representational theory, but geographies based on this concept are painfully few. Thrift's use of the term 'non-representational theory' has been seen as problematic, and Lorimer (2005) *PHG* 29, 1 offers the term 'more-than-representational'. Try Colls (2012) *TIBG* 37, 3, 430 on feminism and non-representational geography.

normative In general, relating to an ideal standard or model; more specifically, describing any academic argument that criticizes current arrangements and calls for the creation of a better future.

North Following the terminology of the Brandt report (1979) a portmanteau term used to describe the advanced economies/more economically developed countries.

North American Free Trade Agreement (NAFTA) A *free trade area, established between Canada and the USA in 1989, Mexico joining in 1993. NAFTA aims to remove import tariffs and *quotas on most raw materials and manufactured goods, and restrictions on trans-border financial services. The agreement also has proposals on health, pollution, and the use of renewable resources. The North American Free Trade Agreement is the most comprehensive free trade pact (short of a common market) ever negotiated between regional trading partners, and the first reciprocal free trade pact between a developing country and industrial countries (G. C. Hufbauer and J. Schott 1993).

Koring (2004, www.Vigile.net) writes that 'the risk is that the trilateral relationship is being further eclipsed by the two bilateral relationships. Even though Canada–Mexico ties and trade have grown over the past 20 years, they remain inconsequential compared to Canada–U.S. relations and the relationship between the United States and Mexico, which is evolving even more rapidly'.

North Atlantic Deep Water (NADW) A water mass largely formed in the Labrador Sea and the Greenland Sea by the sinking of highly saline, dense water. It is part of the 'conveyor belt' model of the *thermohaline circulation. It has been suggested that global warming may cause sufficient freshwater to be released from the melting of ice in the northern latitudes to slow or even to stop the formation of the NADW by reducing salinity to the level where it is insufficiently dense to sink. If this occurred it would stop the Gulf Stream flowing north, and thus cause the oceanic conveyor belt, which is transferring heat from the equator, to cease. This would have a dramatic cooling effect on the climate of north-west Europe (Krom in J. Holden (2012), p. 61).

North Atlantic Drift The Gulf Stream is a warm surface ocean current originating in the Gulf of Mexico, which widens to several hundred kilometres, and slows to less than 2 km/h, and then divides into several sub-currents, one of which is the North Atlantic Drift (Hayward and Williams (2005) *Geol. Today* 21, 4). The intensified westerlies in the northern North Atlantic and strengthened northerlies along the western part of the Nordic Seas force the North Atlantic Drift to follow a more easterly path, giving an asymmetric sea surface temperature response in the northern North Atlantic, and thereby maintaining the properties of the Atlantic Water entering the Nordic Seas (Otterå et al. (2004) *Tellus A* 56, 4). See Rasmussen and Thomsen (2004) *Pal. Pal. & Pal.* 210, 1 on the role of the North Atlantic Drift in millennial glacial climate fluctuations. Bryden et al. (2005) *Nature* 438 calculate that the North Atlantic Drift has slowed by 30% over the past 50 years.

North Atlantic Oscillation (NAO) A large-scale alternation of atmospheric mass between the Icelandic low- and the Azores high-pressure area. When the pressure difference is large, with a deep Icelandic low and a strong Azores high, the NAO is said to be high or positive, and said to be negative when the pressure difference is less (or occasionally negative) as a result of persistent blocking in the Iceland–Scandinavia area. Strong positive phases of the NAO tend to be associated with above-normal temperatures in the eastern USA and across northern Europe, and below-normal temperatures in Greenland and across south-eastern Europe and the Middle East. Opposite patterns are observed during negative phases

(Perry (2000) *PPG* 24, 2—an excellent source). Kingston et al. (2006) *PPG* 30, 2 find generally positive NAO/Arctic Oscillation streamflow relationships in north-west Europe and north-east USA, and positive and negative streamflow relationships in parts of eastern Canada. Kuszmina et al. (2005) *Geophys. Res. Letts.* 32, L04703–L04703 suggest that the NAO may intensify with further increases in greenhouse gas concentrations.

The name **Arctic Oscillation** was introduced to highlight the fact that the pressure anomalies associated with the NAO span most of the Arctic. But 'oscillation' is something of a misnomer, as the northern and *southern annular modes oscillate regularly in time. For this reason, in recent years the NAO/AO nomenclature has been increasingly supplanted in the dynamical literature by the phrase **Northern Annular Mode (NAM)**. Annular denotes the longitudinal scale of the pattern, and annular mode suggests the NAM reflects dynamical processes that transcend a particular hemisphere or, for that matter, planet.

North Pacific Current An eastward current, originating east of the Emperor Seamounts (170° W) and forming the northern part of the North Pacific subtropical gyre. It maintains the Arctic Polar Front with the Pacific Subarctic current. See Dinniman and Rienecker (1999) *J. Phys. Oceanog.* 29, 4.

nuclear family The small family unit of parents and children. 'The institutionalization of the nuclear family household . . . emerges in interaction with the processes of capitalism, as well as in relation to existing (patriarchal) processes' (Lier (2007) *Geog. Compass* 1, 4). D. Lal (1998) claims that 'capitalism is not inevitably connected with the nuclear family or even individualism, although its genesis depended on both'. Phillips (2009) *Canad. Geogr./Géog. canad.* 53, 2, 230, writing of colonial Canada, argues that 'though they often seemed natural, universal, and inevitable, colonial nuclear families were in fact produced through a series of laws and customs that regulated sex and marriage'.

nuclear power A form of energy which uses nuclear reactions to produce steam to turn generators. Naturally occurring uranium is concentrated, enriched, and converted to uranium dioxide—the fuel used in the reactor. This undergoes nuclear fission, which produces large amounts of heat. Some of the highly radioactive spent fuel may be reprocessed while the bulk must be disposed of. Both are costly and hazardous undertakings. The main advantage of nuclear power is the low CO_2 emissions: 'nuclear power is a low-carbon form of electricity generation that can make a significant contribution to tackling climate change. Ruling out nuclear as a low carbon energy option would significantly increase the risk that the UK would fail to meet its CO_2 reduction targets' (Meeting the Energy Challenge: A White Paper on Nuclear Power URN:08/525).

A major disadvantage is the risk of contamination—see Tsuji et al. (2012) *Environ. Health Prev. Med.* 17, 2 124. There are also very high construction and decommissioning costs—'it would be irresponsible for us to benefit from nuclear power and leave it to later generations to deal with the waste' (N. Lenssen, Worldwatch Institute)—and the major problems which may arise with any accident, such as the Chernobyl disaster of April 1986. Furthermore, nuclear power stations have a short lifespan. Parkins and Haluza-DeLay (2011) *ageconsearch.umn.edu/bitstream /.../StaffPaper11-01.* are excellent on the social and ethical considerations of nuclear power development.

nucleated settlement A settlement clustered around a central point, such as a village green or church. 'The problem is to determine where dispersed settlement finishes and nucleated settlement begins' (B. Roberts 1996). Nucleation is fostered by defence considerations, localized water supply, the incidence of flooding, or rich soils so that farmers can easily get to their smaller, productive fields while continuing to live in the village. 'Nucleated settlement does not in itself necessarily induce a sense of community . . . nucleated settlement in itself is only a part of the process of constructing and maintaining a sense of common interest' (C. Dalglish 2003).

nuclei Minuscule solid particles (sing. *nucleus*) suspended in the *atmosphere. Three types of atmospheric nuclei are distinguished: **Aitken nuclei**, of radii less than 0.1 μm (Heggs et al. (1991) *J. Geophys. Res.* 96, 18), **large nuclei**, radii 0.2–1 μm, and **giant nuclei**, with radii greater than 1 μm. See Clarke (1992) *J. Atmos. Chem.* 14.

These nuclei may be scraps of dust, from volcanic eruptions or dust storms, or salt crystals, or given off when bubbles burst at the surface of the sea. Atmospheric nuclei can scatter sunlight enough to lower temperatures, if enough are present, and play an important role in *cloud formation; see Kulmala et al. (2001) *Tellus B* 53, 4.

nuée ardente 'A glowing cloud of volcanic ash, *pumice, and larger *pyroclasts which moves rapidly downslope, typically starting as large blocks, and then fragmenting to a mixture of dust through to house-sized particles' (Mursik et al. (2005) *Rep. Prog. Phys.* 68).

numerical modelling 'Numerical models can be useful for explaining poorly understood phenomena or for reliable quantitative predictions. When modelling a multi-scale system, a "top-down" approach—basing models on emergent variables and interactions, rather than explicitly on the much faster and smaller-scale processes that give rise to them—facilitates both goals. Parameterizations representing emergent interactions range from highly simplified and abstracted to more quantitatively accurate. Empirically based large-scale parameterizations lead more reliably to accurate large-scale behaviour than do parameterizations of much smaller-scale processes. Conversely, purposely simplified representations of model interactions can enhance a model's utility for explanation, clarifying the key feedbacks leading to enigmatic behaviour. For such potential insights to be relevant, the interactions in the model need to correspond to those in the "real" system in some straightforward way. Such a correspondence usually holds for models constructed for predictive purposes, although this is not a requirement' (Murray (2007) *Geomorph.* 90, 3–4).

nunatak A mountain peak which projects above an ice sheet, generally angular and jagged due to *freeze–thaw, contrasting with the rounded contours of the glaciated landscape below. Confirming a former nunatak is generally difficult (Goodfellow (2007) *Earth Sci. Revs* 80, 1–2). See Kleman and Stroeven (1997) *Geomorph.* 19, 1–2.

nuptiality The frequency of marriage within a population, usually expressed as a marriage rate. R. Woods (2002) suggests that the role of nuptiality in influencing fertility during the late 19th century was geographically variable and could not be attributed to some overarching explanatory factor. Aria (2001) *Int. Mig. Rev.* 35, 2 considers *assimilation as a cause of falling nuptiality in Cuban Americans.

nutrient cycle The uptake, use, release, and storage of nutrients by plants and their environments; the transformation of inorganic compounds into organic forms due to biological uptake, and then back to an inorganic state (D. Allen and M. Castillo 2009). The nutrient cycle is the main functional process maintaining the stability and production of an ecosystem, and land cover is the source and sink of the material and energy supporting the biosphere.

Schematic nutrient cycle

Nutrient cycling processes

A. Sources (inputs)
   4. Atmospheric inputs (e.g. dust, aerosol, biological fixation, precipitation)
   5. Weathering from parent material (primary minerals from rocks)
   6. Deposition of minerals through the movement of animals (e.g. movement of ungulates in Serengeti), soil (down slope), and seepage (drainage waters)

B. Storage capacities (pools)
   1. Living biomass
   2. Detritus (eg. litterfall, dead/decaying roots, fallen logs and branches)
   3. Soil

C. Internal ecosystem cycling
   10. Litter inputs (above- and below-ground)
   11. Decomposition and mobilization processes
   12. Uptake, utilization, and redistribution processes by higher plants (eg. trees primary sink for nutrients – $\frac{1}{3}$ of the total annual requirement of nutrients acquired from internal redistribution. Excluding 70% of N removed from the leaves before litterfall)

D. Losses (outputs)
   7. Volatilization to the atmosphere
   8. Harvest transport, and erosion
   9. Leaching below the rooting zone

Nutrient cycle

O

oasis A watered spot in an arid area, occurring when the *water-table reaches the surface. See Woo and Guan (2006) *Permaf. & Perigl. Procs* 17 on a **polar oasis** environment.

object-field A model that attempts to integrate the field- and object-view in a single, combined, integrated data model; see Raper and Livingstone (2001) *TIBG* 26, 2 for a brief explanation and useful references.

objectivity A state of mind free of prejudice; 'an analysis free of values' (Ferber (2006) *AAAG* 96, 1, 176).

occlusion A stage in a *mid-latitude depression where the cold front to the rear catches up with the leading warm front, lifts the wedge of warm air off the ground, and meets the cold air ahead of the warm front. If this overtaking air is colder than the cold air ahead of it, it will undercut it, forming a **cold occlusion**. If it is warmer than the cold air which is ahead of it, it will ride over it, in a **warm occlusion**. See B. Haurwitz (2007); see also Martin (1998) *Month. Weather Revs* 126, 2 and (1999) 127, 10; and Levizzani et al. (1989) *Il nuovo cimento C* 12, 2 (online), on a three-dimensional single radar analysis of an occluded front.

Power and Sidaway (2005) *Soc. & Cult. Geog.* 6, 6 use the term **occluded geographies** in describing Lisbon's Expo '98 to signify a reworking/re-presentation.

occupancy rate The number of people dwelling in a house per habitable room (except for kitchens and bathrooms). In the UK, a rate of −1 implies that there is one room too few for the number of people living in the household. In 2011, England and Wales had 2 million households with an occupancy rate of −1 or lower.

occupational mobility The ease with which a worker can change jobs. The Danish Technological Institute (2008) definition is 'a change in

Occlusion

occupational status, e.g. change of job profile and job content'.

ocean–atmosphere interaction
(ocean–atmosphere coupling) Oceans and the atmosphere constantly interact with each other; surface winds drive the *ocean currents, moving warm water polewards and cold water equatorwards, and evaporation from warm oceans removes latent heat from the atmosphere. This latent heat is released when the vapour condenses with height. See Smith et al. (2006) *J. Climate* 19, 18 for a low-resolution coupled ocean–atmosphere general circulation model, and Ciasto and Thompson (2008) *J. Climate* 21, 6 on the Southern Hemisphere. **Ocean–atmosphere oscillations** are interactions which switch suddenly from one phase to another; the best known being the *El Niño–Southern Oscillation. See McCabe et al. (2004) *Proc. Nat. Acad. Sci.* 101 on ocean influences on drought.

ocean current A permanent or semi-permanent horizontal movement of unusually cold or warm surface water of the oceans, to a depth of about 100 m. Ocean currents are generated by the thrust of the wind, the spin of the Earth, and the Moon's gravity, and are also affected by variations in water temperature and density, and by the *Coriolis force. These currents are an important factor in the redistribution of heat between the tropics and the polar regions; *see* THERMO-HALINE CIRCULATION. The Southern Ocean, for example, is 'a crucial cog in the global heat engine' (anon. (2002) *Australian Arct. Mag.*). Saetra et al. (2008) *Tellus A* 60, 1 believe that an increase in the sea-surface temperatures of 1–2 °C in the North Atlantic current may intensify cyclones. See also Ridgway and Dunn (2007) *Geophys. Res. Letts* 34, 13.
  Cold currents can modify temperatures up to 100 km inland. As tropical air streams move over these currents,

advection *fog forms over the sea, stripping the air streams of most of their moisture; onshore winds are therefore dry; see T. T. Warner (2004). Conversely, **warm currents** originate in tropical waters, bringing unusually warm conditions to coastal areas in higher latitudes. The *North Atlantic Drift, for example, creates a difference of around 4 °C between the winters of coastal Canada and coastal western Europe of the same latitude (Bryden in G. Siedler and J. Church, eds 1999).

oceanic core complexes (OCC)
Massifs in which lower-crustal and upper-mantle rocks are exposed at the sea floor, formed at mid-ocean ridges through slip on detachment faults rooted below the spreading axis. The formation of these large, shallow sea-floor features appears to be an episodic outcome of plate rifting and accretion at slow spreading ridges. An outstanding example of an oceanic core complex has been mapped at the eastern intersection of the Mid-Atlantic Ridge and the Atlantis transform fault; see Planert et al. (2010) *Geophys. J. Int.* 181, 1, 113.

oceanic crust
That portion of the outer, rigid part of the Earth which underlies the oceans. It is made mostly of a basalt layer, rich in iron and magnesium, average density around 2.9, and 5–6 km thick, but see Rodger et al. (2006) *Geology* 34, 2.

oceanic ridge
An underwater mountain range developed where *magma rises up through a cracking and widening ridge in the oceanic crust. Some magma cools below the crust, some forces into fractures, and much flows out to form new crust, which is then pushed away from the ridge. McCarthy (2004) *J. Geophys. Res.* 112, B03410 explains that the disparity in spreading rates between the Northern and Southern

Hemispheres is correlated with the rate of latitudinal circumferential extension.

As new crust is created each side of the ridge, it takes up the prevailing *magnetic polarity of the Earth, which reverses from time to time. As a result, symmetrical bands of crust, with alternating polarity, develop on either side of the ridge. These magnetic patterns are used to calculate the rate of the *seafloor spreading; see Michaud et al. (2006) *Geology* 34, 1. The term **mid-oceanic ridge** properly refers to the ridge at the centre of the Atlantic Ocean, which comes to the surface at points such as Tristan da Cunha and Ascension Island. Other ridges, such as the Pacific–Antarctic ridge, are not truly at the centre of the ocean.

oceanography The study of the oceans. This covers the shape, depth, and distribution of oceans, their composition, life forms, ecology, and water currents, and their legal status. A. P. Trujillo and H. V. Thurman (2007) provide a good general introduction.

ocean trench A long, narrow, but very deep depression in the ocean floor where two adjacent plates converge, and one descends beneath the other. For this reason ocean trenches are also known as convergent plate boundaries. 'The arc like structure of many ocean trenches can be qualitatively understood by the ping-pong ball analogy. If a ping-pong ball is indented, the indented portion will have the same curvature as the original ball, that is, it will lie on the surface of an imaginary sphere with the same radius as the ball. The lithosphere as it bends downward might also be expected to behave as a flexible but inextensible thin spherical shell' (D. L. Turcotte and G. Schubert 2002). An alpinotype orogeny is also termed an **ocean trench orogeny**.

offshore 1. In meteorology, moving seawards from the land, as in an offshore wind.

2. In geomorphology, the zone to the seaward side of the breakers.

offshore bar *See* BAR.

offshore financial centre (OF) Better known as 'tax havens', offshore financial centres—such as the Cayman Islands or Panama—are 'jurisdictions that oversee a disproportionate level of financial activity by non-residents. Financial activity in them is usually dominated by the provision of intermediation services for larger neighbouring countries . . . By providing competition for domestic banks, offshore banks also (inadvertently) lower the costs for anyone using the domestic financial system. For example, simply by being close to Andorra and Monaco, France has a more competitive banking system, which provides more credit at lower interest rate spreads. Such indirect competitive benefits of offshore financial centres more than offset their costs' (Rose and Spiegel (2007) *Econ. J.* 117, 523, which see).

offshoring For businesses, the transference of portions of their work to outside, foreign suppliers rather than completing at home. Frequently, work is offshored in order to reduce labour expenses. See James and Vira (2012) *J. Econ. Geog.* 12, 4, 841 on offshoring and India's service economy. See also NEARSHORING.

ogives Bands of dark and light ice that stretch across glaciers, in the shape of a pointed arch. Multiple shear zones are formed, through which basal ice is uplifted to the glacier surface to produce the dark, foliated ogive bands (Goodsell et al. (2002) *J. Glaciol.* 48, 161).

oil crisis (oil price shock) In 1973, the *OPEC countries, headed by Libya, quadrupled the price of oil, and many member states nationalized the oilfields in their own countries. Oil companies began to develop previously uneconomic alternative resources; attention shifted to alternative sources of energy, energy conservation, and energy-efficient technology; and countries lacking alternative sources of supply, or the technology to develop alternatives, found the cost of their oil imports rising painfully.

okta A measure of cloud cover, from 1 okta (scant) to 8 oktas (complete). Oktas are shown on *synoptic charts by a circle which is progressively shaded in as cloud cover increases.

oligotrophic Poor in nutrients, usually of lakes, soils, and peat bogs.

ombrotrophic Pertaining to vegetation or soils that gain nutrients and water from precipitation only.

onion weathering *See* EXFOLIATION.

ontogeny The origin and development of a phenomenon. Kitchin (2007) *PHG* 31, 3, 331 argues that *cartography is ontogenetic in nature; that its maps emerge through practices and have no secure ontological status.

ontology The study of being. 'Ontology comprises theories, or sets of theories, which seek to answer questions about what the world must be like for knowledge to be possible' (S. Aitken and G. Valentine, eds 2006). Each discipline has its own ontology: 'economists tend to think about labour markets in terms of wages and human capital: workers locate where they receive the highest returns for their skills . . . in contrast, geographers typically see labour location as a result of worker attachments to complex, fixed social institutions and networks' (Ashby and Monk (2007)

Oxonomics 2, 1). See Boschma and Martin (2007) *J. Econ. Geog.* 7, 5 on ontology for evolutionary economic geography. A debate on the ontologies of scale occupies much of the (2007) *TIBG* 32, *passim*.

(((∰))) SEE WEB LINKS
• UK Ordnance Survey ontologies.

ooze A deposit either of algae, or fine inorganic sediments, developed on the sea floor, or at the bottom of lakes (Dasey et al. (2005) *Marine & Freshw. Biol.* 56). See H. Armstrong and M. Brasier (2005). **Calcareous oozes** are made up of the plankton in the surface waters of the oceans.

OPEC The Organization of Petroleum Exporting Countries was established in September 1960 by Venezuela, Saudi Arabia, Iran, Iraq, and Kuwait. Other countries which qualified for membership and joined later were Qatar (1961), Indonesia and Libya (1962), Abu Dhabi (1967, later transferred to the United Arab Emirates, 1974), Algeria (1969), Nigeria (1971), Ecuador (1973), and Gabon (1975). Ecuador and Gabon withdrew in 1992. The San Francisco Bay Area-based collective, Retort, argues that since the 1960s relatively high oil prices have benefited both the oil majors and military contractors and manufacturers, construction companies, engineering and industrial design firms, and the banking and finance sectors as OPEC states recycled and spent their abundant petro-dollars. Watts (2005) *PHG* 29, 360 writes that the removal of Saddam Hussein 'obliterated at a stroke one of the most assertive voices for an aggressive OPEC price policy, and set the stage for clipping the wings of another; Hugo Chavez'. See Bridge (2008) *J. Econ. Geog.* 8, 3.

opencast mining A system of mining which does not use shafts or tunnels and is, hence, cheaper. The layer above the mineral seam is removed and the exposed deposit is extracted using earth-moving machinery. The overburden may then be replaced. See Ghose and Majee (2001) *J. Env. Manage.* 63, 2,193 on air pollution caused by opencast mining and its abatement measures in India, and the 1980 edition of the *New Zealand Geographer* 36, 2, 85 on potential environmental impacts of opencast mining.

open data Data that are freely available to everyone without mechanisms of control, such as copyright or patents.

open field A distribution of farmland associated with *feudalism in Europe. Each manor had two or three large open fields and each farmer was awarded a number of strips within each field, so that no one had all the good or all the poor land. 'From late in the seventeenth century both the open fields and land held in common ... began to be enclosed and to be allocated in compact parcels to each of those who had possessed grazing or other rights in it' (Pounds (2005) *J. Hist. Geog.* 31, 3).

open source software Computer software that is distributed along with the code that is used to create the software, under a special software licence that allows users to use, change, and/or improve the software's source code, and to redistribute the software either before or after it has been modified. If any modifications are made to the source code, other developers must be told what has been changed, and how. Open source software is freely available, but not necessarily *free* in financial terms.

open system A system that is not separated from its environment, but exchanges material or energy with it. Trudgill and Briggs (1977) *PPG* 1 see soil as an open system; see Nordstrom (1994) *PPG* 18 on an open system model of beach change, and Douglas (1981) *PPG* 5 on the city as 'an integrated open system of living things interacting with their physical environment'.

ophiolite suite A sequence of rocks consisting of deep-sea marine sediments overlying a range of igneous phenomena, including basalts and *dykes. Ophiolite sequences are indicators of sea-floor spreading, as tectonic movements pull and thin the crust, allowing magma to rise up and create new crust. When the magma is cooled by the sea water, it forms a variety of igneous rocks and structures. See Saha et al. (2010) *J. Earth Sys. Sci.* 119, 3, 365 on a completely preserved ophiolite suite from South Andaman, India.

opportunist species (opportunistic species) Species which survive by the rapid colonization or recolonization of a *habitat: for example, the roach in Ireland (Caffrey and McLoone, *Environ 2003*). Such species have very good powers of dispersal in order to seek new habitats. Life is short, fecundity and reproduction are high, and these species are able to withstand difficult conditions. The term **fugitive species** is also used. Holling (in C. Folke et al. 1995) sees the proliferation of pioneer and opportunist species as the first stage of the adaptive process.

opportunity cost The second choice that you give up when choosing the best; for example, in choosing a better-paid job further away, commuting time is the opportunity cost. Carlino (1985; cited in Desmet and Fafchamps (2005) *J. Econ. Geog.* 5, 3) argues that, though the cost of moving goods has gone down through improved technology, the cost of moving people—as measured by the opportunity cost of time—has gone up. Baldwin and Okubo (2006) *J. Econ. Geog.* 6, 3 show that regional policies tend to attract the least productive firms since these have the lowest opportunity cost of leaving an agglomerated region; 'the result is a "sorting effect"'. See Willis et al. (1998) *Transp. Res. D* 3, 3 on land prices and **social opportunity cost**.

optimization model A model used to find the best possible choice out of a set of alternatives. A simple example might be finding the most efficient transport pattern to carry commodities from the point of supply to the markets, given the volumes of production and demand, and transport costs. See Ward et al. (2003) *Annals Reg. Sci.* 37 on evaluating urban change. Kuby et al. (2005) *Adv. Water Resources* 28, 8 introduce a new class of combinatorial-optimization models to appraise dam removal.

optimizer A decision-maker who seeks to maximize profits; in choosing an industrial location, the profit maximizer is assumed to know all the relevant factors, including the cost of assembling materials and distributing products, the price of labour, and *agglomeration and *external economies. Among many who find optimizer a static and simplistic concept are Feola and Binder, *Tropentag 2007. Compare with* SATISFICING.

optimum location The best location for a firm in order to maximize profits; 'in many circumstances, the optimum location of the firm and the optimum production relationships can be shown to be inter-dependent' (P. McCann 2002). See also Bradshaw (2001) *Australian Geogr.* 32, 2.

oral history The collection of living people's first-hand knowledge of persons, events, and places. See Riley and Harvey (2007) *Soc. & Cult. Geog.* 8, 3, 345 on oral history and the practice of geography.

ordinal Any phenomenon that can be compared by rank.

ordnance datum (OD) In the UK, mean sea level at Newlyn, Cornwall, from which all other spot heights on Ordnance Survey maps—and hence all *contours—are established.

Ordovician A period of *Palaeozoic time stretching approximately from 500 to 430 million years BP.

(( )) SEE WEB LINKS
• An overview of the Ordovician period.

organic acid Acid compounds of carbon, such as acetic acid, which are produced when plant or animal tissues decompose. Bormann et al. (1998) *Biogeochem.* 43, 2 show that organic weathering plays a significant role in the release of nutrients to plants. See also Jones et al. (2003) *Plant & Soil* 248, 1–2.

organic weathering The breakdown of rocks by plant or animal action or by chemicals formed from plants and animals, such as *humic acids. Pope et al. (2002) *Geomorph.* 47, 211, while not central to the topic, should be useful.

Organization for Economic Cooperation and Development (OECD) Constituted in 1961, this is a group of nations, comprising most of western Europe, together with Australia, Canada, Japan, New Zealand, and the USA, and formed to develop strategies which will boost the economic and social welfare of the member states. See OECD (2005) *Regions at a Glance* (ISBN 9789264018631). An OECD report of October 2007 (ISBN 9789264037809) concludes that 'region-specific advantages—embedded in specialized firms, skilled labour, and innovation capacity—will maintain the competitiveness of OECD regions, thus seeming to contradict the hypothesis that globalization reduces the importance of geographical proximity in business'.

Organization of African Unity (OAU) An association of independent African states constituted in 1963, and designed to encourage African unity, to discourage *neo-colonialism, and to promote development.

SEE WEB LINKS
• The African Union website.

Organization of American States (OAS) An association of Latin American states with the USA. Constituted in 1948, it is designed to encourage American solidarity, to aid cooperation between members, to maintain present boundaries, and to arbitrate in disputes between members.

SEE WEB LINKS
• The OAS website.

orientalism A description of the West's depiction of the cultures of the Eastern Hemisphere, orientalism became a political term through the work of Edward Said (1978), for whom orientalism means European academic and popular discourse about the Orient, expressed through colonial bureaucracies and styles, doctrines, imagery, scholarship, and vocabulary. (In this, he was heavily influenced by Michel Foucault, and his concept of *imagined geographies.) Said argued that 'the Orient has helped to define Europe (or the West) as its contrasting image, idea, personality, experience'. In other words, to Western eyes, orientalism is what the West is not; orientalism is *other. By representing the East as an exotic, bizarre, and inferior attachment, the Orientalists made colonial conquest a natural and logical extension of the rise of the West. Samuel Huntington (1993) *Foreign Affairs* 72, 3, for example, divides huge sections of the Earth into 'civilization groups', with

Western civilizations at the top of the hierarchy, and D. Gregory (2004) shows how the Western powers have reproduced and extended the orientalist scripting of the Middle East in current-day Afghanistan, Palestine, and Iraq.

R. Guha and G. Spivak (1988) criticize Said's work for ignoring voices of resistance within the Middle East, while H. Bhabha (1995) dismisses a clear separation between colonizers and colonized, recognizing 'the ambivalence, *hybridity, and mimicry found in colonial representations of Orientalised places and, at the same time, to the same phenomena in the self-representations of colonized peoples'. See Jouhki (2002) *J@rgonia* 8 on Indo-Orientalist essentialism.

orogeny A movement of the earth which involves the folding of sediments, faulting, and metamorphism. A **cordilleran orogeny** begins with sedimentation at a passive continental margin, ranging from coarse sand and silt near the shore, to limestone reefs in tropical seas. Fine-grained clastic sediments accumulate on the deeper continental slopes, forming shales and greywackes. The deposition of these *geosynclinal deposits is followed by *subduction and compression. Folds, thrust faults, and a volcanic arc form. Lateral growth continues through igneous activity, and metamorphism, uplift, and deformation result from continuing plate convergence. See Şengör (1999) *GSA Spec. Paper* 338.

Initially, a **continental collision-type orogeny** is similar to the cordilleran type. However, when two continental plates collide, both are too thick and too buoyant for subduction. Consequently, the plates are welded together to produce a large mountain chain. See the excellent Andersen et al. (2002) *Abstr. & Procs Norw. Geol. Soc.* 22, 12–14 (available online, 2008).

orographic precipitation (orographic rainfall) Also known as **relief rainfall**, this forms when moisture-laden air masses are forced to rise over high ground. The air is cooled, the water vapour condenses, and precipitation occurs. See Gray and Seed (2000) *Meteorol. Applics.* 7 on the characterization of orographic rainfall, and Minder et al. (2008) *Qly. J. Royal Met. Soc.* 134, 633, 817–39, Part B, on small-scale patterns of mountain precipitation. The step-duration **orographic intensification coefficient method** may be used to estimate the effects of topography on storm rainfall; see Lin (1989) *Atmospheric Deposition, IAHS Publ.* 179.

orthogonal At right angles. Orthogonals plotted through the crests of waves in plan illustrate the process of *wave refraction.

osmosis The passage of a weaker solution to a stronger solution through a semi-permeable membrane. In soils, the more dilute soil moisture passes by osmotic pressure into plant roots. *Salinization causes **reverse osmosis** in soil, where higher concentrations of salt in the soil water draw water from the plant roots, causing crops to wither, and perhaps die. See Kotzer (2005) *Desalinisation* 185, 1–3.

otherness Barnett (2005) *PHG* 29, 1 finds two quite distinct understandings of otherness. One is based upon the empirical analysis of the ways in which identities are constructed through exclusion or denigration. In the second, the movement between identity and difference is understood to be ontologically constitutive of any and all subjectivity (P. Osborne 2000). 'By running these two understandings of otherness together, the uncovering of exclusion becomes the taken-for-granted manoeuvre of critical analysis for

poststructuralized geography, according to which the cultural other is necessarily posited as the unstable foundation of hegemonic identities.' Kuus (2004) *PHG* 28 finds 'a broadly orientalist discourse that assumes essential difference between Europe and Eastern Europe and frames difference from Western Europe as a distance from and a lack of Europeanness'. J. Wedel (2001) observes that the assumption of otherness is not only inscribed on east-central Europe by the West but is also appropriated by east-central Europeans themselves. J. Perlman (2011) argues that 'cultural otherness, or the othering of certain groups within a society as inferior and non-existent, perpetuates social and economic inequalities. Therefore, *favelados* are unable to obtain personhood because they are othered as inferior degenerates within Brazilian society'.

outer space The physical universe beyond the Earth's atmosphere. MacDonald (2007) *PHG* 31, 5 aims to establish outer space as a mainstream concern of critical geography. 'More than half a century after humans first cast their instruments into orbit, contemporary human geography has been slow to explore the myriad connections that tie social life on Earth to the celestial realm. My starting point is a return to an early-modern geographical imagination that acknowledges the reciprocity between heaven and earth.' Warf (2007) *Tijdschrift* 98, 3 reminds us that, although satellites circulate in outer space, their origins and impacts are very much on the ground. See T. Sandler (1980) on the exploitation of outer space.

outlet glacier A fast-moving section of glacier ice bordered by rock, differing from an *ice stream, which is bordered by ice. For an example, see Glasser and Hambrey (2002) *Sedimentol.* 49, 1. An ice stream is a fast-flowing 'river' of ice within more slowly moving ice sheet walls.

outsourcing For businesses, the transference of portions of their work to other companies. See Clark and Ashby (2013) *Econ. Geog.* 13, 2, 279.

outwash (outwash sands) and gravels Sorted deposits which have been dropped by *meltwater streams issuing from an ice front. The material deposited is coarse near the ice front, becoming progressively finer with distance from it. Outwash fabric tends to be well *bedded, and *current bedding is common; *see* SANDUR. The finest elements of the outwash may be deposited in *proglacial lakes to form *varves. A series of streams may produce several alluvial fans which coalesce to form an **outwash plain** (valley train).

overdeepening A phenomenon found in *cirques and in the steps of *glacial troughs, where the middle section(s) of the feature are lower than the mouth. One suggestion is that the over-deepened section is the zone of maximum ice thickness, another that overdeepening coincides with less resistant rocks. Evans (1999) *Annals Glaciol.* 28 ascribes overdeepening to rotational flow, while Hooke (1991) *GSA Bull.* 103, 8 describes a positive feedback process, as overdeepening causes crevassing at the surface, resulting in erosional forces that accentuate that overdeepening.

overland flow The flow of water over a hillslope surface. When water infiltrates a soil and all the pore spaces are full, the soil is saturated and the water table is at the surface. Overland flow will then occur. This is **saturation-excess overland flow**, which can occur at much lower rainfall intensities than those required to generate **infiltration-excess overland flow** (also known as

*Hortonian overland flow). Saturation-excess overland flow can occur at the foot of a hillslope even when it is not raining. Overland flow may be laminar, turbulent, and transitional, or consist of patches of any of these flow states (A. J. Parsons and A. D. Abrahams 1992). Horton (1945) *GSA Bull.* 56 suggests that overland flow is likely to be laminar near watersheds, but with distance down-slope overland flow becomes more turbulent (Smith and Bretherton (1972) *Water Resources Res.* 8). Horton (1945) explains that these changes result from the microtopography of the soil surface: run-off gathers into depressions, which increase in size downslope. As the depressions deepen they capture more flow through the cross-grading of the hillslope surface (Dunne et al. in J. Costa and P. J. Fleisher, eds 1995). Surface run-off deepens as a result of this 'micro piracy'; see Croke et al. (2005) *Geomorph.* 68, 3–4.

The Darcy–Weisbach, Chézy, and Manning equations are used to predict overland flow velocity (Nunnally (1985) *Env. Manage.* 9, 5). Smith et al. (2007) *PPG* 31, 4 review the problems of measuring overland flow, and evaluate some solutions.

overlay At its simplest, a stack of transparent map layers, revealing where things co-occur. See Ahlqvist in R. Kitchin and N. Thrift (2009) for overlays in *GIS.

overpopulation Too great a population for a given area to support. 'What might be termed overpopulation in one context might not in another because of differences in the standard of living aspired to by different populations . . . Overpopulation is often examined in relation to the term "demographic pressure". This is a difficult concept to define since pressure is perceived in various ways, depending on whether countries are developed or less developed. In developed countries, which have reached a high standard of living, demographic pressure . . . is perceived as something that degrades the environment . . . Demographic growth, overpopulation and depopulation cannot be studied in the same way everywhere, because there are countless demographic systems, each of which is nested as a subsystem interlinked with other territorial, economic, cultural and social systems' (Faus-Pujol and Higueras-Arnal (2000) *Applied Geog.* 20, 3).

Perhaps the foremost diagnoses of overpopulation come from P. Ehrlich (with A. Ehrlich 1990). Neo-Malthusians support the argument; see Lempert (1987) *E. Afr. Econ. Rev.* 3, 1. Marxists, however, view overpopulation as the result of the maldistribution and underdevelopment of resources (Marx, *Capital*, vol. iii; see also Y. S. Brenner 1969). In the developed world, some would suggest that pollution and the desecration of the countryside are indicators of overpopulation.

overspill 1. The population which is dispersed from large cities to relieve congestion and overcrowding, and, possibly, unemployment. It occurs with redevelopment in the city where new building is at much lower densities so that some people—the overspill—cannot be housed in the city.

2. The condition of a fluvio-glacial lake when water levels rise above the barrier impounding it; see Kaiser et al. (2007) *Earth Surf. Procs Ldfms* 32, 10.

3. The dispersal—whether spontaneous or planned—of population in response to congestion, overcrowding, and high property prices. Overspill in Canada reflects the ongoing search for cheaper accommodation, by those who have been hardest hit by the crisis in affordable housing that has characterized Canada's largest and fastest-growing cities since

the 1980s (Mitchell et al. (2004) *Canad. Geogr./Géogr. canad.* 48, 2).

overthrust A nearly horizontal fold subjected to such stress that the strata override underlying rocks; see Price (1998) *GSA Bull.* 100, 12 and Hubert and Ruby (1978) *GSA Bull.* 70.

over-urbanization Population growth in an urban area which outstrips its job market and the capacity of its infrastructure; 'urbanization without industrialization' (M. Davis 2006). In 1988, 100 000 migrants from mainland China flooded into Hainan. The provincial government was swamped by more than 180 000 applications for only 30 000 jobs, with the result that those unable to find work turned to street hawking. Meanwhile, in-migrating peasant workers congregated in sprawling urban 'village' settlements that lacked basic amenities. The infrastructure of urban centres in Hainan sagged under these immense pressures (Go and Wall (2007) *Sing. J. Trop. Geog.* 28, 2).

ox-bow lake A horseshoe-shaped lake once part of, and now lying alongside, a meandering river with a narrow 'neck' between meander loops. When the river breaks through this narrow stretch of land, the old meander becomes a temporary lake; ox-bow lakes quickly fill up and become hollows in the landscape. Rasmussen and Ossa (2011) *Phys. Geog.* 32, 6, 497 report that ox-bow lakes are indicators of past channel conditions and can be utilized 'as measuring sticks for geomorphic change'.

oxidation The absorption by a mineral of one or more oxygen ions. Oxidation is a major type of chemical weathering, particularly in rocks containing iron; see Price and Bevel (2003) *Chem. Geol.* 202, 3. Oxidation occurs in soils when minerals take up some of the oxygen dissolved in the soil moisture; see Kayak et al. (2007) *J. Env. Qual.* 36.

oxisol A soil of the *US soil classification. *See* FERRALLITIZATION.

ozone A form of oxygen, and an atmospheric trace gas, made by natural photochemical reactions associated with solar ultraviolet radiation. Ozone has three atoms of oxygen combined in one molecule, rather than two atoms, as in free oxygen. The proportion of ozone in the atmosphere is very small, but it is of vital importance in absorbing solar ultraviolet radiation. The **ozone layer** (**ozonosphere**), is an ozone-rich band of the atmosphere, at 10–20 km above the Earth, but is at its most concentrated between 20 and 25 km. See Schiermeier (2007) *Nature* 449. When the ozone layer thins (the 'hole' over Antarctica is an example) increased solar ultraviolet radiation reaches the surface of the Earth, with consequent damage to human health. See European Space Agency (2004), 5 October.

pacific coast A coastline where the trend of ridges and valleys is parallel to the coast. If the coastal lowlands are inundated by the sea, a coast of interconnected straits parallel to the shore may result.

Pacific Decadal Oscillation (PDO)
A long-term fluctuation in sea surface temperatures in the extratropical north Pacific, waxing and waning every 20–30 years. The 'cool'/'negative' phase is characterized by a wedge of unusually low ocean temperatures in the eastern Pacific and a warmer 'horseshoe' of unusually high ocean temperatures in the north, west, and southern Pacific. In the 'warm'/'positive' phase, the western Pacific cools and the east warms. In western North America, positive phases of the PDO are associated with decreased winter precipitation, snowpack, and streamflow in the northwest, and higher precipitation in the southwest. Conditions reverse during negative PDO phases. Schneider and Cornuelle (2005) *J. Climate* 18, 21 argue that the north Pacific sea surface temperature anomaly and the PDO are 'a response to changes of the north Pacific atmosphere resulting from its intrinsic variability, remote forcing by *ENSO and other processes, and ocean wave processes associated with the ENSO and the adjustment of the north Pacific Ocean by *Rossby waves'. See also Goodrich and Walker (2011) *Phys. Geog.* 32, 4, 295.

pack ice Large blocks of ice on the surface of the ocean, formed when an ice field is broken up by strong waves. Since 2002, the East Greenland pack ice is much less, or even non-existent (Brooks (2005) *Geol. Today* 21, 6).

pahoehoe A type of lava flow which spreads in sheets, associated with highly fluid, basic lava, such as that ejected from *Hawaiian volcanoes. The surface is a glassy layer which has been dragged into ropy folds by the movement of the hot lava below it. Whether the lava is aa or pahoehoe depends mainly on viscosity and strain rate, which in turn depend on crystallinity, dissolved gas content, temperature, bubble content, slope, eruption rate, and lava composition. See Self et al. (2005) at *OpenGeoscience 1* 5, 283.

(⊕) SEE WEB LINKS
• Website of OpenGeoscience, British Geological Survey.

palaeoclimate The climate of a particular period in the geological past, before historical records or instrumental observations. Palaeoclimatology uses evidence from glaciers and ice sheets, geology, sediments, and tree rings; even the micromorphology of leaves (Haworth and McElwain (2008) *Pal., Pal. & Pal.* 262, 1–2) to determine Earth's past climates.

palaeoecology The reconstruction of past environments from fossils evidence. See Seppä and Bennett (2003) *PPG* 27, 4 on progress in palaeoecology, and Meadows (1999) *PPG* 23, 2.

palaeoenvironmentology The study of prehistoric and historic environments. Traditionally, the most widely applied technique in Quaternary studies has been that of *palynology. 'There appears to be no limit to the potential diversity of proxy types of evidence [for palaeoenvironmentology], and Quaternary palaeoecologists are . . . prepared to investigate virtually any organism . . . accumulating in sediments as to its palaeoenvironmental value' (Meadows (1999) *PPG* 23, 2).

palaeohydrology The study of past hydrology. Evidence for hydrological changes during previous periods of Earth history comes from the alteration, deposition, and erosion of rocks from these periods. Palaeohydrology includes changes in flora and fauna assemblages that have been strongly influenced by changes in hydrology. See Willard et al. (2011) *The Holocene* 21, 2, 305 on a pollen-based palaeohydrologic reconstruction from North Carolina.

palaeo-ice streams See Stokes and Clark (2001) *Quat. Sci. Rev.* 20 and Golledge and Stoker (2006) *Boreas* 35, 13 on ice streaming in the British Ice Sheet.

palaeolimnology The reconstruction of the paleoenvironments of inland waters and the changes experienced in these waters due to events such as climatic change, internal ontogenic processes, or human impacts, for example, *eutrophication, or *acidification. Dong et al. (2012) *The Holocene* 22, 107 provide a useful example.

palaeomagnetism *See* GEOMAGNETISM.

palaeosol A soil exhibiting features which reflect past conditions and processes. The presence of a palaeosol may have *palaeoclimatic significance; see Chlachula et al. (2004) *Boreas* 33, 2 for example, on using loess-palaeosols in southern Siberia.

Palaeozoic The era stretching approximately from 570 to 225 million years BP.

(()) SEE WEB LINKS
• An overview of the Palaeozoic era.

palimpsest This is a parchment from a scroll or book from which the text has been scraped or washed off, and which can be used again. The writing was usually washed using milk and oat bran. With the passing of time, the faint remains of the former writing would reappear enough so that scholars could discern the text (called the *scriptio inferior*), and decipher its meaning. Geographers really like the analogy of the palimpsest; Graham (2010) *Tijdschrift Econ.* 101, 4, 422 writes 'All places are palimpsests. Among other things, places are layers of brick, steel, concrete, memory, history, and legend. The countless layers of any place come together in specific times and spaces and have bearing on the cultural, economic, and political characteristics, interpretations, and meanings of place'. In geomorphology, palimpsest landscapes are composed of a mosaic of active and relict (inactive) landforms of different ages. See Kleman (1992) *Geografisk A*, 74, 4, 305.

Palmer Drought Severity Index (PDSI) A drought index that uses temperature and precipitation values from the climate division dataset of the National Climatic Data Center along with other components of the water balance equation to measure the departure of soil moisture supply from normal. The output of the PDSI consists of positive (wet) or negative (dry) values centred on 0 (normal) with values above +4.0 or below −4.0 generally considered extreme (Goodrich (2007) *Geog.*

Compass 1, 4). See Hessel and Graumlich (2002) *J. Biogeog.* 29, 7 on tree-ring reconstructions of this index. Yan et al. (2013) *Mathematical Problems in Engineering* present a modified PDSI.

palsa 'There is no general agreement about the meaning of the word "palsa". Usage and recent suggested definitions indicate that the word is chiefly used for cryogenic mounds covered by peat that were formed by an accumulation of segregation ice in the discontinuous permafrost zone' (Pissart (2002) *PPG* 26, 4). However, Arlen-Pouliot and Bhiry (2005) *The Holocene* 15, 3, 408 are comfortable with this term.

palynology The study of pollen grains as an aid to the reconstruction of past plant environments. Kiage and Liu (2006) *PPG* 30, 5 combine evidence from palynology, lake sediments, and associated records, to provide a more accurate and complete assessment of the paleoenvironmental changes. 'One drawback of palynological data is that lakes accumulate pollen from a large geographic area making it difficult to obtain information on a specific region from lake pollen' (anon. (1997) *Land-Owner Resource Center*). The other weakness is that most of the pollens

found come from wind-pollinated species, and animal-dispersed pollen is under-represented.

pan A large, shallow, flat-floored depression found in arid and semi-arid regions. Pans may be flooded seasonally or permanently. See Viles and Goudie (2007) *Geomorph.* 85, 1–2.

panarchy Global governance or a global assemblage of diverse spheres of governance. Hartzog (2010) *Master's Essay, University of Utah at Panarchy. com* offers this useful equation:

$$\text{Complexity} + \text{Networks} + \text{Connectivity} => \text{Panarchy}$$

See also L. S. Gunderson and C. S. Holling (2001).

pandemic An epidemic over a large area; throughout an entire country, continent, or the whole world. See Chandra et al. (2013) *Int. J. Health Geographics* 12, 9 on the Indian influenza pandemic of 1918–19, where over 15 million died in less than a year.

Pangaea A supercontinent comprising all of Earth's landmass, before it was split by *continental drift. Crowley et al. (1989) *Geology* 17, 5 find that very high summer temperatures and a large

Pangaea

annual range are major features of supercontinents; a finding with implications for *Permian extinctions.

See W. M. Marsh and M. M. Kaufman
(2012), p. 246 on the movements of India
during the break-up of Pangaea.

parabolic dune *See* SAND DUNE.

paradigm The prevailing pattern of
thought in a discipline or part of a discipline. Perhaps the most powerful
Western paradigm has been the 'scientific method'. T. Kuhn (1972) argued that
the evolution of a new paradigm marks a
new stage in thinking. This paradigm
persists until too many anomalies occur,
when it is replaced by a new paradigm
that is able to explain the anomalies. In
geography, for example, the regional
geography paradigm was superseded by
the *quantitative revolution, and then
the *cultural turn.

R. Inkpen (2005), pp. 18–21 is strongly
recommended for paradigms in physical
geography; Lomolino (2000) *Glob. Ecol.
& Biogeog.* 9, 1 offers a new paradigm for
island biogeography, and, in human geography, DeLyser and Shui (2013) *PHG*
are concerned with the open-science
paradigm and the 'fourth paradigm'.

paraglacial processes The non-
glacial Earth-surface processes,
sediment accumulations, landforms,
landsystems, and landscapes that are
directly conditioned by glaciation and
deglaciation (Ballantyne (2002) *Quat.
Sci. Revs* 21). This distinguishes it from
the term 'periglacial' which is defined as
'cold, non-glacial' and is applied to
environments in which frost-related
processes and/or permafrost are either
dominant or characteristic. See Iturrizaga (2008) *Geomorph.* 95, 1–2 on
paraglacial debris landscapes.

parallel drainage *See* DRAINAGE
PATTERNS.

parallel slope retreat The evolution
of a hillslope when the slope angle
remains constant. Ellis et al. (1999) *Basin
Res.* 11, 1 argue that parallel slope retreat
requires some mechanism to remove
weathered material from the foot of the
slope in order to maintain the existing
topography. See Leeder et al. (2002)
Basin Res. 14, 1 on parallel slope retreat
by earthquake-induced collapse failure.
Jackson et al. (2005) *Sedimentol.* 52, 5
find parallel slope retreat on beaches,
when breaking waves on the upper
foreshore are lower; enhanced when
waves approach at large angles to the
beach.

parenting To be or act as a mother
or father to someone. For Aitken in
R. Kitchin and N. Thrift (2009), 'an
emotional and day-to-day practice that
is embedded in larger social and spatial
contexts'. Luzia (2010) *Soc. & Cult. Geog.*
11, 4, 359 writes on the 'actual places and
spaces of parenting'.

parent material The rock or deposit
from, and on, which a soil has been
formed. The nature of the parent rock
will largely determine the nature of the
*regolith, and hence the soil texture—
basalt tends to produce clay soils; granite
sandy soils. S. Smith (1996) *PPG* 20, 4
links parent material with bimodality in
fluvial beds. See also Mahaney (1993)
PPG 19, 3.

Pareto optimality A situation where
it is impossible to improve the economic
lot of some without making others worse
off. Sen (1970) *J. Polit. Econ.* 78 shows
that, within a system of social choice, it is
impossible to have both a commitment
to minimal liberty and Pareto optimality.

park Originally an enclosed area used
for hunting, in the 18th century the term
applied to the grounds of a country
house. It now refers to open land used
for recreation in a town or city. Within

the UK, parks were a feature of 19th-century *urbanization, in order to relieve the 'artificiality' of urban living. However, Brainard et al. (2006) *CSERGE*, ECM-2006–05 argue that parks have usually been sited with little regard for the geography of where different social groups live: within Birmingham, UK, 'even within the most deprived areas, whites have better access to park areas than non-whites'. Conlon (2004) *Sexualities* 7, 462 describes Christopher Park, Greenwich Village, New York, as a specific space that 'highlights the relationships amongst queer theory, identity and material public spaces'.

participatory GIS The gathering of data using traditional methods such as interviews, questions, and focus groups, all using some form of paper maps to allow participants to record spatial details. The key difference between participatory *GIS and community mapping is the way the data are processed after gathering. Dunn (2007) *PHG* 31, 5 describes it as 'a more socially aware type of GIS which gives greater privilege and legitimacy to local or indigenous spatial knowledge ... A Participatory GIS celebrates the multiplicity of geographical realities rather than the disembodied, objective and technical "solutions" which have tended to characterize many conventional GIS applications'. See Elsley and Cartwright in A. Ruas, ed. (2011) for an example.

participatory research (participatory action research) A research strategy that is based on participation and action from the community it is working with; looking for change grounded on local knowledge and local experience. PAR 'seeks to democratize knowledge production and foster opportunities for empowerment by those involved' (Kindon et al. (2008) *Int. Encyclopedia Hum. Geog.*). PRA is thus a bottom-up

approach. Lane et al. (2011) 36, 1, 15 discuss participatory research in comparison with other research methods on flood risk. **Participatory Rural Appraisal (PRA)** uses the same approaches; see Yusuf (2007) *J. Agric. Res. & Dev.* 6, 37.

particulates (particulate matter) Atmospheric particles—dust, pollen, soot, salt, sulphur, etc. See Zhang et al. (1994) *J. Atmos. Scis* 33, 7 on the impact of dust on particulate nitrate and ozone formation.

Partition When the British left India in 1947, they partitioned the country, creating the separate countries of India and Pakistan to accommodate religious differences between Muslims and Hindus. Pakistan has a majority Muslim population and India is primarily Hindu. See M. Hasan (1994) for a collection of essays, extracts, memoirs, and a short story to create an outline of the events preceding and surrounding India's Partition.

( ) SEE WEB LINKS
• Partition in postcolonial studies.

passive margin In plate tectonics, a margin between *continental and *oceanic crust contained entirely within a single *plate. Tectonic activity here is minimal. See Leroy et al. (2007) *Geophys. J. Int.* on the uplift and evolution of passive margins.

pastoralism Raising of livestock; pastoral farmers grow crops to feed their livestock, and pastoral nomads raise their livestock on natural pasture. See Warren (1995) *TIBG* 20, 2, 193.

patch An area of vegetation that has a uniform structure and composition, and differs from the surrounding vegetation. Watson (2002) *J. Biogeog.* 29, 5–6 distinguishes between fragments—remnants of a previously widespread habitat—and islands, which have always been restricted and isolated in their spatial extent. Patch scale is between 0 and 10 hectares.

path In *behavioural urban geography, a channel people move along, within a city.

path-dependent 'Path dependence exists when the outcome of a process depends on its past history, on a sequence of decisions made by agents and resulting outcomes, and not only on contemporary conditions' (Baláž and Williams (2007) *Int. Mig.* 45, 2). 'The future development of an economic system is affected by the path it has traced out in the past' (G. Hodgson 1996). This term seems, to some extent, to resemble *hysteresis, or *contingencies. See also Martin and Sunley (2006) *J. Econ. Geog.* 6. 'The strength of the path dependence model lies in the insistence of its practitioners on the importance of the specific sequence of micro-level historical events' (Ruttan (1997) *Econ. J.* 107, 444). Greener (2002) *Manage. Decisions* 40, 6 explains path dependence as a lock-in (the entry of a system into a trap from which it can't escape without the involvement of an outside force/shock) by historical events. See also Verleysdonk et al. (2011), *Geograf. Series A*, 93, 113; DOI: 10.1111/j.1468-0459.2011.00423.x on the sensitivity and path dependence of mountain permafrost systems.

patriarchy A gendered system of social control that pervaded all aspects of human existence, including politics, industry, the military, education, philosophy, art, literature, and civilization itself (K. Millett 1970). This includes the 'private' spheres of love, sexuality, marriage, and children—the family is politicized as the foundation of patriarchal power; see Bennett (2006) *Gender, Place & Cult.*

13, 2. 'Feminist writings extended the original commonsense meaning of [this] term to apply to a situation of male dominance in a society. This immediately raises the question of whether the term is purely descriptive, or whether it refers to an explanatory theory. It is important to appreciate this difference as many sophisticated feminist arguments use the term in its descriptive sense' (Blackburn et al. (2002) *Brit. J. Sociol.* 53, 4).

'Patriarchy and masculinity are not eternal but (like everything else) constructed products, and open to change' (Massey (2001) *TIBG* 26, 2). Erman (2001) *Int. J. Urb. & Reg. Res.* 25, 1 shows how an urban community reproduces traditional patriarchal authority over inmigrants through the social construction of female labour within the framework of familialism, and the housewife ideology. Adams and Ghose (2003) *PHG* 27, 4 observe that Bollywood idols reflect not only patriarchy but subethnic and colonial power relations. 'It is a sad fact that one of the few profoundly non-racial institutions in South Africa is patriarchy' (A. Sachs, no date). Hern (2006) *Pol. Geog.* 28, 6 defines **transpatriarchies** as forms of patriarchy extending across and between national societies. 'Remasculinization of public rhetoric and privileging economic and military concerns act to defend patriarchy diffusely (Connell (2005) *Signs* 30, pp. 1816–17), and have sharp impacts on diverse women's livelihoods and political spaces' (Radcliffe (2006) *PHG* 30, 4). See Waitt (2006) *AAAG* 96, 4 on white gay patriarchy.

patterned ground Most commonly in *periglacial areas, the arrangement of stones into polygons, isolated circles, and concentrations of circles known as nets, steps, and stripes. Polygons and circles are more common on level surfaces, stripes generally form on slopes;

the patterns are made of coarser stones, separated by much smaller stones (*fines*).

George et al. (1989) *Geophys. & Astrophys. Fluid Dynam.* 46, 3 argue that convection currents within the *active layer produce patterned ground; Grab (1997) *ESPL* 22 cites desiccation and the formation of needle ice as major causal processes. Matsuoka (2003) *Geomorph.* 52, 1–2 attributes patterned ground to the upward injection of slow-freezing, waterlogged silts at particular points. Kessler and Werner (2003) *Science* 299, 5605 note *self-organization in patterned ground.

peace, geographies of Also termed **pacific geopolitics**, this is a way of thinking geographically about world politics in order to promote peaceful and mutually enriching human coexistence (Megoran (2010) *TIBG* 35, 3, 382). Williams (2013) *AAAG* 103, 1, 230 has a practical article on reproducing everyday peace in north India.

peak land-value intersection (PLVI) The point in a *CBD, often, but not always, at a road intersection, where land values are at a maximum.

peasant A farmer whose activities are dominated by the family group. The family provides all the labour and the produce is for the family as a whole; occasional surpluses are sold in the open market. Landholdings are small, sometimes owned by the family, but often leased; see Kull (2006) *AAAG* 96, 3 on the perceived rights of peasants versus state control. 'Peasants are not a declining class but rather are undergoing a variety of transformations. In some settings they will indeed disappear, in others experience stresses that lead to their semi-proletarianization and, in yet others, peasants were, and still are, being created by capitalism' (Johnson et al. (2005)

Antipode 37, 5). Although peasants have been characterized as backward and resistant to change, peasant strategies can be highly rational in a society where there is little margin for error; see Faminow et al. (1999) *World Animal Rev.* 93. See Murray (2006) *J. Peasant Studs* 33, 4 on neo-feudalism in Chile.

peat A mass of dark brown or black, partly decomposed plant material, formed in waterlogged ground, where temperatures are low enough to slow down the decomposition of plant residues. It is characteristically found in cold climates, and as a relict feature in temperate zones. See Turetsky et al. (2007) *Glob. Change Biol.* 13 on climate change and peatland. Peat may be used as fuel, and is widely used by gardeners; so much so that peat bogs are now under threat. See M. Evans and J. Warburton (2007).

ped In a soil, an aggregate of silt, sand, and clay of characteristic shape, resulting from the formation of hydrogen and ionic bonds between soil particles. Peds may be further developed by plant roots, by *polysaccharide gums secreted by soil fauna (Piccolo and Mbagwu (1999) *Soil Sci. Soc. Am. J.* 63), and by alternate freezing and thawing, or wetting and drying (Hart et al. (1985) *Eur. J. Soil Sci.* 36, 3). Ped properties can be reflected in the structure of pore space (Pachepsky and Rawls (2003) *Eur. J. Soil Sci.* 54, 3). *See* SOIL STRUCTURE.

pedalfer Any soil high in aluminium (Al) and iron (Fe), and from which the bases such as calcium and magnesium carbonates have been *leached. Pedalfers generally occur in regions with an annual rainfall of more than 600 mm. H. Jenny (1992) notes that C. Marbut's differentiation of soils into pedocals and pedalfers (1935) applies only to mature soils.

pediment A low, *concave element at the foot of a hillslope (also called a *concave*, or *waning*, slope). See Strudley and Murray (2007) *Geomorph.* 88, 3–4 on the numerical simulation of pediment development.

pedocal A soil high in calcium carbonate and magnesium carbonate.

pedogenesis The formation of soils. **Pedogenic processes** are soil-forming processes. See Foss (2006) *Soil Sci.* 171, 6, Suppl. 1 on milestones in pedogenesis. The chief pedogenic factors are time, *hydrology, *parent rock, climate, fauna and flora, but terrain variables may be ineffective predictors of soil characteristics where a dense canopy hides a non-uniform erosional environment (Williamson et al. (2006) *Soil Sci. Soc. Am. J.* 7). **Pedology** is the science of soils: their characteristics, development, and distribution.

pedon A small sample of a soil large enough to show all the characteristics of all its horizons.

pelagic Of marine life, belonging to the upper layers of the sea.

peneplain An erosional plain, near to base level. 'Despite more than a century of effort, no convincing example of a contemporary peneplain has been identified' (Phillips (2002) *Geomorph.* 45, 3–4).

people's geography project An educational project making geographical insights and skills accessible and useful to non-geographers; 'to popularize and make even more relevant and useful to ordinary people the important, critical ways of understanding the complex geographies of everyday life that geographers have [developed] and continue to develop'.

SEE WEB LINKS
• Website of The People's Geography Project.

perception The manner by which we make sense of the world; a process shaped by a cultural script, or filter, which moulds our take on life. One such filter is the available technology; where once disasters were interpreted as 'Acts of God', today they are perceived as the result of human irresponsibility or malevolence; see Furedi (2007) *Area* 39, 4. Perception also changes with age: Tapsell et al. (2001) *Area* 33, 2 come to the unsurprising conclusion that children's perceptions differ from those of adults.

Of particular interest is **hazard**, or **risk, perception**; 'what science deems to be an acceptable level of risk may not match the social perception of acceptability. When the differences between social and scientific notions of risk become acute, then the outcome is a *social amplification of risk* by the public' (Herrick (2005) *Area* 37, 3). However, Crozier et al. (2006) *Area* 38, 2 find that people in high-risk earthquake zones tend to be fatalist, or at least resigned to the consequences. Perception also shapes economic decision-making; see Farley (2007) *AAAG* 97, 4 on the perception of 'good' forests, and Warren (2007) *PHG* 31, 4 on the clash between perception and ecology; '[the] popular perception of "cultural belonging" conflicts with scientific judgements that such species do not belong ecologically'. *See* ENVIRONMENTAL PERCEPTION.

perched water-table A partly saturated, isolated, confined *aquifer underlain by an impermeable rock, with the main *water-table below the two. See Bryan and Rockwell (1998) *Geomorph.* 23, 1–2 on perched water tables, rill incision, and sediment transport.

percolation The filtering of water downwards through the *bedding planes, *joints, and pores of a permeable rock.

percolines An underground network of water seepage zones: old root channels, soil cracks, and animal burrows enlarged by *interflow. **Percoline flow** is 'laterally concentrated diffuse interflow perhaps including pipeflow' (A. Lerman and M. Meybeck 1988).

perennial stream A watercourse which flows throughout the year.

perfect competition Under conditions of perfect competition, there are many suppliers; a perfectly elastic supply of the *factors of production; no collusion between suppliers; and buyers and sellers are fully aware of the prices being charged throughout the market. This is an unlikely state of affairs; imperfect competition is much more common. Try Mossy (2003) *Chaos, Solutions & Fractals* 18, 3: technical, but interesting.

perforation kame See KAME.

performance The acting out of different identities and roles. Thrift, for example, in I. Cook et al., eds (2000) writes of the possibilities of drawing on 'street theatre, community theatre, [and] legislative theatre'. Eyerman (2004) *Thesis Eleven*, 79, 25 argues that performance implies a notion that action is also symbolic, public as well as social. When we see that social actions and practices are performances we are better able to understand that they are open to interpretation; we can see how people seek to create particular spatialized identities and spaces, and how these might change with different performances. Because of this, Latham (2003) *Env. & Plan. A* 35, 11 argues that reframing the research process as a performance can produce new processes and new insights.

performativity Using speech and gestures to act out/create a situation. J. Butler, who created this concept, describes it as the 'reiterative power of discourse to produce the phenomena that it regulates and constrains' (J. Butler 1993). In other words, performativity indicates a performance that has had its desired effect: what we say and do about something shapes what it is. A good example is given by Waitt (2008) *Soc. & Cult. Geog.* 9, 1, who illustrates the way that aggressive attitudes towards the ocean by Australian shortboard surfriders creates a kind of masculinity. In a **performative perspective**, all subjects are subjects-in-process, and to study a subject is to 'analyse the practices by which it is produced as a self-constituting subject' (Kuus (2007) *TIBG* 32, 1).

peridotite A coarse-grained igneous rock, made mostly of olivine and pyroxene, and thought to be the main constituent of the *mantle.

periglacial Referring to the processes and landforms of any area with a *tundra climate, or where frost processes are active and *permafrost occurs in some form. **Periglacial climates** are arid, with temperatures below 0 °C for at least six months, and summers warm enough to allow surface melting to a depth of around 1 m. See Vandenberghe (2006) *Geomorph.* 54 on the transition from temperate to periglacial climates, and vice versa.

Periglacial processes include *abrasion, *freeze–thaw, *nivation, and *solifluction, and are responsible for the formation of new deposits, the alteration of existing unconsolidated deposits, and the modification of existing landforms by *mass movement; see Millar (2005) *Geomorph.* 72, 1–4 and Braun (1989) *Geomorph.* 2, 1–3. Hall et al. (2002) *PPG* 26, 4 identify snow fall as a crucial factor in periglacial weathering. See Vandenberghe and Thorn (2002) *PPG* 26, 4 for progress in periglacial research.

p

periodicity This refers to recurrences at regular intervals. The major external cycles affecting the Earth are lunar (with daily, monthly, and 18.6 year cycles), planetary cycles, with 20,000- and 400,000-year recurrences (see MILANKOVITCH CYCLES), and solar cycles; Lean (2010) *WIREs Clim. Change*, 1: 111 is excellent on this.

periodic market A market place where buyers and sellers converge at regular intervals (Ghosh (1981) *Prof. Geogr.* 33, 4).

periphery The edge, or margin. Murray and Challies (2006) *Asia Pacif. Viewpt* 47, 3 describe New Zealand as a **resource periphery** and Chile as **semi-peripheral**.

permafrost Areas of rock and soil where temperatures have been below freezing point for at least two years. Permafrost need contain no ice; a sub-zero temperature is the sole qualification. **Continuous permafrost** is present in all *periglacial areas apart from small, localized thawed zones, while **discontinuous permafrost** exists as small, scattered areas of permanently frozen ground. A **frost table** marks the upper limit of permafrost, which is overlain by the *active layer. Ice develops mainly in upper levels in **epigenetic permafrost**; in **syngenic permafrost** it is regularly distributed throughout the whole thickness of the permafrost; go to the US Army Permafrost tunnel for both types.

Permafrost aggradation (growth) decreases the thickness of the active layer, is responsible for the formation of *pingos, and may be caused by the freezing of *taliks. **Permafrost degradation** (decline) plays a key role in the development of *thermokarst; see Jorgensen et al. (2001) *Clim. Change* 48, 4. Permafrost is a very sensitive system; small mistakes in constructing buildings in this environment can have catastrophic effects; see Couture et al. (2000) *Geol. Soc. Canada*, 2000–B2.

Permian The latest period of *Palaeozoic time, stretching approximately from 290 to 248 million years BP. The **Permian mass extinction** occurred between 290 and 252 million years ago, during which 95% of all marine life on Earth was killed, and 70% of all land families became extinct.

((()) SEE WEB LINKS
• Bristol University website.
• An overview of the Permian period.

persistence (geomorphological persistence) The persistence of a landform varies with the magnitude and recurrence interval of the event which brought it about. Calver and Anderson (2004, *TIBG* 29, 1) suggest the terms **static persistence** where 'spatial positioning is constant, [and there is] virtually no change in the feature over time', and **dynamic persistence** where 'the type of feature remains in evidence . . . but individual representations are formed and/or degrade'. See also Wilcock and Iverson, eds (2003) *Geophys. Monog. Ser.* 135.

personal construct theory This suggests that humans are continually constructing and testing their own, individual images of reality, and that investigations should be based on the personal constructs of the people involved, not on those of the researcher. See Pomery and Green in P. Dearden and B. Sadler (1989); see also Hudson (1980) *PPG* 4.

personal space The zone around individuals—usually that reserved for a normal conversational voice and for friendly interaction—which they reserve for themselves. The extent of a personal space around an individual is reckoned to be 1–1.5 m for an Anglo-Saxon (Beaulieu (2004) *J. Applied Social Psych.* 34, 4). Empowerment strategies provide a conduit for women to claim personal space (George (2007) *Gender, Space & Cult.* 14, 6). Barker (2000) *Area* 32, 4 only hints at the 'globalization of personal space', as does McCormack (2003) *TIBG* 28 at the 'personal space of emotion'.

p-forms and micro-channel networks Small-scale landforms of glacial erosion—usually less than 1 m in width and depth—on glaciated bedrock surfaces. Sichelwannen are sickle-shaped bedrock depressions, with an open end pointing in the direction of ice flow, and are the most common (Glasser and Bennett (2004) *PPG* 28, 1). See Shaw (1994) *Sed. Geol.* 91. The term 's-form' applies if a glacial meltwater origin can be established.

phase space 1. Very simply, a continuum representing two variables: A and B. The continuum ranges from 100% A, 0% B, to 0% A, 100% B. See, for example, Hooke (2003) *TIBG* 28, 2 on the phase space of meander movement and bend curvature. Melton (1958) *J. Geol.* 66 describes a mature drainage basin phase space with four variables.
2. Very much less simply, an abstract concept that captures all the possible spaces in which a spatio-temporal system might exist *in theoretical terms*; not just what happens but what might happen; 'the space of the possible' (J. Cohen and I. Stewart 1994). 'The geography of phase space is flexible, but not totally arbitrary: the main possibilities are "already there", constrained by contextual realities' (I. Stewart and J. Cohen 1997). This use of the concept is not totally distinct from 1; there are just many more factors involved. Martin (in Harrison et al. (2006) *Area* 38, 4) draws on Richards in J. Matthews and D. Herbert (2004) to see phase space as 'a middle

road between relational space and more fixed notions of spatiality'.
Hooke (2007) *Geomorph.* 91, 3–4 uses **phase space plots** to help uncover emergent behaviour in meandering rivers.

phenomenon instance In *geomorphology, an entity whose form and behaviour varies over time and space. Thus, for example, a coastal spit may only 'act as such' at certain stages of the tide cycle, at other times more resembling an embryonic *sand dune (Raper and Livingstone (2001) *TIBG* 26, 2). *See* EMERGENCE.

phenotype The outward appearance, physical attributes, or behaviour of an organism, which develop through the interaction of nature and nurture. For **phenotypic variation**, see Robertson and Robertson (2008) *J. Biogeog.* 35, 5.

Phillips curve A negative exponential curve demonstrating the relationship between the percentage change in wages and the level of unemployment. The theory is that high wages cause high inflation, and the lower the rate of unemployment, the higher the rate of inflation. Fujiki and Wall (*St. Louis Fed. W. Papers* 2006–57) adjust the Phillips curve for Japan, by taking account of the geographical dispersion of labour-market slack.

photic zone Those upper levels of a water body that are penetrated by light. See Håkanson (2006) *Glob. Ecol. Biogeo.* 15 on suspended particulate matter and the photic zone.

photochemical smog Nitrogen dioxide (NO_2) is emitted from petrol engines. Ultraviolet light splits this into nitric oxide (NO) and monatomic oxygen (O). The hydrocarbons emitted from the burning of *fossil fuels react with some of the monatomic oxygen to form photochemical smog. See Hart et al. (2004) *13th Joint Conference on the Applications of Air Pollution Meteorology*. Photochemical smog is most common where the sunshine is strong and long-lasting and where car use is high; see Zhang and Oanh (2002) *Atmos. Env.* 36, 26 on photochemical smog in Bangkok. Photochemical smog damages plants (Arbaugh et al. (2003) *Env. Int.* 29, 2–6), and irritates eyes and lungs (Fukuoka (1997) *Int. J. Biometeorol.* 40, 1).

photogrammetry The use of aerial photographs and images from *remote sensing to measure terrestrial features. See *J. Photogrammetry & Remote Sensing*. Photogrammetry employs high-speed imaging and the accurate methods of remote sensing in order to detect, measure, and record complex 2-D and 3-D motion fields; it is 'the art, science, and technology of obtaining reliable information from noncontact imaging and other sensor systems about the Earth and its environment, and other physical objects and processes through recording, measuring, analyzing and representation' (*International Society for Photogrammetry and Remote Sensing*).

(⊕) SEE WEB LINKS

• Website of International Society for Photogrammetry and Remote Sensing.

photography and geography Geography is a visual discipline, working with maps, globes, models, slides, and photographic illustrations; yet 'visual power is simultaneously expressed as something to be wary of, serving historically as the prime tool of the white, western gaze of academia' (Roberts (2012) *PHG* 37, 3, 386). D. Cosgrove (2008) therefore argues that the 'truth' of the image must be sought through its deceptive surface to the 'social practices' beyond it.

phreatic Describing *groundwater below the *water-table. The **phreatic zone** is permanently saturated; see Bekele et al. (2006) *Aus. J. Earth Sci.* 53, 6 on a phreatic aquifer, and Lajura et al. (2006) *Env. Forensics* 7, 4 on a coastal phreatic terrain. A volcanic **phreatic eruption** has *meteoric water mixed with the lava, and may create steam, or a *geyser (Fontaine et al. (2003) *Earth & Planet. Sci. Letters* 210, 1–2). A **phreatophyte** is any desert plant with roots down to the phreatic zone; see Ridolfi et al. (2007) *J. Theoret. Biol.* 248, 2.

phylogenetic Referring to evolutionary development. The **phylogenetic diversity (PD)** of a species is a measure of its taxonomic distinctness and can be estimated by looking at the phylogenetic relationships among taxa. Species-specific metrics on PD can then be used to determine conservation priorities at various biogeographical scales (McGoogan et al. (2007) *J. Biogeog.* 34, 11).

phylum The second highest category (of seven) in the scientific system of classification for organisms (pl. *phyla*), consisting of one or several similar or closely related classes. Hence, **phylogeography**, the study of the historical processes that may be responsible for the contemporary geographic distributions of individuals. This is accomplished by considering the geographic distribution of individuals in the light of the patterns associated with a gene genealogy; J. Avise (2000).

physical geography 'Physical geography focuses upon the character of, and processes shaping, the land-surface of the Earth and its envelope, emphasizes the spatial variations that occur and the temporal changes necessary to understand the contemporary environments of the Earth. Its purpose is to understand how the Earth's physical environment is the basis for, and is affected by, human activity. Physical geography was conventionally subdivided into *geomorphology*, *climatology*, *hydrology*, and *biogeography*, but is now more holistic in systems analysis of recent environmental and Quaternary change. It uses expertise in mathematical and statistical modelling and in remote sensing, develops research to inform environmental management and environmental design, and benefits from collaborative links with many other disciplines such as biology (especially ecology), geology and engineering' (K. Gregory 2002). However, R. Inkpen (2005) makes the soundly based claim that there is not a single history of physical geography.

Between 1850 and 1950, the main ideas that had a strong influence on the discipline were uniformitarianism, evolution, exploration and survey, and conservation (G. P. Marsh 1864). In the 1960s, 'a new type of physical geography began to emerge that accentuated a concern with dynamic processes of earth systems. This new approach, which has evolved to the present, is founded on basic physical, chemical, and biological principles and employs statistical and mathematical analysis. It has become known as the "process approach" to physical geography . . . Over the past fifteen years, physical geographers, who have always acknowledged that the systems they study are complex, have turned to emerging ideas in the natural sciences about nonlinear dynamical systems and complexity to explore the relevance of these ideas for understanding physical-geographic phenomena' (Rhoads (2004) *AAAG* 94, 4). 'Advances in remote sensing, geographical information systems and information technology have enabled a more global approach; a second new development has been the advent of a more culturally-based approach throughout many

branches of physical geography. By 2000 a series of issues can be identified including the increasingly holistic trend, greater awareness of a global approach and of environmental change problems, and of the timely opportunities which can arise from closer links with human geography and with other disciplines' (Gregory (2001) *Fennia* 179, 1).

Harden (2011) *Phys. Geog.* 33, 1, 1 writes that 'while the sub-discipline of physical geography remains firmly grounded in research undertaken to explain Earth's landscapes and its geomorphic, hydrologic, atmospheric, cryospheric, petrologic, and biogeographical processes, which change over time and space, the extent of the human "footprint" on this planet challenges physical geographers to pay greater attention to the role of people in environmental change and the interactions between people and their environments'.

physical quality of life index
(PQLI) An attempt to measure the quality of life or well-being of a country. The numerical value is derived from the adult literacy rate, infant mortality rate, and life expectancy at age 1, all equally weighted. It is one of a number of measures created because of dissatisfaction with the use of GNP as an indicator of development. However, Rodrik and Brodsky DOI:10.1016/0305-750X(81) 90071-1 suggest problems with this indicator.

physiological drought
When soil water, although present, is unavailable, as when it is frozen, or when *evapotranspiration exceeds plants' uptake of water. See Sausalito and Bellinger (2007) *Silva Fennica* 41, 2 on morphological and physiological drought.

phyto-
Of a plant; relating to plants; hence **phytogeography**. See DeForest

Safford (2007) *J. Biogeog.* 34, 10 and
Kuentz et al. (2007) *J. Biogeog.* 34, 10.

piedmont glacier A glacier formed
from the merger of several *alpine glaciers as they emerge from the mountains; see Hall and Denton (2002)
Holocene 12.

**pingo 1. open system pingo, hydro-
static pingo** A large *ice mound
formed under *periglacial conditions,
so called because it is formed from
an unfrozen pocket confined by
approaching *permafrost. Mackay (1979)
Geographie physique et quaternaire 33
argues that open system pingoes
develop on former lakes.

2. closed system pingo An ice
mound, as above, but formed as water,
under *artesian pressure within or below
permafrost, which causes it to buckle
upwards. MacKay (op. cit.) argues that
not all such pingoes form in a truly
closed system. Gurney (1998) *PPG* 22, 3
proposes a third category: polygenetic
(or 'mixed') pingoes.

pioneer community The earliest
*sere developing on a raw site. A

pioneer species is a plant species dominating a community in the first stage of
succession.

pipe 1. A volcanic channel or conduit
filled with solidified *magma. Sometimes the hard pipe rock is exposed after
erosion. See Barnett and Lorig (2007) *J.
Volcanol. Geotherm. Res.* 159, 1-3. For
pipe eruption, *see* CENTRAL VENT
ERUPTION.

2. In *hydrology, a subsurface
channel, often near horizontal,
through which water passes. Pipes
can transfer water underground as
a rapid route for subsurface storm
flow. See Carey and Woo (2000)
J. Hydrol. 233, 1-3.

pipeline *See* GLOBAL PIPELINE.

place Staeheli in J. Agnew et al., eds
(2003) distinguishes five conceptualiza-
tions of place: place as physical location
or site, place as a cultural and/or social
location, place as context, place as constructed over time, and place as a social
process. 'If we make space through
interaction at all levels ... then those
spatial identities such as places, regions,

1. Open system pingo (East Greenland type)

Ice
'Free' water
Talik
Permafrost
Saturated ground

a. Permafrost advances

Pingo

b. 'Free' water freezes and expands, lake floor mounds up, lake water drains away

Surface ruptures

Expanding ice core

2. Closed system pingo (Mackenzie type)

Upward movement of
groundwater through
artesian pressure

3. Collapsed pingo (open or closed system)

Rampart   Previous mound

Ice core
melted   Solifluction

nations, and the local and the global
must be forged in this way, too, as
internally complex, essentially
unboundable in any absolute sense, and,
inevitably, historically changing'
(Massey (2004) *Geografiska B* 86, 1).
H. Raffles (2002) understands places as
'spatial moments [that] come into being
and continue being made at the meeting
points of history, representation, and
material practice'.

'A national ideology about women's
place . . . is filtered through geographical
differentiation at the local level' (Mc-
Dowell (2003) *TIBG* 28, 1). The world's
cultural geography can be viewed as a
'mosaic' (Crang et al. (2003) *PHG* 27).
See also Nijman (2007) *Tijdschrift* 9, 2.
Waterman (1998) *TIBG* 23 comments on
the relationships between the art produced at a specific place and the place
itself: 'place becomes a metaphor for

social trends.' Castree (2004) *Pol. Geog.* 23, 2 has a different take: 'it is useful to reflect on what is lost when we are encouraged to see place-bound movements as tendentially "regressive"; translocal organizing as largely "progressive" (not to mention pragmatically necessary); and an openness to other peoples and places as the mark of a proper (sic) place politics.' Try Larsen and Johnson (2012) *AAAG* 102, 3, 632 for a comprehensive summary of the major geographical work on 'place'—but like most of the references in this entry, there is no reference to any real, existing place or places.

A **relational reading of place** works with the ontology of flow, connectivity, and multiple geographical expression, to imagine the geographies of cities and regions through their plural spatial connections (Amin (2004) *Geografiska B* 84,1).

Narratives of place ascribe meaning to socially and politically bound space, and as Davis (2005) *AAAG* 95, 3, 607 explains, the 'process of transforming empty space into a meaningful construct necessitates first visualizing in the mind's eye what, or more importantly who, occupies that space. The cultural landscape, or social meaning, of space is rendered intelligible only when we attach to it some of the available representations while disregarding others'. When stabilized over time, narrative representations of place have the potential to become deeply imbedded knowledge that has enormous implications for how we understand and interact with the world; see While and Short (2011) *Area* 43, 1, 4.

Martin (2003) *AAAG* 93, 3 argues that **place frames**—the physical conditions of the neighbourhood and the daily life experiences of its residents—are used to unite residents for collective action.

Within physical geography, there remain many possibilities that are determined by local, specific, factors linked to particular places and times— that is, by *contingencies (Phillips (2007) *Geomorph.* 84, 3–4). Contingencies are not necessarily complications that can eventually be described by general laws. They may be irreducibly place dependent, and as important as, or even more important than, general laws in determining how the world works (Phillips (2004) *Prof. Geogr.* 56, 1).

place dependent Temporally and spatially bound, by, for example, fixed *capital, institutions, labour skills, and shared conventions. 'Products and commodities take on the qualities of the places from which they come. In their tendency to persist even in the context of geographically homogenizing forces, place differences permeate the artefacts whose creation they stimulate' (Molotch (2002) *Int. J. Urb. Reg. Res.* 26, 4).

place identity 'Place identity is subjected to an imposed, created and manufactured place construction, while sense of place remains personal, found and grounded in the lived experiences of many residents' (Carter et al. (2007) *Soc. & Cult. Geog.* 8, 5). See Hall (1997) *Ecumene* 4 on repositioning place identity; see McGuirk and Rowe (2001) *Australian Geog. Studs* 39, 1 and Schwartz (2002) *Pol. Geog.* 25, 1.

placelessness E. Relph (1976) claimed that, with mass communication, and increasingly ubiquitous high technology, places become more and more similar, so that locations lose a distinctive 'sense of place'. 'Placelessness dehumanizes the world and because dehumanized places have less or no human attachments, the people in these placeless places become even more vulnerable to more dehumanization' (I. Vogeler 1996).

Dicken in J. Peck and H. Yeung, eds (2003) argues that 'geography and place are profoundly important to the ways firms are produced and behave. Firms are embedded in specific places, both in the place(s) in which they operate, and more so in their place of origin.' Dicken (2004) *TIBG* 29, 1 thinks the claim that TNCs are becoming 'placeless' is 'a nonsense'. And Gertler in Peck and Yeung (op. cit) demonstrates the limited ability of head offices to impose uniformity across their operations overseas.

SEE WEB LINKS
- Ingolf Vogeler's lecture notes about placelessness.

place-making The way we collectively shape our public realm to maximize its value; the planning, design, management, and programming of public spaces, rooted in community-based participation. Pierce et al. (2011) *TIBG* 36, 1, 54 define place-making as the: 'set of social, political, and material processes by which people iteratively create and recreate the experienced geographies in which they live. Place-making is an inherently networked process, constituted by the socio-spatial relationships that link individuals together through a common place-frame', which is the same definition, but in loftier language.

place marketing 'Branding' a location in order to 'sell' it, that is, to make it more attractive; a tool to promote the attractiveness of a locality in such a way that target groups know its distinctive features and are persuaded to go there. See Hospers (2011) *Tijdschrift* 102, 3, 369. Hospers (2010) *Place Branding and Public Diplomacy* 6, 280 is also useful on the limits of place marketing.

place names The study of the early forms of present place names may indicate the culture which gave the name together with the characteristics of the site. In the UK, for example, *ey* meaning a dry point and *ley* meaning a forest, wood, glade, or clearing appear in many place names such as Chelsea and Henley-in-Arden.

Place names label, define, and represent places and people; 'a place name sometimes fills up its territory with sense of place and homogenizes it. Fixing a boundary on a map reinforces a real differentiation and might segregate the dwellers' (Okamoto (2000) *PHG* 24, 328). In the Hawaiian Islands, place names change 'from being reflections of Hawaiian geographic discourse, to being encoded within Western approaches to knowledge, commodification of the environment, and control of territory' (Hermann (1999) *AAAG* 89, 1). Place names mark the spatiality of power relationships (Myers (1998) *Tijdschrift* 87, 3); see also Usher (2003) *Canad. Geogr./Géog. canad.* 47, 4 on aboriginal land claims in Canada. Kearns and Berg (2002) *Soc. & Cult. Geog.* 3, 3 consider pronunciation to be an important element of the cultural politics of place naming within post-colonial societies.

It is common for revolutionary regimes to create new state symbols to 'remove evidence of the deposed regime and to establish an identity for the usurper'. During the Soviet period a special framework of laws and instructions was created to regulate a renaming policy. With the fall of the Soviet Union, renaming began in earnest; in Armenia, for example. By 1988 some 598 out of 980 place names in the Armenian SSR had been renamed.

SEE WEB LINKS
- Arseny Saparov on the alteration of place names in Soviet Armenia.

places of memory Sites where the symbolic imaginings of the past interweave with the materialities of the present;

places of memory are both constructed and contested (Rose-Redwood (2008) *Soc. & Cult. Geog.* 9, 4). K. Till in J. Agnew et al., eds (2003) explores places of memory, topographies and sites of social memory, and conflicts over national places of memory. See Forest and Johnson (2002) *AAAG* 92, 3 on Soviet memorials, and Alderman (2003) *Area* 35, 2 on street names.

place utility The significance a place has to a particular function or people. Adams and Adger (2013) *Environ. Res. Lett.* 8, 1 investigate the attachment people form to place, and the role of the environment in creating that attachment. Lin-Yuan and Kosinski (1994) *Int. Migr.* 32, 1 argue that when people evaluate places, they don't necessarily look for the one with the highest overall utility.

planation surface *See* EROSION SURFACE.

planetary boundary layer The lowest 500 m of the troposphere; the layer most strongly influenced by the land, or sea, beneath it. See Amarydis et al. (2007) *Atmos. Chem. Phys. Discuss* 7, 13537–660.

planetary winds The major winds of the Earth such as the westerlies, *trades, etc., as opposed to local winds. See Civiet et al. (2005) *ADS* on measuring planetary winds.

planeze One of a series of triangular facets facing outward from a conical volcanic peak. The planezes are separated by radiating streams which run down the flanks of the cone. Try Coltorti et al. (1999) *Geol. Soc. Special Publication* 162.

planform The outline of an object viewed from above.

plankton Minute organisms which drift with the currents in seas and lakes. Plankton *algae, various animal larvae, and some worms; the animals are **zooplankton** and the plants are **phytoplankton**. See W. Lampert and U. Sommer (1997) for a discussion of the 'top-down' versus 'bottom-up' plankton controversy.

planning As practised by local or national government, the direction of development. See S. Ward (2004) for a well-presented and easily read history of planning in the United Kingdom. S. Campbell and S. Fainstein, eds (2003) provide a valuable stimulant for constructive debate and reflection.

Planning blight is the adverse effect of a proposed development, such as a motorway, which could cause a drop in house prices. If the landowner cannot dispose of the property, or cannot make as much use of it as was previously possible, he or she may serve a purchasing notice on the planning department of the local authority. See R. Eddington (2006).

plant 1. In a system, the buildings, machinery, and land that inputs enter, and output exits.
2. In industrial geography, an individual factory producing power or manufactured goods.

plantation A large agricultural system, generally a monoculture, employing labour on a large scale to produce tropical and subtropical crops. Early stage processing often takes place on site. Plantations may be seen as a spatial expression of *imperialism and *capitalism (B. Warren 1980). See Caney and Voelks (2003) *PHG* 27, 2 on landscape legacies of plantations in Brazil, and P. Curtin (1990) on the rise and fall of the plantation.

plant community An assembly of different species of plants growing together in a particular habitat; the floral component of an *ecosystem. The concept of community can be applied to a range of scales from a small pond to the Amazon rain forest. Kent et al. (1997) *PPG* 21, 3 summarize developments in the conceptualization of the plant community. Burns (2007) *J. Biogeog.* 34, in a study of the assembly of an island plant community off the west coast of Vancouver Island, BC, finds that his results provide evidence for both the deterministic and individualistic viewpoints on community structure, and suggests that both chance and determinism contribute to the assembly of this island plant community. Duckworth et al. (2000) *PPG* 24, 4 argue that plant functional types can aid in the understanding of the assembly and stability of plant communities.

plant recruitment The establishment of new seedlings and small plants.

plant succession The gradual evolution of a series of plants within a given area. This occurs in a roughly predictable order, while the habitat changes. **Primary succession** is 'ecosystem development on barren surfaces where severe disturbances have removed most biological activity' (L. Walker and R. de Moral 2003). **Secondary succession** is the replacement of a community after a disturbance; see Halpern et al. (1997) *Ecology* 70 for a summary of the variables influencing secondary succession. In **autogenic succession** the plants themselves are the genesis of change (Ellis and Coppins (2006) *J. Biogeog.* 33, 9); in **allogenic succession** the changes are driven by forces outside the ecosystem. The two are not mutually exclusive; see Francis (2006) *Area* 38, 4.

plastic flow Movement of material, especially rocks and ice, under intense pressure, when it flows like a very *viscous substance and does not revert to its original shape when pressure is removed. As it moves, it *shears. In ice, plastic flow is due to pressure at depth; a thickness of at least 22 m is needed for plastic flow *temperate glaciers. Plastically flowing ice will flow around and over an obstacle, which may cause deposition in the lee of the obstruction: this is **plastic moulding**; see R. LeB. Hooke (2005). **Plastic deformation** is an irreversible change in the shape of material, without fracturing, resulting from compression or expansion. See Moeyersons et al. (2006) *Geomorph.* 76, 3–4 on plastic deformation in a *vertisol.

plate A rigid segment of the Earth's crust which can 'float' across the heavier, semi-molten rock below. The plates making up the continents—**continental plates**—are less dense but, at up to 35 km deep, are thicker than those making up the oceans—the **oceanic plates**—which are up to 5 km deep. A plate is a part of the *lithosphere which moves over the plastic *asthenosphere. The boundary of a plate may be a *constructive, *destructive, *conservative, or, more rarely, a *collision margin. The theory of **plate tectonics** submits that the Earth's crust is made up of six large plates: the African, American, Antarctic, Eurasian, Indian, and Pacific plates, and a number of small plates, the chief of which are the Arabian, Caribbean, Cocos, Nasca, Philippine, and Scotia plates. Hansen (2007) *Geol.* 35, 12 postulates that the ability of a terrestrial planet to evolve plate tectonics is a function of planet size.

The movement of plates causes global changes, such as *continental drift and a remodelling of ocean basins and the creation of major landforms: *oceanic ridges, *fold mountains, *island arcs, and

PACIFIC
PLATE

EURASIAN PLATE

INDO-AUSTRALIAN PLATE

AFRICAN PLATE

ANTARCTIC PLATE

NORTH AMERICAN PLATE

SOUTH AMERICAN PLATE

Collision zone

NAZCA PLATE

Plates

1 Juan de Fuca
2 Cocos
3 Adriatic
4 Aegean

5 Turkish
6 Arabian
7 Iranian

→ movement of plates

② rate of movement in cm/year

– – – uncertain plate boundary

ⴷⴷⴷ destructive margin

━━━ constructive margin

▬ ▬ ▬ collision zone

*rift valleys, together with *earthquakes and *volcanoes, which occur at a destructive plate boundary where one plate plunges below another. See Cloos (1993) *GSA Bull.* 105, 6 on lithospheric buoyancy and subduction of oceanic plateaus, continental margins, island arcs, spreading ridges, and seamounts. Plates are driven, at least partly, by forces from the convecting mantle. Oceanic driving forces are exerted when a lithospheric plate is being carried by a faster-moving *asthenosphere. *Ridge-push* is the gravitational sliding of a plate off a mid-ocean ridge, in response to the *ridge-push force*, caused by the horizontal spreading of the near-surface asthenosphere at constructive margins; see Wilson (1993) *J. Geol. Soc.* 150. Anderson (2001) *Science* 293 holds that only upper mantle convection may be driven by plate tectonics; Bokelman (2002) *Geology* 30, 11 finds that, in the case of the North American plate, the mantle plays an important role in driving the plates. The sinking of a cold, dense slab of oceanic plate provides a *slab-pull force* (see Schellart (2004) *Geophys. Res. Letters* 31, L07611). Conrad and Lithgow-Bertelloni (2002) *Science* 298 estimate the relative importance of 'pull' versus 'suction', finding that the pull from lower mantle slabs is not coupled to the surface plates. Scoppola et al. (2006) *GSA Bull.* 118, 1 observe that slab pull is 'notoriously considered one order of magnitude larger than ridge push'.

Eagles et al. (2009) *Tectonophysics* offer an animation of the Southern convergent margins, and M. Summerfield, ed. (2000) has a collection of papers on global tectonics and macroscale landform development. N. Oreskes, ed. (2001) brings together the reminiscences of the major players in development of plate tectonics theory, and C. Ollier (2000) analyses current problems related to mountain tectonics.

plateau An extensive, relatively flat upland. Some plateaus are formed structurally, from resistant, horizontal rocks; some are *erosion surfaces; some made of **plateau lavas**; see Bondre et al. (2004) *Procs Ind. Acad. Scis, Earth Planet. Sci.* 113 on the Deccan Plateau.

playa A flat plain in an arid area found at the centre of an inland drainage basin, such as the Salinas Grandes, Jujuy Province, Argentina. Within such an area, lakes frequently form. Evaporation from the playa is high and alluvial flats of saline mud form. The term is also used to describe a lake within such a basin. The deepening of the basins by deflation is favoured by salt weathering processes. See Goudie and Wells (1995) *Earth-Sci. Revs* 38; and Gutiérrez-Elorza et al. (2005) *Geomorph.* 72, 1–4.

Pleistocene An *epoch of the Quaternary period, also known as the Ice Age, from 1.8 million to ~10000 years ago.

(⊕) SEE WEB LINKS

• An overview of the Pleistocene epoch.

plinian eruption A highly explosive volcanic eruption with dense clouds of gas and *tephra propelled upwards for many kilometres. *Fissure eruptions are common, as are the destruction of the crater walls, and the fragmentation of solidified lavas. 'The magma mass discharge rate governs the eruption magnitude of plinian eruptions, and depends strongly on conduit geometry' (Varekamp (1993) *J. Volc. & Geothermal Res.* 54, 3–4). Arce et al. (2003) *Geol. Soc. Amer. Bull.* 115, 2, 230 describe the Plinian eruption of Nevado de Toluca volcano, Mexico.

plinthite A hard capping or crust at the surface of an unconsolidated soil. See D. Steila and T. Pound (1998) on

p

the stages of plinthite development, and Yaro et al. (2006) *Int. J. Soil Sci.* 1, 171.

plucking *See* QUARRYING.

plume An upwelling of molten rock through the *asthenosphere to the lower *lithosphere. The *hot spot thus formed shows up in volcanic activity at the surface. As continents move over the hot spot, there forms a chain of volcanoes at the surface, since the plume is stationary with respect to the *mantle. See Constantini et al. (2007) *J. Volc.*

plunge pool A pool at the base of a waterfall, formed by *eddying, *hydraulic action, *cavitation, and *pothole erosion. Nott and Price (1999) *Earth & Planet. Letters* 171, 2 use plunge-pool palaeo-flood records to predict the frequency and magnitude of extreme floods.

pluralism 1. Any situation in which no particular political, cultural, ethnic, or ideological group is dominant (R. Dahl 1961). When individuals affected by a given issue in the same way band together, the group empowers the individual by aggregating resources. 'Each group competes, negotiates, and compromises with other groups for influence, resources, such as money, or public support' (Eisenberg (1996) *Soc. Sci. Info.* 35, 2).
2. The cultural diversity of a plural society. Clifford (2006) *Area* 38, 4 speaks of 'pluralism, which has been much talked about but rarely practised', and Smith (2005) *Antipode* 37, 5 has a merry tilt at pluralism. Ethington and Meeker in M. Dear and S. Flusty (2002) suggest that the urban is itself plural, such that 'any attempt to view the city other than through a tolerance of multiple views cannot suffice' (Moore et al. (2003) *Area* 35, 2). R. Eversole and J. Martin (2005) observe the underlying assumption in Australia that 'a recognition of pluralism

may be able to overcome potentially conflicting political interests in the regions, or actually sustain some communities that may be becoming economically and environmentally non-viable or socially dysfunctional'. See Nash (2005) *Antipode* 37, 2 on pluralism in Northern Ireland. Mouffe uses **agonistic pluralism**—in which *the political* refers to the antagonism that is constitutive of society and *politics* refers to practices, discourses, and institutions that order and organize society but that are always open to the possibility of antagonisms because they are embedded in the political (Mouffe (2000) IHS). M. Kenny (2004) discusses the **new pluralism**, finding that the preservation of difference tends to become a condition of any group's inclusion in the political community in a way that 'dethrones the principles of democratic citizenship'.
3. An academic stance that does not focus on one approach only; see Clare and Siemiatycki (2014) *Prof. Geogr.* 66, 1, 4.

pluralist geography An 'academic debate that allows distinct approaches to engage with one another without presuming either that one must eventually "win" by dominating the other, or that anything goes' (Plummer and Sheppard (2006) *J. Econ. Geog.* 6). Smith (2005) *Antipode* 37, 5 sees pluralism as 'the core politics of neo-critical geography'.

pluton A solidified, underground, igneous mass, varying in size from *batholiths to *sills and *dykes. See McNulty et al. (2000) *GSA Bull.* 112, 1 on the Mount Givens pluton, Calif., and Stevenson et al. (2007) *J. Geol. Soc.* 164, 1 on the Eastern Mourne pluton, Ulster. A **plutonic rock** is an igneous rock which has cooled and crystallized slowly, at great depth.

pluvial A time of heavier precipitation than normal. See Pinson (2008) *Geoarchaeology* 23, 1 on evidence of more than one pluvial during the last million years. Pluvials have created current desert landforms, including wadi accretion terraces (A. Alsharhan 1998).

podzol (podsol) A soil characteristic of the coniferous forests of Russia and Canada. **Podzolization** occurs when severe *leaching leaves the upper *horizon virtually depleted of all soil constituents except quartz grains. Clay minerals in the A horizon decompose by reaction with *humic acids and form soluble salts. The leached material from the A horizon is deposited in the B horizon as a humus-rich horizon band or as a hard layer of sesquioxides. See Sauer et al. (2007) *J. Plant Nutrit. & Soil Sci.* 170, 5 for a review of the present knowledge on the development, functions, and threats of podzols in 2007.

poetics The linguistic, literary, and theoretical resources used in *representation; hence **geographical poetics**— the varieties of language in which geography has been, continues to be, and ought to be, written. See Kong (2001) *PHG* 25, 2 for an example.

poikilotherm An organism which has its body heat regulated by the temperature of its surroundings. See Lindsey (1966) *Evolution* 20 on latitudinal variations in the body sizes of poikilotherm vertebrates.

point In a *GIS, a zero-dimensional feature. Because every object takes up space, no geographic feature can really be without length or width. Instead, points can be used to represent geographic features on maps; only its name and location are needed. See M. N. DeMers (2009).

point-bar deposit The accumulation of fluvial sediment at the *slip-off slope on the inside of a *meander. Van de Lageweg et al. (2014) *Geology* 42, 4, 319 explain the formation of point bars.

point pattern analysis For an examination of the method, potential, and limitations of two classes of point pattern analysis, see Bishop (2007) *Area* 39, 2.

polar Applying to those parts of the Earth close to the poles. The **polar area**, in particular the Arctic, is considered to be the forerunner of climate change (Schiermeier (2006) *Nature* 441, 146). **Polar air masses (P)** originate in the mid-latitudes (40°–60°) and are characteristically cold.

polar front The discontinuous, variable *front which forms over the North Atlantic and North Pacific, where polar maritime air meets tropical maritime air. The formation of *mid-latitude depressions at the polar front is connected with the development of troughs in the **polar front jet stream**, a band of high-velocity winds in the wider *Rossby waves. The polar front *jet stream moves southwards in winter and northwards in summer. During an *El Niño event, it shifts south, bringing cooler air, which might suppress tornado activity in the US Midwest. Mid-latitude aridity can result from the effect of mountains on the polar jet stream; see W. G. Whitford (2002). See Dupont et al. (2005) *Geol.* 33, 6 on the sensitivity of the Namibian upwelling to latitudinal shifts in the Southern Ocean polar front zone; Ingvaldsen (2005) *Geophys. Res. Letter* 32, LI6603 on the link between the polar front and the North Cape Current; and Hayes and Zenk (1977) *NOAA Technical Report ERL PML* 28 on the Antarctic Polar Front.

    Polar easterlies blow east–west, from *polar highs to sub-polar lows. A **polar**

p

high is a mass of cold, heavy air, centred at the poles, and produced by downward, vertical air currents from the *polar vortex, bringing high pressure at high latitudes. For **polar low**, *see* COLD LOW.

polar vortex A rotating polar high pressure system, formed as a distinct column of cold air in the middle to lower stratosphere, developed during the long polar night; wind speeds may reach 100 metres s^{-1}. Upper *troposphere air is sucked into the vortex, then sinks to cause *polar highs. See Chshyolkova et al. (2007) *Angeo* 25 on the Arctic Polar vortex.

polder Land reclaimed from the sea, lakes, or river deltas; the land is bounded by a dyke, drained, and is maintained by pumping. **Calamity polders** are overflow areas, to be flooded during extreme discharges; see Roth and Warner (2007) *Tijdschrift* 98, 4.

political ecology The study of the relationships between nature and society: how political, economic, and social factors affect, and are affected by, environmental issues. These issues may address: 'mutually causal' interactions between society and land-based resources, for example that poverty is brought about by poor management, which leads to environmental degradation, which, in its turn, deepens poverty. However 'it cannot be assumed that the poor have an intrinsic propensity to degrade environmental resources' (*International Fund for Agricultural Development); the influence of political, cultural, or economic systems; for example, the cultural *eutrophication of Lake Erie (D. Kaufman and C. Franz 1993); the role of external structures; for example, the deepening poverty of Masai herders when blocked from traditional grazing (P. Blaikie 1985); the

reinforcement and misrepresentation of existing economic, gender, and racial inequalities: see Moore (2000) *Jour. Political Ecol.* 7 for an analysis of 'law's patriarchy' confronting global inequalities; see King (2010) *PPG* 34, 1, 38–55 on political ecologies of health, and Jewitt (2008) *Asia Pac. Viewpt* 49, 1 on the political ecology of the Jharkhand region of India.

Walker (2005) *PHG* 29, 73 notes that 'while political ecology has thrived, its coherence as a field of study and its central intellectual contributions remain the subject of sometimes contentious debate'. One of the recurrent, and unresolved, questions has been 'Where is the ecology in political ecology?' Vayda and Walters (1999) *Hum. Ecol.* 27 argue that 'some political ecologists do not even deal with literally the influence of politics in effecting environmental change but rather deal only with politics'. **Global political ecology** concerns itself with the globalization of environmental problems and responses through processes of financialization, generalized environmental collapse, and the mobility of expertise.

(⊕) SEE WEB LINKS

• An overview of the potential violent conflict in pastoral communities.

political economy The study of the principles that govern the production and distribution of goods under capitalist systems, which includes the roles that individuals and political entities such as states play in the production and consumption of goods. In a useful article, Sheppard (2010) *J. Econ. Geog.* 11, 2, 319 explains that geographical political economists consider a wide-ranging set of theories of the capitalist space economy, connecting agency with socio-spatial structure, through extensive multi-method case study research, as well as theoretical analysis.

political geography The study of the geographical and spatial aspects of politics; the spatial expression of political ideals and the consequences of decision-making by a political entity. Of particular interest are the geographical factors which influence voting patterns; for example, Morrill et al. (2007) *Pol. Geog.* 26, 5 on the red (conservative/republican) and blue (liberal/democrat) divide in the United States, or Buhaug (2006) *Pol. Geog.* 25, 3 on local determinants of African civil wars. M. Castells (2000) concludes that power 'is no longer concentrated in institutions (the state), organizations (capitalist firms), or symbolic controllers (corporate media, churches)... Power, as such, is the fulcrum of identity.' Agnew (2005) *AAAG* 95, 2 agrees: 'we cannot meaningfully apply the orthodox conception of sovereignty to the conditional exercise of relative, limited, and partial powers that local, regional, national, international, and non-territorial communities and actors now exert.' However, Houston (2005) *Pol. Geog.* 24, 1, writing on Turkey, finds that the production and signification of space by the nation-state is not cancelled out by the global logic of capital. For clear reviews, see Sparke: (2004) *PHG* 28, 6; (2006) *PHG* 30, 3; (2008) *PHG* 32, 3; and Staeheli (2010) *PPG* 34, 1, 67, on the role of disorder in political geography.

Studies on quantitative methodologies and spatial analyses based on GIS include Buhaug and Lujala (2005) *Pol. Geog.* 24, 4, who find that the relation of a conflict affects its duration. For critical, feminist, and popular geopolitics, see Hyndman (2004) *Pol. Geog.* 23, 3; for peace and conflict studies, see Sene (2004) *African Geopol.* 15–16, on the federating role of soccer, and territoriality.

politics of place An emphasis on the importance of place and locality as arenas through which the forces of capitalism and the state structures that support it are challenged; 'far from being bounded, homogenous, and stable entities, localities are actively produced through a politics of place that may simultaneously involve political economic processes (capital, and localized modes of political regulation), and cultural processes (contested meanings and senses of place)' (Elmhirst (2001) *Sing. J. Trop. Geog.* 22, 3; a clear and helpful paper); see also D. Massey (2005) and (2004) *Geografiska B* 86. Amin (2004) *Geografiska B* 86, 1 (unclear and verbose) detects a new politics of place, characterized by demand for localized decision-making; 'a restricted democracy'. See also Cumbers et al. (2008) *PHG* 32, 2.

polje In *karst terminology, a flat-floored, steep-sided, enclosed basin up to 65 km long and 10 km wide. Streams can be ephemeral or permanent; usually the water drains into *streamsinks. Most poljes are aligned with underlying structures such as folds, faults, and troughs. See Nicod (2003) *Acta Carsalogica* 32, 2 and Miller (2006) *GSA Sp. Paper* 406.

pollen analysis *See* PALYNOLOGY.

pollution The presence in, or introduction into, the environment of a substance that has harmful or poisonous effects. 'As far as the geographical distribution of pollution is concerned: pollution is concentrated in agglomerations, as in [major] Third World cities... environmentally harmful activities are located in peripheral regions, as is the case with nuclear power stations in industrialised countries... [and] in some cases... pollution is rather uniformly distributed over the country' (Rauscher and Bouman siti.feem.it/worldcongress/abs2/rausch2.html). See Lange and

Quaas (2007) *BEJ Econ. Analys. & Policy* 7, 1 on the effect of environmental pollution on agglomerations. **Diffuse pollution** is pollution from widespread activities with no one discrete source; for example from pesticides or from urban run-off.

((∰)) SEE WEB LINKS
• Definition of diffuse pollution.

Biodegradable pollutants like sewage, cause no permanent damage if they are adequately dispersed; see Zagorc-Koncan and Somen (1999) *Water Sci. & Tech.* 39, 10 on biodegradable industrial pollution. Land cover change and the expansion of modern agricultural practices has significantly increased leaching of chemicals to the surface waters, leading to the degradation of aquatic ecosystems worldwide (Donner (2003) *Glob. Ecol. & Biol.* 12, 4).

Non-biodegradable pollutants such as lead, may be concentrated as they move up the food chain; this is *biological magnification*—see J. Nathanson (2002). Within western Europe, air pollution, associated with basic industries such as oil refining, chemicals, and iron and steel, as well as with motor transport, is probably the principal offender, followed by water and land pollution. Other forms include noise, and thermal pollution.

Present-day problems of pollution include *acid rain and the burning of *fossil fuels to produce excessive carbon dioxide. The **pollution haven** (or ecological dumping) hypothesis states that if free trade occurs between countries with different environmental standards, countries with lower standards will develop a comparative advantage in relatively 'dirty' industries (Rauscher (1994) *Oxf. Econ. Papers* 46).

pollution dome A mass of polluted air in and above a city or industrial complex which is prevented from rising by the presence of an *inversion above it.

See Victor et al. (2004) *Annal. NY Acad. Scis* 1023. Insolation responds to pollution domes—Tomalty and Brazel (2000) *Procs Urb. Envs*, Conf. AMS, Davis, record inner-city values as 15% lower than outside values. Winds may elongate the dome into a pollution plume; see Nemecek et al. (1995) *Int. J. Rock Mech. &c.* 32, 7 on pollution plume containment. The term has been extended to describe **light pollution domes**.

polygon 1. In geomorphology a sand wedge polygon is a micro-relief landform, 7–14 m in diameter, 'devoid of vegetation and . . . outlined by textural changes in the soil' (Péwé (1959) *Am. J. Sci.* 257). Ice-wedge polygons are formed after a temperature drop so extreme that 'the ground contracts and cracks, forming a narrow void. During the winter, snow blowing across the tundra surface fills in the crack and freezes to form a thin ice vein. The following spring the ice vein thaws, but it refreezes the following winter and the ground preferentially cracks again along the same ice vein. New snow blows into the crack and freezes, gradually "wedging" the ground open' (Morgan (1972) *Canad. J. Earth Scis* 9, 6).
2. In *GIS, a closed area bounded by a connected sequence of paired x and y coordinates, the first and last coordinate pairs being the same, and all other pairs being unique. More commonly known as an area.

polysaccharide gum The sticky by-product of the decomposition of roots by micro-organisms, which can bind soil minerals into aggregates, such as *peds.

polythermal glacier A thermally complex glacier with both warm and cold ice. Warm ice usually occurs where the ice is thickest, as a result of geothermal heating, while the snout and margins of the glacier are frozen to the bed. See Roberson et al. (2011) *Geograf. Series A* 93, 71 DOI: 10.1111/j.1468-0459.2011.00420 on the physical properties of a polythermal valley glacier in Norway.

pool and riffle The alternating sequence of deep pools and shallow riffles along the relatively straight course of a river. The distance between the pools is 5–7 times the channel width. 'If alternate bars are present, the pools deform into meander bends and the riffles become points of inflection . . . however, in this paper it is suggested that pools correspond to the points of inflection in a meander system' (Leopold in R. D. Hey et al. 1998). See also Wohl and Merrit (2008) *Geomorph.* 93, 3–4.

population A group of individuals of the same species within a *community. Life tables provide a picture of survival and mortality in populations. Cohort life tables—whereby individuals are tracked from birth—are appropriate for plants and sessile organisms. Static life tables—which record age at death within a certain period of time—are appropriate for mobile and long-lived organisms. For the laws of population ecology, see Turchin (2001) *Oikos* 94, 1.

population density The ratio of a population to a given unit of area. Lagerlöf and Basher (2006) *MPRA Paper* 369 make the bold claim that population density correlates with per capita income in Canada and the United States. On firmer ground is Sato (2007) *J. Urban Econ.* 61, 2, who argues that migration from areas of lower to higher population density in Japan maintains regional variations in fertility. **Crude population density** is simply the number of people living per unit area and is not necessarily a development indicator.

population dynamics The study of the numbers of populations and the variations of these numbers in time and space. The main determinants are birth and death rates and migration; which, in turn, reflect levels of development, and socio-economic and cultural factors. See Millington et al. (1999) *App. Geog.* 19, 4 on population dynamics and socio-economic change; Bhandari and Grant (2007) *J. Rural & Comm. Dev.* 2 on population dynamics and land use; Ahn in H.-R. Kim and B. Song (2007) on population dynamics and urbanization in Korea. See also H. Macbeth and P. Collinson, eds (2002); try L. M. Hinter (2000) on the environmental implications of population dynamics. For geolinguists, see N. Komarova and M. Nowak (2002) on the population dynamics of grammar acquisition.

population ecology The study of how the population sizes of species living together in groups change over time and space. It includes population dynamics and distribution, evolutionary ecology, ecological genetics, theoretical models, conservation biology, agroecosystem studies, and bioresource management. See the *Journal of Population Ecology*.

population equation The future size of a population depends on a range of functions. Thus:

$$P_{t+1} = P_t + (B - D) + (I - E)$$

where P_t and P_{t+1} are the sizes of population in an area at two different points in time, t and $t + 1$ are those points, B is the birth rate, D is the death rate, I is the immigration, and E is the emigration. For an example of its use, see Roos (2005) *J. Econ. Geog.* 5, 616.

population geography The study of human populations; their composition, growth, distribution, and migratory movements with an emphasis on the last two. It is concerned with the study of demographic processes which affect the environment, but differs from demography in that it is concerned with the spatial expression of such processes. *Population, Space and Place* is the journal of the UK Population Geography Research Group.

population growth The rate of increase of the world's population has become ever faster:

year	population in billions
1850	1.2
1900	1.6
1950	2.55
2000	6.1
2050	(estimated) 9.2

'Future population growth is highly dependent on the path that future fertility takes. In the medium variant, fertility is projected to decline from 2.6 children per woman today to slightly over 2 children per woman in 2050. If fertility were to remain about half a child above the levels projected in the medium variant, world population would reach 10.6 billion by 2050. A fertility path half a child below the medium would lead to a population of 7.6 billion by mid-century. That is, at the world level, continued population growth until 2050 is inevitable even if the decline of fertility accelerates (Population Division, Dept. Econ. & Soc. Affairs, UN Secretariat, *2004 Revision*). *See* POPULATION PROJECTION.

Wolfgang Lutz et al. (2004) discuss expert argument-based probabilistic forecasting of population; Bongaarts and Bulatao (1999) *Pop. & Devt Rev.* 25 quantify the relative contributions of fertility, mortality, migration, and age structure in projections for all countries of the world.

population potential The accessibility of people from a given point.

population potential,

$$V_1 = \sum_{j=1}^{n} \frac{P_1}{d_{ij}} = \frac{P_1}{d_{i1}} + \frac{P_2}{d_{i2}} + \frac{P_3}{d_{i3}} \cdots + \frac{P_n}{d_{in}}$$

where the population potential (V_1) at point i is the summation (Σ) of n populations (j) accessible to the point i divided by their distance (d_{ij}) to that point. Transport costs may be used instead of distance.

population projection The prediction of future populations based on the present age–sex structure, and with the present rates of fertility, mortality, and migration. Wilson and Rees (2005) *Pop. Space Place* 11 review developments in population projection methodology.

population pyramid *See* AGE–SEX PYRAMID.

pore In geomorphology, a minute opening in a rock or soil, through which fluids may pass. Porous rocks allow water to pass through or be stored within them. **Pore water pressure** is the pressure applied by water in the pores to soil and rock particles, reducing internal friction; when the rock or soil is saturated, pore water pressure can be so great that slope failure results; see Leroueil et al. (2009), *Proc. 1st Italian Workshop on Landslides*, on slope instability due to pore water pressure increase. However, when the material dries out, only the smallest pores are water-filled, resulting in **negative pore-water pressure**. The water is held in these pores through strong suction, so that negative pore water pressure can increase the

Population projection. Population of the world 1950–2050 by projection variants

pressure, especially with clays. Negative pore water pressure in soil is referred to as soil moisture tension (or suction).

porosity The ratio of the volume of pores to the volume of matter within a rock or soil, expressed as a percentage. Iverson et al. (2000) *Science* 20 conclude that relatively small variations in porosity can influence landslide behaviour profoundly. Koulouri and Giourga (2007) *Catena* 69, 3 use soil porosity as one variable in determining soil erodibility. Chemical weathering increases rock porosity under virtually all circumstances (Gordon (2005) *Physical Geogr.*). Damody et al. (2008) *Geomorph.* 94, 1–2 use the porosity of granite to date tors. Porosity is one of the best indicators of soil structural quality.

positionality Where you are coming from must reflect your own personal viewpoint. Thus, positionality is the 'recognition and declaration of one's own position in a piece of academic work' (A. Rogers, N. Castree, R. Kitchin 2013). Gold (2002) *Ethics, Place & Env.* 5, 3 looks useful.

positive discrimination A policy designed to favour some deprived region or minority and to redress, at least in part, *uneven development. Policies of positive discrimination have been criticized for treating the effects of inequality rather than tackling its causes. Other criticisms made are that not all of the minority or region needs help, and that many deprived people are outside the catchment area. Nonetheless, the EU, for example, still uses schemes of positive discrimination, through the European Regional Development Fund; see Roberts (2003) *Prog. Planning* 59, 1. Bosher et al. (2007) *Dev. & Change* 38, 4 find examples of positive discrimination in Andra Pradesh, India.

positivism A positivist method is a philosophy of science of simply sticking to what we can observe and measure. It uses repeatable research methods, so that the same tests can be performed again as a quality control.

possibilism The theory that, although the physical environment may promote certain actions or rule them out, humans also play their part in choosing their strategies; such choices are not pre-determined by that environment. See J. Anderson (2010) pp. 42–43.

post-colonialism A sub-discipline of geography which awaits a consensus on its definition; some definitions and observations follow: 'The geographically dispersed contestation of colonial power and knowledge' (A. Blunt and C. Mc-Ewan, eds 2000). 'Much of what passes for postcolonial theory in British geography reinforces new forms of colonial epistemologies and colonial hierarchies, while destabilizing their older forms' (Gilmartin and Berg (2007) *Area* 39, 1). 'Post-colonialism studies might best be regarded now as a term for a body of diverse and often contesting formulations of the cultural production of colonized people rather than as a discipline or methodology *per se*' (B. Ashcroft et al. 2002). 'Post-colonialism has two key meanings: temporal—the period after colonialism; and critical—the study of colonialism and its aftermath' (Gilmartin (2009) in C. Gallaher et al.). J. P. Sharp (2008) provides an excellent introduction for students of colonialism and post-colonialism. **Post-colonial urban theory** is the study of the conditions of post-imperial places, particularly the examination of colonial inscriptions on cities and the reinscriptions that follow decolonization. See Short (2012) *AAAG* 102, 1, 129.

postdevelopment A political acknowledgement of the injustices and unequal power relationships of the practice of development, and a view that development cannot offer sufficient tools for transformation. Instead, it suggests resistance to *globalization provided by grassroots organizations and non-governmental organizations, alternative forms of livelihood, and local forms of knowledge and practice. See A. Ziai, ed. (2007). See also Sidaway (2007) *PHG* 31, 3.

post-Fordism A system of production and *accumulation characterized by flexibility of both labour and machinery, the vertical break-up of large corporations, small-batch production, better use of links between firms so that subcontracting is increasingly used, *just-in-time production, niche consumption, commodified culture, and the popularity of monetarist ideologies among governments. Post-Fordism is associated with *agglomeration, which will simplify interaction between linked forms of economic activity. See A. Amin (1994). Phelps (2002) *Antipode* 34, 2 considers the lineage of post-Fordist work practices. Linda McDowell (2010) *TIBG* 16, 4, 400 is somewhat brisk on 'the new gender order of post-Fordism'. *See* FLEXIBLE ACCUMULATION; FLEXIBLE SPECIALIZATION.

post-humanism A definition of humanity in which the 'human' is merely one life form located on the technological and the biological or 'green' continuum. Post-humanism rejects the definition of the human subject as 'separate and liberated from nature and fully in command of self and non-human others' (Castree and Nash (2006) *Soc. & Cult. Geog.* 7, 4, 501). Braun (2004) *Geoforum* 35, 269 provides perhaps the clearest available discussion on post-humanism in geography.

post-industrial city A city exhibiting the characteristics of a *post-industrial society. Service industries dominate with a strongly developed *quaternary sector and *footloose industries abound, often on pleasant open space at the edge of the city. Post-industrial cities are also characterized by large areas of office blocks and buildings for local government administration. These cities often exhibit marked inequality of income distribution because of the contrasts between those who are appropriately skilled—professionals, managers, administrators, and those in high-technology service industries—and the poorly paid service workers who look after their needs, together with the unemployed. The former can afford high house prices, and, in fact, contribute to them; the latter cannot. Phelps and Ozawa (2003) *PHG* 27, 5 present a taxonomy of forms of agglomeration in proto-industrial, industrial, and post-industrial urban contexts, commenting that contemporary, post-industrial agglomerations are larger than preceding forms, with an increasingly complex pattern of specialization within and between urban areas. O'Connor and Go (2013) in T. Flew (ed.) have an interesting paper on Manchester as a post-industrial city. *See also* ENTREPRENEUR; INFORMATION CITY.

post-industrial society A post-industrial society has five primary characteristics: the domination of service, rather than manufacturing, industry, the pre-eminence of the professional and technical classes, the central place of theoretical knowledge as a source of innovations, the dominating influence of technology, and levels of urbanization higher than anywhere else in the world. Mahon (2006) *TIBG* 31, 4 argues that childcare arrangements are central to post-industrial societies.

postmodernism A philosophical stance which claims that it is impossible to make grand statements about the structures of society or about historic causation, because knowledge is partial and situated, and no one interpretation is superior to another. Postmodernism in geography is expressed in the recognition that space, place, and scale are social constructs, not external givens.

Of particular interest is the way that *time and space have been 'compressed' by modern transport systems, especially by jumbo jets; hence **postmodernity**: the ever swifter movements of capital and information in postmodern society. This development has reduced the *friction of distance for most *advanced economies. D. Harvey (1989) sees *time–space compression as a fundamental social feature of postmodernism. Some geographers claim that postmodernism challenges the dominance of time and history in social theories; instead stressing the significance of geography and spatiality (see E. Soja 1989). The postmodern tradition also stresses, and indeed champions, difference.

Ley (2003) *PHG* 27, 5 notes that interest in postmodernism is waning; he approaches postmodernism as a form of local knowledge, characterized as knowledge inevitably framed through the 'here' of our collective presence and the 'now' of our collective interests. He then extends his argument to modernism and post-modernism in the built environment.

post-productivist transition A transformation in rural economies since the mid-1980s characterized by the production of food within an increasingly competitive international market; the progressive withdrawal of state subsidies for agriculture; reductions in food output; growing concerns over food quality; the growing environmental regulation of agriculture; and the creation of a more sustainable agricultural system. Wilson and Rigg (2003) *PHG* 27, 6 ask whether post-productivist notions can be applied to the developing world, concluding that the concept needs to be adapted and developed to address specific conditions in the rural South, whilst Holmes (2002) *TIBG* 27 judges that Australia's rangelands are moving from a post-productivist agricultural occupancy to a multifunctional rural occupancy.

p

In the UK the *Countryside Steward-ship Scheme attempts to bring about a transformation from a *positivist to a more post-positivist approach to agriculture. Amsted in L. Landmark and C. Sandstrom, eds (2013) takes issue with this concept. *See* THIRD FOOD REGIME.

post-socialist city Discussions of the post-socialist city can founder on the definition of post-socialism (see Andros (2001) *Anthropoids*). Lin (2004) *Eurasian Geog. & Econ.* 45, 1 argues that the interaction of the socialist legacy of industrialization 'with the new forces of marketization and globalization has given rise to a peculiar pattern of simultaneous industrialization and tertiarization, differing from the Western norm of linear progression'. Tsenkova (2008) *Urb. Geog.* 29, 78 is richly detailed; see also Hirt (2007) *Urb. Geog.* 28, 8, and try K. Stanilov (2007).

poststructuralism Poststructuralists believe that knowledge production is political, and that economic structure could not fully capture the human experience; therefore, they focus on the way the social categorization of dominant and weaker groups is normalized and rationalized through language; the study of dominance and 'otherness', for example, is often based on the way people are defined in society and how they maintain, resist, subvert, or nuance those identities. For geographers, it has confirmed the *spatiality of social life and the social construction of *space. See J. Munroe (2006) on the amalgam of socio-biological-physical processes that generate the 'geographies' that are commonly discussed as 'geography', and M. A. Doel in P. Cloke et al., eds (2004); also try M. Doel (1999) with a sound review of the different poststructural writers.

potential evapotranspiration (PET) The maximum continual loss of water by evaporation and transpiration, at a given temperature, given a sufficient supply of water. PET often outstrips actual evapotranspiration. Trudgill and Briggs (1981) *PPG* 5 consider the significance of potential evapotranspiration in soil and land potential.

pothole Loosely, a vertical cave system; more precisely, a more or less circular hole in the bedrock of a river. The hole enlarges because pebbles inside it collide with the bedrock as the water swirls. Potholes carved into streambeds can be important components of channel incision. Springer et al. (2005) *Geomorph.* 82, 1–2 demonstrate that the pothole dimensions of radius and depth are strongly correlated.

poverty 'A pronounced deprivation in well being.' A. Sen (1985, 1993) argues that what is important to well-being is the 'capability to function in society'. These capabilities include access to food, shelter, clothing, education, security, and good health: capabilities which the poor lack. The poor also lack a voice in society, are vulnerable to risk, and are powerless.

'Over the past 30 years, the world has undoubtedly registered impressive gains in the average life expectancy of its population—from 60 years at birth, to 67 years. Infant mortality has decreased substantially—from 96 to 56 per 1,000 live births' (Bidwai (2006) *Development Dialogue* 47, 31). The 2005 UN Human Development Report explains that these changes are partly a product of falling child death rates, improvements in access to water and sanitation, and global immunization. There is also evidence that literacy levels have risen.

There have been setbacks: eighteen countries registered lower scores on the HDI in 2003 than in 1990. Twelve of

Percentage of people living on less than $1 and $2 per day

	\$1.08 per day at 1993 PPP							
	1981	1984	1987	1990	1993	1996	1999	2001
East Asia	57.7	38.9	28.0	29.6	24.9	16.6	15.7	14.9
of which China	63.8	41.0	28.5	33.0	28.4	17.4	17.8	16.6
Eastern Europe and Central Asia	0.7	0.5	0.4	0.5	3.7	4.2	6.3	3.7
Latin America and Caribbean	9.7	11.8	10.9	11.3	11.3	10.7	10.5	9.5
Middle East and North Africa	5.1	3.8	3.2	2.3	1.6	2.0	2.6	2.4
South Asia	51.5	46.8	45.0	41.3	40.1	36.6	32.3	31.3
of which India	54.4	49.8	46.3	42.1	42.3	42.2	35.3	34.7
Sub-Saharan Africa	41.6	46.3	46.8	44.6	44.0	45.6	45.7	46.9
Total	**40.4**	**32.8**	**28.4**	**27.9**	**26.3**	**22.8**	**21.8**	**21.1**

	\$2.15 per day at 1993 PPP							
	1981	1984	1987	1990	1993	1996	1999	2001
East Asia	84.8	76.6	67.7	69.9	64.8	53.3	50.3	47.4
of which China	88.1	78.5	67.4	72.6	68.1	53.4	50.1	46.7
Eastern Europe and Central Asia	4.7	4.1	3.2	4.9	17.2	20.6	23.7	19.7
Latin America and Caribbean	26.9	30.4	27.8	28.4	29.5	24.1	25.1	24.5
Middle East and North Africa	28.9	25.2	24.2	21.4	20.2	22.3	24.3	23.2
South Asia	89.1	87.2	86.7	85.5	84.5	81.7	78.1	77.2
of which India	89.6	88.2	87.3	86.1	85.7	85.2	80.6	79.9
Sub-Saharan Africa	73.3	76.1	76.1	75.0	74.6	75.1	76.0	76.6
Total	**66.7**	**63.7**	**60.1**	**60.8**	**60.1**	**55.5**	**54.4**	**52.9**

Source: Chen & Ravaillon (2004)

these are in sub-Saharan Africa and the former Soviet Union accounts for the other six (ibid.). Chuhan (*Glob. Monitoring Re. 2006*) adds that 'over three-quarters of a billion people, many of them children are malnourished. Each year nearly 11 million children die from malnutrition or disease before reaching their fifth birthday . . . and over half a million women die every year in childbirth from lack of appropriate health care, malnourishment, or disease. More than 100 million children of primary school age, the majority of them girls, do not attend school.'

Poverty is not restricted to the developing world: in inner London in 2006, 51% of children were living in poverty and US Census Bureau records show that the proportion of Americans living in poverty rose between 1974 and 2004, from 11.2% to 12.7%. For the period 2000–2005 the Cuban infant mortality rate was 8/000; the same as for the USA (*World Mortality Report*, 2005). 'The only positive interpretation that can be given to this picture of poverty and inequality in the developed world is that many of the poor are recent arrivals from even poorer places in Africa, Asia and Latin American countries' (Gilbert (2007) *Geog. Compass* 1, 3).

SEE WEB LINKS
- World Bank Poverty and Equity data.
- Official website of the United Nations Development Programme.
- Oxford Poverty and Human Development Initiative.
- The Human Development Report.

power The power structure of a society is reflected in its social organization and in its economy: *capitalism is controlled by a minority who dominate the *factors of production. In a *centrally planned economy it is the state that dominates, and these have their own spatial expression; see Boudreau (2007) *Env. & Plan. A* 39, 11 on making new political spaces.

Geographies of power study the spatial distribution of power across a multiplicity of geographical scales, from the body to the globe. The mix of scales changes dynamically, as territories are continually subjected to resistance, contestation, renegotiation, *deterritorialization, and reterritorialization; see, for example, Mawdsley (2007) *Geog. Compass* 1, 3 on China and Africa as emerging challenges to the geographies of power.

Themes include state formation, where power is transmitted through social channels to territorial ones; see, for example, Jessop (2007) *Pol. Geog.* 26, 1 and the relationship of minority cultures and empowerment, as in Mavroudi (2008) *Pol. Geog.* 27, 1, on Palestine. See Kobrin (2001) in A. Rugman and T. Brewer (2003).

power law A mathematical relationship between two quantities. When the frequency of an event varies as a power of some attribute of that event (e.g. its size), the frequency is said to follow a power law. Thus, Hack's law (Hack, (1957), *US Geol. Surv. Prof. Pap.* 294-B:45) states that the length *l* of the main stream of a river basin scales like a power *h* of the basin's area *a* :

$$l \propto a^h$$

and Li et al. (2011) *Geomorph.* 130 (2011) 221 find a power law relationship between rainfall and the frequency of landslides in Zhejiang province, China.

prairie A large area, found outside the tropics, with grassland and occasional trees as natural vegetation, as in the prairies of North America, the South American pampas, the Russian *steppes, and the South African veld. A **prairie soil** is a soil of the wetter prairies, resembling *chernozem in its high humus content and its development under grassland. Prairie soils are *leached of calcium, and

are slightly acid. See Barrett and Randall (1998) *Soil Sci.* 163, 6.

praxis Often designating accepted practice, at its most simple, praxis is practice as opposed to theory; Kitchin et al. (2013) *Dialogues Hum. Geog.* 3, 1, 56 argue that new social media produces new forms of public geography and digital praxis.

Pre-Boreal A warmer division of the *Flandrian. For the rapid climate change at the Younger Dryas–Pre-Boreal transition, see Starkel (2003) *Glob. & Planet. Change* 35, 1.

Precambrian The oldest *era in Earth's history dating from about 4600 million years BP. 'To speak of "the Precambrian" as a single unified time period is misleading, for it makes up roughly seven-eighths of the Earth's history' (UCMP).

(())) SEE WEB LINKS
• The divisions of Precambrian time.

precautionary principle In 1992, the Rio Declaration on Environment and Development, stated that 'where there are threats of serious or irreversible damage, lack of full scientific certainty shall not be used as a reason for postponing cost-effective measures to prevent environmental degradation'. In other words, if a project might well have adverse effects, these can be evoked as a reason for abandoning it, even if the scientific evidence is not totally convincing.

precession The wobbling of the Earth on its polar axis, which has a return period of 26000 years. This change in movement and orientation alters the amount and location of solar radiation reaching the Earth.

precipitation In *meteorology, the deposition of moisture from the atmosphere onto the Earth's surface in the form of *rain, *hail, *frost, *fog, sleet, or *snow. Initially, cloud droplets grow around nuclei through condensation and diffusion. In warmer clouds the larger droplets then grow by collision and coalescence with the smaller ones. In colder clouds the *Bergeron–Findeisen mechanism is thought to operate, probably in conjunction with the growth of ice crystals: through *accretion, as supercooled water droplets freeze on impact with the ice; and aggregation, as smaller ice crystals stick to larger ones. Much precipitation begins in the form of ice crystals, but melts as it falls, to become rain. Variations in the intensity, amount, timing, duration, and frequency of precipitation, whether frozen or liquid, have important implications on the physical, chemical, and biological processes on Earth. See Zhang et al. (2007) *Nature* 448 on *anthropogenic precipitation; Holmer (2007) *Geograf. Annal.* 89, 4 on rainfall change; Halfon and Kutiel (2007) *Eur. Geosci. Union* 9 on precipitation mapping; and Malby et al. (2007) *Hydrol. Scis J.* 52, 2 on long-term variations in *orographic rainfall.

predator–prey relationships In theory, there should be an equilibrium between predators and prey. Thus, when predators are scarce, the numbers of prey should rise. Predators would respond by reproducing more and, possibly, by changing their hunting habits. As the population of predators rises, more prey is killed and their numbers fall. Many of the predators then die; thus numbers of predators and prey oscillate between two extremes. See Gilpin and Rosenzweig (1972) *Science* 177, 1358 for the classic model. Gilg et al. (2003) *Science* 302, 866 should be useful.

The **keystone predator effect** predicts that prey that are strong resource competitors are expected to dominate at

low levels of potential productivity; augmenting prey resources permits less efficient competitors but more predator-resistant prey to invade and dominate. The underlying trade-off among prey can be termed the 'keystone trade-off' (Leibold (1996) *Am. Nat.* 147). See Steiner (2003) *Oikos* 101, 3.

pre-industrial city A model of the pre-capitalist city, proposed by G. Sjöberg (1960). The city centre is occupied by a small elite. The lower classes occupy the concentric zone surrounding the centre, and the outcasts are consigned to the outer edges of the city. Abbott (1974) *Amer. Soc. Rev.* 39, 4 finds that the distribution pattern of socio-economic status in pre-revolutionary Moscow was consistent with Sjöberg's model.

preservation The protection of human features in the landscape, as opposed to *conservation which is concerned with the protection of the natural landscape. This distinction is not always made; see, for example, Shafer (2004) *Lands. & Urb. Plan.* 66, 3 on the voluntary preservation of America's natural heritage.

pressure gradient The rate of change in *atmospheric pressure between two areas, providing a force which moves air from *high to *low in an effort to even up the unequal mass distribution of the air. On a global scale, the most powerful pressure gradients are in a *meridional direction, caused by meridional disparities in *insolation. See Bacon and Carter (1993) *Int. J. Climat.* 13 on the connection between mean wave height and atmospheric pressure gradient in the North Atlantic, and McPherson (2007) *PPG* 31, 3. The **pressure gradient wind** is the movement of air in response to pressure differences, blowing from high to low. It is modified, however, by the action of the *Coriolis force. The term is also used in fluvial geomorphology; see Powell (1998) *PPG* 22, 1.

pressure melting point In glaciology, the temperature, often well below 0 °C, at which ice under pressure will melt. In basal ice, flow at the pressure melting point is partly from basal sliding and partly from shear deformation (Zwally et al. (2002) *Science* 297). Warm glaciers have bases at or above the pressure melting point of their ice.

pressure release The expansion of a rock formed under pressure when that pressure is released (also known as *dilatation*). A glacier may remove the overburden, and the revealed rock 'bursts' open; see the classic Lewis (1954) *J. Glaciol.* 2. Twidale (1973) *Rock Mech. & Rock Eng.* 5, 3 attributes sheet jointing to pressure release.

price mechanism The regulation of economic activity by supply. Hayek (1945) *Amer. Econ. Rev.* 35 described the price mechanism as 'that division of labour on which our civilization is based'. The spatial impossibility theorem tells us that price mechanism does not work in a homogeneous spatial economy (Ottaviano and Thiesses (2005) *Env. & Plan. A* 37, 10. **Pricing policies** are the arrangements whereby prices of commodities to the consumer are determined; see, for example, Arentze and Timmermans (2007) *J. Transp. Geog.* 15, 1.

primary industry (primary activity) Economic activity, such as fishing, forestry, and mining and quarrying, concerned with the extraction of natural resources.

primate city The largest city within a nation which dominates the country not solely in size—being more than twice as large as the second city, as in London and Birmingham, UK—but also in terms of influence. Wani Rashid and Kairkhar (2010) *Int. J. Env. Sci.* 2, 4 are downbeat on Srinagar, which they describe as a primate city. Tammarau (2000) *Tijdschrift* 91, 1 explores this concept in centrally planned, developed, and developing economies.

principal components analysis (PCA) Probably the oldest and best known of the techniques of multivariate analysis, this procedure reduces the dimensionality of a data set in which there are a large number of interrelated variables, while retaining as much as possible of the variation present in the data set. This is achieved by transforming to a new set of variables, the principal components, which are uncorrelated, and which are ordered so that the first few retain most of the variation present in all of the original variables. Demšar et al. (2013) *AAAG* 103, 1, 106 provide an excellent overview of PCA in spatial settings.

private space Spaces, such as private properties, that are not open to all. Bondi (1998) *Urb. Geog.* 19, 2 shows how divisions between public space and private space operate at different scales and take different forms in different neighbourhoods. These forms illustrate how gender and class are interwoven in demarcations between and connotations of public and private spaces.

process model 1. A way of looking at the world based on the processes involved, rather than starting with the object. 'These phenomena include migration, erosion, urban sprawl, percolation, and development, which form the basis of a geographer's education in processes as opposed to objects. Modelling such phenomena as a collection of objects that undergoes change upon a spatial plain ignores that change defines the process. An alternative basis for modelling spatio-temporal phenomena is proposed that involves developing a conceptual framework that takes change as its core, a framework defined by process' (Femke Reitsma, Department of Geography University of Wisconsin, Milwaukee). This means that a process model is not based on a research question. Hollyer explains this approach very clearly at Geosoft.com.

   2. A process-based software program model. Wergen et al. (2012) *Geomorph.* use a 3-D morphodynamic, process-based numerical model (Delft3D) to investigate the processes governing the evolution of major morphological features in alluvial estuaries as they evolve through the interaction between tidal movement, the available sediments, and the geometry of the tidal basin.

(⊕) SEE WEB LINKS

- Geosoft website with Hollyer explanation of process model approach.
- Delft3D Open Source Community website will give all you need for fluvial, marine, and sedimentary processes.

process–response system A natural system in which energy and matter flow through a set of entities arranged in a specific manner. The flows interact with the entities; there is a two-way feedback between process and form. Castedo et al. (2011) *Geomorph.* 177–78, 128 give an excellent example of a process–response coastal recession model of soft rock cliffs.

producer An organism which can fix energy from the sun and transform it by photosynthesis into food. See Olff et al. (2000) *Glob. Ecol. & Biogeog.* 9, 4 on producers and consumers. **Producer services** are services such as accountancy, marketing, advertising, market testing, and commercial lending provided by one business to another. *See* OUTSOURCING.

production In ecology, the increase of body mass as food is converted into new living material.

production, geographies of The study of global divisions of labour; production as instituted process; regional worlds of production, and the dynamic network/territory interface. 'Changing geographies of production are a product of the inter-play of (inter alia) corporate, state, and trades union strategies, as companies pursue profitability, trades unions and workers seek new employment and/or protect existing jobs, and states attempt to balance the pursuit of accumulation in their territory with the claims of equity and socio-spatial justice' (Hudson *TIBG* (2002) 27, 262). Coe, *PHG* 36, 3, 389 provides a really useful A-Z of the distinguishing features of **Global Production Network** analysis.

productivist Describing an agricultural system which is *intensive, expansionist, and based on the expansion of world trade in food, ever-increasing farm sizes, and the use of technology to increase output; factory farming and *agribusinesses are both examples. Walford (2003) *J. Rural Studs* 19, 4 reports that commercial farmers continue to intensify and specialize; in Brittany, for example, the removal of hedges has decreased *infiltration and *groundwater storage (Douguet and Schembri (2000) CRITINC *W. Paper* 13). Annually, some 30000 tonnes of nitrates are leached into British rivers (Harvey (1998) *TIBG* 4), and slurry from livestock has added to pollution; see Willis (2001) *Asia-Pac. Viewpt* 42, 1. See also Jay (2007) *Food Policy* 32, 2.

productivity The output of an economic activity, in terms of the economic inputs. Fischer et al. (2007) *SSRN*, illustrate the 'central link between productivity and knowledge capital', and

Broersmar and van Dijk (2007) *J. Econ. Geog.* 8 find that investing in highly dense regions increases congestion, thereby reducing productivity. Some argue that social and environmental 'costs' must also be considered; see Compassion in World Farming Trust (1999) on factory farming, for example, and the cost of raw materials is also the depletion of finite resources—see Ehui et al. (1990) *J. Env. Econ. & Manage.* 18, 2. A **productivity rating** is an estimate of an area's ability to support plant growth; see Yang et al. (2005) *Rev. Agric. Econ.* 27, 1 on conservation reserve enhancement.

proglacial Situated in front of a glacier. A **proglacial lake** is formed between the terminus of the ice and the higher ground which is often in the form of a terminal *moraine. Glacial Lake, Wisconsin, was a large proglacial lake. Glacial recession caused the ice dam to fail catastrophically and the lake drained rapidly (Clayton and Knox (2008) *Geomorph.* 93, 3).

progradation The accumulation of beach material when there is an excess in the supply of sediment; see Goy et al. (2003) *Geomorph.* 50, 1–3. Progradation is a feature of, for example, *delta and *mangrove coasts; Hori et al. *Geomorph.* 41, 2 find that the average progradation rate of the Yangtze delta rose from 50 to 80 km/year about 2000 years BP.

project In economic geography, 'a temporary system with institutionalised termination' (Lundin and Söderholm (1995) *Scand. J. Manage.* 11); a 'one-off' activity, such as shipbuilding, film-making. See Grabner (2002) *Env. & Plan. A* 34, 11 and Kloosterman (2008) *J. Econ. Geog.* 8, 4 on **project-based organizations**.

projection *See* MAP PROJECTION.

protalus rampart A thin ridge, about a metre high and many metres long, composed of rock fragments, running along the foot of a mountain face and thought to be produced from rock fragments sliding over a snow surface (Curry et al. (2001) *Procs. Geols. Ass.* 112, 4). Also known as **pronival rampart**: Margold et al. (2011) *Geograf. Series A*, 93, 2, 137.

protection 1. The imposition of *quotas or *tariffs on foreign imports by a government: to become self-sufficient, to protect new industries, or as a bargaining tool. See Bai et al. (2003) *W. Davidson W. Paper* 565, and Young (2000) *Quart. J. Econ.* 115, 4 on China.

2. A form of copyright; European legislation has provided legal protection to designated regional foods. See Parrot et al. (2002) *Eur. Urb. & Reg. Studs* 9, 3, 241.

protest, geographies of A protest registers an objection, either expressed individually or organized collectively: in India, for example, *satyagraha* describes the practice of non-violent protest that also involves a quest for truth. These days, internet-based protests are adding to and extending more traditional forms of objection. Geographies of protest are concerned with: mapping the distribution of protests; the detailed disposition of protestors and authorities; and the protest camps and occupations themselves.

proximity Nearness in space, time, or relationship. Oerlemans et al. (2001) *Tijdschrift* 92, 1 explain the **proximity effect** as benefiting from localized ties. Abramovsky and Simpson (2011) *J. Econ. Geog* 11, 6, 949 find evidence of co-location (within 10 km) of R&D facilities in pharmaceuticals with high research-rated chemistry departments, consistent with geographically localized knowledge spillovers and the importance of accessing academic knowledge for pharmaceutical firms.

proxy (proxies) Datum/data on areas that do not permit direct measurement; for example, in studies of past climates where tree-rings, borehole temperatures, and pollen abundance are used. Different proxies preserve the climate information in different ways and are sensitive to climate variables at different time scales.

psychic income The enjoyment, which cannot be measured in financial terms, that people derive from location or particular place (G. Becker 1964). See Reardon and McCorkle (2002) *Int. J. Retail & Distribution Manage.* 30, 4 on psychic income and shopping. Garnham (1996) *Festival Manage. & Event Tourism* 4 argues that major events improve self-image and raise morale, but Olds (1998) *Festival Manage. & Event Tourism* 11 suggests that the psychic income of a hosting community is not always positive.

psychogeography The study of specific effects of the geographical environment, consciously organized or not, on the emotions and behaviour of individuals; the hidden landscape of atmospheres, histories, actions, and characters that charge environments. See I. Sinclair (2003).

public administration, geographies of The study of the roles of space and place in contemporary public administration; the implications for public policy-making of the reconfigurations of spaces and places; and the use of information and communications technology to redesign organizations (for example, the e-Europe Action Plan). See Baltodano (1997) *J. Pub. Admin. Res.* 7, 6. Historically, the organization of government activity has been based on co-location of actor (public administrator), act (administration), and geographical setting (administrative unit); see Painchaud (1977) *Publius* 7.

Globalization brings about new spatial forms of organization and processes in a 'virtual world', and these new spatial forms of organization and processes represent new 'public spaces'; see Toonen (1998) *Public Admin.* 76, 2. The key criteria in determining the appropriate sizes of administrative units may well be a sense of geographical identity for the 'citizens', and equity in resource distribution and access to services. See S. Crate (2003) and Ahas et al. (2005) *Futures* 37, 6.

public choice theory The behaviour of public-sector bureaucrats is at the heart of public choice theory. Hospers (2005) *Intereconomics* 40, 6 discusses two types of policy-maker: offensive and future oriented, or traditional and preserving, and plumps for the latter. Parlow (2008) *U. Colo. L. Rev.* argues that public choice theory is flawed.

public geography Geography for, by, or with the non-academic public. This may include making geographical research more accessible, making geographical knowledge together with members of the public, and perhaps giving greater recognition to 'lay' geographical knowledge in its own right. But Fuller and Askins (2011) *PHG* 33, 4, 654 explain that definitions of this term can be more complicated than the one here.

public goods Goods freely available, either naturally, like air, or from the state, like education in most developed countries. Krause and Krause (2011) *Urb. Studs* 49, 11, 2399 argue that climate protection is a global public good. See also R. Baldwin (2005).

publics Small groups of people who follow one or more particular issue very closely. Different kinds of publics occupy different kinds of spaces: 'for example, a public space that would foster a liberal-economistic version of publicity might be one that facilitates state functioning and makes it easier for the state to safeguard and regulate individual rights and to promote economic development and growth' (Staeheli and Mitchell (2007) *PHG* 31, 6). 'Publics are deeply embedded in social and machinic complexes involving the mobilities of people, objects, and information' (Sheller (2004) *Env. & Plan. D* 22, 1). For Fuller and Askins (2007) *Antipode* 39, 4 publics, while multiple and in flux, are also created.

public sector That part of a national economy owned and controlled by the state, including nationalized industries, national and local government services, and public corporations. Mrinska (2007) *Northern Economic Agenda*, paper 2, concludes that public intervention is necessary to revive the economy of economically depressed regions, resulting in increased employment and spending, which should stimulate growth in the private sector.

public space The space that the public values: 'space to which it attributes symbolic significance and asserts claims. Citizens create meaningful public space by expressing their attitudes, asserting their claims, and using it for their own purposes' (Goheen (1998) *PHG*, 22, 4). These purposes include getting drunk and urinating/vomiting; see Jayne, Holloway, and Valentine (2006) *PHG* 30, 4, 451. At a more elevated level, see Springer (2011) *Antipode* 43, 2, 525, and Ramadan (2013) *Eur. Urb. & Reg. Studs* 20: 145 on Tahrir Square, Cairo.

Public spaces, including parks, squares, and markets, are 'co-produced', and the key principles for their development include leaving room for self-organization, diversifying activities to encourage diverse people to participate, and maintaining access and availability (M. Mean and C. Tims 2005). K. Worpole and K. Knox (2007) conclude that strategies intended to 'design out crime', such as cutting down bushes, installing vandal-proof street furniture, and closing public toilets, affected their attractiveness and damaged the usefulness of public spaces; see also Klauser (2007) *Eur. Urb. & Reg. Studs*, on using CCTV to regenerate public space. Veronis (2006) *Env. & Plan. A* 38, 10 finds that the use of public space is central to the cultural politics of migrants in Toronto; Ortiz et al. (2004) *GeoJournal* 61, 3, report on women's use of public space in Barcelona; and Mackintosh (2007) *J. Hist. Sociol.* 20, 1–2 writes on the engagingly titled bourgeois geography of bicycling.

pull factor A positive factor exerted by a locality that people move to. These have included: new land grants for farmers (the Great Plains; Gutmann (2000) *Climatic Change* 44, 3), assisted passages and other government inducements (for Australia, see P. Burroughs and A. J. Stockwell 1998), freedom of speech or religion (see N. Philbrick 2006), and material inducements (Hong Kong; see Lin (2002) *Asia Pac. Viewpt* 43, 1).

pumice A very light, fine-grained, and cellular rock produced when the froth on the surface of lava solidifies. See Calcaterra et al. (2007) *Geomorph.* 87, 3. A **pumice raft** comprises pumice fragments floating on the sea; one was spotted in 2006 (*Geol. News*, Nov. 2006).

pumped storage scheme Electricity cannot be stored, so when demand is low, at night, some can be used to pump water from a lower to a higher reservoir. At peak demand, the water is allowed to fall back to the lower level, passing through turbines which turn generators. See Bruch (2007) *UNEP Dams and Development*, 92–3 on the Palmiet Pumped Storage station, South Africa.

push factor In *migration, any adverse factor which causes out-migration. Examples include: famine, changes in land tenure (the Highland Clearances, 1790–1850, A. MacKenzie 1999), political persecution (Tamil separatists, Sri Lanka, Stokke (2000) *Growth & Change* 31, 29), and mechanization which made agricultural workers redundant and which made factory products cheaper than those of cottage industry (see Pisani and Yaskowitz 2002, *Soc. Sci. Quarterly* 83, 2 on rural depopulation in Portugal). Relatively few migrations are spurred by push factors alone. See Barnett and Adger (2007) *J. Polit. Geog.* 15, 4 on climate-change induced migration.

push moraine *See* MORAINE.

puy The French term for a volcanic neck, revealed by differential erosion. The type location is the Puy de Dôme. The Hohentweil, Hegau, Germany, is a further example.

pyramidal peak Synonym for horn or aiguille.

pyramid of numbers (ecological pyramid) A diagram of a *food chain which shows each *trophic level as a horizontal bar, drawn in proportion to the *biomass. There is a large fall from *producers and primary *consumers; thereafter the decreases are smaller. The animals on the higher levels are generally larger and rarer than animals lower down the pyramid. See Kent (2000) *PPG* 24.

pyroclast A fragment of solidified lava, ejected during explosive volcanic eruptions. Classification of pyroclasts is by size. Fragments less than 4 mm across are ash; compacted ash is *tuff*. Material between 4 and 32 mm is *lapili* and fragments larger than 32 mm are *blocks*. Collectively, these fragments are *tephra. Pyroclasts formed from lava produce volcanic bombs and volcanic breccia. **Pyroclastic flows** are also known as *nuées ardentes. They result from the bursting of gas bubbles within the magma, which fragments the lava. Eventually, a dense cloud of fragments is thrown out to form a mixture of hot gases, volcanic fragments, crystals, ash, pumice, and shards of glass. See Adas et al. (2006) *GSA Bull.* 118, 5.

p

Q

qualitative Concerned with meaning, rather than with measurement. The emphasis is on subjective understanding, communication, and empathy, rather than on prediction and control, and it is a tenet that there is no separate, unique, 'real' world. All qualitative researchers are positioned subjects; as such, the rigour of their research depends not only on the suitability of the methodology, the use of multiple methods, and the inclusion of verbatim quotations, but also on its credibility and transferability. Thomas (2007) *Prof. Geogr.* 59, 4 suggests that researchers must avoid psychoanalysing research subjects: 'this limits the ways in which scholars can "read" personal narratives for unconscious processes.' Bradshaw (2001) *Area* 33, 2 outlines 'some choices which might help address possible difficulties in qualitatively researching the powerful'. For methods in qualitative research, see Davies and Dwyer (2007) *PHG* 31, 2; 2008, *PHG* 32, 3; and for a checklist for evaluating qualitative research, see Baxter and Eyles (1997) *TIBG* 22, 4. See Suchan and Brewer (2000) *Prof. Geogr.* 52, 1 on qualitative methods in cartography.

quality of life The well-being of individuals and societies; 'either perceived or as identified by observable indicator' (Pacione in R. Kitchin and N. Thrift 2009). See the UN Development Index and K. C. Land, A. C. Michalos, and M. J. Sirgy, eds (2013).

quantification The numerical measurement and analysis of processes and features.

quantitative revolution In geography, the intellectual movement beginning in the 1950s that explicitly introduced to the discipline scientific forms of theorizing and techniques of empirical verification (Barnes (2004) *PHG* 28, 5), transforming geography into an analysis-oriented scientific discipline. The quantitative revolution is generally considered to have emerged from a general dissatisfaction with regional geographic study, and a consequent shift in focus towards more systematic and specialized approaches (Keylock (2003) *TIBG* 28, 2). Barnes (1998) *Env. & Plan. A* 30 argues that the success of this new geography was due to the manner in which it was able to collapse an initially disparate set of concerns into a single framework (the regression or multiple regression equation), to which Bracken and Wainwright (2006) *TIBG* 31, 2 would add the use of concepts 'the Western modernist psyche finds easy to accept'. See Johnston and Sidaway (2004) *PHG* 28. Quantification began, flourished, and was, in turn, criticized: 'the dominant paradigm at the time was that the western experience of urban-industrial development held the key to progress everywhere, and the challenge was to overcome any obstacles to the diffusion of ideas and innovations that would hinder this transformation . . . much that characterised the quantitative revolution was about theorising and employing statistical procedures in the office to generalise about change everywhere' (Bedford (2005) *Asia Pac. Viewpt* 46, 2). Quantification was attacked for being unrealistic and bloodless, turning humans into automata, for being too deterministic, and for ignoring the importance of subjective experience. T. J. Barnes (1998) attributes the reduction in emphasis upon quantification in more recent human geography to the growing awareness that techniques built upon an assumption of statistical independence are not strictly applicable to situations where spatial interdependence is apparent. Was the quantitative revolution a conspiracy? No, says Johnston (2007) *TIBG* 32, 3.

quarrying 'The removal of pre-loosened bed material and/or material resulting from bed failure' (Rea and Walley (2006) *Earth Surf. Procs & Landfms* 19, 7). Since the tensile strength of ice is low, *glacial quarrying is not possible unless the rock is shattered.

Glasser and Bennet (2004) *PPG* 28, 1 observe that there is now general agreement that quarrying is favoured beneath thin, fast-flowing ice, and Drake and Shreve (1973) *Procs Royal Soc. London, Ser. A* 332, propose a heat pump effect: water that is melted in high-pressure areas flows away and does not refreeze at the immediately adjacent low-pressure area, and this leaves cold patches that advect downglacier. Fluctuations in basal water pressure also play an important role in the formation of glacially quarried landforms; see Glasser and Bennet (2004) *PPG* 28, 1. Landforms of glacial quarrying include roches moutonnées, rock basins, and zones of *areal scouring.

Quaternary (Pleistogene) The most recent period of geological time,

covering the last 2 million years, and including the *Pleistocene and the *Holocene.

quaternary industry (quaternary activity, quaternary sector) Economic activity concerned with information: its acquisition, manipulation, and transmission. Into this category fall law, finance, education, research, and the media. See D. Ley (1996) on the middle classes and the expansion of quaternary industry.

queer geographies A major concern of queer geographies has been the critical role of place and space in the production of sexual identities, practices, communities, subjectivities, and embodiments, and the scholarship of queer space includes its location, nature, and definition, its history, memories, events, and subcultures, and the relationship between space and social justice. This is examined by looking at spaces such as closets, communities, and cruising grounds; 'finding or creating spaces in which to know ourselves and become known to others' (Sember (2003) *Space and Culture* 6, 3). 'Through repeated acts of "coming out", gay men and lesbians map where and how to be 'out' and when and where not to be. These environments are largely concealed and require careful reading' (Space, Sexualities and Queer Working Group of the RGS). Correspondingly, there is a focus on the centrality of sexualities in constituting social space. See Gorman-Murray (2007) *Soc. & Cult. Geog.* 8, 1 on rethinking queer migration through the body.

Podmore, in a study of Montreal lesbian territory (2006, *Soc. & Cult. Geog.* 7, 4), asserts that, although gay men have often produced clearly visible territorial enclaves in inner-city areas (as in Manchester or San Francisco), lesbian spaces have been comparatively invisible 'since

q

their communities are constituted through social networks rather than commercial sites'. Further, Nast (2003, *Antipode* 34) argues that 'gay white patriarchies' exist, which depend 'structurally and implicitly upon white supremacy and heteropatriarchy'. Oswin (2004) *Acme* 3, 2, also, warns against an empiricist, cognitive reading of the affluent, white, gay male. Binnie and Valentine (1999) *PHG* 23, 2 believe that queer geographies need to move away from simple mapping of lesbian and gay spaces towards a more critical treatment of the differences between sexual dissidents. Knopp (2007) *Prof. Geogr.* 59, 1 focuses on the potential of feminist-inspired and allied queer geographies to rethink a variety of spatial, and other, ontologies. See also Brown and Knopp in Anderson et al., eds (2003, *Antipode* 34). Hubbard (2000) *PHG* 24, 2 writes that 'everyday space' is experienced as aggressively heterosexual by lesbians and gay males; heterosexuality has served to create and justify other forms of oppression and confinement in Western cities.

A. Tucker (2009), on queer geography in the Republic of South Africa, is recommended.

Queer theory challenges the idea of the preconstituted sexual subject and understands power as productive rather than simply oppressive: 'yet critical geographers generally depict queer spaces as spaces of gays and lesbians or queers existing in opposition to and as transgressions of heterosexual space' (Oswin (2008) *PHG* 32).

quickclay Clay whose structure collapses completely on remoulding, and whose shear strength is thereby reduced to almost zero (Eilertsen et al. (2008) *Geomorph.* 93, 3–44).

quickflow That part of a storm rainfall that moves quickly to a stream channel. Amit et al. (2002) *Ground Water* 40, 5 compare quickflow—probably through cracks—with *base flow—the slow flow through the porous medium. 'Quickflow and slowflow refer to relative time constants for characteristic components of observed basin-scale response' (Schwartz (2007) *J. Am. Water Resources Ass.* 43, 6).

quota A limit on the import or export of a particular product imposed by a government. Import quotas may be imposed as protectionism; export quotas may be imposed in countries that depend on the export of a particular raw material, as a means of stabilizing prices. Quotas are usually controlled by the issue of licences. Tokatli (2007) *J. Econ. Geog.* 7, 1 explains that *transnationals can relocate production from one country whose annual quota is exhausted to another whose quota has not been filled or that is not bound by quotas, in the process playing off different producers and governments against each other. The Multi-Fibre Arrangement, which has governed the international trade in textiles since 1974 through sets of trade restrictions and quotas, terminated in 2005. In consequence, Hale and Burns in A. Hale and J. Wills, eds (2005) expect a large-scale shift in manufacturing to the lowest-cost areas of production, such as China and India.

race A classification system used to categorize humans into large and distinct populations. Race is not to be confused with *ethnicity; the answer to the question 'does race exist?' is, biologically no, socially yes, for race is a social construct which, on occasion, can be a matter of life and death. Winlow in R. Kitchin and N. Thrift (2009) is excellent on this. **Geographies of racism** range from examining geographies of segregation and racism, to exploring the cultural politics and social practice of racism, to everyday geographies of identity and experience. See C. Dwyer and C. Bressey, eds (2008).

radial drainage *See* DRAINAGE PATTERNS.

radiation Energy travelling in the form of electromagnetic waves: X-rays, ultraviolet, visible, infra-red, microwaves, or radio waves. In urban areas, a reduction in the *sky view factor decreases the loss of terrestrial radiation thereby contributing to the creation of *urban heat islands (Grimmond (2007) *Geog.* 30, 3).

radiation fog *See* FOG.

radiative forcing The increase in the trapping of outgoing *terrestrial radiation by greenhouse gases. In a study of the capacity of agriculture to reduce the emission of greenhouse gases, Robertson et al. (2002) *Science* 289 find that only soil carbon accumulation in non-tillage systems came close to mitigating radiative forcing. **Radiative cloud**

forcing is the change of the radiative energy flux caused by the presence of liquid water and ice in the atmosphere.

radical geography An analysis of the processes by which inequalities— in race, class, gender, or age—are produced and maintained. Radical geographers hold that studying the visible geography of spatial relationships is not enough; the power relationships, specifically the political and economic structures, are fundamental. For example, any study of industrial location would be worthless, unless the operations of *transnational corporations, and tariff and trade agreements, are taken into account (Byrne (1995) *Capital & Class* 56). Radical geography includes attempts to change the situation on the ground for everyday people in everyday places. See D. Fuller and R. Kitchin (2004) for an exploration of the role of the radical academic in society, and Lawson (2007) *AAAG* 97, 1 on geographies of care and responsibility. **Radical cartography** is the practice of mapmaking to subvert conventional notions and thus actively promote social change. See A. Bhagat and L. Mogel (2008).

radioglaciology The use of ice-penetrating radar to investigate the thickness, roughness, motion, and debris of a body of ice, and to measure and detect crevasses and sub-ice lakes. See Plewes and Hubbard (2001) *PPG* 25, 2 for a detailed account of the investigative processes.

radiometer A passive remote sensor, sensitive to *terrestrial radiation of one or more wavelengths of the visible and infra red. Advanced Very High Resolution Radiometer currently flies aboard many US and European satellites, and its data may be analysed using the AVHRR Hydrological Analysis system, together with ILWIS.

radiosonde A free-flying balloon carrying meteorological instruments. The balloon climbs to a height of 20–30 km above mean sea level, sending information from these sensors to ground stations, whereupon it bursts, returning the equipment to the ground.

rain A form of *precipitation consisting of water droplets ranging from 1 to 5 mm in diameter. The type of rain produced reflects the circumstances in which it formed. A mass of warm air rising at a warm front will develop layered clouds and produce steady rain. Air forced to rise quickly at cold fronts will produce heavier rain. These are both examples of frontal rain. *Convection rain occurs when warm, unstable air rises rapidly. Air forced to rise over mountains may form *orographic (relief) rain. *See also* BERGERON–FINDEISEN THEORY; COALESCENCE THEORY.

raindrop erosion The dislodging of soil particles by large drops of rain. See Wainwright et al. (1995) *ESPL* 20, 277–91 on raindrop erosion and desert pavements. See also Patrick (2002) *PPG* 26.

rainfall intensity The rate at which rain falls, usually measured in millimetres per hour. Intense rainfall is associated with **convectional rain**, notably in thunderstorms and tropical regions, where intensity may be over 100 mm per hour. (British rainfall intensity is normally of the order of 2 mm per hour.) The intensity of rainfall is normally inversely proportional to its duration. Denault et al. (2006) *J. Am. Water Resources Ass.* 42, 3 examine the potential impacts of increases in short duration rainfall intensity on urban infrastructure.

rainfall run-off The overland and downslope flow of rainwater into channelled flow when the rock or soil is saturated. In **rainfall run-off modelling**, hydrological models are categorized as *lumped*—which treat the catchment area as a single unit—or *distributed*—which represent the catchment as a system of interrelated subsystems, both horizontally and vertically. For **deterministic rainfall models**, see Ramos et al. (1995) *Water Resources Res.* 31, 6. For **fuzzy conception rainfall models**, see Ozelkan and Duckstein (2001) *J. Hydrology* 253, 1. For a very clear guide through a **parametric rainfall-run-off** simulation model see Dawdy, Lichty, and Bergmann (1972) *US Geol. Surv., Prof. Paper* 506-B. For the use of *artificial neural networks to relate meteorological variables to run-off, see Antar et al. (2006) *Hydrol. Procs* 20, 5.

rain forest An area of luxuriant forest, found where rainfall exceeds 1000 mm yr^{-1}. Although rain forest does develop in temperate latitudes, most of the world's rain forest is tropical. Hill and Hill (2001) *PPG* 25, 3, 326 explain why tropical rain forests are so species-rich, and Puyravaud et al. (2003) *J. Biogeog.* 30, 7 applaud the use of vegetation thickets as restoration tools in rain forests.

rain shadow An area of relatively low rainfall to the lee of uplands. The incoming air, forced to rise over the high land, causes precipitation on the windward side. The descending air is subject to *adiabatic warming, which increases its capacity to hold much of the remaining water vapour, further reducing rain on the lee side. See Ralph et al. (2002) *Conf. Mount. Meteo.* on the impact of a prominent rain shadow in California.

rainsplash The impact of raindrops on the soil may break down soil *peds, loosen soil particles, and cause *turbulence in the *sheet wash of water flowing downslope. 'Rainsplash . . . is not a single process, but a combination of several discrete but interacting soil particle detaching and [soil particle] transport mechanisms' (Terry (1998) *Austr. J. Soil Res.* 36, 3). As overland flow becomes deeper, the effect of raindrop impact on the soil decays significantly (Mosley (1973) *Zeitschrift* 18). Furbish et al. (2007) *U. Vanderbilt* argue that soil grain transport by rainsplash is largely an *advection–dispersion process.

raised beach A former *beach, recognizable by beach deposits and marine shells, often accompanied by rock platforms and dead *cliffs, which now stands above sea level some metres inland. At sites of *isostatic rebound, several raised beaches may be seen at different levels; the raised beaches of Scotland range from 6 to 14 m OD. Raised beaches result from tectonic uplift; see van Vliet-Lanoe et al. (2000) *J. Geodynam.* 29, 1. Gallagher and Thorpe (1997) *Irish Geog.* 30, 2, 68 explain how to date a raised beach.

ramparts As *ice wedges grow, they push up the soil at the edges into ramparts; see M. Shahgedanova (2002). **Protalus ramparts** are coarse, angular, non-stratified thin ridges of rock fragments, about 1 m high, lying along the base of a mountain face. They are associated with persistent snow banks (Sass (2006) *Geomorph.* 80, 1–2). Shakesby et al. (1999) *Geografiska A* 81, 1 show that they can be formed by snow-push.

range of a good or service The maximum distance/travel time an individual will travel to obtain a given good or service; rare speciality shops have a larger range than shops selling *convenience goods.

ranker An *intrazonal soil, not yet fully developed. This soil is shallow, with a fibrous O *horizon, and an A horizon directly on top of loose, non-calcareous rock.

ranking of towns Various methods have been used to devise a *hierarchy of towns, including comparing the number and type of functions, using multivariate analysis, measuring the interaction between a settlement and its field, ranking settlements by the size of their *spheres of influence; and using graph theory. Recent rankings have been based on quality of life (Rogerson in F. T. Seik et al., eds (2000); Berger et al. (1987) *Soc. Sci. Quart.* 68) or on urban competitiveness (Rodinelli and Vastag (1997) *Econ. Dev. Qly.* 11). The 2007 China Urban Competitiveness Study (Hong Kong) found, unsurprisingly, that Hong Kong was at the top of the Chinese hierarchy. A. Christopher (2012) has an investigation of the *rank-size rule in Delta state, Nigeria.

rank-size rule This 'rule' predicts that, if the settlements in a country are ranked by population size, the population of a settlement ranked n will be $1/n$th of the size of the largest settlement. When settlement size is plotted against rank, on normal graph paper, a concave curve results; plotted on logarithmic scales, a straight line emerges—this is the *rank-size pattern*.

Many developing countries show *primacy*: a sharp fall between the largest, *primate city and the other cities. The *binary pattern*, found mostly in federal countries such as Australia, shows a concave curve. The *stepped order* pattern shows a number of settlements at each level, with each place resembling others in size and function. Fonseca (1989, Institute of Mathematical Geography) explains why the rank-size rule 'works'. Taylor (2007) *GaWC Res. Bull.* 238 notes that national urban hierarchies assume that the national urban system is a closed one; relations between cities in different countries are factored out. 'However, vibrant dynamic cities are always cosmopolitan, [and] to treat New York as US only is to severely under-estimate its economic significance . . . the national bounding of cities was [always] a critical weakness . . . it is even more nonsensical today.'

Raoult's law This states that the presence of a solute will lower *saturation vapour pressure. *See* CONDENSATION NUCLEI.

rapids Areas of greatly disturbed water across a river, rapids have a continuous and relatively gentle slope, rather than a sudden vertical drop.

raster data These data represent the landscape as a rectangular matrix of square cells, such as the pixels that make up a digital photograph. *See* VECTOR.

rational choice theory A theory arguing that individual self-interest is the fundamental human motive, and that individual actors pursue their goals efficiently (see Clarke in M. Tonry and N. Morris, eds 1983). Strauss (2008) *J. Econ. Geog.* 8, 2 notes that economic rationality finds expression in rational choice theory. Cornish in R. Clarke and M. Felson, eds (1993) argues that criminals make travel decisions based on rational choice theory: everything else being equal, criminals choose closer targets rather than those that are further away; see also Herbert and Brown (2006) *Antipode* 38, 4.

rationality The process of using reason or logic to solve a problem. 'The insights of behavioural economists have allowed social scientists to acknowledge certain unique traits specific to humans as a species as rationality without succumbing to an ontological position that posits *utility maximisation as an *a priori* foundation of human behaviour . . . In the UK, for example, strong assumptions are being made about the economic rationality and decision-making competence of individuals in areas of government policy such as health, pensions and education' (Strauss (2008) *J. Econ. Geog.* 8, 2).

Ravenstein's 'laws' of migration
E. Ravenstein's laws of migration, 1889 (1995, *J. Royal Stat. Soc.* 48) introduced the notion that people move in order to better themselves economically. In this view, migration is considered as the individual's response to regional differentials in economic development. Ravenstein's fifth 'law' of migration suggests that women are more mobile than men, at least across short distances—a contention supported by Boyle and Halfacree (1995) *Reg. Studs* 29, 1 and Faggian et al. (2007) *J. Reg. Sci.* 47, 3.

reach (river reach) A short length of river channel. Unstable river reaches that display rapid planform change are described as active reaches; see Martin and Pavlowsky (2011) *Phys. Geog.* 32, 5. However, Parker et al. (2012) *River Res. & Applications* 28, 10, 1708 state that there is a notable lack of consistency in its definition. Even so, they explain how to establish the boundaries of a river reach.

reaction time The time between any kind of change and the response it elicits in a system. In geomorphology it is the interval between a disturbance or change in controls and observable morphological change. In hydrology, the bigger a catchment, the longer its reaction time. Harrison (2003) *Am. Geophys. Union* C22A-01 extends this term to glaciers, but Oerlemans (2007) *J. Glaciol.* 53, 182 thinks that reaction time is not a well-defined concept, 'because the reaction time may depend on the glacier/climate history in a non-transparent way'.

realism At its most simple, realism describes the belief that what is 'real' is what is immediately present to and open to the apprehension of the basic human senses, especially sight and touch; see Pratt in N. Thrift and R. Kitchin, eds (2009). 'One of the characteristic realist ideas is that there is a difference between appearance and reality, and that it is the task of inquiry to find access beyond what appears to be the case to what is really the case' (Mäki (2004) *Env. & Plan. A*, 36, 1719). In considering realism in geography, Sack (1982), *TIBG* 7, 4 504 argues for the ideas of natural necessity, and the existence in nature of underlying structures, potentials, or tendencies. He continues: 'the features that are essentially realist point to the necessity of deeper structures and tendencies in nature than are observed on the sensory surface. Realism, however, does not tell us what the particular structures look like. This is the task of science'. Try Sayer in S. Aitken and G. Valentine (2006), p.154, who is refreshingly honest.

recession The decline in river flow after a storm event has passed. On a *hydrograph, the **recession limb** records the fall in *discharge after the river has reached peak flow due to a storm event. The slope of the recession limb reflects the amount of water stored in the basin and the way it is held in the *catchment area. See Stravs et al. (2008) *Geophys. Res. Abstr.* 10, EGU2008–A-02978 on recession limbs and flow forecasts.

recessional moraine *See* MORAINE.

reclamation The process of creating usable land from waste, flooded, or derelict land. Wilson (2011) *AAAG* 101 presents a highly political analysis of the reclamation of the American West and the location of Japanese 'concentration camps' during the Second World War.

recoding Placing an item of data into a different classification class. This is often done when two different data sets have different category boundaries.

recovery rate The time taken for a *diffusion wave to re-form having been blocked by a diffusion barrier. The recovery rate of a wave front is directly related to both the type and size of the barrier it encounters. See Cliff and Hagget (2006) *J. Geogr. Syst.* 8, 3.

recreation Activities which are undertaken in leisure time. C. M. Hall and S. J. Page (1999) is well reviewed. See also *tourism. *See* LEISURE, GEOGRAPHY OF.

   Recreation carrying capacity is the amount of recreation which a site can take without any deterioration of its qualities. See, for example, W. Theobold (2005).

rectangular drainage *See* DRAINAGE PATTERNS.

recurrence interval The length of time between events of a given magnitude. Anderson and Calver (2004) *TIBG* 29, 1 note that 'landscape form depends on the overall effect of past frequency distributions of sizes of geomorphic events, and on the weather conditions' return'. In hydrology, the return period has an inverse relationship with the probability that the event will be exceeded in any one year. See Keylock (2005) *Adv. Water Resources* 28, 8 on flood recurrence intervals. See also Srikanthan and McMahon (1985) *Hydrol.*

r

Scis 30, 2 on calculating the recurrence interval of drought events through the stochastic analysis of rainfall and streamflow data.

recycling The reuse of *renewable resources in an effort to maximize their value, reduce waste, and reduce environmental disturbance. A *cycle economy*, based on the economical and responsible treatment of limited resources (Seliger et al. (2001) *CIRP Anns.—Manuf. Ind.* 51, 1) is a key concept of today's Japanese environmental policy (Kenichi (2002) *Mining & Materials Processing Inst. Japan* 118, 9). Williams and Kelly (2003) *Resources, Conserv. & Recycl.* 38, 2 evaluate public perception of local authority recycling, and Asomi-Boateng (2007) *J. Plan. Ed. & Res.* 27, 2 reports that waste-based urban cultivation in Ghana could minimize waste collection and disposal costs.

REDD+ (removals of greenhouse gas emission by forests) In developing countries, a programme for lowering emissions by stopping deforestation and forest degradation with an international payments scheme, the slowing of habitat fragmentation, and the conservation of forest biodiversity and soil. See Lu et al. (2012) *Chin. Geogr. Sci.* 22, 4, 390.

redevelopment The demolition of old buildings and the creation of new buildings on the same site. Redevelopment can solve existing problems of congestion and poor design but, for residential areas in particular, it is seen to be wasteful of resources, destroying communities. Wu and He (2005) *Tijdschrift* 96, 1 find that the root cause of such problems was the way local officials and local developers 'utilised their privileges inherited from a planned economy to benefit from the emerging real estate markets'. Fang and Zhang (2003) *Asia Pac. Viewpt* 44, 2 also detect 'the hidden agenda of local elites' during urban redevelopment.

redistricting The process of establishing electoral district boundaries for electing members to a legislative body. Redistricting is conducted periodically for legislative bodies in countries that elect candidates from discrete geographic districts; for example, Canada, the UK, and the USA. See Leib, in B. Warf (2007).

redlining Limiting or charging more for things such as banking, insurance, health care, or even supermarkets, to people who live in what are seen as high-risk districts. 'The practice originated in the 1930s, when federal agencies encouraged lenders to rate neighbourhoods for mortgage risk. Since the 1960s, especially in the US, it has been associated with disinvestment, racial discrimination, and neighbourhood decline. It has always been viewed as a feature of the inner city. Historical evidence indicates that across Canada the first areas to be redlined were the less-desirable suburbs' (Forrester (2003) *Urb. Studs* 40, 13 2661. Harris (2003) *Canad. Geogr./ Géogr. canad.* 47, 3 argues that the suburban origin of redlining in Canada 'perpetuated social class diversity in Canadian suburbs.' Eisenhauer (2001) *GeoJournal* 53, 2 describes the location of out-of-town superstores as 'supermarket redlining'.

red rain The washing out of fine dust particles over mid-latitudes. For Europe, the major sources of dust are the Sahara and the fringes of the Sahel. This dust is picked up by air streams which rise into the upper *troposphere, and carried north and west until it is washed out by precipitation. McCafferty (2008) *Int. J. Astrobiol.* 7 examines historical and mythical accounts of red rain.

reduction The loss of oxygen from a compound. For example, the sesquioxide ferric oxide can be reduced to the monoxide ferrous oxide by bacteria. *See* GLEY SOILS.

reductionism Explaining any phenomenon in terms of only one causal factor.

reflective beach A beach which reflects a major part of the incoming wave. A reflective beach has a convex beach profile, without nearshore bars, is steeper than the wave it reflects, and is typically caused by *accretion. Reflection may generate *edge waves on beaches. See Miles and Russell (2004) *Cont. Shelf Res.* 24, 11.

reflexivity By a researcher, the habit of reflecting upon her/his own identity or standpoint in relation to her/his scholarship; self-transformation and self-examination in the pursuit of knowledge, and between the researcher and the phenomena studied; 'reflexivity that aims, even if only ideally, at a full understanding of the researcher, the researched, and the research context' (Rose (1997) *PHG* 21, 3, 305).

reforestation Replanting a previously wooded area that has been felled; the implicit assumption is that the restoration will succeed reforestation. See Vallauri et al. (2002) *Restoration Ecol.* 10, 1. Bhagwat et al. (2014) *Forest Ecol. & Manag.* (in press) see the reforestation in the Western Ghats of India as a cultural matter.

refugee 'A person who, owing to a well-founded fear of persecution for reasons of race, religion, nationality, membership of a particular social group or political opinion, is outside her or his country of nationality and who is unable or unwilling to return' (UN Protocol 1976). Australia's refugee programme seeks to provide a humanitarian response and protection to individual refugees; participate responsibly in the international community; honour its Convention obligations; further the interests of the people of Australia; meet high standards of administration; and acknowledge as much as possible changes in refugee populations (King (2001) *Int. Mig.* 39, 1). 'Though attempts to de-naturalize the relationship between people and places have been important for how the refugee experience is conceptualized, there has been too much focus on imagination accompanied by a neglect of the local perspective of migrants and displaced people' (Brun (2001) *Geografiska B* 83, 1). Ramadan (2013) *TIBG* 38, 1, 65 has an interesting paper on the refugee camp as a distinctive political space. *See also* ASYLUM MIGRATION.

refugium Refuge areas (pl. *refugia*). Haffer (1969) *Science* 165:131 argues that 'during several dry climatic periods of the Pleistocene and post-Pleistocene, the Amazonian forest was divided into a number of smaller forests which were isolated from each other by tracts of open, nonforest vegetation. The remaining forests served as "refuge areas" for numerous populations of forest animals, which deviated from one another during periods of geographic isolation. The isolated forests were again united during humid climatic periods when the intervening open country became once more forest-covered, permitting the refuge-area populations to extend their ranges (Baker *Amer. Scientist*, nd). This is the **refugia hypothesis** which T. T. Veblen, A. Orme, and K. Young, eds (2007), p. 125, cast doubt on.

reg A North African term for *desert pavement.

regelation Within a moving glacier, the melting of the basal ice at points of increased pressure and refreezing at points of decreased pressure. **Regelation slip** occurs when a glacier moves over an obstacle. Pressures build up on the up-glacier side, causing localized pressure melting. The meltwater flows around the obstacle and then refreezes on the down-glacier side where pressure falls.

regime The seasonal variation in river flow which tends to be repeated each year. A river is said to be **in regime** if the morphological characteristics of a given channel, such as its size, fluctuate around a mean condition over time.

regime theory A set of empirical equations relating channel shape to discharge, sediment load, and bank resistance. The theory proposes that dominant channel characteristics remain stable for a period of years and that any change in the hydrologic or sediment regime leads to a quantifiable channel response (such as erosion or deposition). Valentine and Haidera (2005) *Water Manage.* 158, 2 propose a modification to the Wallingford rational regime theory. See also Eaton and Millar (2004) *Geomorph.* 66, 1–2. Eaton and Church (2007) *J. Geophys. Res.* 112, F03025 find that classical hydraulic geometry represents only an approximation of the variation in channel forms.

   urban regime theory A theory that attempts to locate a phenomenon, such as a city, in its wider spatial, political, and economic context, arguing that growth and development results from groups of actors, working cooperatively together. See Mossberg and Stoker (2001) *Urb. Affairs Rev.* 36, 6, 810 on the development of urban regime theory.

region Any tract of the Earth's surface with either natural or man-made characteristics which mark it off as being different from the areas around it. 'The formation of any given regional map is reflective—and indeed constitutive—of an unevenly developing, often overlapping and superimposing mosaic' (Martin and MacLeod (2004) *TIBG* 29, 4).

   Paasi (2001, *Eur. Urb. & Reg. Studs*) writes of regions as 'collective institutional structures . . . regions are not independent actors; they exist and "become" in social practice and discourse'.

   Many geographers have attempted to distinguish regional boundaries; for example, Slaymaker (2007) *Sing. J. Trop. Geog.* 28, 1 discusses whether or not South-East Asia is a legitimate physical geographical region. However, as Shields (2007, Putting Region in its Place Conf., U. Alberta) points out, 'topography is only one component of geographical identities . . . a place might be said to realize or to embody a regional character, but these statements are always made with a view to contrasting some quality of the region with the qualities of other regions.' Jones (2006) *AAAG* 96, 2 finds international region building 'a messy, problematic, and highly contested activity for parcelling, regulating, and representing geopolitical space'.

regional climate model Any software system designed to model climates between global and local level. Relationships are derived from large-scale variables (predictors) and local/regional variables (predictands). For example, RegCM3 is a three-dimensional, hydrostatic, compressible, primitive equation, σ-coordinate regional climate model which is maintained at MIT.

regional development 1. The provision of aid and other assistance to regions which are less economically developed.

   2. The differential in economic outcomes. Ezcurra and Rapún (2006) *Eur.*

Urb. & Reg. Studs 13, 4 explore the relationship between regional *inequality and economic development level in fourteen western European countries for the period 1980–2002, finding that, beyond a given level of per capita GDP, regional inequality can be seen to decrease. McCoe et al. (2004) *TIBG* 29, 4, in a study of BMW in Germany and Thailand, argue that the 'strategic coupling' of the global production networks of firms and regional economies ultimately drives regional development. Liard-Muriente (2007) *Area* 39, 2 highlights the potential opportunities and pitfalls when designing incentive strategies.

Regional Development Agency

(RDA) Eight regional development agencies were established in England in 1999. These are:

- the North West RDA—North-West
- OneNorthEast—North-East
- Yorkshire Forward—Yorkshire and Humberside
- East Midlands Development Agency— East Midlands
- Advantage West Midlands—West Midlands
- East of England Development Agency—East of England
- South West of England Development Agency—South-West
- South East England Development Agency (SEEDA)—South-East

The London Development Agency was established in 2000. Their remit is to promote sustainable economic development. Each has stakeholders from the local public, private, and voluntary sector community, and each RDA produces a 'regional economic strategy' which outlines the planned economic development and regeneration of the respective region.

regional geography The study in geography of *regions and of their distinctive qualities. A precondition of this study is the recognition of a region, its naming, and the delimitation of its boundaries. One approach has been to identify 'natural' regions while another was to establish economic regions based on agriculture and/or industry. Often there was an intimation of a link between the two types of region. Once the keystone of geography, the status of regional geography has been in decline: 'regional geography survives—if not in most of our major universities, then at least in the public eye. Believe it or not, the informed public still thinks that geographers have particular and unusual perspectives and insights on foreign areas as well as domestic issues, and that geographers can explain "the way the world works" because they have a comprehensive view of it' (Murphy et al. (2005) *PHG* 29, 2). 'The great number of feasible ways of dividing space into regions and time into periods opens the door for alternative narratives, including those which challenge conventional Eurocentric interpretations' (Wishart (2004) *PHG* 28, 3).

regional inequality A disparity between the standards of living applying within a nation. It is difficult to quantify the prosperity or poverty of a region, but there are two basic indicators: unemployment (which has been used in Britain as a symptom since the 1920s), and per capita income, which in England generally falls to the north and west: the North–South divide was 'produced essentially by the . . . metropolitan South as a way of differentiating itself from northern "others"' (A. Baker and M. Billinge 2004). Some would assert that economic development brings about regional inequality, but Fan and Sun (2008) *Euras. Geog. & Econ.* 49, 1 reveal that both interregional and intraregional inequalities in China have declined since 2004. *See also* UNEVEN DEVELOPMENT. M. Amin, A. Ash, and N. Thrift (2003) write that 'One of the most persistent characteristics of the geography of the United Kingdom is the wide inequality that exists between its constituent regions. In the present period, in spite of many stated intentions and much government rhetoric to the contrary, it has on many measures grown considerably worse.'

regional innovation system Very generally, a set of interacting private and public interests, formal institutions, and other organizations that promote the generation, use, and dissemination of knowledge. 'The basic argument is that this set of actors produce pervasive and systemic effects that encourage firms within the region to develop specific forms of capital that is derived from social relations, norms, values and interaction within the community in order to reinforce regional innovative capability and competitiveness' (Yu and Jackson (2001) *Growth & Change* 42, 2, 111).

regionalism Regions may be seen as 'sub-national or transnational areas differentiated from their larger host state(s) by regional claims to distinctive historical, cultural, political and/or economic experiences' (Dahlman (2009) in C. Gallaher et al.). Regionalism, then, is a set of claims or an identity made on the basis of the region. Giordano (2000) *Pol. Geog.* 19, 445 distinguishes between **institutional regionalism**, which relates more to the processes of 'regionalization' that have taken place within and between European states, and **autonomist regionalism**, which refers to the forms of minority, separatist, and ethnic regionalisms that have gained increasing exposure in recent decades. Also, the two categories are not mutually exclusive. Jonas and Pincetl (2006) *Pol. Geog.* 25, 482 describe a new civic regionalism, based on participatory, inclusive, and

partnership models of governance in California. Jones and MacLeod (2004) *TIBG* 29, 4 write on **spaces of regionalism**, which (re)assert 'national and regional claims to citizenship, insurgent forms of political mobilization and cultural expression and the formation of new contours of territorial government'; see J. Agnew (2002), for example.

regionalization The demarcation of regions. E. Ginzburg (1977) suggests that regionalization is 'a slippery concept that varies from writer to writer', but that, regardless of definition, the intention of regionalization programmes is to improve access, quality, cost, and equity. Parr (2002) *Int. Reg. Sci. Rev.* 25 sees the advantages of economic activity being organized around a well-defined localization, urban concentration, or activity complex (or some combination of these) realized with a regional spatial structure which is more dispersed. Wishart (2004) *PHG* 28, 3 thinks regionalization should be seen as part and parcel of the art of *representation. See Wang et al. (2008) *Nat. Hazards* 44, 169 on the regionalization of urban natural disasters in China.

regional multiplier In the framework of a simple regional multiplier model, the operation of a new enterprise creates additional regional income due to its regional exporting activity, which may be considered as the first round of impact. In a second round, the additional expenditures of the firm in the local economy will create more income (Skuras et al. (2005) *Eur. Urb. & Reg. Studs* 12, 4). See Keet in J. Adésínà et al. (2006) and A. Hirsch (2005), both on African topics, and Bathelt (2007) *Geog. Compass* 1, 6 on a knowledge-based multiplier model of *clusters.

regional policy A policy, adopted by government, aimed at redressing *uneven development within a country. Pritchard (2005) *Geog. Res.* 43, 1 argues that the neoliberal approach to regional policy 'advances a relatively simplified conception of regional economic and social life that pays little regard to their richness, complexity and grounded realities'. Gleeson (2003) *Austral. Geog. Studs* 41, 3 critically analyses the European Union's regional policy framework, particularly the advocacy of 'economic normalization' as a leading aim. 'While regional policy traditionally draws upon regional assets to produce competitiveness, the value of this becomes questionable in the era of globalization' (Bathelt (2006) *PHG* 30, 2).

regional science An interdisciplinary study which concentrates on the integrated analysis of economic and social phenomena in a regional setting. It seeks to understand regional change, to anticipate change, and to plan future regional development, and draws heavily on mathematical models. See Bailly and Gibson, and Boyce, both (2004) *Papers Reg. Sci.* 83, 1. See also R. Florax and D. A. Plane (2004).

regolith Derived predominantly from rocks, regolith consists of detritus and chemically altered residues of physically and/or chemically altered components of rocks, as well as many new minerals precipitated during the chemical alteration of parent rocks (Taylor in A. S. Goudie and K. A. Gregory 2011). The depth of the regolith varies with the intensity and duration of the weathering process; within the tropics it may be hundreds of metres deep. Soil is simply regolith with added organic material.

regression analysis A technique for identifying the relationship between a dependent variable y, and one or more independent variables x. While correlation 'makes no assumption as to which is the dependent and which the independent variable, it simply assesses the degree of association between the variables. Regression attempts to describe the dependence of a variable on one or more explanatory variables. It assumes that there is a one-way causal effect from the explanatory variable(s) to the response variable. *Principal components analysis is a regression technique.

(⊕) SEE WEB LINKS

• University of Leicester website of online statistics.

regulation theory The regulation school, developed in France by M. Aglietta (1976), holds that:

• capitalism is always prone to periods of crisis but occasionally displays periods of long-run stability known as regimes of accumulation
• regimes of accumulation are composed of a structural coupling between an accumulation system (form of production) and a mode of social regulation (MSR)—extra-economic dimensions, such as institutions, habits, customs, and norms
• a structural coupling involves co-evolution mediated by class struggle and is a chance discovery.

Two examples of regulation theory are the *Fordist post-war period of capital expansion (which broke down in the early 1970s); and the current post-Fordist (or after-Fordist) capitalist period based on several suggested capitalist forms—flexible specialization, flexible accumulation, and reflexive accumulation.

'Because capitalism is prone to destabilizing and self-generated crises, and because the system has nevertheless persisted, there must be some complex of economic, social, cultural, and political "regulation" that works to soothe,

mute, and partially resolve the contradictions and crises of capitalism' (Purcell
and Nevins (2005) *Pol. Geog.* 24, 211).

Reilly's law The principle that the
flow of trade to one of two neighbouring
cities is in direct proportion to their
populations and in inverse proportion to
the square of the distances to those cities
(W. Reilly 1929). See Brodsky (2003)
Prof. Geogr. 55, 2.

rejuvenation The renewed vigour of
a once active process. The term is
generally applied to streams and rivers
which regain energy due to the uplift
of land through *isostasy or by a fall
in the base level (Schlunegger and
Schneider (2005) *Geomorph.* 69, 1–4).
Rejuvenation may also apply to tectonic movements (Valdiya (1993)
Current Sci. 64).

relationality Relational means indicating relation(s) or relationship; concerning the way in which two or more
people or things are connected. It would
follow, therefore, that relationality
means 'being related to'; 'in a relationship'; 'connectedness'; but try Yeung
(2005) *TIBG* 30, 1, who describes relationality as an 'essential quality embedded in an iterative process of drawing
interconnections between two or more
discrete categories and phenomena that
may not necessarily be binaries'.
 Smith et al. (Open University) shed
some light: 'one of the ways that geographers explore relationality is not
simply by making assertions or pro-
nouncements, but by actually tracking
the practices through which geographically extensive linkages are established
and sustained. Following flows through
specific networks and nodes, researchers
"flesh out" the causal chains that impli-
cate distant lives.' Hopkins and Pain
(2007) *Area* 39, 3 also help: 'relationality
does not just pose interesting questions

about age, but marks a fundamental
change in the way we approach and
think about it, focusing more widely on
families, generations and interactions';
and Sparke (2008) *PHG* 32, 3 writes
that local regions and places should
be studied 'with attention to the ways
in which they are interarticulated
with global space-spanning ties and
traces'.

(((⊕))) SEE WEB LINKS

• Smith et al. on interdependence.

relational space A perception of
space as unbounded and formed
through non-territorial flows and
networks. Ash (2004) *Geografiska B* 86, 1
sees regions as 'places of overlapping—
but not necessarily connected—relational networks'. Hudson, too (2005,
PHG 29, 5), sees regions as 'constituted
from spatialized relations stretched out
over space and materialized in various
forms, and representational narratives'.
Jones (2005, *Reg. Studs Ass. Conf.*)
points out that regions and networks
should not be seen as non-spatial and
without 'geographical anchors', nor
should territories and scales be viewed
as closed and static. There have been
criticisms. Yeung (2006) *W. Paper Series*
2006–15 stresses that regions 'are not
some kind of autonomous actors capable of effecting spatial change', and
Tomany (2007) *Env. & Plan. D* 25, 2
challenges those advocates of the relational region 'who see places as simply
the local articulation of global flows and
who present any concern with local
culture and identity as atavistic and
archaic'.

relative humidity (U) The ratio of
the actual vapour density (which indicates the amount of water vapour
present in the air) to the theoretical
maximum (saturation) vapour density at

the same temperature, expressed as a percentage. This may be expressed as:

$$U = 100\, e/e'_w$$

where e is actual vapour pressure and e'_w saturation vapour pressure with respect to water at the same temperature.

Saturated air has a relative humidity of 100%. Air with a relative humidity in excess of 100% is said to be **supersaturated**. Relative humidity varies both diurnally, with a dawn maximum and an afternoon minimum (Priante-Filho et al. (2004) *Glob. Change Biol.* 10, 5), and, less conspicuously, annually, both variations being in opposition to the pattern of temperatures.

relaxation time The time taken by a geomorphological system to adjust to a sustained change in the nature/intensity of an external process. For the Yellow River, Wu et al. (2008) *Geomorph.* 100, 3–4 find a mean relaxation time of about five to six years for channel adjustment.

relict landform A geomorphological feature which was formed under past processes and climatic regimes but still exists as an anomaly in the changed, present-day conditions. Only a few very young landforms are the result of currently operating geomorphic processes (Bloom (2002) *Geomorph.* 47, 2–4). See Stroeven et al. (2002) *Geomorph.* 44, 1–2 on tors, boulder fields, and weathering mantles in north-eastern Sweden, and Gutiérrez-Elorza et al. (2005) *Geomorph.* 72, 1–4 on blow-outs and playas in the Duero Depression, Spain. Relict surfaces contain information on past surface processes and long-term landscape evolution (Goodfellow et al. (2008) *Geomorph.* 93, 3–4).

relief rain *See* OROGRAPHIC PRECIPITATION.

religion, geographies of The study of the impact of place and space on religious belief and vice-versa. This includes: the mapping, distribution, and diffusion of religions; the study and analyses of the different sites of religious practice, and the ways in which religion shapes human response to them; and the significance of religions for different populations at different spatial scales.

relocalization The preservation of local values and services; an alternative to *globalization. See C. Hines (2000). DuPuis and Goodman (2005) *J. Rural Studs* 21, 8 hold that 'resistance to the agro-food distanciation' is in some key ways at the core of relocalization appeals within local food systems; see also Feagan (2007) *PHG* 31, 1. See Hendrickson and Heffernan (2002) *Sociol. Ruralis* 42 on opening spaces through relocalization. See the special issue on food relocalization, *Cambr. J. Regions, Econ. & Soc.* 3, 2.

rematerialization The grounding of geographical analysis in the concrete world of actual physical (Jackson (2004) *TIBG* 29, 2); 'the way that the "material" and the "social" intertwine and interact material in all manner of promiscuous combinations' (Bakker and Bridge (2006) *PHG* 30, 1). B. Olsen, M. Shanks, and T. Webmoor have witnessed in the last decade or so something of a rematerialization. However, Kearnes (2003) *Soc. Cult. Geog.* 4, 2 takes issue with the rematerialists, particularly with Jackson's assumption that matter is a universally undifferentiated conditionality. He argues, instead, that social and cultural geography must account for the 'wayward expressiveness of matter'.

re-memory *See* MEMORY.

remote sensing The scanning of the Earth by satellite or high-flying aircraft in order to obtain information about it.

'Remote sensing technology is deemed to include photo interpretation, photogrammetry, aerial photography, image processing, and other related technologies in the field of survey, planning, and management of natural resources, and other areas of application where the technology is considered to be appropriate' (*Indian J. Remote Sensing*).

rendzina An *azonal soil rich in humus and calcium carbonate, developed on limestone. The A *horizon is dark and calcareous, but usually thin; the B horizon absent, and the C horizon is chalk or limestone.

renewable resource A recurrent *resource which is not diminished when used but which will be restored, such as *wind energy. Renewable resources may be consumed without endangering future consumption as long as use does not outstrip production of new resources, as in fishing. In principle, wood is a renewable resource, but in the absence of well-planned management, short-term exploitation can induce environmental impacts or conversion to other uses yielding results better likened to mining than sustainable use (Berlik et al. (2002) *J. Biogeog.* 29). The EU proposes a mandatory target of 20% for renewable energy's share of energy consumption by 2020 and a mandatory minimum target of 10% for biofuels. It also proposes creating a new legislative framework to enhance the promotion and use of renewable energy.

⊕ SEE WEB LINKS
- The 2013 update of the Renewable Energy Roadmap.

rent gap The gap between the actual rent paid for a piece of land and the rent that could be collected if the land had a 'higher' use. The idea is central to, but does not entirely explain, *gentrification; see N. Smith (1996).

rent gradient The decline in rents with distance from the city centre, reflecting the cost of transport from the outlying districts to the centre. See Wu in L. Ma and F. Wu (2005).

replacement rate The fertility rate required to hold the population constant in the absence of immigration or changes in longevity (Greenspan (2004) *Pop. & Dev. Rev.* 30, 4).

representation The ways in which meanings are formed, conveyed, and shared among members of social groups. These representations can be defined as *culture, and cultural forms, notably language, and to some extent shape the reality they represent. Given that it is a working, creative, language-driven process, representation cannot be neutral or without power relations. 'Representation is linked to studying the formal and informal relationships between discourse, text, and imagination [for] place and the human landscape are neither blank nor neutral but laden with meaning, symbolism, and contradiction' (C. Gallaher et al. 2009). *Landscapes may also be representations, from which people choose, rearrange, and specify their meanings; 'representation frequently draws on cultural imaginings of places and a past that may never have existed . . . such representations are responsive to the geographical and historical specificities of the context within which they are created and consumed' (Markwick (2001) *Geog.* 86, 1).

representational See NON-REPRESENTATIONAL THEORY.

reptation A method of sand transport whereby grains are set into a low motion due to the high-velocity impact of a descending saltating grain. In this way, grains are set in motion at wind velocities lower than those required to move them by wind alone.

research and development
(R and D) Research on new discoveries for industrial processes. Rutherford (2004) *PHG* 28, 4 observes that, while R and D *transnational corporations remain home-nation oriented, and in many cases locally embedded, there are forces eroding this orientation. 'TNCs may innovate domestically, but the need for maximizing returns on investment means that R+D will spill over internationally. Home base has become insufficient to generate technological advantage.' In 2004, 1.9% of GDP was spent on research and development in the EU-25 (1.95% in the EU-15), compared to 2.6% of GDP (NFS data) for the USA (Crescenzi et al. (2007) *J. Econ. Geog.* 7, 6).

reserves The proportion of resources, notably mineral resources, which can be extracted using the prevailing technology.

residential differentiation (residential segregation)
The evolution of distinct neighbourhoods, recognizable by their characteristic socio-economic and/or ethnic identity. *See also* SEGREGATION; ETHNIC SEGREGATION.

residential mobility Newbold and DeLuca (2007) *Urb. Geog.* 28, 7 suggest that whereas mobility is high, new arrivals primarily remain in their initial destination with little difference in the overall distribution. Residential moves are associated with various individual and household characteristics, along with neighbourhood effects and the type of housing initially occupied.

resilience 1. In *geomorphology, the ability to recover towards a predisturbance state, which is directly related to dynamic stability. Long et al. (2006) *Geomorph.* 82, 309 explain the way Dungeness Foreland—a large sand and gravel barrier located in the eastern English Channel—has, for 5000 years, demonstrated remarkable geomorphological resilience in accommodating changes in relative sealevel, storm magnitude and frequency, variations in sediment supply, as well as significant changes in back-barrier sedimentation.

2. Of an economy, the ability to withstand an economic shock, such as the closure of a factory, or the loss of a key market, or more global events such as the global financial crisis of 2007. Resilient economies are able to bounce back quickly from the negative effects of the shock. See Martin (2012) *J. Econ. Geog.* 12, 1, 1.

resistance The ability of a system to avoid or minimize responses to externally imposed changes. Resistance has two main aspects: some indication or measure of strength, chemical or mechanical stability, or susceptibility to modification; and absorption, generally assessed as some measure of capacity (A. Goudie and K. J. Gregory 2011). In hydrology, resistance is the friction force counterbalancing the downslope gravitational force (L. B. Leopold et al. 1964). Resistance is a function of the surface area in contact with the flow, and so flow becomes more efficient when the ratio of the flow cross-sectional area A to the wetted perimeter P is high (defining the *hydraulic radius R). See Smith et al. (2007) *PPG* 31, 4 on applying flow resistance equations to overland flows.

resource An available supply of something that is valued because it can be used for a particular purpose, usually to satisfy particular human wants or desires (C. Park 2007). Giordano (2003) *AAAG* 93, 2 develops a typology classifying common resources into one of three categories—open access, fugitive, and migratory—based on spatial relationships between resources and resource users. *See also* NATURAL RESOURCE.

resource allocation The assessment of the value of a resource or of the effects of exploiting a resource. Spatial decision support systems (SDSS) are designed to make complex resource allocation problems more transparent and to support the design and evaluation of allocation plans (Jeroen et al. (2003) *Trans. GIS* 7, 2). K. Kyem et al. (2001) *Trans. GIS* 5, 2 describe a decision support algorithm for managing resource allocation problems involving competing interests.

resource-frontier region A newly colonized region at the periphery of a country which is brought into production for the first time. Hacquebord and Avango (2009) *Arctic Anthrop.* 46, 1–2 use a *core–periphery model in order to understand the general trends in the history of natural resource exploitation in an Arctic Resource Frontier Region.

resource management The allocation and conservation of *natural resources. The main emphases are on: an understanding of the processes involved in the exploitation of resources; the analysis of the allocation of resources; and the development and evaluation of management strategies in resource allocation. Sustainable development and environmental protection are major goals; see Barton (2006) *Asia Pac. Viewpt* 47, 3 on fishery resource management in Chile and New Zealand. See Johnson et al. (2004) *Nat. Res. Forum*, 28, 3 on gender analysis in resource management. *See also* NATURAL RESOURCE MANAGEMENT.

response time The speed of a reaction to a change in environment. See Dyke and Peltier (2000) *Geomorph.* 32, 1–4 on response times and sea-level

changes. For land ice, the dynamic response time is proportional to the size of the ice mass (W. Paterson 1994). J. Bamber and A. Parsons (2004) have a table of estimated response times for various components of the climatic system. *See also* REACTION TIME; RELAXATION TIME.

restructuring A change in the economic make-up of a country. It may involve: reordering production to achieve *economies of scale; a switch of investment from one sector to another (*see* DEINDUSTRIALIZATION); a change in the spatial distribution of industry; or a change in the economic system; see D. Harvey (2005) on the restructuring of capitalism. 'The tendential "ecological dominance" of neoliberal restructuring strategies . . . has profoundly disrupted inherited scalar hierarchies, while setting in train new rounds of creative institutional destruction across multiple scales' (Peck and Theodore (2007) *PHG* 31, 6; see also N. Brenner et al. (2005) and Jessop (2000) *Int. J. Urb. & Reg. Res.* 24). 'Firms emphasizing the social at the expense of the material may require dramatic restructuring during times of crisis when debt, for example, becomes unwieldy' (Ettlinger (2008) *PHG* 32, 1). See Clarke et al. (2006) *Env. & Plan. A* 38 on retail restructuring and consumer choice. Restructuring may be imposed by an authority, like the World Bank, to improve the ailing economy of a nation which is in debt to that authority. *See also* STRUCTURAL ADJUSTMENT.

retailing, geography of The study of spatial patterns of retail and consumer behaviour. This includes the analysis of retailing within a city; hierarchies of retail centres based on *central place theory, and the relationship between out of town malls and city shopping centres. Models are used to forecast retailing and consumer decisions, mostly at the intra-urban scale. These methodologies treat space as neutral, or independent.

In contrast, the '**new economic geographies of retailing**' (N. Wrigley and M. Lowe eds, 1996) view the relations between space and retail activity as mutually constitutive. In this way, retail capital can structure spaces, defining our urban spaces with shopping streets, markets, and malls, but is also configured by socio-spatial processes. Key themes include: the reorganization of working corporate structures, retailer–supplier interfaces and labour; the social relations of production; the organization and technology of retail distribution; and the workings of retail capital. 'The potential of retail geography is that categories such as "economy" or "culture" are constantly being shattered. The two seem mutually implicated' (Nick Blomley, Simon Fraser University).

In the developed world, it is becoming increasingly difficult to distinguish between consumption and entertainment: 'not only is shopping melting into everything, but everything is melting into shopping' (S. T. Leong 2001). Retailing is now part of the expression, construction, and contestation of identity; consumption may define who we are. At the same time, much of current thinking is that consumption is intrinsically evil, and morally corrupting.

Perhaps the supreme incarnation of shopping as leisure is the out-of-town shopping centre/shopping mall. Malls were built on a large scale and pioneered by Victor Gruen (1973), who aimed to 'restore the lost sense of commitment and belonging . . . [to] counteract the phenomenon of alienation, isolation and loneliness and achieve a sense of identity'. Shopping malls offer consumers a safe and climate-controlled alternative to the perceived dangers and unpredictability of city centre shopping,

transforming our public spaces into sanitized, privatized, serially reproduced zones. G. Ritzer (2004) argues that 'these malls, corridors, consumers and shops could be almost anywhere—Los Angeles, Singapore, Moscow, Rio or Johannesburg', and M. Sorkin (1992) claims that 'the globalisation of retailing is creating the ageographical city, a city without place attached to it'. 'The mall is essentially a pseudo-place which works through spatial strategies of dissemblance and duplicity' (J. Goss (1991) *AAAG* 81, 1). See M. Birkin, G. Clarke, and M. Clarke (2002) and try P. Scott (2007).

re-territorialization The restructuring of a territory/place that has experienced *deterritorialization. See Bunnell et al. (2006) *Int. J. Urb. Reg. Res.* 30, 1 on the complex re-territorializations on the Indonesian island of Bintan.

return flow Water which has seeped through the soil as *interflow but which backs up the hillslope when it has reached a saturated layer. See Guebert and Gardner (2001) *Geomorph.* 39, 3–4.

return period The length of time between events of a given magnitude.

reurbanization The development of new homes, businesses, and community facilities within existing urban areas. Ogden and Hall (2001) *TIBG* 29, 4 provide strong evidence for re-urbanization in ten French cities, 1990–99. Pietry (2000) *Cybergeo: écon. et statis.* 336 attributes some of this to the preference of French young people for city-centre living. See Champion in R. Pattison and W. Leaver, eds (2001).

revegetation Re-establishing and developing plant cover, either by artificial or natural means; see, for example, Li et al. (2007) *Geomorph.* 88, 3–4. Owen et al. (2001) *J. Biogeog.* 28, 5 report that most revegetation following anthropogenic and/or environmental interference is through vegetative reproduction. One of the key components of any restoration programme is to adequately and efficiently identify potential sites for revegetation. As with any restoration activity, economic cost plays a major role (Lawson et al. (2007) *Geogr. Res.* 42, 4).

Reynolds number (R$_e$) Four factors combine to determine whether the flow of water within a channel is *turbulent or *laminar: the density, velocity, and viscosity of the water, and the hydraulic radius of the channel. Since the density of water is 1, the Reynolds number expresses this combination as:

$$R_e = \frac{VR}{\mu}$$

where V = velocity of the liquid, R = hydraulic radius, and μ = viscosity of the liquid. The Reynolds number is a dimensionless quality; in streams, the maximum number for laminar flow is between 500 and 600, depending on temperature, and at high Reynolds numbers, above 2000 to 25000, flow is turbulent. Smith et al. (2007) *PPG* 31, 4 note that the flow is laminar for Reynolds numbers below 2300 (in circular pipes), and above 4000 in turbulent. 'An ill-defined region between these two limits is known as the transitional zone. The critical Reynolds number transition to turbulent flow depends on the exact flow configuration and should be determined experimentally.'

ria The seaward end of a river valley which has been flooded as a result of a rise in sea level. The name is from the type location in Galicia, Spain.

ribbon development A built-up area along a main road running outwards from the city; a location that combines cheaper land with high accessibility, and the chance of attracting trade from passing traffic. See Milward (2003) *Canad. Geogr./Géog. canad.* 46, 1 and Murdoch and Lowe (2003) *TIBG* 28, 3 on the campaign by the Council for the Preservation of Rural England (CPRE) against ribbon development.

Richter scale A scale of the *magnitude of earthquakes, ranging from 0 to (in theory) 10. Richter (1935) *Bull. Seism. Soc. Am.* 25 equated the magnitude of an earthquake to the base 10 logarithm of the ground motion in millimetres, measured on a certain type of seismograph, plus a correction factor related to the distance of the earthquake. The distance is calculated from the difference in arrival time for different types of waves that travel at different speeds. So, for a constant distance between an earthquake's hypocentre and the seismograph, the ground motion has to increase by a factor of 10 to cause an increase of 1 on the Richter scale. Professor Richter was trying to characterize the energy of a seismic event, not damage. Newer measurements, referred to as the 'seismic moment' and 'moment magnitude', have been developed to address some of the Richter scale's shortcomings (Johnson (2004) US Dept Energy). *See also* MERCALLI SCALE, for measurements of earthquake *intensity.

ridge and furrow A set of parallel ridges and depressions formed during the period of strip cultivation in the Middle Ages. As the land was ploughed always to the same pattern, the plough threw up earth to make ridges which often survive in the present landscape. See Whittington and Brett (1979) *J. Hist. Geog.* 5, 1. Ridge and furrow landscapes are still clearly visible in the landscapes of midland England.

rift valley A long strip of country let down between normal faults, or between a parallel series of step faults (also called a *graben*). There is a maximum rift valley relief, controlled by the competition between isostasy and lithospheric geometry (Quin and Buck (2005) *J. Geophys. Res.* 110, B03404). The biggest terrestrial rift valley system is the East African system, at 3000 km long. *Plate tectonic theory suggests that rift valleys are the result of large-scale doming above a mantle *plume, followed by fracturing along the crest of the dome as plates diverge (Moucha et al. (2008) *Geology* 36, 6). Many rifts, like the Rhine rift valley, have the Y-shaped pattern characteristic of a triple junction, indicating that they arise from plate separation; see Horner-Johnson et al. (2003) *Eos* (*Trans. Am. Geophys. Union*) 84, 46.

right to the city Writings on 'the right to the city' stress the intrinsic worth of cities to their inhabitants, not merely their instrumental or economic value. While Painter (2005) ICRRDS, U. Durham, comments that 'the right to the city is the right to make full use of the city and to live a richly urban life'. Attoh (2011), *PHG* 35, 5, 669 critically examines the concept of the right to the city.

rills Small *channels, between 5 and 2000 mm in width, and very closely spaced. They develop well in areas with heavy rainfall, especially upon weaker rocks, such as volcanic ash. Sediment yields increase with slope gradients and flow rates (the former having more effect than the latter) and greater flow rates cause more rill erosion and soil loss under the same slope gradient (Lei et al. (2006) *Eur. J. Soil Sci.* 57, 3). Torri et al. (2006) *Geomorph.* 76, 3–4 find a power relationship exists

between width and total flow discharge in rills. Rills may widen and deepen to form *gullies.

rime *See* FROST.

ring dyke *See* DYKE.

riparian Relating to a river bank. The riparian zone is composed of the area of land adjacent to streams and rivers. This zone is often a narrow strip between the channel and hill slope in headwater areas but also occurs as a more extensive floodplain adjacent to larger rivers (Naiman and Decamps (1997)). A river and its surrounding riparian vegetation are two dynamic systems that interact through several hydrological, geomorphological, and ecological processes; see Perucca et al. (2007) *River Resources Res.* 43, 3 on the role played by vegetation on meandering river morphodynamics. For an exhaustive treatment of the topic, see Hill in J. Jones and P. Mulholland, eds (2000). Riparian vegetation dynamics are driven by allogenic hydrogeomorphological factors, with autogenic (plant-induced) influences affecting both plant dynamics and the river environment from the earliest stages of plant establishment, and becoming more important as landform stability is achieved (Francis (2006) *Area* 38, 4).

rip current A strong current moving seawards in the near-shore zone. Munk (1949) *Ann. NY Acad. Sci.* 51, 3 suggested that rip currents were formed by onshore wave mass transport and that water piled up against the beach, which provided a hydraulic head for a rip current (an offshore return flow). Brander and Short (2001) *J. Coast. Res.* 17, 2 explain velocities in a rip channel by the continuity equation, in which the discharge out of the rip channel is equal to incoming wave-induced transport over the transverse bar. See also MacMahan et al. (2005) *Marine Geol.* 218, 1–4.

rising limb The section of a *hydrograph from the start of increased discharge to the maximum flow.

risk The likelihood of possible outcomes as a result of a particular action or reaction. Technically speaking, the likely outcomes of risks can be assessed as a series of different odds, while there is no calculation of probabilities in *uncertainty. 'Social scientists have long argued that risk is socially constructed . . . the interpretation of physical threats is not just a subjective process engaged in by individuals but is also strongly affected by mores, norms, values, institutions, and other influences on choice that are held in common by members of social groups. By these means the almost infinite number of physical risks that inhabit our world is prioritized to facilitate collective action' (Mitchell (2007) *AAAG* 97, 2). See J. Kasperson and R. Kasperson (2005). Herrick (2005) *Area* 37, 3, in a study of risk perception and GM food, writes that 'what science deems to be an acceptable level of risk may not match the social perception of acceptability. When the differences between social and scientific notions of risk become acute, then the outcome is a "social amplification of risk" by the public.' Wakefield and Elliott (2003) *Prof. Geogr.* 55, 2 suggest that risk messages are chosen and shaped by journalists on the basis of the pressures they themselves are under, and that, while newspapers were a major source of risk information, their impact was mitigated by readers' distrust, and access to their own personal information networks. Watson and Stratford (2008) *Soc. & Cult. Geog.* 9, 4 recognize three socio-spatial orderings of risk (displacement, replacement, and reorientation).

Risk analysis identifies the level of hazard in an area, and estimates the probability of occurrence of future risk (Chung and Fabbri (2008) *Geomorph.* 94, 3–4). See Hungr et al. (2008) *Geomorph.* 96, 3–4 on **quantitative risk analysis**. *See also* UNCERTAINTY.

river 'Rivers are truly perceived not as "things in space" but as "processes through time"' (Goodwin (1999) *Stream Notes*).

river basin management 'The process of coordinating conservation, management and development of water, land and related resources across sectors within a given river basin, in order to maximise the economic and social benefits derived from water resources in an equitable manner while preserving and, where necessary, restoring freshwater ecosystems' (Global Water Partnership (2000) *Backgr. Paper* 4). Restoration, rehabilitation, river basin management, and their derivatives are practised in at least 21 different countries in response to the exploitation and subsequent deterioration of the riverine environment (Wheaton (2006) *Area* 38, 2). 'Interventions on hydro/ecological systems by different categories of stakeholders characterized by different political, decision-making, and discursive power, and varied access to resources, tend to generate costs, benefits, and risks that are distributed unevenly across spatial and temporal scales and across social groups. This is due to the interconnectedness of users through the hydrologic cycle entailed by their dependence upon the same resource' (Molle (2007) *Geog. J.* 173, 4). See Hirsch and Wyatt (2004) *Asia Pac. Viewpt* 45, 1 on the Se San River.

river capture *See* CAPTURE.

riverine Relating to, similar to, or formed by, a river. Ward et al. *Freshwater*

Biol. 52, 1 argue that hydrological connectivity (the exchange of matter, energy, and biota via water) plays a major role in sustaining **riverine landscape diversity**.

river restoration Assisting the recovery of ecological integrity in a degraded watershed system by re-establishing hydrologic, geomorphic, and ecological processes and replacing lost, damaged, or compromised biological elements (Wohl et al. (2002) *Landsc. Ecol.* 17). Restoration activities include: cutting a new, meandering course, creating stepped banks to provide diverse ecological niches, removing dams to restore fish movement, reconnecting flow by cutting linking channels, setting back or breaching *levees, and dredging. McDonald et al. (2004) *TIBG* 29, 3 argue that the purpose of restoration has shifted from simple utilitarian needs for flood and erosion/sedimentation control towards the incorporation of ecological concerns.

See Millington and Sear (2007) *Earth Surf. Procs Landforms* 32, 8 on the geomorphic and ecological effects of reintroducing woods in river restoration schemes. See Downs and Kondolf (2002) *Environ. Manage.* 29, 4, 477 on post-project appraisals in adaptive management of river channel restoration.

(⊕) SEE WEB LINKS
- The UK River Restoration Centre has case studies.
- The European Centre for River Restoration has a useful bibliography.

River Styles framework A method of river maintenance based on four key principles: respect for river diversity; working with river dynamic and change; working with the linkages of biophysical processes; and using geomorphology as an integrative physical template for river management activities. Brierley et al.

(2011) *Applied Geog.* 31, 1132 provide a case study.

(⊕) SEE WEB LINKS
- Website of River Styles® framework.

river terrace A bench-like feature running along a valley side, roughly parallel with the valley walls. Bridgland and Westaway (2007) *Geomorph.* argue that uplift is essential for the formation of river terraces; 'their cyclic formation, however, almost invariably seems to have been a response to climatic fluctuation'. Bridgland (2002) *Quat. Sci. Revs* 19, 13 suggests a six-stage model:

1. the incision phase in which terrace generation occurs, occurring at the transition to interglacial conditions (discharge is high as a result of melting permafrost;

2. an aggradation phase, again at the glacial–interglacial transition, seen mainly in the lower reaches of valleys;

3. the interglacial phase, in which fine-grained sedimentation (rarely preserved) is predominant;

4. a further phase of incision at the interglacial–glacial transition;

5. the main aggradational phase, at the interglacial–glacial transition, as a result of considerable sediment being liberated by the decline of vegetation;

6. a phase of glacial climate during which there is relatively little activity, much of the potential discharge being locked up in permafrost.

river training The attempt to 'flood-proof' a river/channel. This may include the straightening, narrowing, and shortening of a river course, removing river gravels, installing new flood banks and river-bank walling, and improving floodplain drainage (McDonald et al. (2004) *TIBG* 29, 3). See Korpak (2007) *Geomorph.* 92, 3–4 on the geomorphological impacts of river training in mountainous channels.

road pricing A strategy to reduce *urban *congestion, first used in Singapore in 1978. An electronic road pricing (ERP) system can electronically monitor, and track, vehicles entering a restricted zone to control the flow of inbound traffic. The system is capable of automatically imposing a demand-sensitive congestion toll on every vehicle without requiring them to slow down or stop, when the congestion level in the restricted zone exceeds a preferred threshold level.

The ERP system was first tried out in Hong Kong from July 1983 to March 1985, but not implemented due to public rejection, arising from concerns over the privacy of movement. In Singapore, on the first day of its implementation (1 April 1998), the usual morning rush hour traffic along one of the heavily congested highways decreased by 17% (Goh (2002) *J. Transp. Geog.* 10, 1). Road pricing varies in coerciveness depending on households' economic resources, and the public perceive road pricing as unfair, given that wealthy households are not forced to adapt (Jones in J. Schade and B. Schlag, eds 2003). Jakobsson et al. (2000) *Transp. Policy* 7, 2 find that lower-income groups find road pricing less acceptable than high-income groups because it affects their perceived freedom and justice negatively. A. Mahendra, ed. (2010) is useful on the impacts of road pricing on business.

rock avalanche The mass movement of a minimum volume of 500000 m^3 of rock (Scheidegger (1973) *Rock Mechanics* 5) derived from the collapse of competent masses of coarse-grained, hard bedrock. Hewitt (2006) *PPG* 30, 3 examines rockslide-rock avalanches in mountain landscapes, and the landforms associated with them, backed up with useful pictures and really clear explanations.

rock cycle The conversion of rocks to other forms; under pressure and heat, *igneous and *sedimentary rocks can become metamorphic rocks. All rock types can erode to form layers of sedimentary rocks and all rocks can be completely melted, cooling and hardening at or near the Earth's surface to form igneous rock.

rock fall The detachment and downward fall of rock from a cliff face to its base. Rock fall may be triggered by permafrost degradation, seismicity, and *glacial debuttressing, but it is its geological structure that makes the rock face more or less prone to rock fall. See Ravanel and Deline (2011) *Holocene* 21, 357.

rock flour Silt- and clay-sized particles of debris formed from grinding due to *abrasion within and at the base of a glacier, and carried by *meltwater streams.

rock glacier A very slowly moving river of angular rock debris, characteristic of high mountain permafrost and made up of debris and ice that creeps on hillslopes at a typical rate of a few decimetres per year (Serrano et al. (2006) *Geomorph.* 74, 1-4). Ikeda and Matsuoika (2006) *Geomorph.* 73, 3-4 distinguish between **boulder rock glaciers**, with an active layer of matrix-free boulders derived from crystalline rocks, and massive limestone; and **pebbly rock glaciers**, made of matrix-supported debris derived from less resistant shales, and platy limestone.

Rossby waves Long ridges and troughs in the westerly movements of the upper air, with a wavelength of around 2000 km, discovered by C. J. Rossby in 1939. Four to six waves girdle the Northern Hemisphere at any one time. Some are a response to relief barriers, like those east of the Rockies, and east of the Himalayas (O'Kane and

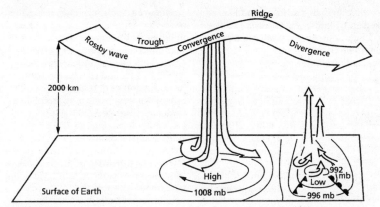

Rossby wave

Frederiksen (2005) *ANZIAM J.* 46 (E), C704–C718).

Rossby waves are thought to be a reaction to the unequal heating of the Earth's surface, and are intimately connected with the formation of *cyclones (Li et al. (2006) *J. Atmos. Scis* 63, 5), *anticyclones (Chen and Newman (1998) *J. Climate* 11, 10), and *mid-latitude depressions. As air in the middle latitudes of the Northern Hemisphere travels west–east into a trough, it slows down, and piles up, causing *convergence just ahead of the ridge which follows. Convergence in the upper air causes a downflow to the ground, creating high pressure systems at ground level, below and just ahead of troughs in the Rossby waves. As air leaves the trough, and its passage straightens out, air speeds pick up, and the air moves very fast as it swings round the outer arc of the ridge. Air then diverges just ahead of the next trough. *Divergence in the upper air causes low pressure systems at ground level below and just ahead of ridges.

At times, the waves are few, and shallow; a pattern known as a high zonal index. In other cases, the flow becomes markedly *meridional; a pattern known as a low *zonal index. Upon occasion, the waves break down into a series of cells (**Rossby wave-breaking**; *see* BLOCKING). Woollings and Hoskins (2008) *J. Atmos. Scis* 65, 2 suggest that the low-frequency variability of the *North Atlantic Oscillation results from variations in upper-level Rossby wave-breaking events.

rotational slip The semicircular motion of a mass of rock and/or soil as it moves downslope along a concave face. Evidence for rotational slip in *cirques comes from the dirt bands observed in cirque ice, which become progressively steeper from the back wall, but then flatten towards the cirque mouth.

r-selection, k-selection Two major strategies may be adopted for the survival of a species. r-selected plants (where r stands for maximum increase) respond swiftly to favourable conditions with most of their energies devoted to rapid maturity and reproduction. *See* OPPORTUNIST SPECIES. k-selected plants survive by putting their energies into persistence. *See* EQUILIBRIUM SPECIES.

rugosity Roughness, wrinkliness; see Donovan (2002) *Geol. Today* 18, 4 on rugosity in karst landscapes, and Knudby et al. (2007) *PPG* 31, 4 on coral reefs. In ecology, more rugosity will provide more available habitat for colonization, foraging, and shelter.

SEE WEB LINKS

• Monitoring of coral reef ecosystems.

run-off The movement over ground of rain water.

rural In, of, or suggesting, the country. Bell (2007) *J. Rural Studs* 23, 4 distinguishes between 'first rural' (basically, low population density), and 'second rural'—a rural of associations. 'It calls upon the connections we have long made between rural life and food, cultivation, community, nature, wild freedom, and masculine patriarchal power, and the many contradictions we have also so long associated with the rural, such as desolation, isolation, dirt and disease, wild danger, and the straw-hatted rube.' Halfacree (1995) *J. Rural Studs* 11, 1 makes the case for defining 'the rural' as 'an abstract social representation; a set of rules and resources existing out of space and time which are drawn upon in both discursive and non-discursive actions. The precise form that this representation takes in these actions is highly contextualised and depends upon its precise usage'. Rurality, through its strong cultural entanglement with 'nature', is seen as a repository of more embedded and authentic values (see R. Mabey 2005). McCarthy (2006) *PHG* 29, 6 argues that 'rurality' 'remains at least as much a product of divisions of labor within the academy and social contexts as a category defined by particular sectoral mixes, land uses, densities, or other empirical descriptors'. M. Woods (2005) concludes that 'it seems that the real power in the British countryside is the very idea of rurality itself'.

Halfacree in P. Cloke et al., eds (2006) notes that *productivist agriculture was to be the essence of 'modern' rurality. Tonts and Grieve (2002) *Australian Geog. Studs* 40 examine the destructive tendencies associated with the **commodification of rurality** in some of Australia's more scenic and accessible rural areas. See Bell (2000) *Rural Sociol.* 65, 4 and Little (2002) *PHG* 26, 5 on rurality and sexuality.

rural depopulation The decrease in population of rural areas, by migration, or through falling birth rates as young people move away. In 2004, Ioffe et al. (*AAAG* 94, 4) reported that, as a result of farmland abandonment, no fewer than 20 million hectares of arable land were already deserted in European Russia. However, Stockdale (2002) *Int. J. Pop. Geog.* 8 asserts that the historical trend of rural depopulation is being replaced by a repopulation of the countryside.

rural geography The study of the rural landscapes of the developed world. It includes the origin, development, and distribution of rural settlement, *rural depopulation, the causes and consequences of agricultural change, patterns of recreational use of the countryside, tourism, planning, and the growing influence on rural areas of urban dwellers. For a review of the sub-discipline, see Roche (2005) *PHG* 29, 3. McCarthy (2006) *PHG* 29, 6 argues that recent developments 'call into question the geographic categories and concentrations of rural geography'.

rurality The values and beliefs of 'rural folk' as constructed (by both rural and urban residents) as distinct and different, and defined by particular mixes of economic exploitation, land uses, and densities; Halfacree in P. Cloke, M. Goodwin and B. Mooney, eds (2006) suggests a threefold and entwined structure for analysing rurality: rural localities (how places are shaped by spatial practices and processes, often linked to production or consumption); formal representations of rurality (often dominated by hegemonic interests); and everyday lives of the rural (how agents perform, practise, and experience rurality). Try Woods (2010) *PHG* 34, 6, 835.

rural–urban continuum The belief that between the truly rural and the truly urban are many 'shades of grey'; if we actually look along a scale from the single isolated farm all the way to the *megalopolis, we do not find any clear boundaries between hamlets, villages, towns, and cities. Sheppard and Nagar (2004) *Antipode* 36, 4 state that 'it is important to reconceptualize the urban as rural–urban continuum in attending to processes of domination, co-optation and resistance, especially in Asia and Africa where populations are predominantly rural and historical patterns of colonial and postcolonial capitalist development and labor migration connect the lives in the urban, peri-urban, and rural areas in intricate and inseparable ways'.

rural–urban fringe The transition zone between the city and its suburbs, and the countryside. Certain types of land use are characteristic of this zone: garden centres, country parks, riding stables, golf courses, sewage works, and airports are common, and these are neither truly urban nor truly rural uses. They do, however, give an urban air to the countryside, which can be cited as an argument for further development—

since the zone is not really 'countryside', it need not be preserved. Thus, C. Bryant et al. (1982) define the **inner urban fringe** as 'land in the advanced stages of transition from rural to urban uses—land under construction, land for which subdivision plans have been approved—in short, land where there is little doubt over much of its area about its urban oriented function and ultimate conversion to urban uses'.

Irwin and Bockstaehl (2002) *J. Econ. Geog.* 2, 1 think that fragmented patterns of development in rural–urban fringe areas could be due to negative externalities that create a 'repelling' effect among residential land parcels. Lagarias (2007, *Cybergeo, systèmes, modél. géostat.*, 391) presents a fractal analysis of urbanization at the periphery of Thessaloniki, Greece. Bunker and Houston (2003) *Australian Geog. Studs* 41, 3 report that the fringe is becoming increasingly complex due to multifaceted demographic change, a broadening economic base, and demands for better environmental management, 'all within the context of an evolving understanding of sustainability'. Qviström (2007) *Geografiska B* 89, 3 objects to the characterization of the rural–urban fringe as a *process* rather than a *place*; questions representations informed by a dichotomous way of thinking; and hopes to 'bring forward more nuanced accounts of fringe landscapes'.

See Nengroo (2011) on the rural–urban fringe of Srinagar.

rustbelt *See* SNOWBELT.

S

Sahel The **Sahelian zone** borders the southern Sahara. Vegetation—scattered grasses, shrubs, and trees—increases in density towards the southern margin. Annual rainfall is 200–400 mm, but unreliable. Severe droughts bring losses of livestock, crop failures, and famine. See *Glob. Env. Change* (2001) 11, 1, special Sahel issue. The work by D. Hummel, M. Doevenspeck, and S. Samimi (2001) *micle (Migration, Climate & Environment) working paper #1* is a superb resource on climate change, environment, and migration in the Sahel.

saline Salty. A **saline soil** is a non-*sodic soil containing enough salt to adversely affect crop growth. Saline soils get their salts from natural or anthropogenic *salinization. **Salinity** is the degree to which water contains dissolved salts, usually expressed in parts per thousand.

salinization The accumulation of unusually high concentrations of dissolved sulphates and chlorides of sodium and calcium, often as a result of large-scale irrigation in semi-arid areas; as irrigation water is evaporated or transpired, the salt ions are largely left behind to accumulate in the soil (Saysel and Barlas (2001) *Ecol. Model.* 139, 2–3). Salinization can reduce plant growth rate and yield, and, in severe cases, cause total crop failure. In certain circumstances, excessive salinity in soils cannot be reduced over time by routine irrigation and crop management practices (Qadir et al. (2000) *Land Degrad. & Dev.* 11).

In coastal areas, salinization can be associated with the over-extraction of groundwater, which can lower the water table and lead to the invasion of sea water (Cardonna et al. (2004) *Env. Geol.* 45, 3). Tsunamis can cause severe salinization problems.

(((●))) SEE WEB LINKS
• European Land Management and Natural Hazards Unit on soil salinization.

saltation The bouncing of material from and along a river bed or a land surface. The impact of a falling sand grain may splash other grains upwards so that a chain of saltating particles may be set up. Saltation upwards is the result of *lift forces (see J. S. Bridge 2003); the downward movement occurs when lift is no longer effective and the particle is subject to drag and gravity. See Stout (2004) *ESPL* 29, 10 on the relationship between saltation activity, and relative wind strength; see also Bauer et al. (2004) *Geomorph.* 59, 1–4 on shear velocity. The **saltation–abrasion model** represents river incision by saltation; see Sklar and Dietrich (2006) *Geomorph.* 82, 1–2.

salt dome A large mass of evaporate minerals that has pierced, and risen through, denser, overlying *sedimentary rock, forming a dome-shaped arch with salt at its core. See Autin (2002) *Geomorph.* 47, 2 on salt domes in Louisiana.

salt marsh Tidal salt marshes are found along the low-energy coastlines of mid- to high-latitude regions of the major continents (except Antarctica), and lie at the *ecotone between terrestrial, freshwater, and marine systems (Greenberg et al. (2006) *BioSci.* 56, 8). As the marsh develops, *halophytes (such as marsh samphire and sea aster in Britain) pave the way for less hardy species, and the marsh becomes part of the coastland; see van Wijnen et al. (1997) *J. Coastal Cons.* 3, 1. See Santos et al. (2013) *PLoS ONE* 8, 11 e78989. DOI:10.1371/journal.pone.0078989 on salt marsh vegetation stress and recovery after the oil spill in Barataria Bay, Gulf of Mexico.

salt water intrusion The movement of saline water into fresh water aquifers. It happens to a restricted extent in most coastal aquifers, because, due to its mineral content, salt water is denser than fresh water, and can push inland beneath the fresh water. Certain human activities have contributed to salt water intrusion, which has risen in many coastal areas, as a result of pumping ground water from coastal fresh water wells. Asfahani and Abou Zakhem (2012) *Acta Geophysica*, DOI: 10.2478/s11600-012-0071-3 review all aspects of this problem, from surveying to corrective action in the Khanasser Valley, Syria.

salt water weathering (salt weathering) A form of weathering, especially important in hot deserts, which combines crystal growth with *hydration. See Smith and Warke in D. Thomas et al., eds (1997) on the importance of the nature of salt and moisture delivery to weathering in arid areas, and Buj et al. (2011) *ESPL* 36, 3, 383 on the role of salt in the weathering of limestone. Salt weathering is closely linked to groundwater controls in enclosed arid depressions (Aref et al. (2002) *Geomorph.* 45, 3–4). See Viles and Goudie (2007) *Geomorph.*

85, 1–2 on salt weathering in coastal salt pans. Smith et al. (2005) *Geomorph.* 67, 1–2 evaluate recent simulations of salt weathering. Landforms associated with arid salt weathering include columnar weathering networks (Goudie et al. (2002) *Zeitschrift Geomorph.* 46) and *tafoni.

sand Particles of rock with diameters ranging from 0.06 mm to 2.00 mm. Most sands are formed of the mineral quartz. Sandy soils are loose, non-plastic, and permeable, and have little capacity to hold water.

sand dune A hill or ridge of sand accumulated and sorted by wind action. Sand will settle on a dune because the friction of the sandy surface is enough to slow the wind, which then sheds some of its *load. Dunes formed in the *lee of an obstacle are **topographic dunes**: *lunettes* form in the lee of a *deflation hollow, *nebkhas* in the lee of bushes, and **wind shadow dunes** in the lee of hills and plateaux. **Crescentic dunes**—sand mounds, barchans, barchanoids, and transverse ridges—are restricted to areas with unidirectional high winds and minimal vegetation (Bishop (2001) *ESPL* 26).

Transverse dunes form when sand supply is abundant, and wind direction constant (Walker and Nickling (2002) *PPG* 26, 1). As the sand supply is reduced, the dunes transform into **parabolic dunes**—hairpin-shaped with the bend pointing downwind (Mitasova et al. (2005) *Geomorph.* 72, 1–4). Where the direction is very changeable, **star dunes** form. The linear **seif dunes** form when two prevailing winds alternate, either daily or seasonally; the leeward slope is a zone of deposition, as well as a zone of erosion (Tsoar et al. (2004) *Geomorph.* 57, 3–4). When sand is limited, **barchans** form with the horns pointing downwind;

Sand dunes Arrows indicate wind direction

see Katsuki et al. (2011) *ESPL* 36, 3, 372 on a collisional simulation of barchans. The height of the slipface of a barchan is proportional to the width of the horns (Wang et al. (2007) *Geomorph.* 89, 3–2). Barchans may be changed into dome dunes, and vice versa; completely gradational forms exist between the two

S

(Fryberger et al. (1984) *Sedimentol.* 31). **Barchanoid** dunes are undulating, continuous cross-wind dunes which may grade into long transverse dunes; see R. Cooke et al. (1993). Dune fields have recently come to be recognized as self-organizing systems that can be seen progressing from states of disorganization or randomness to uniformity; see Wilkins and Ford (2007) *Geomorph.* 83, 1–2.

sand dune dynamics The processes of dune movement; see Wiggs (2001) *PPG* 25, 1.

sand dune stabilization Techniques designed to prevent the erosion and deposition of sand include: the establishment of shelterbelts; mass tree revegetation using a synthetic water absorbing polymer; developing live hedges; and mechanical fencing. Shelterbelts were the most effective; mechanical fencing the least (Raji et al. (2004) *Env. Model. & Assess.* 99, 1–3). Schwendiman (1997) *Int. J. Biomet.* 21 describes the three-stage approach in temperate climates, and Gadgil and Ede (1999) *Land Deg. & Dev.* 9, 2 outline sand dune stabilization methods in New Zealand. See also Lomba (2008) *J. Coastal Res.* on monitoring sand dune stabilization in Israel.

sand flux The flowing of sand. Dong et al. (2004) *Geomorph.* 57, 1–2 find that sand flux over a sandy surface increases with height in the very near surface layer, but then decays exponentially; see also Wang et al. (2008) *Geomorph.* 96, 1–2. Yang et al. (2011) *J. Arid Land* 3, 3, 199 estimate a sand flux in the Taklimakan Desert, China, to be 258.67×10^{-4} kg/m˙s.

sandur A sheet, or gently sloping fan, of *outwash sands and gravels (pl. *sandar*); see Marren (2004) *Sed. Geol.* 164, 3–4, and Magilligan (2002) *Geomorph.* 44, 1. Sandar fans are commonly found in glacial troughs (Thomas and Chiverrell (2006) *Quat. Sci. Revs* 25, 21–2).

sand-wedge polygons *See* FROST CRACKING.

Santa Ana *See* LOCAL WINDS.

sapping The breaking down and undermining of part of a hillslope, triggering small slips. Sapping is significant in valley formation (Nash (1996) *ESPL* 21, 9). See Twidale (2007) *Revista C. & G.* 21, 1–2 on sapping at a river bluff. For **sapping erosion**, see Fernández et al. (2008) *Geomorph.* 95, 3–4.

SAPRIN The Structural Adjustment Participatory Review International Network is a global network established to strengthen the organized challenge to *structural adjustment programmes.

(((●))) SEE WEB LINKS
• The SAPRIN website.

saprolite Weathered rock *in situ* formed by deep weathering to depths of tens or hundreds of metres (Ollier et al. (2007) *Geomorph.* 87, 3).

sapropel An unconsolidated mud in estuaries, lagoons, lakes, and shallow seas, its formation depending on heavier precipitation (Rohling and Hilger (1991) *Geol. Mijnbouw* 70), and to variations in the Earth's orbit—they are thus useful in reconstructing palaeoclimates (Larrasoaña et al. (2006) *Phys. Earth & Planet. Interiors* 156, 3–4).

saprophyte A largely surface-dwelling, primarily free-living, organism, using enzymes to break down dead organic matter. Most are fungi or bacteria—fungi play vital roles in the nutrient cycles of forest ecosystems, acting as mutualists, parasites, and saprophytes of virtually all plants (Czederpiltz et al. (1999) *McIlvainea* 14).

sastrugi Ridges of ice particles lying across an ice sheet, formed by strong winds, and oriented at right angles to them. The sastrugi alter the surface roughness, and this alters the resistance to wind. Consequently, more drift snow builds up; this self-enhancing process thus forms larger sastrugi (Herzfeld et al. (2003) *Hydrol. Procs* 17, 3).

satellite town Satellite towns are established around big cities to attract population surplus and control population growth. See Ziari (2006) *Cities* 23, 6 on satellite towns in Iran. Try V. Vasudehan (2013) on life in an Indian satellite town.

satisficing The setting of sub-optimal targets to which people aspire. Satisficing has been widely observed among residential developers: 'for developers, the most important rule is to choose locations that meet a minimum profit criterion—a rule that results in satisficing . . . The main consequence of satisficing is that developers prefer projects on greenfield sites that take a shorter time to build and sell' (Mohamed (2006) *J. Planning Ed. & Res.* 26).

saturated adiabatic lapse rate (SALR) *See* ADIABAT.

saturated mixing ratio lines Lines of constant humidity mixing ratio plotted against height and pressure on a *tephigram.

saturated zone overland flow (saturation overland flow) *See* OVERLAND FLOW.

saturation (saturated air) There is a constant exchange of water molecules between liquid water, or ice, and the air, as evaporation and *condensation take place. **Saturation vapour pressure** (es) depicts a balance in the air between condensation and evaporation. es is greater in water droplets than in sheet water, and lower in impure water than in pure. Where es in an air parcel is greater than e (the *ambient *vapour pressure) there will be net evaporation; where e is greater than es, there will be net condensation.

savanna Broad belts of tropical grassland flanking each side of the equatorial forest of Africa and South America. Furley (2006) *PPG* 30, 1 reviews the determinants of savanna: climate and water, fire, soil–vegetation relationships, grazing. See also Boone et al. (2002) *Afr. J. Ecol.* 40. The savanna belts are associated with the sinking of high-level equatorial air on its return to the *intertropical convergence zone. Rainfall is therefore slight, and trees are modified to minimize water loss, with small leaves, and often thorny (Chidumayo (2001) *J. Veg. Sci.* 12).

The boundaries of the savanna are far from clear; there is a gradual change from tall grasses, 1–3 m high with scattered trees, to grassy woodland, and finally to the rain forest. However, where savanna vegetation has been repeatedly burnt (Balfour (2002) *Afr. J. Range & Forest Sci.* 19, 1) there can be a sharp division between this and the equatorial forest, which is less easily fired. Laris (2011) *AAAG* 101, 5, 1067, explains that anthropogenic burning differs from the one based on ecological theory in that its spatiotemporal pattern is relatively consistent from year to year.

scalar In *GIS, **scalar data** are defined by measurements based on a common standard. A particular data set may use a measurement system that may only apply to itself. 'GIS allows you to establish a scalar description for features you can't really measure any other way' (M. N. DeMers 2009) as in a ranking of landscapes by beauty.

scale A level of representation; in cartography, the ratio between map distance and distance on the ground. Geographical difference is expressed at all scales, from the inter-personal to the institutional, and from the national to the international. However, plotting geographical information on too small a scale can conceal information which only becomes apparent at larger scales and higher resolutions. The map of US persistent poverty below purports to show poverty at county level; and shows no poverty in Oregon. The map of Oregon on p. 433, also at county level, tells a different story.

Conversely an examination of geographical information at too large a scale may conceal the bigger picture. Cash and Moser (2000) *Glob. Env. Change* 10, 2 advise on matching the scale of the assessment with the scale of management, thus avoiding **scale discordance**. Vincent (2007) *Glob. Env. Change* 17, 1 describes the development of two empirical adaptive capacity indices for use at different scales of analysis.

'Scale is both a methodological issue inherent to observation—its epistemological moment—and an objective characteristic of complex interactions within and among social and natural processes—its ontological moment' (Sayre (2005) *PHG* 29, 3). 'Every scale brings a new set of problems that coincide with a new nature/society interface' (Dolidon (2007) *Cybergeo: Env. Nat. Paysage* 363). Sassen (2000) *Pub. Cult.* 12, 1 observes that the national and the global are not mutually exclusive scales but 'overlap and interact in ways that distinguish our contemporary moment', and Biggs et al. (2007) *Ecol. & Soc.* 12, 1 produce a dialogue on multi-scale scenarios. A. Herod (2011) provides an integrated single-volume introduction to scale in human geography.

The **politics of scale** describe the ways in which scale choices are constrained overtly by politics, and more subtly by choices of technologies, institutional designs, and measurements; see Lebel et al. (2005) *Ecol. & Soc.* 10, 2, and Silvey (2004) *PHG* 28, 4. Haarstad and Fløysand (2007)

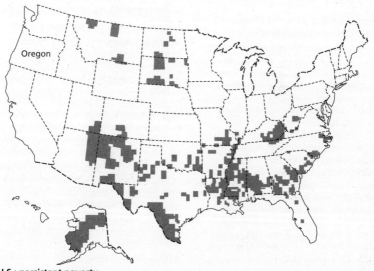

U.S.: persistent poverty

Scale. Oregon: persistent poverty

Pol. Geog. 26, 3 criticize the debate on politics of scale 'for leaving several central questions relatively unexplored'.

For modelling scale in GIS, see E. Wentz (2003) and Florinsky and Kuryakova (2000) *Int. J. GIS* 12.

'Scale is also understood to be dynamic, created or at least shaped by political contests over control of space. It also helps relate political geography to economic, social, transnational, and natural phenomena' Gallaher (2009) in C. Gallaher et al.

scale effect (scale-dependence, spatial-scale dependence) At different scales, certain different processes prevail. The simplest example of this is found in sediments: small particles, such as clays, are 'stickier' as cohesive forces are relatively more important when the particles get smaller; *see* HJULSTROM DIAGRAM. In biogeography, a major scale-dependence notion is that different ecological patterns and processes are dominant at different spatial scales. This notion in turn implies that changes in scale, be it the size of unit plot or the size of study area, involve a change in the correlation structure among biotic and abiotic components; see Kim et al. (2012) *AAAG* 102, 2, 276. Wei et al. (2012) *Chin. Geogra. Sci.* 22, 2, 127 show that the driving forces and mechanisms of water erosion variations are quite different at different scales.

scaling law If something is **scalable**, a change in its size will not change its nature. Thus, a triangle is always a triangle, however large or small it is. In contrast, an elephant shrew could not be scaled up into an elephant; its skeleton would not be strong enough, its metabolism unworkable; do see http://www.av8n.com/physics/scaling.htm. If there is a relationship between two scalable entities which can be expressed as a law, such as the ratio between the length of one side of a square to its area, then that relationship will hold good whatever the scale they are both at, and that law is a scaling law. With a scaling law, it is possible to compare information collected at one scale with information at a different scale, and to predict outcomes at different scales.

SEE WEB LINKS
• Introduction to scaling laws.

scarp retreat A synonym for *parallel retreat.

Schmidt hammer (Swiss hammer, rebound hammer) This device can measure the strength of a rock. See Goudie (2006) *PPG* 30, 6, 703 on the Schmidt hammer in geomorphological research.

science park An area of industrial development set up in collaboration with, and in close proximity to, a centre of higher education, in order for industry to capitalize on academic research, for jobs to be created locally, and for the educational institute to generate income. Knowledge from foreign companies is also important; see Lefner et al. (2006) *Env. & Plan. A* 38, 1 on the Zhongguancun Science Park, Beijing.

Scirocco *See* LOCAL WINDS.

scoria A volcanic rock made of sharp rock fragments and full of air pockets once occupied by gases. **Scoria cones** are constructed by the accumulation of ballistically ejected clasts from discrete, relatively coarse-grained Strombolian bursts, and subsequent avalanching (Valentine et al. (2005) *Geology* 33, 8).

scour Erosion; see, for example, Saynor et al. (2004) *Geografiska A* 86, 2 on scour holes and gullying. The depth of a scour hole in a river bed is strongly correlated with *discharge, with scouring on the rising limb of the hydrograph, and filling as the rate of transport through the pool declines (Borg et al. (2007) *Geomorph.* 89, 1–2). See also Leopold in L. Leopold et al. (1964). **Scour and fill** is the erosion and subsequent filling of a channel; see Xu (2004) *Geomorph.* 57, 3–4, and Fuller and Hutchinson (2007) *NZ Geogr.* 63, 3 on measuring scour and fill. For **river bed scours**, see Eilertsen and Hansen (2008) *Geomorph.* 93, 3–4.

scree Shattered rock fragments which accumulate below and from free rock faces and summits. Scree is formed by *freeze–thaw; see Statham (1973) *TIBG* 59. Its formation requires jointed rocks that allow water to penetrate.

sea-floor spreading The creation of new crust as *magma rises up at a *constructive *plate margin. Sea-floor spreading can be detected through the presence of magnetic anomalies; see Roest and Srivastava (1989) *Geology* 17. A **spreading axis** is a zone of active eruptive fissuring at a mid-oceanic ridge; see Wise et al. (1992) *Procs Ocean Drilling Program* 120 on the south-east Indian ridge.

sea level Modern analyses of sea-level changes due to glacial *isostatic adjustment are based on the classic sea-level equation by Farrell and Clark (1976) *Geophys. JR Astr. Soc.* 46; Mitrovica and Milne (2003) *Geophys. J. Int.* 154, 2 present a further equation. See Kendal et al. (2005) *Geophys. J. Int.* 161, 3 on theoretical approaches to computing gravitationally self-consistent sea-level changes due to ice growth and *ablation.

For something like 227 000 of the last 250 000 years, sea levels were around 10 m below their current stand (Voris (2000) *J. Biogeog.* 27). Nunn (2000) *NZ Geogr.* 56, 1 demonstrates the 'possible and/or likely' effects of a fall in sea level around AD 1220–1510 on Pacific islands: falling water-tables; the emergence of reef surfaces and the consequent reduction of nearshore water circulation; the emergence of reef islets; and the conversion of tidal inlets to brackish lakes. Heany et al. (2005) *J. Biogeog.* 32, 2 argue that late Pleistocene low sea-level shorelines, rather than current shorelines, define patterns of genetic variation among mammals on oceanic Philippine islands.

N. Bindoff et al. (2007) conclude that sea levels rose at an average rate of 1.7 ± 0.5 mm yr^{-1} during the 20th century. They estimate future sea-level rise of between 0.18 and 0.59 m for the 21st century. 'These simple figures conceal substantial spatial and temporal variability' (Church et al. (2006) *Glob. & Planet. Change* 53); see Edwards (2007) *PPG* 31, 6. For many atoll island states, sea-level rise is more than just a threat to their tourism; it also determines their survival (Wong (2003) *Sing. J. Trop. Geog.* 24, 1). These countries also have low adaptive capacity to sea-level rise, heavily dependent as they are on fragile environmental goods and services, such as coral reefs and fisheries (Nurse and Moore (2007) *Nat. Resources Forum* 31, virtual issue). Sea-level rise in Tuvalu is likely to contribute to saltwater intrusion, long-term loss of land, and damage to ecosystems, agriculture, and livelihoods (J. McCarthy et al., eds 2001). See Dawson in C. Gopalakrishnan and

N. Okada, eds (2008) on sea-level rise and the Thames estuary. For a thorough review of this complex subject, see J. A. Church et al., eds (2010).

seamount A relatively small-scale (up to 250 km across) rise, sub-circular in plan view. The date a seamount is formed, and the tectonic setting they erupt into (on-ridge or off-ridge), reflect the way the lithosphere interacts with the fluid mantle beneath. For the radiometric dating of seamounts, see Hillier (2007) *Geophys. J. Int.* 168. McClain (2007) *J. Biogeog.* 34, 11 challenges the view that seamounts are oases/biodiversity hot spots.

search behaviour The way an individual, or entity, selects a location. The choice depends on the information available, and on the decision-maker's ability to evaluate the information. Try McColl-Kennedy and Fetter (2001) *J. Services Marketing* 15, 2.

seasonal forecasting A forecast of weather conditions for a period of three to six months, based on, in particular, sea surface temperatures. See A. Troccoli et al., eds (2005).

seawater incursion (seawater intrusion) In coastal aquifers, lighter fresh water lies above a wedge of more dense salt water, and is separated from it by a mobile transition zone. If aquifers are pumped at rates exceeding their natural capacity to transmit water, seawater is drawn into the system to maintain the regional groundwater balance. See Qahman and Larabi (2006) *Hydrogeol. J.* 14, 5 on Gaza.

secession The transfer of part of the territory and population of one state to another, whether pre-existing, or newly created. 'The last act of the European geopolitical drama has yielded the greatest "advantage" for the former minorities in Central and Eastern Europe,

but, as calculations show, the nation-state model became a less effective tool to solve ethnic problems, since each new secession engenders the appearance of new minorities' (Kotossov and Treivish (1998) *Polit. Geog.* 17, 5). Writing on Biafra's abortive secession in 1967, Diamond (2007) *Dialec. Anthrop.* 31, 1–3 observes that 'Nigeria is not a nation but a political idea imposed by force of foreign arms. The variety of peoples within its borders will continue to seek accommodation.' See Le Breton and Weber (2001) IMF WP/ 01/176, on 'how to prevent secession'.

secondary air mass An *air mass which has been modified by the passage of time or by its movement to a different area. A *k* air mass is colder than the surface it moves over, and is usually unstable (*see* INSTABILITY; a *w* air mass is warmer, and usually stable).

secondary industry (secondary sector) The creation of finished products from raw materials; that is, manufacturing.

second home A property occasionally used by a household whose normal residence is elsewhere. Second homes are usually found in rural areas; the 'compensation hypothesis' describes the association between urban density and the consumption of second homes (Módenes and López-Colás (2007) *Tijdschrift* 98, 3). Second homes can bring increased custom to rural areas, but may drive up house prices beyond the pockets of local residents. Stressing that he is referring to Sweden only, Marjavaara (2007) *Tourism Geogs.* 9, 3 observes that second homes become a convenient scapegoat for rural decline: second home owners are often targeted as holders of alien values; locals are seen as fragile victims.

sector theory The view that housing areas in a city develop in sectors along the lines of communication, from the

Sector theory

*CBD outwards. High-quality areas locate along roads; industrial sectors develop along canals and railways, away from high-quality housing; working-class housing locates near industry. 'Basic urban pattern such as concentric zone, sectors and multi-nuclei . . . can be interpreted as "ideal" conditions or basic structure at local scale, and the real pattern could be viewed as a composition of multiple structures in the urbanization process' (Xu et al. (2007) *Landsc. Ecol.* 22, 6).

securitization The practice of pooling various types of debt, such as mortgages, car loans, or credit card obligations and selling these together. This process of 'increasing institutional and geographic dispersion of financial asset ownership, feeds the development of securities industry; an industry links issuers with investors, sellers with buyers, and offers services to address the information asymmetry and agency problems arising from securitization. Securities firms benefit from co-location' (Wójcik (2011) *J. Econ. Geog* 11, 6, 925).

sedentary Fixed, not moving, as in **sedentary agriculture** where the farmer and the fields are permanently settled.

sediment 1. Material which has separated and settled out from the medium—wind, water, or ice—which originally carried it. For fluvial sediments the ability of a river to carry sediment depends on particle size as well as the river discharge. *See also* LOAD. A **sediment slug** is a body of *clastic material associated with disequilibrium conditions in fluvial systems over time periods above the event scale. Slugs range in magnitude from unit bars to sedimentary features generated by basin-scale sediment supply disturbances; see Nicholas et al. (1995) *PPG* 19, 4, 500.

2. A **sediment budget** is the tally of inputs and outputs for a specific *open system, over a given time period; the classic paper is Trimble (1983) *Am. J. Sci.* 283. A **balanced sediment budget** is an equation of *mass conservation in which the sediment fluxes related to the sources, sinks, and storages are balanced. The concept is important in environmental management: see Nordstrom et al. (2002) *Geomorph.* 47, 2–4.

sedimentary rock A rock composed of sediments, usually with a layered appearance (*bedding plane). **Sedimentation** is the deposition of sediment from a state of suspension in water or air, also called *siltation*.

sediment cell A length of coastline, and its associated nearshore area, within which the movement of sand and shingle is largely self-contained. The littoral sediment cell concept 'has become synonymous with pro-active coastal management practices in England and Wales' (Cooper and Pontee (2006) *Ocean & Coastal Manage.* 49, 7–8).

Sediment yield is the total mass of sediment that reaches the exit of a drainage basin. Tamene et al. (2006) *Geomorph.* 76, 1–2 find that pronounced terrain steepness, easily detachable slope material, poor surface cover, and gullies promote high sediment yields; de Vente and Poesen (2005) *Earth Sci. Revs* 71, 1–2 propose a general relation between basin area, active erosion processes, sediment sinks, and total sediment yield; and Verstraeten et al. (2003) *Geomorph.* 50, 4 develop an index model to explain variability in area-specific sediment. A **specific sediment yield** is the sediment yield per unit area of the drainage basin, usually per square kilometre, and is calculated from:

specific sediment yield $(t\ km^{-2}y^{-1})$
= sediment yield $(t\ y^{-1})/$
drainage basin area (km^2)

Holeman (2010) *Water Res.* 4, 4, 737 provides a table of the sediment yield of major rivers of the world.

Sedimentation is a term often used to describe the blocking of an aquatic system by the deposition of sediment; see Bennett et al. (2005) *Water Resources Res.* 41, W01005, and Wang et al., *Water Resources Res.* 41, W09417. See Flemming (1965) *J. Sed. Res.* 35, 2 on **sedimentary particles**, their transport history, and their depositional environments.

segmented labour market A *dual labour market may be composed of skilled and unskilled segments. 'Welfare, tax and labour-market institutions continue to facilitate and maintain a segmented labour market with a wide range of employment systems and practices, making it virtually impossible for the growing proportion of the labour force in low-wage, service-dominated jobs to achieve independent living on an individual basis' (McDowell (2004) *PHG* 28, 2). See also Hedberg, SULCIS *W. Paper* 2008:1.

segregation The separation of a large population into subgroups—based on grounds of income, race, religion, or language—located in distinct residential areas. Ellis et al. (2004) *AAAG* 94, 3 write on prejudice and racial segregation, while Blumen and Zamir (2001) *Env. & Plan. A* 33 find spatial segregation to be intrinsic to the capitalism-patriarchy. In a similar vein, Smith (2012) *TIBG* 37, 3, 461 outlines ways in which distinct subpopulations of British society are becoming more transient and segregated. His work is on British coastal towns, but his conclusions apply throughout the UK. Schnell and Yoav (2001) *AAAG* 91, 4 suggest a socio-spatial isolation index of segregation. Goldhaber (2007) *Urb. Geog.* 28, 6 presents an empirical model for testing ethnic segregation in the context of spatial perception. *See also* INDEX OF SEGREGATION.

seiche A stationary or *standing wave in an enclosed body of water, resulting in major shifts in wind magnitude or direction (Luettich et al. (2002) *C. Shelf Res.* 22, 11–13).

seif *See* SAND DUNE.

seismic Of an *earthquake. The **seismic focus** is the point of origin of the earthquake within the crust; see Deji (1999) *Eng. Geol.* 51, 3 on seismic foci around the Three Gorges Project. The velocity of **seismic waves** varies with the density of the material through which they travel. This property is used in **seismic tomography**—the investigation of the interior of the Earth; see W. K. Lee (2003).

selective logging The felling, at intervals, of the mature trees in a forest of mixed age. Selective logging (combined with large tree retention) is recommended for minimizing ecological impacts; see Rüger et al. (2007) *Forest Ecol. & Manage.* 252, 1–3. Selective logging can, 'together with forest management practices, significantly delay or accelerate the transition of a forest landscape to a deforested state and its return to a recovered state' (Rondon et al. (2012 *Int. J. Forestry Res.* DOI:10.1155/2012/170974).

selective migration This may be spontaneous, as when a particular age–sex group constitutes most of the migrants; 'age-selective migration tends to dichotomize areas between younger and older, with higher and with lower natural increase' (Morrill (1995) *Annal. Reg. Sci.* 29, 1). 'Age-selective migration is systematically draining the entry

labour pool in rural America' (W. Kandell and D. L. Brown 2006). Boyle (2004) *PPG* 28, 6 argues that selective migration 'will alter our interpretation of health and mortality inequalities'.

A leading cause of selective migration is the increasing skill-focus of immigration policy in a number of industrialized countries—'a trend that is likely to intensify as rich countries age, and competitive pressures build in knowledge-intensive sectors' (Kapur and McHale (2005) Center for Global Development). Vigdor (2002) *Rev. Econ. & Sta.* 84, 4 considers selective migration and economic outcome. See Hjort (2009), *Gerum Kulturgeografi* (2009) 1, notably p. 18.

self-organization A process whereby some sort of order arises from an initially disordered system, a process arising from the interactions within the system. For example, distinctive zones of land use may arise within a city though they have not been formally designated as such. 'Morphodynamic systems often display a sequence of positive feedback driving the system towards a new state, followed by negative feedback, which stabilizes the system, resulting in equilibrium. This is referred to as self-organization. The result of this process is a rather orderly arrangement of sediments and landforms' (Masselink in J. Holden (2012) p. 447). Examples include the formation of evenly spaced, nearly uniform ripples on sand dunes or stream beds, or of *patterned ground in periglacial landscapes. 'Nonlinear, dissipative interactions among the small- and fast-scale constituents of a system give rise to order at larger spatial and longer temporal scales' (Phillips (2003) *PPG* 27, 1). 'Those not convinced by arguments of self-organisation and inherent behaviour as an explanation in itself commonly ask the mechanism of the changes. Even for those who support the theory, questions still arise of how the self-organisation takes place and how the changes are communicated from one part of the system to another such that order emerges' (Hooke (2007) *Geomorph.* 91, 3–4).

Self-organized criticality describes a system in dynamic equilibrium near a threshold condition. The concept arises from Bak et al. (1987) *Phys. Rev. Letts* 59, 4, who develop a simple *cellular automaton model in which sand is added, grain by grain, to a surface to form a pile. When local slopes become too steep a collapse occurs, moving sediment to neighbouring cells, which too can collapse if the adjusted slopes are too steep. See Coulthard and Van De Wiel (2007) *Geomorph.* 91, 3–4.

self-similarity The whole of a **self-similar** object has the same shape as one or more of the parts; thus, self-similarity is a typical property of *fractals. A state of constant proportion, or self-similarity, is called *isometry*. In geography, these similarities are not exact, but statistical; it's therefore necessary to agree upper and lower limits to the scales involved, and sometimes the nature of the similarity will depend on the scale, since different processes may operate at different scales; *see* SCALING LAW. Mandelbrot (1967) *Science* 156, 636 saw self-similarity in the curves of coastlines, but Andrle (1996) *ESPL* 21, 955 finds that the west coast of Britain is not statistically self-similar over the range of scale of measurement. Baas (2002) *Geomorph.* 48, 309 notes that theories of self-similarity are controversial, 'especially with regard to their application to real physical systems' and that 'the peak of popular research interest into chaos, fractals, and self-organization has passed'. Nonetheless, Gaucherel et al. (2011) *ESPL* 36, 10, 1313 argue that self-similar dimension has been checked using either the whole channel network

or a single channel link, and uses a new approach to check self-similarity at intermediate scales.

semiotics The ways in which signs and meanings are created, decoded, and transformed; 'cultural landscapes . . . are a collection of symbols to be deciphered—an exercise in semiotics' (Knudsen et al. (2007) *Tourism Geog.* 9, 3). Semiotics are 'commonly understood as abetting our understanding of the reflexive relationship between people and their environment' (Boogart (2001) *Middle States Geogr.* 34, 38).

sense of place Either the intrinsic character of a place, or the meaning people give to it, but, more often, a mixture of both. 'Sense of place . . . is an umbrella concept that includes all the other concepts—attachment to place, national identity, and regional awareness . . . The messages transmitted are not neutral, but rather they reflect the subjective senses of the beholder or the perceptions of "society"' (Shamai (1991) *GeoForum* 22, 3). Robinson (2006) *UNESCO Univ. & Heritage, 10th Int. Seminar*, argues that it is 'the intangible characteristics of places which are central to our recognition of the more perceptually and socio-culturally founded notion of sense of place'. Planners consciously create or preserve memorable and singular structures to make a space distinctively different—such as the UK Angel of the North. Guthy et al. (2014) *J. Manage. Inq.* DOI: 10.1177/9780199680856511, find—not entirely convincingly—that a sense of place is linked with sustainability.

serac An ice pinnacle, or series of pinnacles on the surface of a glacier resulting from tensional failure in the more rigid upper crust when the glacier: moves over a slope; spreads out over a plain; or passes round a bend in its valley.

sere One of a sequence of developmental stages in plant succession. A primary stage is a **prisere**; a **hydrosere** develops in water; a **psammosere** on a sand dune; and a **xerosere** in an arid location—see Dimeyeva (2007) *Basic & Appl. Dryland Res.* 1. If a plant succession is interrupted, the resulting secondary stage is a **sub-sere**.

service industry Those economic activities (also called *tertiary industry) concerned with the distribution and consumption of goods and services. Proximity to clients is the major locational factor. Other factors promoting the *clustering of service firms and workplaces are: agglomeration economies, supply of infrastructure, supply of labour, and new firm formations.

Some services—particularly less specialized household services—are fairly evenly located in relation to settlement patterns. Those less evenly located include specialized household services, central government functions, and professional or business services, which concentrate in major city regions, or within regional clusters. With information and communication technology, service activities can decentralize from urban cores to the suburbs, and to more peripheral sites, both nationally and internationally (Hermelin and Rusten (2007) *Geografiska B* 89, 1).

settlement Any form of human dwelling—from a single house to the largest city.

settlement hierarchy A division of settlements into ranks, usually by population size.

settlement pattern The spatial distribution of settlements. Some settlement patterns may be seen as a reflection of cultural traditions: a dispersed settlement pattern in Norway, for example, has acquired a symbolic

meaning (Cruickshank (2006) *Norsk Geograf. Tidssk.* 60, 3). *Nearest neighbour analysis may be used to test for regularities in settlement patterns.

sexuality, geographies of The spatialities of sexualities, intersecting discussions of sexualities with issues such as development, race, gender, and other forms of social difference (K. Browne et al., eds 2007). For example, C. Hemmings (2002) examines bisexual spaces as places that are defined by both geographical boundaries and cultural significance, showing how and why safe places have developed for gay, lesbian, and bisexual communities. In 1999, Binnie and Valentine (*PHG* 23, 2) argued for a move away from a simple mapping of lesbian and gay spaces towards a more critical treatment of the differences between sexual dissidents. Consult Hubbard et al. (2008) *PHG* 32, 3 and Oswin (2008) *PHG* 32, 1 to see if their call has been answered, while Del Casino (2007) *Health & Place* 13, 4 discusses geographies of sexual health. *See also* QUEER GEOGRAPHIES.

shadow price The shadow price of any good is the increase (or decrease) in social welfare brought about by providing one more (or one less) unit of that good. Thus, the shadow price of carbon is defined as the social cost of emitting a marginal tonne of carbon (or the social benefit of abating a tonne). See Dietz (2007) DEFRA. This concept is used in *cost–benefit analysis to value intangible items like *amenity.

shanty town *See* SQUATTING.

share-cropping A farming type where the tenant pays his rent in produce rather than in cash. In the USA, lack of capital left the cropper dependent upon others for subsistence until market time. Storekeepers furnished the cropper with food and clothing during the year, the accumulated debt to be repaid, with interest, at harvest time. The store owner was often also the buyer of the crops (M. P. Conzen 1994). 'Share-tenant relations are . . . compromises between risk and incentive . . . Given a choice, better-off workers would seek tenant relations, while poorer workers or recent arrivals would have to enter sharecropping arrangements' (Brannstrom (2001) *Polit. Geog.* 20, 7).

shear (shearing) The deformation of a material so that its layers move laterally over each other. Shearing bends, twists, and draws out rocks along a fault or *thrust plane. A **shear plane** is the face along which shearing occurs. **Shear strength** is the ability of a rock or soil to withstand shearing; soil shear resistance is proportional to the tangent of the friction angle and therefore soils with larger friction angles exhibit a higher shear strength. Two major parameters of soil shear strength are cohesion, which is boosted by plant roots and peaks when soil moisture content is 12% to 14%, and internal friction, which is the resistance of a soil mass to sliding, itself inversely related to the amount of moisture in the soil. Slope failures are the consequence of loading conditions that exceed soil shear strength (Simoni et al. (2008) *Hydrol. Procs* 22, 4). **Shear stress** happens when the applied force acts tangentially to the surface of the body (the down-slope component). It is opposed to a normal stress (the slope-normal component) which is applied perpendicularly.

See Namikas (2000) *J. Sed. Res.* 72, 2 on measuring bed shear stress during aeolian saltation; Sime et al. (2007) *Water Resources Res.* 43, W03418 on estimating river bed shear. Within a glacier, the level of shear stress (τ) experienced at any point is dependent on the ice thickness and the surface slope. It can be expressed as:

$$\tau = \rho g(s - z) \sin \acute{\alpha}$$

where ρ is the density of ice, g is the acceleration due to gravity, s is the surface elevation, z is the elevation of the point within the glacier, and $\acute{\alpha}$ is the surface slope of the glacier.

sheet erosion A very slow-acting form of erosion whereby a thin film of water—

Shear Stress = W sin⊖   Normal Stress = W cos⊖

Shear Stress

Normal Stress

W   ⊖

⊖

Shear stress

sheet wash—transports soil particles by rolling them along the ground. Sheet erosion is a major process in the denudation of land surfaces, but R. Morgan (2005) claims that it is only really effective on steep and relatively unvegetated slopes. See Chaplot and Bisonnais (1998) *Soil Till. Res.* 46 and A. Goudie (2004).

sheet flow At the base of a glacier, a thin film or sheet of water at very high water pressure, a thousandth of a millimetre to a millimetre in thickness, being thinnest over bumps in the bed because of the higher pressure upstream. Sheet flow is thought to be dominant only beneath ice masses where the majority of basal water is derived from subglacial melting due to geothermal heating, sliding, or bed deformation.

sheeting The separation and peeling away of the outer layers of a rock mass (Wakasa et al. (2006) *Hydrol. Procs* 31, 10). *See also* EXFOLIATION.

shield The very old, rigid core of relatively stable rocks within a continent. See Stein (2003) *Precamb. Res.* 123, 1–2 on the formation of the Arabian-Nubian Shield, and Lidmar–Bergström (1995) *Geomorph.* 12, 1 on the Baltic Shield.

shield volcano A volcano formed of successive eruptions of free-flowing *lava which create a gently sloping, broad dome tens of kilometres across and more than one kilometre high. Mount Mazama, Oregon, is composed of several overlapping composite and shield volcanoes (Bacon and Lanphere (2006) *GSA Bull.* 118, 11).

shifting cultivation A farming system based on the clearance of land, cultivation until soil fertility declines, followed by fallowing for up to twelve years, in order to restore the vegetation to its initial condition. Vasconcelos et al. (2002) *Appl. Geog.* 22, 2 argue that shifting cultivation is ecologically sustainable for the subsistence of small groups. 'Shifting cultivation is a truly *sustainable system, perhaps the only one for humans. It can maintain agricultural production indefinitely with little reliance on external inputs when population levels are low—roughly no more than one person per 2 hectares of agricultural land' (C. A. S. Hall 2000). Uhlig et al. (1994) *Ecol. Studs* 101, 145 find that shifting cultivation alters carbon-release estimates.

shift-share analysis A method of estimating the relative importance of different elements in any growth or decline of regional industrial employment. See Fotopoulos (2007) *Growth & Change* 38, 1 for the methodology and for the application of shift-share analysis to demography, and Chen and Xu (2007) *World J. Models & Simulation* 3, 2, 90 on using shift-share analysis to gauge differences in economic strength.

shingle Pebbles on a beach, rounded by *abrasion, and reduced in size by *attrition. The abrasion rate of shingle is linked to, in rank order: mean wave height, pebble type, and pebble weight (Dornbusch et al. (2002) *J. Coastal Res.* SI36).

shit, geographies of The studies of spatial inequalities in access to sanitation and the consequences of this for human, environmental, and economic health, and of spatial and temporal differences in cultural attitudes towards excrement. Sanitation continues to be largely neglected in many developing countries, and government-backed sanitation initiatives, where they exist, often prioritize flush and discharge systems over low-cost or community-based alternatives. In 2012, the Census Commissioner of India reported that about 600 million Indians practise open-air defecation.

• Online version of 'Economic Impacts of Inadequate Sanitation in India'.

shopping, geographies of *See* RE-
TAILING, GEOGRAPHY OF.

shopping centre M. Ibrahim and P. McGoldrick (2003) examine the relations between the use of shopping centres and the transport modes used or available, together with policies that attempt to reduce the negative impacts of decentralizing retail facilities.

shopping goods In the process of selecting and buying things, those goods which the customer compares on such bases as suitability, quality, price, and style—such as furniture and clothing (AMA Report Definition Committee (1948) *J. Marketing*, 13, 2). Makers of shopping goods can limit the number of wholesalers and retailers in a geographic area in order to maintain channel control and enhance the image or exclusivity of their products (Gable et al. (1995) *J. Retail. & Consum. Services* 2, 4).

shore The land adjoining a large body of water/the sea. The **backshore** is normally above the high water mark, but still influenced by the sea; the **foreshore** is the area between high and low tide marks and is exposed at low tide. The **nearshore** is seaward of the foreshore, ending at the breaking point of the waves; the **offshore** is the zone seaward of the breakers, but in which material is moved by the waves.

shore platform A very gently sloping platform extending seaward from the base of a cliff, and subject to *salt weathering, alternate wetting and drying, *water-level weathering, and *quarrying, *hydraulic action, pneumatic action, and *abrasion. Trenhaile (2005) *ESPL* 30 finds that platform gradient increases with tidal range; with slow

downwearing, platform width increases with tidal range (but decreases with fast downwearing); and platform gradient decreases and width increases with wave energy, decreasing rock resistance, and platform roughness. Stephenson and Kirk (2000) *Geomorph.* 32 argue that waves on platforms are not capable of causing erosion because of their high loss of energy in these shallower waters. Kennedy et al. (2011, *ESPL* 36, 5, 686) conclude that a New Zealand shore platform is a function of the relative balance between wave energy and weathering, while the width of the platform is related to the ability of waves to dissect the platform along vertical joint lines.

sial The material of the continental crust, mostly of silica- and aluminium-rich minerals.

silcrete A *duricrust cemented with silica. See Thiry (2006) *J. Sedim. Res.* 61, 1, 111.

sill An *intrusion of igneous rock which spreads along bedding planes in a nearly horizontal sheet. Sills are mostly concave, and can be fed from dykes, or the steep climbing portions of deeper sills. Both sills and dykes can provide magma to overlying volcanic fissures, and sills can feed shallow laccoliths (Thomson (2006) *Bull. Volc.* 66, 4).

silt Fine-grained minerals, 0.002–0.06 mm in diameter, ranging in size between *clay and *sand in size. Silts are often deposited by rivers when the flood water is quiet.

Silurian A period of *Palaeozoic time stretching approximately from 430 to 395 million years BP.

• An overview of the Silurian period.

sima The material of the lower part of the continental crust and the oceanic crust, dominated by silica and magnesium.

sink A point, line, area, or reservoir where energy or mass is removed from a system. For example, soil and trees tend to act as natural sinks for carbon.

sink hole *See* DOLINE.

sinter Mostly silica and sulphate minerals, precipitated in layers from volcanic activity. Sinters provide evidence of past environments (Lynne et al. (2006) *Geology* 34, 9). **Snow sintering** is the bonding of snow particles, either by the diffusion of water molecules, or by pressure-induced creep; see D. McClung and P. Schaerer (1993).

sinuosity The extent to which a river bends. The **sinuosity ratio** is the distance between two points on the stream measured along the channel, divided by the straight line distance between those points; it is used to determine whether a channel is straight or meandering (Aswathy et al. (2008) *Env. Monit. Assess.* 138, 1–3).

site The position of a structure or object in physical, local terms, such as a river terrace.

Site of Special Scientific Interest (SSSI) A site in the UK which is of particular importance because of its geology, topography, or ecology. SSSIs are graded in terms of importance from 1 to 4.

(((🌐))) SEE WEB LINKS
• The Board of Natural England's SSSI website.

situated knowledge All knowledge is produced by people who have their own viewpoints, which reflect their time, place, and biases; their knowledge is *situated*. D. Haraway (1991) was the first to argue that knowledges are always situated, always produced by positioned actors working in/between all kinds

of locations, working up/on/through all kinds of research relation(ships). 'All these make a huge difference to what exactly gets done by whom, how, and where it's done, how it's turned into a finished product, for whom' (Cook et al. in D. Atkinson et al. 2008).

situation The location of a phenomenon, such as a town, in relation to other phenomena, such as other towns. *Compare with* SITE.

skid row An inner-city district characterized by a resident population that is in some way disadvantaged: many are 'poor, street-entrenched, addicted, alcoholic and/or mentally ill, the presence of a local drug market and a concentration of bars, low-rent hotel rooms and social service facilities' (Huey and Kemple (2007) *Urb. Studs* 44, 12).

sky view factor A measure commonly used by urban climatologists to quantify the openness of a site within an urban setting. It has major implications for incoming and outgoing solar and terrestrial radiation (Grimmond (2007) *Geog.* 30, 3).

slaking In geomorphology, the disintegration of fine-grained rocks. Slaking is one of the main rock-weathering mechanisms (Stefano and Ferro (2002) *Biosys. Engineer.* 81). **Slake durability** is the rate and amount of strength reduction in a rock after soaking (Morgenstern and Eigenbrod (1974) *J. Geotech. Eng. Div., ASCE* 101), and is an important parameter in slope stability (Yilmaz and Kalacan (2005) *Env. Geol.* 47, 7).

slash and burn An agricultural system whereby the trees are felled, the land cleared of most of the trunks, and the remaining vegetation fired. See Tschakert et al. (2007) *Ecol. Econ.* 60 on carbon offsets in small-scale slash-and-burn agriculture.

slickenside A set of linear marks on a fault or bedding plane caused by the frictional movement of one rock body against another. See Ree et al. (2003) *Island Arc* 12, 1.

slide A form of mass movement in which material slides in a relatively straight plane, generally breaking into many blocks as it moves. Slides usually have a length far greater than the depth of the moving material, and are triggered by high water pressure (Dai et al. (1999) *Eng. Geol.* 53, 3–4), but the competence of the rock is also significant; see Warburton et al. (2003) *ESPL* 28, 5.

slip-off slope The relatively gentle slope at the inner edge of a meander.

slope *See* HILLSLOPE.

slope wash The downslope transport of sediment by an almost continuous film of water. It can be seen as three erosional processes: sheet flow, rill erosion, and gullying.

slope winds *Anabatic and *katabatic winds.

slum An area of poor housing, characterized by multi-occupance and over-crowding; see McFarlane (2008) *Env. & Plan. A* 40, 1 on sanitation in Mumbai's informal settlements. Rao (2006) *Int. J. Urb. & Reg. Res.* 30, 1 rails against the use of 'slum' as shorthand for 'the distortion of urban substance into a dysfunctional stage for violence, conflict, and the iniquitous distribution of resources'; although places of marginalization, slums can also be places of great creativity and community survival (Appadurai (2002) *Public Culture* 14, 21; P. Chatterjee 2004). For example, slums often play an important role in a city and region's economy, with many slum residents

taking part in vital informal production, services, and recycling-based services that support more affluent sections of the city; see Pyati and Kamal (2012) *Area* 44, 3, 336. **Slum clearance** is the demolition of substandard housing, usually accompanied by *rehabilitation and *redevelopment. At the time of writing, history's most ambitious slum-clearance project is nearing completion, with the elimination of Dharavi, a Mumbai slum that houses between 600 000 and a million people.

slump A form of *mass movement (also called a single rotational slide) where rock and soil rotate backwards with movement down a concave face: 'a more or less rotational movement, about an axis that is parallel to the slope contours, involving shear displacement along a concavely upward facing failure surface' (D. Varnes 1978). Slumps are most common in thick *regoliths (Nwadije (1998) *Nat. Hazards* 1, 2) and large mudstone rock units (Plint (1986) *Sedimentol.* 33, 3).

slushflow A linear, channelled downslope movement of water-saturated snow on glaciers and in alpine stream channels during the beginning stages of the spring break-up. See Roberson et al. (2011) *Geograf. Series A*, 93, 2, 72.

small and medium enterprises (SME) Definitions vary across countries, but commonly enterprises with fewer than 500 employees are considered small. Ojala and Tyrväinen (2007) *Jyväskylä Univ. Archive* investigate the influence of cultural distance, geographical distance, and market-size variables to the target country preference of SMEs in the software industry.

smog A combination of smoke and fog. The fog occurs naturally; the 'smoke' is introduced into the atmosphere by human agency. See R. Kosobud (2006) on the cost-effective control of urban smog. *See also* PHOTOCHEMICAL SMOG.

SMSA A Standard Metropolitan Statistical Area in an urban area of the USA. This can be a town of 50 000 people, or two towns, each with more than 15 000 people, and together totalling more than 50 000; or a county with more than 75% of its population working in industry. Areas which seem, by employment, commuting, or population density, to be urban rather than rural are also included.

snow Frozen atmospheric water. A snow crystal is ice, up to 5 mm across, and variously shaped as a prism, plate, star, or needle. Snow crystals fall from stratiform clouds when the low-level air is degrees below freezing point, and the air above it is colder (Mason (1998) *QJ Royal Met. Soc.* 124, 545). When the low-level air is near 0 °C, snow crystals aggregate to form snowflakes (Fujiyoshi and Wakahama (1985) *J. Atmos. Scis* 42, 15).

Snow cover plays an important role in the hydrological cycle; see Holko et al. (2011) *Geog. Compass* DOI: 10.1111/j.1749-8198.2011.00412.x.

Snowball Earth This theory posits that, several times in Earth's history, oceans froze, ice sheets covered the tropics, and global temperatures plummeted to −50°C, so that the entire Earth may have actually frozen over, resembling a 'snowball', and potentially causing some of the most severe crises in the history of life on the planet. Such episodes may have been triggered by a reduction in greenhouse gases in the atmosphere, principally CO_2 (carbon dioxide) and CH_4 (methane). This would have made the global climate colder, creating larger areas of ice and snow. This ice and snow reflects more solar radiation than does bare ground or liquid water, which creates a 'positive feedback'. If the Earth ever became approximately half-covered by ice or snow, the feedback would become self-sustaining and glacial ice would rapidly spread to the equator.

SEE WEB LINKS
• Snowball Earth website.

snowbelt The states of the north-east USA, such as Michigan, Illinois, Indiana, Ohio, New Jersey, New York, and Pennsylvania. Discussion of a population shift from the snowbelt to the *sunbelt was frequent in the 1970s, but Essletzbichler (2004) *AAAG* 94, 3 maintains that the snowbelt/sunbelt dichotomy collapsed in the 1990s.

snowpack An area of naturally formed, packed snow that usually melts during the warmer months; see Caine (1995) *Geog. J.* 161, 1, 55 and *Field Guides* (2009), 15, 471.

social area analysis The analysis of a city to define **social areas** (urban areas which contain people of similar living standards, ethnic background, and lifestyle). 'Social area analysis tends to view the city as constituted of a mosaic of socially homogeneous containers' (Schnell and Benjamini (2005) *Urb. Studs* 42, 13).

social capital The relations that exist between individuals within both families and communities, together with the educational, cultural, and social resources and expertise of a society (J. Coleman and F. Thomas 1987). 'At the macro level, formal institutions, such as state constitutions and legal norms, make up the social capital of a society . . . at the micro level, social capital can be conceptualized as personal trust which exists in a particular community or social network' (Batheld and Glückner (2005) *Env. & Plan. A* 37). However, Clarke (2008) *Pol. Geog.* 27, 1 claims that 'many scholars would agree that an emphasis on community as the site for generating social capital and citizenship may be misleading', and Antoci et al. (2007) *J. Socio-Econ.* 38, 1 emphasize the negative welfare consequences of social capital. For an overview of geographical accounts of social capital see Holt (2008) *PHG* 32, 2.

social cleavage The spatial division of society into distinct groups; as manifestations of divisions within society, social cleavages are held to be linked to geographic patterns of electoral politics (Secor (2001) *Pol. Geog.* 20, 5; see also Perepechko et al. (2007) *Pol. Geog.* 26, 2). However, Blondel (*U. Siena & Eur. U. Inst., Occ. Paper* 16/2005) argues that 'the time has come to consider whether it is realistic to continue to regard as correct the view that social cleavages are the paramount . . . factor accounting for the relationship between "modern" parties and their supporters'.

social cohesion The interdependence between the members of a society, shared loyalties, and solidarity (Durkheim, quoted by Jensen (1998) *Canad. Policy Res. Netwk. Stud.* F03). 'A cohesive community is a community that has naturally many cross-links, where people from different race, age, background, feel free and happy to mix together in housing, in education, [and] in leisure facilities' (UK Housing, Planning, Local Government and the Regions Sixth Report). Nagel and Staeheli (2008) *Soc. & Cult. Geog.* 9, 4 see the social cohesion agenda as 'marked by the assertion that the host society should define the terms of integration, and that the primary responsibility for integration lies with immigrants and minorities'. Yuval-Davis et al. (2005) *Ethnic & Racial Studs* 28 see the concern with social cohesion as an attempt to limit the exclusion of groups.

((⊕)) SEE WEB LINKS
• Council of Europe website.

social costs In economics, the total costs of any action. **Private social costs** are met by the individuals concerned; Williams and Balaz (2002) *J. Ethn. & Mig. Studs* 28, 4 note that Slovaks working in Austria reduce their social costs through commuting rather than migrating.

social Darwinism The application of the concept of evolution to the development of human societies over time, emphasizing the struggle for existence of each society, and the survival of the 'fittest' of them. Such pernicious ideas have been used to justify power politics, imperialism, and war; see Weikart (2003) *German Studs Rev.* 26, 2. Dorling (2010) *Geog. J.* 176, 3, 186 argues that the UK's post-1970s rise in social and economic inequalities contributed to a revival of social Darwinism ('a kind of growing racism against the poor'), which had been rejected in the more equitable post-war era.

social distance The perceived distance between social strata (different socio-economic, racial, or ethnic groups), usually measured by the amount of social contact between groups; see White et al. (2005) *Sociol. Methods & Res.* 34, 2. Agrawal et al. (2006) *J. Econ. Geog.* 6, 5, analysing knowledge flows, find that geographic proximity works to overcome social distance. Autant-Bernard et al. (2007) *Papers Reg. Sci.* 83, 3 define this term as the number of links between any two firms: 'firms with a higher number of partners and a smaller social distance are more likely to be engaged together in a research project . . . in this context, social distance matters more than geographical distance.'

social economy A term with various meanings, but one definition is that the social economy 'includes those organizations which are animated by the principle of reciprocity for the pursuit of mutual economic or social goals, often through social control of capital. This definition would include all cooperatives and credit unions, non-profit and volunteer organizations, charities and foundations, service associations, community enterprises, and social enterprises that use market mechanisms to pursue explicit social objectives' (BALTA, B.C.–Alberta Social Economic Alliance).

social exclusion The exclusion of part of society from 'any of the social, economic, political and cultural systems which determine the social integration of a person in society' (A. Walker and C. Walker 1997). Kenyon et al. (2003) *J. Transp. Geog.* 10, 3. see social exclusion as 'the denial of access, to an individual or group, to the opportunity to participate in the social and political life of the community, resulting not only in diminished material and non-material quality of life, but also in tempered life chances, choices and reduced citizenship'.

Social exclusion is a more complex concept than poverty: the Transport Research Group notes that 'it is possible to be excluded without experiencing poverty; and it is possible to experience poverty, yet feel included'. Exclusion may come about through a combination of problems such as income poverty or unemployment; geographical isolation or reduced accessibility to social networks, facilities, goods, and services unemployment; poor housing; high crime environments; family breakdown; or the denial of citizenship rights and freedoms. See the DFID Policy Paper *Reducing Poverty by Tackling Social Exclusion.* Try the *Social Exclusion Literature Review 2008.*

social formation The prevailing pattern of class structure which goes hand

in hand with a particular *mode of production; try Bruneau (1981) *Antipode* 13, 3. J. Osterhammel and N. Petersson (2005) list qualifying criteria: the social interaction between two people; the longevity of these interactions; and their reinforcement by institutions.

social geography Originally this was defined as the study of the spatial patterns of social, as distinct from political and economic, factors; see, for example, Meijering et al. (2007) *Tijdschrift* 98, 1. Social geography may now be subdivided into three categories. The first is the spatial expression of capitalism: 'private property . . . is a spatial expression of power and surveillance that we have obediently internalized to make the actual exercise of power unnecessary' (Gruenewald (2003) *Am. Educ. Res. J.* 40, 3). The second stresses the 'alternative' view of human geography; for example, studying the economically disadvantaged rather than the successful: D. Darton et al. (2003) point out that the distribution of disadvantage in the UK has a strongly geographical dimension. A third category emphasizes *welfare geography. See Pain (2003) *PHG* 27, 5, 649.

socialism A social system based on equality and *social justice, once linked with common ownership of the *means of production and distribution. Hensman (2001) *Antipode* 33, 3 argues that the opposition to globalization is 'national socialism, socialism or state socialism [but] still a form of capitalism'. 'In the mid-1970s, Zygmunt Bauman, 1976, described socialism as intrinsically utopian. Contemporary critical geographers, however, are better at mapping current dystopias than imagining utopic alternatives' (Blomley (2007) *PHG* 31, 1). 'There is a moral responsibility to take ownership of what we propose, implicitly or explicitly, when we counter capitalism in the name of socialism or (today) postdevelopmentalism' (Corbridge (2005) *PHG* 29, 5).

social justice 'Putting it simply, geographers interested in justice focus on who gets what, who misses out and where all this occurs' (Bauder (2006) *PHG* 30). These are questions that can be asked on global, national, regional, and local scales. 'Social justice requires the removal of the obstacles that inhibit individuals and groups from developing their potential . . . contrary to the orthodox view, increasing social equality improves economic efficiency' (Kitson et al. (2000) *Camb. J. Econ.* 24, 4). Also of interest to geographers is the way in which moral systems vary spatially; 'humans seek not only to improve technical means but also to use these means to reflect upon, regulate, and place value on their own conduct' (Collier and Ong (2003) *Current Anthrop.* 44, 3). *See also* JUST CITY.

social physics The use of concepts from physics to illuminate aggregate human behaviour; for example, the *gravity model. In general, 'new social physics falls well short of sociological adequacy' (Crossley (2005) *Soc. Rev.* 53, 2).

social polarization The results of *segregation within a society; 'a widening gap between individuals, households or groups of people in terms of their economic and social circumstances and opportunities' (Dorling and Woodward (1996) *Prog. in Plan.* 45, 2). 'Differentiation as such is not necessarily a problem, but polarization between the haves and the have-nots is. It becomes even more of a problem if polarization increases' (Brattbakk et al. (2000) *Paper, 2nd Int. Conf. Critical Geog.*).

social reproduction The processes through which the fundamental structures and relations of society continually

recreate the existing mode of production—a mode in which the structures of dominance of one group over another are maintained (Althusser (1971) cited in Mitchell et al. (2003) *Antipode* 35, 3). Katz (2001, *Antipode* 33, 4) writes that 'social reproduction is the fleshy, messy, and indeterminate stuff of everyday life . . . secured through a shifting constellation of sources encompassed within the broad categories of the state, the household, capital, and civil society . . . It is also a set of structured practices that unfold in dialectical relation with production, with which it is mutually constitutive and in tension . . . the production and reproduction of the labour force calls forth a range of cultural forms and practices that are also geographically and historically specific, including those associated with knowledge and learning, social justice and its apparatus, and the media.'

'Distinctive national, regional and local patterns in the ways in which men and women divide paid labour and caring work result in an uneven geography of the total work of social reproduction' (McDowell (2003) *TIBG* 28, 1). Even childcare arrangements are integral to social reproduction in post-industrial economies (Mahon (2006) *TIBG* 31, 4), while in developing countries 'traditional means of social reproduction, including children's participation from an early age in family businesses and agricultural endeavours, is giving way to oversight by federal and state government agencies' (Jennings et al. (2006) *Area* 38, 3).

social space The combined use and perception of space by distinct social groups, as opposed to *personal space. 'Social space is produced by societies according to the spatial practices that exist within the society. The produced space is a set of relations between objects within the space' (Carter (2004) *Am. Ass. Geogrs, Mid. Atlantic Div.*

Annual Meeting). Social space provides an environmental framework for the behaviour of the group; it is flexible/networked (Peck and Tickell (2002) *Antipode* 34, 3), and multi-layered—'vulnerability is a multi-dimensional and multi-layered social space defined by political, social and institutional capacities' (Watts and Bohle (1993) *PHG* 17).

social well-being A situation where income levels are high enough to cover basic needs, where there is no poverty, and where there is easy access to social, medical, and educational services. To achieve this in developing countries, El-Ghannau and Ragab (2002) *Perspec. Glob. Dev.* 1, 1 conclude that population growth must slow down.

socionature (socio-nature) An *ecosystem that has been socially created; produced nature. 'The concept of socionature asserts that social relations are inherently ecological and that ecological relations are inherently social' (Alkon (2013) *Antipode* 45, 3, 663). See Swyngedouw's conference paper of 1997 and J. Anderson (2010) p. 144.

sodic soils Soils that are traditionally defined as having an exchangeable sodium percentage greater than 6%—'but there are many soils with lower sodium levels which exhibit similar behaviour' (S. Raine and R. Loch 1995).

soil The naturally occurring, unconsolidated, upper layer of the ground, made of humus and weathered rock. Major factors affecting soil formation are: climate, relief, parent material, vegetation, and time; see Svoray and Shoshany (2004) *Rem. Sens. Env.* 92, 2. Soil is the medium on which we produce 99% of our food; a major store of carbon; and regulator of climate, holding two to three times more carbon than exists in the atmosphere (Davidson et al. (2000) *Nature* 408, 789). Soils are also a regulator

of water resources, attenuating hydrological responses and removing contaminants from percolating water (Haygarth and Ritz (2009) Land Use Policy 26S, S187–97).

soil association A cluster of soils dominated by a single soil taxon (US Dep. Agric. (1993) *Agric. Handbook* 436).

soil classification A naming system to describe and identify soils.

(⊕) SEE WEB LINKS
• National soil classifications.

soil creep *See* CREEP.

soil erosion The removal of the soil by wind and water and by the *mass movement of soil downslope. The wind erosion is by *deflation; water erosion takes place in gullies, *rills, or by *sheet wash; downslope mass movement ranges from soil *creep to *landslides. For methods in soil erosion budgeting, see Stolz et al. (2012) *Holocene.*

Accelerated soil erosion is erosion increased by human activity. Causes include: wind, tillage (*NSR Report for DEFRA*); and continuous cropping (Jaiyeoba (2003) *Soil & Till. Res.* 70, 1). However, Ives (1991) *New Scientist* 1764 disputes a link between deforestation and soil erosion, and Rowntree et al. (2004) *Land Degrad. & Dev.* 15, 3 doubt the claim that overgrazing leads to soil erosion. War (as a result of agricultural intensification and the use of marginal land, or of 'scorched earth policies'), fire, urbanization, and mining also accelerate soil erosion; see Robichaud (2005) *Int. J. Wildland Fire* 14 and Limartin in M. Haigh, ed. (2000) on mining. Loughran and Balog (2008) *Geogr. Res.* 46 record a drop in soil erosion through changing from intensive tillage to none at all. However 'there is still some way to go before the cause and effect links are sufficiently understood to form a basis of tools for the management of grazing in a way that would minimise erosion and maximise the benefits' (Thornes (2008) *Geogr. Res.* 46).

SEE WEB LINKS
• Erosion data and erosion models.

soil flux The flow of soil over a slope, generally downslope. See Furbish and Haff (2010) *J. Geophys. Res.* 115, F03001.

soil heave The cracking and lifting of soil material on a slope which occurs when shrunken clay, in rewetting, returns to its original volume, or through the growth and melting of ice crystals. See R. Charlton (2008).

soil horizon *See* HORIZON.

soil moisture Moisture is held in the *capillary soil pores. Soil moisture controls many important processes in seedling emergence, evapotranspiration, mineralization of the soil organic fraction, surface run-off, leaching, and crop yield; the extent of these processes cannot be quantified without knowing the moisture status of the root zone (Akinrem and McGinn (1996) *Can. J. Soil Sci.* 76).

The **soil moisture budget** is the balance of water in the soil, resulting from precipitation and *potential evapotranspiration in combination. Mintz and Serafini (1992) *Clim. Dynam.* 8 give a simple water budget model. *See also* MOISTURE INDEX.

soil pipe A horizontal tube-like subsurface cavity within the soil; *macropores greater than 1 mm in diameter and up to a metre or more in diameter. They are continuous in length such that they can transmit water, sediment, and solutes through the soil. The classic paper is Wilson and Smart (1984) *Catena* 11, 145.

soil pore Any open space within the soil. **Soil porosity** is the percentage of pore space. Water will not drain freely through the fine **capillary pores** (average

diameter < 0.03 mm), which retain water through surface tension, but drains freely through the larger non-capillary pores. With 2–5% soil porosity, most water flows directly to the subsurface.

soil profile A vertical series of soil *horizons from the ground surface to the parent rock. A soil is classified according to the arrangement of its horizons. The Soil Profile Analytical Database of Europe of Measured Profiles was created to provide a common structure for storing harmonized information on typical soil profile properties of European soils (Hiederer et al. (2006) *Geografisk Tidsskrift* 106, 1).

soil science 'Historically, soil science has followed a circuitous path in its evolution from a discipline with foundational roots in geology, to an applied agricultural and environmental discipline, and now to a bio- and geo-science through the Earth's Critical Zone investigations. This closes the loop or spiral, but along the way, soil science has become more comprehensive, extensive, integrative, analytical, and quantitative . . . now is a golden era for soil science to integrate its expertise more closely with other bio- and geo-sciences. This will significantly enhance the opportunity to obtain extramural funding and public support, as well as the advancement of soil science. As such, soil science needs to vigorously become more interactive and extend its role beyond traditional agriculture. The knowledge of spatial soil diversity and landscape dynamics is a fundamental underpinning critical to the success of this venture. Pedology, as a unique sub-discipline of soil science, contributes inordinately to Earth science, including, for example, elucidation of field variability, surficial weathering processes, Earth system dynamics, and vadose zone flow and transport. With the blooming of hydrogeosciences, hydropedology is a timely addition in this era of interdisciplinary, multidisciplinary, and systems approaches for developing comprehensive prioritization of science and applications in Earth science. Hydropedology has a niche in this march with other bio- and geo-sciences in addressing global Earth science priorities. Soil scientists support changing paradigms and favour closer linkages with the bio- and geo-sciences community. In this regard, hydropedology has a unique role to play' (Wilding and Lin (2006) *Geoderma* 131, 3–4).

soil series A group of soils formed from the same parent rock and having similar *horizons and *soil profiles, but with varying characteristics according to their location.

SEE WEB LINKS
• Soil Survey Staff, Natural Resources Conservation Service, United States Department of Agriculture: Official Soil Series Descriptions.

soil structure The arrangement of soil particles into larger particles, or clumps. Platey structures are formed of thin, horizontal layers; prism-like structures are called columnar where the tops are rounded, and prismatic where the tops are level; blocky peds are bounded by flat or slightly rounded surfaces; and spheroidal structures are called crumbs if highly water absorbent, and granular if only moderately so. See Levine et al. (1996) *Ecol. Model.* 92, 1.

soil texture The proportions of sand, silt, and clay in any soil. Twelve different textural classes are recognized, and the structure of the soil can be determined when the percentage of these three soil constituents is plotted on a ternary diagram. TAL is a group of programs to determine soil texture classes based on major classification schemes.

SEE WEB LINKS
• TAL software.

Soil texture

solar constant The rate per unit area at which *solar radiation reaches the outer margin of the Earth's atmosphere. This fixes the energy supply for the *atmospheric heat engine. Gribben (1991) *New Scientist* 132, 1796 discusses variations in the solar constant.

solar energy 1. Of all the sunlight that passes through the atmosphere annually, only 51% is available at the Earth's surface to do work. This energy is used to heat the Earth's surface and lower atmosphere, melt and evaporate water, and run photosynthesis in plants. Of the other 49%, 4% is reflected back to space by the Earth's surface, 26% is scattered or reflected to space by clouds and atmospheric particles, and 19% is absorbed by atmospheric gases and clouds. The geography of a place determines how much solar energy is absorbed by the earth, how much is stored and how readily it is released to the atmosphere; see R. Thomas and R. Huggett (1980) and Martin et al. (1998) *Geophys. Res. Lett.* 25, 23 on the distribution of solar energy at the Earth's surface.

2. Any energy source based directly on the sun's radiation. See Zweibel et al. (2008) *Scientific American* for a wide-ranging review.

solar radiation The electromagnetic waves emitted by the sun, varying in wavelength from long-wave radio waves, through infrared waves and visible light, to ultraviolet waves, X-rays, and gamma radiation. Earth gets only 0.0005% of the sun's radiation Most solar radiation passes straight through the *atmosphere without warming it, but it is received and absorbed by the Earth. See Allen (2006) *QJ Royal Met. Soc.* 84, 362. Changes in solar radiation are known as solar forcing.

sol brun lessivée A type of *brown earth from which some clay has been *leached, forming an E *horizon.

solfatara A vent through which steam and volcanic gases are emitted.

solifluction The slow, downslope movement of water-saturated debris in *periglacial regions and other areas with cold climates. Rates of movement may be 0.9 cm yr^{-1} on gentle slopes, 12–25 cm yr^{-1} on steeper slopes. Temperature is the most important factor controlling morphometry of solifluction landforms, explaining about 22% of the variability (Ridefelt (no date) ISSN 1650–6553, 70). Solifluction may develop in areas without *permafrost; winter freezing of the sub-surface layer may be sufficient (Matsuoka (2001) *Earth-Sci. Revs* 55, 1–2). **Solifluction terraces** are step-like, linear, patterned-ground formations, made of fine-grained material (Walsh et al. (2003) *Geomorph.* 55, 381). Heights and widths are about 1 m, and, with very fluid material, terraces show rough bedding (*tumultuous bedding*). **Solifluction lobes** are bulges in the slope profile that do not break the surface (Matsumoto and Ishikawa (2002) *Geografiska A* 84, 3–4). The major environmental controls on solifluction lobes are gradient and soil moisture (Ridefelt and Boelhouwers (2006) *Perm. & Periglac. Procs* 17).

solonchak An *intrazonal saline soil found in hot, arid climates. Evaporation of soil moisture brings saline groundwater to the surface where it, too, evaporates. The sodium and calcium chlorides and sulphates which have been translocated remain as a grey surface crust.

solonetz An *intrazonal, formerly saline soil. Periodic rainfall has leached the salts from the surface layer and these accumulate in the B *horizon. Deep ploughing destroys the solonetz horizon (Bogomolova (2008) *Geophys. Res. Abstr.* 10, EGU2008–A-10603).

solum In a soil, the layer above the parent material. The solum is strongly influenced by vegetation (Schaetzl (1991) *Eur. J. Earth Sci.* 42, 3) and climate. Alvarez and Lavado (1998) *Geoderma* 83, 1–2 find precipitation as a key influence in Pampa and Chaco soils.

solution In geomorphology, the picking up and dissolving of particles by a fluid, usually water or carbonic acid. *See also* CARBONATION; AGGRESSIVITY.

sorting In geomorphology, the deposition of sediments in size order. 'Grain diameter indicates the strength of the transporting environment' (Potocki and Angiel (2004) *Miscellanea Geographica* 11 on *aeolian sorting). River beds become finer grained downstream through a combination of grain abrasion, size-selective transport, and deposition (Ferguson (2003) *Geology* 31). *See also* VARVE.

soundscape The landscape of sounds in which we live. Saldhana in R. Kitchin and N. Thrift, eds (2009), uses this term 'to conceive how sound gives meaning to spaces and places'. He continues 'sounds such as music help orientate the lives of all humans . . . [but] to be sensed, sound has to be embodied'.

Southerly Burster *See* LOCAL WINDS.

southern annular mode (SAM)
Also known as the Antarctic Oscillation, this term describes the north–south movement of the westerly wind belt that circles Antarctica, dominating the middle to higher latitudes of the Southern Hemisphere. It is a zonally symmetric or annular structure, with a large low pressure anomaly centred on the South Pole and a ring of high pressure anomalies at mid-latitudes.

In a positive SAM event, the belt of strong westerly winds contracts towards Antarctica. This results in anomalously dry conditions over southern South America, New Zealand, and Tasmania, and wet conditions over much of Australia and South Africa. The stronger westerlies above the Southern Ocean also increase the insulation of Antarctica. As a result, there is less heat exchange between the tropics and the poles, leading to a cooling of Antarctica and the surrounding seas.

sovereignty Conventionally, political authority exercised by a state over a given territory. Steinberg (2009) *AAAG* 99, 3, 467 is among several commentators who argue that 'although the concepts of sovereignty and territoriality frequently are used interchangeably in discussions of the modern polity, the two terms in fact reflect different organizing principles that can exist independent of each other'. **Domestic sovereignty** includes both authority and control but only within a state, while **international legal sovereignty** concerns the authority (government) of the state as recognized by other sovereign states. This legal personality can survive territorial and internal changes. **Westphalian sovereignty** implies the territorial organization of the state as an inviolate realm, free from intervention by other states, while **interdependent sovereignty** relates wholly to the state's control of its borders and its exposure to external influences. 'Sovereignty in the modern era of multi-tiered governance is often murky and shifting' (Rodden (2004) *PSOnline*, on **fiscal sovereignty**). Ong (2000) *Theory, Cult. & Soc.* 17, 4 defines **graduated sovereignty** as the state's differential treatment of the population based on ethno-racial differences and the dictates of developmental programmes; see also Park (2005) *Pol. Geog.* 24, 7.

S

sovereign wealth fund (SWF) An investment fund that is owned and controlled by a national government. SWFs are mainly located in Asia and in the Middle East. In Asia, they are generally 'non-commodity' funds whose resources originate from current and financial balance of payments surpluses and/or from fiscal surpluses. In the Middle East and Africa, they are primarily 'commodity' funds managing the incomes of oil and mineral exporting countries. However, the world's largest fund is in fact European—the Norwegian government pension fund—and it is considered one of the most transparent sovereign funds in the world (*Banque de France, Franc Zone Annual Report* (2012).

space The extent of an area, usually expressed in terms of the Earth's surface. The term **spatial** derives from this meaning; and spatial relationships are at the heart of geography. '"Space" refers to the operational dynamics of a network: across space, over time, at a variety of scales, from the body to the neighbourhood, to the region, and across nations' (Ettlinger and Bosco (2004) *Antipode* 36, 2). D. Massey (2005) sees space as lived experience: 'one important aspect of space is that it is the dimension of things (and people) existing at the same moment. If *time* is the dimension of change then *space* is the dimension of simultaneity. Space is the dimension of all these things happening at once. Space is the dimension, then, that presents us with the existence of others.'

Y.-F. Tuan (1997) contends that a space requires a movement from a place to another place. Similarly, a place requires a space to be a place. 'Hence, the two notions are co-dependent.' Thrift in S. Holloway et al. (2003) writes that 'place refers to the process whereby spaces are ordered in ways that open up affective and other embodied potentials'. See also Sanders (2008) *PHG* 32, 2.

'**Absolute space** is fixed and we record or plan events within its frame . . . the **relational view of space** holds there is no such thing as space or time outside of the processes that define them. . . . Processes do not occur *in* space but define their own spatial frame' (D. Harvey 2006). Raper and Livingstone (1995) *Int. J. GIS* 9 distinguish between absolute space, which is geometrically indexed, and **relative space**, which is object oriented.

Mark and Freundschuh (1997) *Trans. GIS* 2, 4 identify 'at least five types of spaces . . . because for most if not all human cultures, there are different kinds of spaces, applied to different situations or phenomena that are conceptualized in different ways'. Every society uses its space differently, both artistically and technologically: 'Europeans have a notion of time and space that is generally assumed by them to be universal. This gratuitous assumption is naive, arrogant, and wrong' (Hopgood (1993) *AGOCG Tech. Report* 24). 'Space is used as a site and mechanism for gendered relationships of power' (Corroto (2001) *J. Arch. Educ.* 55, 2).

space-for-time substitution Certain processes in geomorphology are too slow to be wholly investigated in one individual's lifetime, so the development of such a process may be observed by finding other sites where that process is exhibited at a different developmental stage. In ecology this means that past or future trajectories of ecological systems may be inferred from contemporary spatial patterns (Blois et al. DOI: 10.1073/pnas.1220228110).

J. D. Aber and J. M. Melillo (2001) give an important caveat: 'this approach assumes that the sites only differ in time and not disturbance or other underlying factors unrelated to time'.

space of flows The communicational space created by new techno-economic networks: 'a new type of space that allows distant synchronous, real-time interaction' (M. Castells 1989). See Reades and Smith (2014) *Regional Studs* 48, on global business telecommunications and employment specialization in the London mega-city region.

space–time 'Space and time are treated as sticky concepts that are difficult to separate from each other' (Dodgshon (2008) *Geografiska B* 20, 1). We need to be able to understand *time* to tell the story of how an individual place developed, and to understand *space* to understand the complexity of the way different places develop. Galton (2004) *Spat. Cogn. & Comput.* 4, 1 explains that in the real world there is rarely a clean separation between (spatial) objects and (temporal) events. 'For example, a meteorologist tracking a tropical cyclone will naturally view it as an object which comes into existence at a certain place and time and moves along a well-defined path, eventually fading away and ceasing to exist. From the point of view of an inhabitant of a town in its path, the cyclone is much more like an event: the sudden onset of strong winds and rain, bringing destruction to the town and then dying away leaving the inhabitants to pick up the pieces.'

Massey (1999) *TIBG* 24, 3 argues that space 'is not static, not a cross-section through time; it is disrupted, active and generative. It is not a closed system; it is constantly, as space-time, being made.' Lane (2001) *TIBG* 26, 2, following Massey's argument (op. cit.) that the sort of space–time model that we adopt

needs to be informed by the entities that we study, concludes that some of the space–time models that Massey critiques 'may still be fundamental to what we do, and in no sense necessarily ahistorical'.

For *geomorphologists, Raper and Livingstone (1995) *Int. J. GIS* 9 propose a **relational space–time**, since the identity of a geomorphological phenomenon is generated by the spatio-temporal relations between causal processes and the local nature of the environment. 'The relational notion of space-time implies internal relations; external influences get internalized in specific processes or things through time' (D. Harvey 2005). Neutens et al. (2007) *Int. J. GIS* 21, 10 visualize space and time in three dimensions using a hybrid (CAD/GIS) system.

spatial To do with geographic space; with distribution or location across a landscape or surface. See B. Warf and S. Arias (2008) on the **spatial turn**.

spatial analysis (spatial data analysis) A geographical analysis which seeks to explain patterns of human behaviour and their spatial expression in terms of mathematics and geometry *See also* GEOCOMPUTATION.

spatial autocorrelation A clustering pattern in the spatial distribution of some variable which seems to be due to the very fact that the occurrences are physically close together, that is, that they are in geographical proximity. Spatial autocorrelation is widespread: rich people move to areas where other rich people live; people only go to parties because other people go, and so on. In any exercise using sampling, the problem with spatial autocorrelation is that if observed values are spatially clustered, samples will not be independent; see T. Bailey and T. Gatrell (1995) and Ehls-chlaeger (2000) *Computers Env. & Urb.*

S

Sys. 24, 5 on measuring spatial autocorrelation. Getis (2008) *Geog. Analysis* 40, 3, 297 provides a history of spatial autocorrelation.

spatial differentiation A. Knowles (2002) is probably not alone in defining geography as 'the study of spatial differentiation'. Within geography, spatial differentiation is seen as:

- a process (see Krevs (2004) *Dela* 21 and Li and Wu (2006) *Housing Studs* 21, 5),
- a causal mechanism (see Walker (1978) *Rev. Radical Polit. Econs* 10, 3),
- a response (see Spark and Burt (2002) *Soil Sci. Soc. Amer. J.* 66).

The identification and measuring/ mapping of spatial differentiation is fraught with difficulties. The European Union has proposed seven criteria for the differentiation of the European territory (Study Programme on European Spatial Planning—Final Report, 2001). There you will find the list of methods, together with the problems.

spatial diffusion A phenomenon spreading through geographic space, such as a language. Silvestre and Campoy (2002) *Studia Anglica Posnaniensia* 38, conclude that models may give adequate statistical explanations for the volume, distance, and direction of geographical diffusion flows, but they do not reveal causal factors. Comparato and Mitchell (2010) http://dx.doi.org/10.2139/ssrn.1670344 employ a social learning model to examine the role of geography in the diffusion of US Supreme Court precedent in State Supreme Courts, and find that both the ideological distance between State Supreme Courts and the geographic proximity of states influences the likelihood of adoption re: Miranda v. Arizona.

spatial-interaction theory The view that the movement of persons between places can be expressed in terms of the attributes, such as population or employment rates, of each place. See Greenwood and Hunt (2003) *Int. Reg. Sci. Rev.* 26, 1 for an excellent review.

spatiality The effect of space on actions, interactions, entities, and theories. 'Spatiality is a social construct, not an exogenously given, absolute coordinate system . . . but a product of the political economic system' (Sheppard (2004) *Antipode*, 36, 3). 'Spatiality is constitutive of the particular ways in which the different modalities of power take effect. The two major dimensions of spatiality are reach and intensity' (Clegg (2006) *Area* 38, 1). 'From the perspective of spatiality, space and society do not gaze at each other but rather are mutually embedded' (Ettlinger and Bosco (2004) *Antipode* 36, 2). The term is also used as a synonym for distribution, or spatial expression.

spatial monopoly The monopoly of a good enjoyed by a supplier over a marked area where no competitor exists. 'In the literature on spatial monopoly, it is common to assume that the firm produced (or sold) only one good. This assumption, however, departs from the real world' (Peng (2004) *Southern Econ. J.* 1 Jan.). 'Spatial monopoly leads to the spatial impossibility theorem: no competitive equilibrium involving trade across locations exists in a homogeneous space' (Stafford (2003) *AAAG* 93). Spatial monopolies happen when firms agree to split the market spatially, or through the privatization of public utilities, as in the case of the British Water Boards.

spatial ontology The many ways people make geography. Hayes and Laforte (IHMC, U. W. Florida) identify some particularly 'geographical' ontological concepts: continuant—physical entity with space-like part; occurrent—

physical entity with time-like parts; location—a piece of physical space; terrain—a piece of geographical space (consisting of locations suitably related to each other); and history—the spatio-temporal region. Martin (2013) *PHG* DOI: 10.1177/9780199680856209 writes of knowledge systems: 'Each contains its own spatial ontology, its own rhythms, its own theories of matter, systems of circulation, and rules of exchange'.

spatial preference The choice of one spatial alternative, such as a housing area, or a shopping centre, rather than another. Meester and Pellenbarg (2006) *Tijdschrift* 97, 4, in a study of Dutch entrepreneurs, find that their criteria are potency, activity, and evaluation, and that spatial preference does not change much over time.

spatial randomness A distribution pattern of some variable where the observed spatial pattern is no more, nor less, likely than any other; 'complete spatial randomness is a neutral condition often thought to reflect the dominance of external, stochastic processes, for example disturbance in ecosystem development' (Cutler et al. (2008) *J. Ecol.* 36, 1).

spatial science The measurement, management, analysis, and display of spatial information describing the Earth; its 'natural' and socially constructed features. It emphasizes the study of aggregate spatial patterns, including spatial behaviour, within clear theoretical frameworks, using *quantitative methods to evaluate models and hypotheses. As such, it uses *cartography, *geodesy, *geographic information science, *hydrography, photogrammetry, remote sensing, or surveying.

spatio-temporal fix Capitalism is an imperfect system, which creates problems that need fixing. For example, over-

*accumulation can result in surpluses, *dumping, idle capacity, lack of investment opportunities, and unemployment. The capitalist system fixes these problems by developing new (*temporal*) sites (*spatial*) that can absorb surplus labour and/or capital. In other words, spatio-temporal fix is 'a metaphor for a particular kind of solution to capitalist crises through temporal deferral and geographical expansion' (D. Harvey 2003).

special economic zone (SEZ) An area within the territory of a *state where normal regulatory practices, such as foreign exchange and remittance restrictions, and export levels, do not apply. By 2002, there were more than 5000 SEZs in China (Chen (2008) *Urb. Studs* 45, 2). Li et al. (2005) *Pacif. Econ. Rev.* 10, 4, find that foreign investment has played an important role in Chinese SEZs, but that the future of the Chinese open-door policy will depend on improving the Chinese political system. See Tantri (2012) *India Qly.: A. J. of Int. Affairs* 68, 3, 231 on China's Policy for SEZs. *See also* FREE TRADE ZONE.

speciation The concept of ecological speciation invokes Darwinian natural selection as a means of generating biodiversity and, specifically, reproductive isolation (Rundle and Nosil (2005) *Ecol. Letts* 8). Patten (2008) *J. Biogeog.* 35, 1 argues that speciation and specialization, 'long studied as distinct disciplines, one evolutionary, the other ecological', could well be 'one and the same thing'.

species–area relationship 'Along a gradient of ecosystems of increasing size, the numbers of species inhabiting those ecosystems increases; rapidly at first, but then more slowly for the larger ecosystems' (Lomolino (2001) *J. Biogeog.* 27, 1). Losos and Schluter (2000) *Nature* 408 explain the relationship as the outcome of the effect of area on immigration and extinction rates. Báldi (2008) *J. Biogeog.* 35, 4 argues that habitat heterogeneity overrides the species–area relationship.

specific humidity The actual mass of water vapour present in a kilogram of moist air; in general terms, the mixing ratio. See Chapman and Thornes (2003) *PPG* 27, 3 on the use of GIS to compare specific humidity with wind flow.

spectacle A striking visual presentation. For example, Cinquegrani (2010) looks at the ways in which early films, as much as photography and colonial exhibitions, transported the spectators to a version of India existing in great part as the creation of unchecked British imagination; the 'Imperial Spectacle'. Thomas (2007) *Cult. Geogs* 14, 3, 369 on the clothing (the 'lived garment') of Lady Curzon, Vicereine of India 1898–1905 is strangely fascinating.

speleothem A collective noun for depositional features such as *stalactites and *stalagmites. Most are made of calcareous rock. See Mattey et al. (2008) *Geophys. Res. Abstr.* 10, EGU2008-A-00461 on seasonal controls on speleothem growth.

sphere of influence The field/hinterland, or area of control; see Bilgim (2007) *Pol. Geog.* 26, 7 on Turkey's 'natural sphere of influence'. Referring to the Ukraine crisis of 2014, Thomas Donilon, former US national security adviser, said 'with respect to Putin my observations [are] . . . that for him, these concepts of balance of power, spheres of influence, are quite real concepts. And quite relevant to how he views the world . . . he regards it as an element of the Russian sphere of influence' (*Brookings.edu*).

The term is also used for the *urban field of a city; see Ma (2005) *Pol. Geog.* 24, 4.

spheroidal weathering The weathering of jointed rocks by water along the joints, such that shells of decayed rock surround isolated, unaltered corestones. Sarracino and Prasad (1989) *Geojournal* 19, 1 see the concentric patterns formed by spheroidal weathering as evidence of *self-organization.

spillover In economic geography, an overflow, or spreading, of information from elsewhere; an indirect, unpaid-for effect of economic activity (also called an *externality), such as spillovers between university *research and development, and firms (although Boders (2004) *J. Econ. Geog.* 4 reports that only regions with low R&D density benefit from interregional spillovers). **Knowledge spillovers** mean that firms can acquire knowledge and information created by others through expenditure on industrial R&D without payment. Movements of labour between firms, informal contacts, business meetings, close customer-producer relations, and face-to-face contacts are all mechanisms fostering spillovers; see Engelkstoft et al. (2006) *Papers Reg. Sci.* 85, 1. Adams (2002) *J. Econ. Geog.* 2 finds that academic spillovers are more localized than industrial spillovers. *See* PROXIMITY.

spit A ridge of sand running away from the coast, usually with a curved seaward end. Breaking waves produce a turbulent flow of water that dislodges sediment from the beachface, and create the *longshore currents that transport the sand along the beachface. 'Without breaking waves, the littoral current and the drift of sand ceases' (Oertel and Overman (2004) *Geomorph.* 58, 1–4). Since the water beyond the end of the spit is usually deep, waves do not break, but are refracted and diffracted around the spit, curving it. Clemmensen et al. (2001) *Holocene* 11, 3 find that spit growth is closely linked to sea-level variation.

Ciavola (1997) *Catena* 30 concludes that 'the strategy of coastal management that seems more appropriate to the spit-tidal flat sedimentary system is a "do-nothing option". The sea has to be left free to overwash the spit, supplying sand to the tidal flat. The spit will start again to migrate uniformly westwards, even if this means that for some time it will be transformed into an island.' Marghany (2012) *Acta Geophysica* DOI: 10.2478/ s11600-012-0061-5 presents a new approach to 3-D spit simulation using differential synthetic aperture interferometry (DInSAR).

spodosol A soil of the *US soil classification. *See also* PODZOL.

spontaneous settlement An unsatisfactory term for a *squatter settlement, since it implies that squatter settlements are unplanned. They often are planned, although rarely legally. For instance, between 2003 and 2005, landless families organized 552 settlements in the Brazilian Amazon (INCRA 2007), and these settlements represent one-third of all new settlements formed throughout Brazil during 2003–2005 (INCRA 2007), pointing to the significance of what has been identified as direct action land reform (DALR) (Caldas et al. (2010) *J. Latin Amer. Geog.* 9, 1).

spread (lateral spread) The *slumping of clayey sediment at the edge of *ice sheets.

spread effect The filtering of wealth from central, prosperous areas, to *peripheral, needier areas. The spread effect is the spatial equivalent of trickle-down economics. Spread-backwash effects were outlined by G. Myrdal (1957) and A. Hirschman (1959) in the late 1950s. 'Generally, spread effects are the positive effects of urban proximity for communities, and backwash effects are the negative consequences of proximity. According to Hirschman, the most important spread effect is the purchase and investments of the affluent region in the outlying region' (Ganning 2010).

spring tide *See* TIDE.

squall A storm characterized by sudden and violent gusts of wind. A **squall line** is a cluster of storm-bearing *convection cells along a non-frontal line, or belt; see Wang and Carey (2005) *Monthly Weather Rev.* 133, 6. Squall lines form along an *occlusion, along an *inversion created by a flow of maritime air over a cool surface, or below an upper-air trough (López (2007) *Atmos. Res.* 83, 2–1). Mohr et al. (2003) *J. Hydromet.* 4, 1 define squall line as a convective line with a trailing stratiform cloud anvil.

squatting The unlawful requisition of housing or land. **Squatter settlements** are illegally occupied, which necessarily involves insecurity of tenure, irrespective of the quality of the buildings and the prevailing physical conditions. Squatter settlements are not necessarily *slums because not all houses in a squatter settlement have to be substandard for it to be a classical squatter settlement (Matovu (2000) Municipal Dev. Prog., Eastern and Southern Africa). *Compare with* SLUM.

Squatter settlements grow because demand for cheap housing outstrips supply. Houses are often flimsy, and sanitation is grossly inadequate (in Indian urban areas, scarcely 20% of the population has access to water/flush toilets; see Raghunaathan (2006) *Hafta Magazine*), power may be unavailable, roads are not metalled, and education and medical facilities are severely limited. Brazil's squatter settlements are characterized by a lack of access to credit and funding for urban infrastructure,

S

and alarming levels of violence, particularly in poor areas; see the movie *City of God*.

It's heartening that Klaufus (2000) *J. Housing & Built Env.* 15, 4 focuses on the inhabitants of squatter settlements as individuals; similarly, Datta (2007) *Gender, Place & N. Cult.* 14, 2 describes the way different women in a Delhi squatter settlement experience different forms of spatial control, using multiple modes of resistance to different spatial controls. Patel and Arputham (2007) *Env. & Urbs.* 19, 2 warn of 'the disruptions that the slum dwellers will bring if they are not involved in the planning and implementation of such redevelopments'.

stability The state of a parcel of air which, if displaced vertically, will return to its original position. Thus, if a parcel of air cools more on rising than the air which surrounds it, it becomes denser than its surroundings and therefore sinks. The atmosphere is **absolutely stable** when the environmental *lapse rate is less than both the dry and saturated *adiabatic lapse rates. Atmospheric stability is reinforced by *inversions. See Peppier (1988) *SWS Miscellaneous Publication* 104.

stable ecosystem An ecosystem that will maintain or return to its original condition after any disturbance. D. A. Perry defines a stable ecosystem as one that is capable of constraining its own fluctuations within certain bounds and, if not maintaining constancy in its species composition and productivity, at least maintaining constancy in certain potentials. However, 'Bowker in A. Ong and S. Collier, eds (2005) argues that 'the stable ecosystem in equilibrium is a myth'.

stable population A population where fertility and mortality are constant. A stable population will show an unvarying age distribution and will grow at a constant rate. Stable population theory is based upon measuring the reported age–sex distribution against that of an appropriately chosen stable population, in order to validate census information (*UN Publication* ST/ESA/STAT/SER.M/67/Rev. 2).

stack An isolated islet or pillar of rock standing up from the sea bed, close to the shore. Initially, caves develop; some ultimately meet to form an *arch. With the collapse of the roof of the arch, a stack is left.

stadial A time when glaciers advanced and periglacial conditions extended, but not as significantly as in a *glacial. Renssen et al. (2005) *Vrije Universiteit Amsterdam* maintain that stadial–interstadial transitions in the North Atlantic were most probably caused by changes in the thermohaline circulation in the North Atlantic Ocean.

stalactite A limestone column hanging from a cave roof. It grows as an underground stream deposits its dissolved load of calcium carbonate, and may extend far enough to meet a *stalagmite, thus forming a continuous column. Stalactites retain a record of environmental conditions; see Mayer et al. (2003) *Geophys. Res. Abstr.* 5, 12952.

stalagmite A limestone column, formed on a cave floor when underground water evaporates, depositing the calcium carbonate dissolved within it. R. Boch (2008) provides useful detail on stalagmite growth dynamics.

standard industrial classification A grouping of industries classified by a government.

(((◉))) SEE WEB LINKS

- British government standard industrial classification.

standardized mortality ratio The ratio of observed to expected deaths. Julious et al. (2001) *J. Public Health Med.* 23, 1 criticize this statistic.

standard of living, index of An assessment of living standards using indicators such as access to health care, standard of education, house ownership, car ownership, take-home pay, and employment rates. As the standard-of-living index is a constructed measure, it does not have an absolute interpretation, so it's better to use it hierarchically (Subramanian et al. (2006) *Am. J. Public Health* 96).

standing crop The *biomass present at a given time in a given area.

standing wave A wave where the surface oscillates vertically each side of a fixed, motionless node, with alternating up and down motion each side of it. The wavelength of a standing wave is the length of the water body it develops in (R. Davis and D. FitzGerald 2004). The most common standing wave is the *seiche.

standpoint The way in which something is thought about or considered. *See* SITUATED KNOWLEDGE; see also McDowell (1992) *TIBG* 17, 4, 399 on feminist standpoint theory.

star dune *See* SAND DUNE.

state A territorial unit with clearly defined and internationally accepted boundaries, an independent existence, and responsibility for its own legal system. 'Most social science has taken state territoriality for granted, equating it with homogeneous bounded space neatly dividing the domestic from the foreign' (N. Brenner et al., eds 2003). 'The state thrives through simplification, making social relations legible so they can be controlled' (Taylor (2007) *TIBG* 32, 2).

B. Jessop (2002) argues for a new state form: 'Schumpeterian competition replaces Keynesianism, workfare substitutes for welfare, postnational multiple scales are imposed upon the national mosaic, and a governance regime blurs the private–public division that defines the state.' Wacquant (2001) *Eur. J. Criminal Policy* 9, 4 argues that neoliberal ideology has brought about the erasure of the economic state, the dismantling of the social state, and the strengthening of the penal state: 'an integral part in the consolidation of the penal state in France has been the state's practices that produce spaces.' 'Processes of globalization increasingly breach state borders and have led to questioning of the degree to which "fixed" territorial structures of the modern state are adapted to organizing, describing and analysing the contemporary world . . . The spatial foundations of the state are also undermined by formal international integration processes, most notably those of the supranational European Union' (Wise (2006) *Area* 38, 2).

State-theoretic approaches incorporate one or more theories of the state; these include liberal democratic theories, social democratic theories, and critical theories; the first part of Baily (2006) *J. European Public Policy* 13, 1, 16.

steady state landscape A landscape whose input, output, and properties remain constant over time. Riggins et al. (2011) *Geomorph.* 128, 1–2, 84 show that over long time scales a hilltop can indeed be in a geomorphic steady state; see Willet and Brandon (2002) *Geology* 30, 2 on steady states in mountain belts. *See also* DYNAMIC EQUILIBRIUM.

steam fog A shallow, wispy, smoke-like fog, formed when cold air passes over warmer water and is rapidly heated. Convection currents carry moisture upwards, which quickly recondenses to form fog. *See also* ARCTIC SEA SMOKE; FOG.

stem flow *See* INTERCEPTION.

step An abrupt, but short, break of slope, especially in a *glacial trough. Steps tend to be smoothed and striated upstream, as the ice grinds the bedrock, and craggy downstream, because of *quarrying; see R. Huggett (2003).

steppe The wild grasslands of Central Eurasia stretch from the Danube, through modern Ukraine and southern Russia, to the Caspian; P. B. Golden (2003) provides an interesting early history. The natural vegetation has by now been removed or much altered by cultivation and grazing.

stepped leader *See* LIGHTNING.

stepwise migration A type of migration which occurs in a series of movements, for example, moving to a town larger than the home town, but not directly to one of the city regions (Nilsson (2003) *Urb. Studs* 40, 7).

sticky 'Referring to those cities and regions which can "anchor" income-generating activities; localized' (Bathelt and Glückler (2003) *J. Econ. Geog.* 3, 117); unresponsive (Cheshire and Magrini (2008) SERC discussion paper #10). For the conditions under which some places manage to remain 'sticky', see Markusen (1996) *Econ. Geog.* 72, 3.

sticky spot An area of high drag at the glacier bed; see Mair et al. (2001) *J. Glaciol.* 47, 9–20, and Kavanaugh and Clarke (2001) *J. Glaciol.* 47, 158 on basal motion precipitated by the gradual failure of a sticky spot.

stillstand A time of tectonic inactivity: see Everest and Kubik (2006) *J. Quat. Sci.* 21.

stochastic Governed by the laws of probability. Frenken and Boschma (2007) *J. Econ. Geog.* 7 reformulate Simon's model of stochastic growth.

stock An irregular igneous *intrusion, cutting across the strata of the *country rock. A stock is less than 10 360 hectares in area (Buddington (1969) *GSA Bull.* 70).

stocking rate The number of livestock per unit area, a rate used to measure the ecological pressure of livestock (K. Zimmerer and T. Bassett 2003).

stone pavement In a *periglacial landscape, pavements of large boulders on saturated land in valley floors. Wells et al. (1995) *Geology* 23 argue that stone pavements remain at the surface because of vertical inflation caused by windblown dust.

stone stream A linear arrangement of rocks, seemingly developed through *freeze–thaw; see Hall (2002) *S. African J. Sci.* 98, 78.

stone stripes *See* PATTERNED GROUND.

stoping The assimilation at depth of *country rock by an igneous *intrusion. The heat of the intrusion melts the country rock which then mingles with the *magma. Žák et al. (2006) *Int. J. Earth Sci.* 95, 5 hold that roof pendants, stoped blocks, and discordant intrusive contacts are evidence of magmatic stoping.

storm beach A beach ridge situated well above the normal limit of high tides, formed as unusually large waves fling ashore shingle, cobbles, and boulders.

storm surge (sea surge) Also referred to as the residual sea levels, these surges occur when a *depression passes over the sea, subjecting the water to lower *atmospheric pressure than its surroundings. This causes the water level to rise and the surrounding water to sink; a fall in pressure of 1 mb will produce an increase in height of almost 1 cm. Such rises in the sea surface can be compounded by high winds.

Storm surges associated with *hurricanes are common in the relatively confined locations of the Gulf of Mexico and the Bay of Bengal. Abeysirigunawardena et al. (2011) *AAAG* 101, 5 find that climatic variability is a fundamental element in explaining the changing frequency and intensity of sea surges in coastal British Columbia.

stoss and lee Glacial landforms with a pronounced asymmetric profile—an abraded slope on the up-ice side (**stoss**), and a steeper rougher quarried slope on the down-ice side (**lee**) that has both abraded and quarried surfaces. Individual examples are commonly known as roches moutonnées (Glasser and Bennet (2004) *PHG* 28, 1).

The term may be applied to sand ripples (Zheng in R. Allen, U. Förstner, and W. Salamons 2009).

strategic environmental assessment The consultation and monitoring of the likely impacts of any strategies on the environment. It is a key component of sustainable development, establishing important methods for protecting the environment, and extending opportunities for public participation in decision-making. 'SEA has been promoted as a promising instrument expected to be able to provide better informed, more credible and more broadly beneficial strategic initiatives, as well as more timely and clearer guidance for subsequent undertakings' (Kirchoff (2011) *UWSpace*). *See* ENVIRONMENTAL IMPACT.

stratified Showing distinct layers. Glacial till is often stratified through

sorting and redeposition by *meltwater. *See also* SORTING.

strato-, stratus Layered cloud. *See also* CLOUD.

stratosphere A layer of the Earth's atmosphere, above the *troposphere, 50 km thick. See Cohen et al. (2007) *J. Climato.* 20, 21 on stratosphere-troposphere interaction events.

stratovolcano A cone-shaped volcano with a layered internal structure; see Hildreth et al. (2003) *GSA Bull.* 115, 6.

streamer Aeolian sand transport occurs primarily in the form of streamers—'wriggles' of sand—that are driven by distinct eddies travelling down toward the surface through the boundary layer and scraping across the bed while exciting saltation along their trail (Baas (2007) *Geomorph.* 91, 3–4).

streamflow Discharge that occurs in a natural stream channel. Streamflow velocity varies with distance from bed; with distance from banks, downstream, and over time. Channel geometry affects turbulence, and thus velocity, and hydraulic jumps and drops affect velocity and depth. Variations in streamflow affect the amount and type of geomorphic work a river can do. The driving force is gravity, which varies with channel gradient; the resisting forces are the viscosity of the water; the *Reynolds number; *Manning's roughness coefficient; and the *Froude number.

See Rantz et al. (1982) *US Geol. Surv. Water-Supply Paper* 2175, 631 on the measurement and computation of streamflow.

stream gauge (stream gage) A location equipped to measure the water surface elevation and/or the *discharge of running water. Equipment used includes: weirs, flumes, and cableways from which to suspend a hydrographer and/or current meter over a river.

stream order A method of classifying the different segments of a drainage system in order to investigate any possible relationship between these segments. Horton (1945), *Geol. Soc. Amer. Bulln.* 56, 3, 275 promoted stream ordering, but it is Strahler's modification of Horton's ordering system (Strahler (1957), *Trans. Amer. Geophys. Union* 38, 6, 913) that has been most widely adopted. Gilvear and Jefferies in J. Holden (2005) find that 'despite the variable size and nature of stream patterns, each network has been found to have an ordered internal composition'. *See also* BIFURCATION RATIO.

Horton   Strahler

Stream order

stream power This describes the energy available, measured in watts, in a particular area of channel. It is calculated at any given cross-section in a river by the equation:

$$\Omega = \rho g Q S$$

where Ω is stream discharge, ρ is the density of water (1000 kg/m^3), g is acceleration due to gravity (9.8 m/s^2), Q is discharge (m^3/s), and S is the channel slope.

Total stream power can be related to rates of bank or bed erosion.

Unit stream power (P) describes the energy available per unit width of channel, and provides a more detailed insight into morphological processes:

$$P = \frac{\rho g Q S}{w}$$

where w is water width in metres and the other variables are in the earlier stream power equation.

Unit stream power can be used to predict when particles are entrained and the rate at which they move downstream. It may also be used to assess the likely morphological impact of altering one or more independent fluvial variables.

⊕ SEE WEB LINKS

- EPA public access website with RIVERMorph program, which may be used to determine changes in unit stream power and thence to predict any resulting responses by the channel.

streamsink *See* DOLINE.

strength In geomorphology, the resistance of a rock mass to rupture under stress. Strength varies with, in order of importance: the spacing of joints, the cohesion and *frictional force of the rock, the *dip of any fissures, the state of weathering of the rock, the width of fissures, the movement of water in or out of the rock mass, the continuity of the fissures, and the amount of infilling of soil within the fissures. See Jackson (2002) *GSA Today* 12, 9 on the 'jelly sandwich' view of the continental lithosphere; see also Townend and Zoback (2000) *Geology* 28, 5 on the way faulting keeps the crust strong.

stress The force applied to a unit area of a substance measured in newtons per square metre. **Compressive stress** crushes the rock; see Bunds (2001) *GSA Bull.* 113, 7. **Tensile stress** is a force which tends to pull a rock or soil apart and which may cause fractures and pores to open; see Wiprut and Zoback (2000) *Geology* 28, 7.

striations Long scratches on a rock, a few millimetres across, formed by glacial abrasion, usually in clusters, often well developed on the stoss side of roches moutonnées. Striations can show the direction of glacier movement (Fjellanger (2006) *Geomorph.* 82, 3–4).

strip mining The removal of the overburden to extract minerals. See E. Reece (2006) on strip mining and the devastation of Appalachia.

Strombolian eruption Volcanic activity, relatively frequent and mild, in which gases escape at intervals, producing small explosions.

structural adjustment Changes to a nation's economy usually imposed upon debtor nations by their institutional creditors; for example the *International Monetary Fund. Changes include: slimming down of government employment, cuts in taxation to stimulate investment, a reduction in public services, privatization of the economy, the promotion of exports; liberalization, and a reduction in government subsidies in order to bring domestic prices more in line with world prices (Clarke and Howard (2006) *Geogr. J.* 172, 2) and combat inflation (with great success in Bolivia, for example: Kohl (2006) *Antipode* 38, 2). However, in countries that have undergone structural adjustment programmes, the majority of citizens have seen their standard of living fall (M. Kurtz 2004). 'It is clear that structural adjustment policies have negative human rights consequences for loan recipients, [but] these bad outcomes probably have been unintended' (Abouharb and Cingarelli (2006) *Int. Studs Qly.* 50, 2). Clarke and Howard (loc. cit.) note mixed effects in Jamaica: while formal employment has fallen in Kingston, the investment in

education and housing undertaken by the middle and upper classes has helped them to sustain the dislocation introduced by structural adjustment.

Structural adjustment policies may be adopted unilaterally; see Wong (2007) *China & World Econ.* 15, 2.

structuralism This is based on the thesis that behind local cultural variations there are deep-laid foundations (structures): 'all manner of practices, objects, events, and meanings, are taken to exist not as discrete entities, but as parts relationally embedded within, and constituted by, underlying wholes' (Dixon and Jones in J. S. Duncan, N. C. Johnson, and R. H. Schein (2004)). *See* POSTSTRUCTURALISM. For example, one society might kill to stop schoolgirls being educated while another bans female popes. In both cases, the deep structure is *patriarchy.

structuration theory A theory of the creation and reproduction of social systems that is based in the analysis of both structure and agents, with neither being more important than the other. This is the **duality of structure**, where both the agent and structure interact to bring about social change. Structure is not the main object of social research, but acts of structuration are. Lippuner and Werlen in N. Thrift and R. Kitchin, eds (2009) explain at length, but the abstract should be enough.

subaerial Occurring on land, at the Earth's surface.

subduction The transformation into *magma of a denser *plate as it dives under another, less dense, plate. A **subduction zone** occurs where rocks of an oceanic plate are forced to plunge below much thicker continental crust. As the plate descends it melts and is released into the *magma below the Earth's crust. Tectonic erosion of the overriding plate by the downgoing slab is believed to occur at half the Earth's subduction zones; see Vannucchi et al. (2008) *Nature* 451 who provide excellent diagrams. The subduction of features such as ridges and seamounts increases basal erosion and the subsidence of the accretionary wedge (Morgan (2008) *Geophys. Res. Abstr.* 10, EGU2008–A-00560). See also Chiarabba et al. (2008) *Earth and Plan. Sci. Letts* 268, 3–4.

subglacial Formed or deposited beneath a glacier. When an ice sheet flows over unfrozen sediment, it may cause **subglacial bed deformation**, as the water in the sediment pores increases enough to reduce the resistance between the individual grains. **Subglacial water flows** are maintained by high water pressure, rather than by melting of the subglacial channel walls. See Siegert (2005) *Ann. Rev. Earth & Plan. Sci.* 33 and Boulton et al. (2007) *Qly. Sci. Rev.* 26, 7–8. More than 70 **subglacial lakes** exist in Antarctica; the largest, L. Vostok, lies beneath 3 km of ice and is 230 km long—see M. M. Bennett and N. Glasser (2009).

subgraph In *network analysis, the graphs, or networks, forming an unconnected part of a whole graph or network.

sublimation Sublimation is *not* the direct conversion of solid ice to vapour—as the liquid on the surface of the solid lump has a high partial pressure it rapidly evaporates, so that it appears that the solid phase is changing directly into the gas phase (Holland (2004) *Weather* 59, 5). **Latent heat of sublimation** is released in this way.

sublittoral zone The zone from the lowest mark of ordinary tides to the end of the *continental shelf.

submarine slide A form of underwater *mass movement. 'Earthquakes are among the most frequently suggested

triggering mechanisms for submarine slides . . . sliding in various forms is a major agent in the evolution of passive continental margin morphology' (Bugge et al. (1987) *Geo-Marine Letts* 7). See Yamamoto et al. (2007) *Island Arc* 16, 4; and Dawson and Stewart (2007) *PPG* 31, 6 on tsunamis generated by submarine slides.

sub-optimal location A satisfactory, but not optimal, location.

subsequent streams Rivers running down the strike of usually weak *strata, or along the line of a fault, at right angles to *consequent streams.

subsistence farming A farming system where almost all of the output feeds and supports the household. See Pandi et al. (2007) *Ag. Sys.* 92, 1–3 for a detailed, practical analysis.

subsurface flow *See* THROUGHFLOW.

subtropical anticyclones Areas of high pressure formed when air, which has risen in the tropics, subsides in subtropical areas. The air is warmed adiabatically as it descends; therefore rainfall is unlikely. The subtropical anticyclone over the North Atlantic, often called the 'Azores' or the 'Bermuda high', influences weather and climate over the eastern United States, western Europe, and north-western Africa (Davis et al. (1997) *J. Climate* 10, 4). Miyasaka and Nakamura (2005) *J. Climate* 18, 23 suggest a local land–sea–atmosphere feedback loop associated with a subtropical high.

suburb A one-class community, located at the edge of the city, with low rates of housing per hectare. 'Suburbs were originally characterised by housing types not entirely different from those found closer to the urban core, but the "garden suburb" tradition came to dominate

Western cities during the twentieth century . . . Individual suburbs became more regulated, more "planned" and more middle class. The widespread adoption of the garden suburb model transformed streetscape into landscape' (Sharpe (2005) *Canad. Geogr./Géog. canad.* 49, 4). Suburb-making in the USA is 'the most geographically extensive version of suburbanization' (Martin (2007) *Area* 39, 1). Rudel et al. (2011) *AAAG* 101, 3 observe that 'more recently, suburbs have adopted land use controls that promote upper class sprawl by reserving large areas for the construction of small numbers of expensive homes on spacious lots. This regulatory shift can be explained in several ways: a home voter hypothesis that derives the new controls from the economic interests of suburban homeowners, and a regional spillover hypothesis that attributes the adoption of new controls to desires by planning commissioners, consultants, and non-governmental organizations to do as other communities are doing'. Peach (2000) *PHG* 24, 4 reports on immigrants in 'leafy suburbs': 'instead of the traditional central concentration of the first generation, the new pattern achieves instant suburbanization but remains fairly segregated.'

succession (plant succession) A series of complexes of plant life at a particular site. Plant succession is viewed as the development of plant life on originally bare earth, in a definite sequence; see Leal and Lorscheitter (2007) *Acta Bot. Bras.* 21, 1. The term **successional dynamics** applies to the shifts between successional stages of an *ecosystem.

Suess cycle *See* DE VRIES CYCLE.

sunbelt In the USA, the southern and western states, such as New Mexico, Arizona, and Florida. Guhathakurta and Stimson (2007) *Int. Plan. Studs* 12, 2

report on sunbelt coalitions that form to save the environment, 'supported by the same businesses that had earlier supported new development at the expense of existing neighbourhoods'. *See* SNOWBELT.

sunspot A dark patch on the surface of the sun, usually occurring in clusters, and lasting about two weeks. Sunspot activity fluctuates in an eleven-year cycle. It has been suggested that the sun is 1% cooler when it has no spots, and that this variation in solar radiation affects the climates of the Earth; see Schrier and Vesteeger (2001) *Geophys. Res. Letts* 28, 5.

superadiabatic lapse rate A *lapse rate over the 9.8 °C/1000 m dry *adiabatic lapse rate, rare in the free *atmosphere, but common just above land surfaces emitting strong terrestrial *radiation. When the environmental lapse rate is superadiabatic, the density of the atmospheric parcel relative to its surroundings results in a buoyancy force in the same direction as original vertical movement (Slonaker (1996) *Weather & Forecast.* 11).

superimposed drainage If a drainage pattern develops on a non-resistant mass covering a buried, resistant bedrock, and if base level falls enough, the drainage will cut into the bedrock in the established pattern (Douglass and Schmeeckle (2007) *Geomorph.* 84, 1–2).

superimposed ice Water that refreezes at the base of the snowpack. See King et al. (2002) *Annals Glaciol.* 34, 1, 335 on the detection of superimposed ice in northern Norway.

Super Output Areas (SOAs) A new geographic hierarchy designed to improve the reporting of small area statistics in England and Wales. The layers of SOA are:

S

Lower layer	Minimum population 1000; mean 1500. Built from groups of OAs (typically 4 to 6) and constrained by the boundaries of the Standard Table (ST) wards used for 2001 Census outputs.
Middle layer	Minimum population 5000; mean 7200. Built from groups of lower layer SOAs and constrained by the 2003 local authority boundaries used for 2001 Census outputs.
Upper layer	To be determined; minimum size c.25000.

- Names and codes for Super Output Area geography.
- Set of slightly revised Super Output Areas (SOAs) on NISRA Geography website.

superterrane *See* TERRANE.

supraglacial On the surface of a glacier. See Benn et al. (2000) *Int. Ass. Hydrol. Scis* 264 on the early stages of **supraglacial lake** growth.

surf The waves of the sea as they break on a shore or reef. On beaches, surf is formed between the breaker and *swash zones. See S. H. Sweeny (2008) on the geography of surfing.

The **surf similarity parameter** is based on a consideration of the wave characteristics and the gradient of the beach (Battjes 1974) may be used to predicted breaker type via the formula:

$$\xi = \frac{\tan \beta}{\sqrt{H_b/L_o}}$$

where tan β is the gradient of the beach and the subscripts $_b$ and $_o$ indicate breaker and deep-water conditions respectively. Small values for ξ are attained when the beach has a gentle gradient and the incident wave field is characterized by a large wave height and a short wave length (or a short wave period). Large values of ξ are found when the beach is steep and the incident wave field is characterized by a small wave height and a long wave length (or a long wave period). Spilling breakers occur

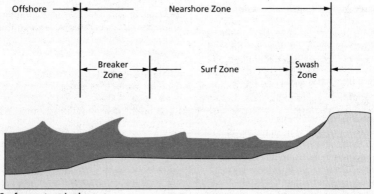

Surf zone terminology

for $\xi < 0.4$; plunging breakers for $\xi = 0.4$ to 1; and surging breakers for $\xi > 1$ (Masselink in J. Holden 2012).

surface In a *GIS, a surface has three dimensions—width, length, and a third dimension determined by the character of the surface.

surface boundary layer *See* BOUNDARY LAYER.

surface process model *See* LAND SURFACE PROCESS MODELS.

surge phenomenon *See* EDDY.

surging glacier A faster-flowing glacier. Normally, glaciers flow at 3–300 myr^{-1}, whereas a surging glacier may reach 4–12 $kmyr^{-1}$. Some glaciers always flow at surging speed; others surge periodically, suddenly accelerating to 10–100 times their normal velocity. Surging glaciers are closely crevassed, with compressional flow at the leading edge and *extending flow at the rear. The net result is a deepening of the long profile of the glacier. Kamb et al. (1985) *Science* 227 and Kamb (1987) *J. Geophys. Res.* 92 suggest that surging is initiated by a switch from a tunnel system to a distributed system that facilitated sliding. The surge will continue as long as high water pressure sustains rapid sliding.

susceptibility The degree to which an environment, landform, or organism is sensitive to a particular factor. A landslide zonation map divides the land surface into zones of varying landscape susceptibility, based on the estimated significance of causal factors in inducing instability. Such maps identify and delineate potentially unstable areas, so that suitable mitigation strategies can be used, and help planners choose favourable locations for siting development schemes (Tangestani (2004) *Australian J. Earth Scis* 51, 3). See Sandercock et al.

(2007) *PPG* 31, 2 on a **vegetation-susceptibility index**, defining the force required to bend vegetation to 45° from the vertical; see Märker et al. (2008) *Geomorph.* 93, 1–2 on **land degradation susceptibility**.

suspension The state in which small, insoluble particles are evenly distributed within a fluid. Particles are carried upwards when *turbulence outstrips the force of gravity.

sustainable development Article 9 of the UN Declaration of Human Rights (1948–66) states that all people 'should promote sustainable development all over the world to assure dignity, freedom, security and justice for all people'. Sustainable development 'meets the needs of the present without compromising the ability of future generations to meet their own needs' (1987 UN Convention on Environment and Development). 'Sustainable development requires a balance between economic prosperity, social equity and the environment to ensure quality of life now without damaging the planet for the future' (UK DEFRA). For the economic geographer, development is sustainable 'if it does not decrease the capacity to provide non-declining per capita utility for infinity' (Neumayer (2003) *Env. & Res. Econ.* 15).

Advocates of '**strong sustainability**' argue that natural capital—the range of functions the natural environment provides for humans and for itself—should be afforded special protection, while those who espouse '**weak sustainable development**' hold that natural capital can be substituted by other forms of capital, especially produced capital. 'The weak interpretations are bound to an anthropocentric world-view, and have typically phrased the essence of sustainable development in terms of an environmental question, rather than also a

S

developmental one . . . The strong interpretations refuse to treat humans and nature as separate from each other and wish to maintain life opportunities by reconciling sound human development and environmental integrity . . . while the weak interpretations tend to have faith in authoritative and coercive structures, such as market forces, the strong interpretations hold that more participatory, transparent and democratic processes are required' (Kallio et al. (2007) *Sust. Dev.* 15).

*Agenda 21 concludes that 'indicators of sustainable development need to be developed to provide solid bases for decision-making at all levels and to contribute to a self-regulating sustainability of integrated environment and development systems'.

'A **sustainable community** is one in which improvement in the quality of human life is achieved in harmony with improving and maintaining the health of ecological systems; and where a healthy economy's industrial base supports the quality of both human and ecological systems (Indigo Development). A **sustainable drainage system** aims to mimic as closely as possible the natural drainage of a site to minimize the impact of urban development on the flooding and pollution of waterways (Jones and MacDonald (2007) *Geoforum* 38, 3).

(⊕) SEE WEB LINKS
- UK government sustainable indicators.
- Indigo Development.

Sustainable Urban Drainage Systems (SUDS) Programmes designed to reduce the risk of small and medium-sized floods, and the pollution of urban rivers. Strategies include the creation of ponds, wetlands, and porous paving that can increase the storage of runoff in the catchment, and the improvement of water quality by physical, chemical, and biological processes.

sustained-yield resource A view of resource extraction as a mutually advantageous long-term partnership with nature (N. Langston 1995). Forestry may be managed as a slow-growing but renewable resource of fuel and timber and a sustained-yield recreational resource. McCarthy (2006) *AAAG* 96, 1 chronicles the failures of forestry sustained-yield policies to achieve their stated goals. The implications of sustainable yield resource input depend on the relative importance of capital and resource inputs in production (M. Common and S. Stagl 2005).

swale A long, narrow, usually shallow trough between ridges on a beach, running parallel to the coastline.

swallow hole *See* DOLINE.

swash The water moving up a beach from a breaking wave. The **swash zone** involves a moving land–water boundary travelling across the beach; at least some part of the beachface is usually unsaturated. Water depths in the swash zone are very shallow, and sediment is often transported as complicated, single-phase, granular-fluid flows (Horn (2002) *Geomorph.* 48, 1–2). **Swash-aligned beaches** are parallel to the incoming wave crests and net *longshore drift is at a minimum.

swidden cultivation *Shifting cultivation.

symbiosis An association of two participants whereby both partners benefit, as when flowering plants rely on insects for pollination and the insects feed on their nectar. **Industrial symbiosis** describes organic relationships between dissimilar industries; see Lyons (2005) *Local Env.* 10, 1 on the use of waste products from one industry as input to another. See Wu and Wall (2005) *Int.*

J. Tourism Res. 7 on economic and environmental symbiosis in tourism.

symbolic capital The resources owned by an individual that serve as a marker of her or his status; 'the material display of social status and how it is structurally constrained' (Sweet (2011) *Am. J. Public Health* 101, 2, 260). Forset and Johnson (2002) *AAAG* 92, 3 discuss the role of symbolic capital in Moscow in the transformation of national identity. See Matos-Wasem (2005) *J. Alpine Res.* 93, 1 on 'good' air—a slightly diagonal usage of the term 'symbolic capital'.

symbolic places Places invested with meaning: 'symbolic place is the keeper of national consciousness and promoter of communality. The national consciousness and self-esteem are fed by tradition and ethnic symbols' (G. Taylor and S. Spencer 2004). Missingham (2002) *Urb. Studs* 39, 9 describes Government House, Thailand, in a time of popular protest by the poor: 'the streets outside Government House are both particular and symbolic. The symbolic and metaphorical meanings of mobilising on the streets and pavements served to highlight the poor's exclusion from the places of authority and power within the walls of Government House.'

sympatric Living in the same region; a term used in ecology to specify separate species whose territories overlap. **Sympatric speciation** is the emergence of two species from a single, mixed-parent population, without the aid of geography. Jiggins (2006) *Current Biol.* 16, 9 finds evidence of sympatric speciation in isolated island habitats.

synchronic analysis The study of the internal linkages of a system at a given time, as in cross-sections in *historical geography (Conedera et al. (2007) *J. Hist. Geog.* 33, 4).

syncline and synclinorium *See* FOLD.

synekism The geographical relationships that create and give importance to cities; the causal power of urban agglomeration (E. Soja (2000)). See Blake (2002) *J. Soc. Anthrop.* 2, 2 for Soja's explanations.

synoptic Overall, or relating to a large-scale or broad view. Synoptic meteorology is primarily concerned with large-scale weather systems, such as extratropical cyclones, but not on a global scale. See C. N. Hewitt and A. Jackson (2003).

system Any set of interrelated parts; an abstraction that is assumed to exist in reality, unlike a *model which does not and is not intended to mirror reality (R. Inkpen, 2005).

Systems can be classified as open, closed, or isolated. **Open systems**—such as the ocean—allow energy and mass to pass across the system boundary. 'Capitalism emerges as an inherently open system which, through abstract and concrete labour, is constantly infused by its putative "exteriors"; differences of nationality, gender, sexuality, geographical location and so on are constantly gathered together in the domain of concrete labour and . . . forcibly articulated into a global system' (N. Castree 1999).

A **closed system** allows energy but not mass across its system boundary. The Earth system *as a whole* is a closed system (the boundary of the Earth system is the outer edge of the atmosphere). The Japanese *keiretsu* model represents a relatively closed system of vertically networked firms; see Gross et al. (2005) *J. Japanese Int. Economies* 19.

In a **cascading system**, a series of small sub-systems are linked from one system to another. Thus, the drainage basin system is made up of the weathering, slope, channel, and network subsystem, where the output from one is the input of other, and one subsystem affects another subsystem.

'Terrestrial hydrology involves a dynamic, nonlinearly lagged and cascading system consisting of complex feeds forward and feedbacks, with any perturbations to inputs or stores interacting' (Schultze (1997) *PPG* 21). In geomorphology, a **morphological system** is a subsystem of morphological components.

systematic geography The study of a particular element in geography, such as agriculture or settlement, seeking to understand the processes which influence it and the spatial patterns which it causes. Ye and Cai (2012) *Geog. Res.* 31, 5, 771 claim that 'systematic geographers stress that geography is a science pursuing the general rules, but regional geographers regard the region as the centre of geographical research'. This is worthy of debate.

systems analysis The study of the way that the component parts of a system work within it, and the way they interact. R. Inkpen (2005) points out that 'systems are still, however, a simplification of reality, not reality as it really is'. See R. Hugget (2005) for an excellent summary.

S

tacit knowledge The hidden, unspoken, unwritten, knowledge that we all possess, which we cannot pass on through writing, like riding a bike, or recognizing a face. This is knowledge that can only be transferred by face-to-face contact (Sternberg in R. Kitchin and N. Thrift, eds 2009). Howells (2002) *Urb. Studs* 39, 5–6, 872 is very helpful on this.

tafone A hollow on a sheer face, produced by localized *weathering, mainly through *granular disintegration (pl. *tafoni*). Canopy-shaped, overhanging cave roofs are typical. In semi-humid to semi-arid regions, tafoni weathering is among the most effective weathering processes operating on surfaces of plutonic rocks such as granite (Brandmeier et al. (2011) *ESPL* 36, 6, 839). Data gathered by Roqué et al. (2013) *Geomorph.* 196, 94 support a subterranean origin of some tafoni. Norwick and Dexter (2002) *ESPL* 27 find that the rate of tafoni enlargement decreases over tens of thousands of years. Hejl (2005) *Geomorph.* 64, 1–2 describes a prehistoric wall painting which displays stages of tafoni development.

taiga The predominantly coniferous forest, south of the *tundra in northern continents, also known as boreal forest. See Eugster et al. (2000) *Glob. Change Biol.* 6, 1 and Kuris and Ruskule (2006) *Baltic Env. Forum* on management of boreal forest habitats.

talik Within a *permafrost zone, the layer of unfrozen ground that lies between the permafrost and the seasonally thawed *active layer. Talik most often occurs below rivers and lakes, or where groundwater upwelling occurs along fractures (Delisle (1998) *J. Quat. Scis* 13). See Grosse et al. (2007) *Geomorph.* 86, 1–2. For the role of talik in the formation of pingoes, see Gurney (1998) *PPG* 22, 3.

talus A *scree slope formed of *frost shattered rock *debris which has fallen from the peaks above. A talus slope is usually straight, and at an angle of 34–35°. Slope angle in a talus cone reflects the balance between the supply of sediment from upslope (usually by physical weathering of bedrock) and the processes moving the sediment down-slope while spreading it out. Lithology is also an important control. Grain flows are frequent occurrences; see Croasta et al. (2007) *Geophys. Res. Abstr.* 9, 07610 on grain size, and Pech and Kotarba (2002) *Studia Geomorph. Carpatho-Balcanica* 36, on talus slopes in the high Tatra. Obanawa and Matsukura (2006) *Computers & Geosci.* 32, 9 propose a mathematical model for the topographic change of a cross-section of a talus landform. See van Steijn et al. (2002) *PPG* 26, 4 on **cliff-talus systems**.

The slow, downslope movement of talus is **talus creep**, initiated either by the shock of new fragments falling on the scree or by the movement of individual particles resulting from heating and cooling.

taphonomy The study of the processes impacting an organism, especially fossilization, between the time of its death and its later discovery. See, for example, Lyman (2010) *Can. J. Earth Scis* 47, 11, 1367.

tariff A list of duties or customs to be paid on imports. Tariffs may be imposed to increase the cost of imported goods in relation to domestic production, thereby reducing the volume of imports and keeping the *balance of payments in credit, or to protect domestic industry from foreign competition. **Preferential tariffs** reduce import duties on products of a certain type or origin. The impact of tariffs may be less than expected: Behrens (2006) *Papers Reg. Sci.* 85, 3 shows that decreasing transport costs are more likely to trigger agglomeration than decreasing tariffs, and Medvedev (2006) *World Bank Policy Research Working Paper* 4038 finds that preferential trade agreements do not much increase trade flows. In a highly technical paper, Mai et al. (2006) *CIRJE-F-435* show that, when the transport cost is sufficiently small, tariff competition coupled with firm migration leads to a core–periphery economy.

tax havens and offshore financial centres (THOFC) The term 'tax haven' applies to a small, low-tax jurisdiction specializing in providing corporate and commercial services to non-resident offshore companies, and for the investment of offshore funds, together with connotations of financial secrecy and tax avoidance. Despite the use of the acronym THOFC, a tax haven is *not* the same as an offshore financial centre, since many of the latter have no statutory banking secrecy ('FACTBOX-Status of bank secrecy protection in Europe', *Reuters*, 11 August 2009) and most of which have adopted tax information exchange protocols to allow foreign countries to investigate suspected tax evasion (*OECD Report*, 2 April 2009).

taxocene *See* TROPHIC LEVEL.

taxon *See* TROPHIC LEVEL.

taxonomy A classification system, especially of organisms; the method by which scientists, conservationists, and naturalists classify and organize 'the vast diversity of living things on this planet in an effort to understand the evolutionary relationships between them' (Fenneman, E-Flora BC).

(●) SEE WEB LINKS
• Electronic Atlas of the Flora of BC.

Taylorism A system of production characterized by the division of factory work into the smallest and simplest jobs, while closely coordinating the sequence of tasks in order to achieve maximum efficiency, as, for example, on a production line. Skilled managers and technicians oversee semi-skilled or unskilled workers who are engaged in simple, repetitive chores. Ellem (2005) *Asia Pacif. J. Hum. Resources* 43, 2 observes that 'most people would say that the biggest change in work towards the end of the last century was the abandonment of Taylorism'. However, Lorenz and Valeyre (2005) *J. Indus. Relations* 47 observe that traditional Taylorist forms of work organization are holding their own in certain nations and sectors of the EU.

tear fault A *fault characterized by lateral movement, transverse to the strike of the rocks. See Huang et al. (2004) *Marine Geophys. Res.* 25, 1–2.

technological revolution The development in the last 25 years in information and telecommunications technology. It has made possible the worldwide integration of economic processes while allowing organizational structures to respond rapidly to changes in markets and technology. Information technology has pushed forward globalization and its overall effects (anon. (2006) *Pop. & Dev. Rev.* 32, 4). Unilateral advantages do not last long; backwardness can sometimes be advantageous if it creates opportunities for technological 'leapfrogging', as 'late-comer' institutions and capital are not locked into older technologies, enabling quicker uptake and diffusion of new ones (Carmody (2008) *Geog. Compass* 2, 1). S. Amin (2003) even refers to the 'obsolescence of capitalism' due to the technological revolution, making labour redundant. See Harris in S. Danzigler and R. Haveman (2001) on anti-poverty policy after the technological revolution. **Technological determinism** 'simply put, is the idea that technology has important effects on our lives. This idea figures prominently in the popular imagination and political rhetoric, for example in the idea that the Internet is revolutionizing economy and society' (Adler in S. Clegg and J. R. Bailey, eds 2008).

technopole A global, rather than local, *growth pole, which M. Castells (2014) calls a key feature of national economic development. Technopoles include three major types: industrial complex of high-tech firms, emerging spontaneously, and linking research and development with manufacturing. See Wang et al. (1998) *Asia Pac. Viewpt* 39, 3 on the development of technopoles in China, and Fontan et al. (2004) *Canadian J. Urb. Res.* 13, 2 on the Angus Technopole in Montreal: 'a new type of community initiative which combines endogenous and community development with business oriented projects.'

technoscience The marriage of technology and science, with each of these using the other for support. A. Nordmann (2013) refers to a 'growing number of fields such as nanotechnology, synthetic biology, and climate studies where it is not possible to separate the scientific from the technological'. **Geographies of technoscience** consider the way that space matters in the production of science and technology; the 'implications of the circulation of expertise and materials in the situating of science and technology; as well as the importance of locating the effects of technoscience in their social, cultural, economic and political context' ('Locating Technoscience').

(((●))) SEE WEB LINKS
• UCL's Locating Technoscience website.

tectonic Of, or concerned with, the processes acting to shape the Earth's crust.

telecommunications The electronic transmission of information over long distances. 'Geographers have long found telecommunications of interest because of the importance of such technologies in the folding of space and their impacts on various economic, political, and social phenomena' (B. Warf (nd) *Oxford Bibliographies Online*). See Rimmer in F. C. Lo and Y.-M. Yeung, eds (1998) on transport and telecommunications among world cities.

teleconnections Causal links between patterns of weather in two locations, or between two atmospheric occurrences, which are far apart, such as the *El Niño–Southern Oscillation, where unusually high sea surface temperatures off the west coast of South America can be connected with droughts in Indonesia (M. E. Mann, R. S. Bradley, and M. K. Hughes 2000), and the blocking-like pattern over the British Isles in summer which co-occurs with a low over southeastern Europe (Trouet et al. (2012) *Holocene* 22, 8, 887). Teleconnections are usually expressed statistically by the amount of shared variance between the two locations or systems. See Simpson and Erkal (2004) *Papers Applied Geog. Confers* 27 on using teleconnections to predict temperatures.

teleology The study of purposes, goals, or ends; the theory that acts, objects, states of affairs, and so on can only be justified in terms of the ends—'rightness' is not intrinsic in an action or process, but is dependent on the consequences of the action or process. A. Memmis (1957) describes the teleology of colonial power and its resistance. H. Obenzinger (1999) claims that the field of 'Holy Land' studies has been controlled by colonialist teleology in much American and Israeli scholarship. S. Elden and E. Mendieta (2011) is worth a try, particularly ch. 10.

temperate Describing those locations and climatic types falling between subtropical and subarctic.

temperate glacier A glacier with basal temperatures of around 0 °C. Warmth is provided by friction with the bedrock, by shearing within the ice mass, and from *geothermal heat. Glacier flow is rapid because warm glaciers are lubricated at their base, and velocities of 20–30 m per day have been recorded. Warm glaciers are the most effective agents of *glacial erosion. See D. Evans (2003) on active temperate glacier margins, and Lingle and Fatland (2003) *Annal. Glaciol.* 36, 1 on temperate glacier surges.

temporality In contrast to the measurable and calculated notion of time/chronology, temporality is concerned with the way in which a sequence of events, a kind of history, is physically experienced by those who live through them or experience them. Thus the passing of time is treated not as a neutral dimension but rather as being

constituted by social practices (T. Darvill 2002). Academic geographers do not express—or do not want to express—this concept very clearly: 'time and space do not exist singly, but only as a hybrid process term' (J. May and N. Thrift 2001); 'the predominant poststructuralist understanding of temporality is in terms of a series of successive moments of pure contingency, tied together by nothing other than the force of an imposed convention or act of vitalistic will' (Barnett (2004) *Pol. Geog.* 23, 5); 'to take on board the coevality of space is to refuse that flipping of the imaginative eye from modernist singular temporality to postmodern instantaneity' (Massey in O. Eliasson 2003), and Brown (2007) *Canad. Geogr./Géogr. canad.* 51, 1 writes of 'the implicit linear temporality of the two-dimensional planes that define the Western agricultural landscape'.

tephigram An *aerological diagram, of two axes aligned at 45° to each other. The nearly horizontal lines show *atmospheric pressure, together with the heights at which they are found; the lines at 45° to the isobars show temperatures, running from bottom left to top right. Superimposed upon these two are three sets of guidelines: those running from bottom right to top left show the dry *adiabatic lapse rate (DALR); those running from bottom left to top right, at about 60° to the base, show constant humidity (mixing ratio lines); and convex, curved lines indicate the saturated adiabatic lapse rate (SALR). With the help of these lines, the behaviour of a rising parcel of air may be predicted. First, the readings from a *radiosonde are plotted to show the environmental lapse rate. Next, from the same source, dew point temperatures are plotted against height. From the temperature and height of the air parcel under investigation, a plot is made of its temperature fall at the DALR, parallel to the

DALR guidelines. Similarly, the dew point change of the air parcel with height is plotted, parallel to the mixing ratio lines. Where these two plots intersect, the parcel will begin to cool at the SALR, and a plot is made of the parcel parallel to the SALR guidelines. At the point where this plot intersects the environmental lapse rate line (above), the height of the top of any cloud formed by the parcel may be established.

tephra A deposit made of fragments of rock—from so-called 'bombs' to dust and *ash—shattered by an explosive volcanic eruption. The coarser, heavier particles fall out close to the volcano vent, while the finer dust may be carried hundreds of kilometres. **Tephrostratigraphy is** the study of sequences of tephra layers and related deposits and their relative ages. It may be used as a tool for correlation in the study of

Tephigram

stratigraphic sequences. See Hall and Mauquoy (2005) *Holocene* 15 on tephra dating, and Xia et al. (2007) *Env. Geol.* 51, 8 on correlating tephra layers.

terminal moraine *See* MORAINE.

terms of trade The ratio of the index of export prices to the index of import prices. If export prices rise at a faster rate than those of imports, then there has been an improvement in the terms of trade.

'Since the early 1980s, all developing countries taken together have been experiencing a downward trend in their net barter terms of trade . . . From a development perspective; the use of the additional income resulting from terms-of-trade changes is of crucial importance. For example, if terms-of-trade gains resulting from higher export prices accrue in the form of higher company profits, and if these are reinvested, the medium-term impact on growth will be much greater than in a situation where the gains accrue to the government through transfers from State-owned enterprises, which are used to service the public debt, or in a situation where they accrue to workers in the form of higher wages that are spent for consumption. Similarly, a deterioration in the terms of trade resulting from higher import prices or lower export prices can lead, *inter alia*, to either a reduction of investment, an increase in government indebtedness or higher unemployment and wage compression if it is not counterbalanced by productivity and export volume growth' (UNCTAD Trade and Development Report 2005).

Chung (2007) *Int. Econ. J.* 21, 1 finds that labour-augmenting technological progress turns the terms of trade against the growing country while capital-augmenting technological progress shifts them in favour of the growing country. Broda (2002) *Staff Report, Federal Reserve Bank of New York*, 148, on terms of trade and developing countries, is useful.

terrace *See* RIVER TERRACE.

terracette A small staircase-like landform generally less than 1 m in tread width and riser height, and as long as 300 m, running transversely along slopes. Explanations of terracette formation include animal disturbance, soil creep, solifluction (gelifluction), slumping and rotational slippage, regolith and vegetation control, subsidence, and anthropogenesis or tectonism; see Brandon and Scroder (2011) *Zeitschr.* 55, 1, 45.

terracing The construction of a series of horizontal levels built on a hillside, especially in upland or piedmont zones, in order to retain water and reduce soil erosion (C. Barrow (1999), Chen et al. (2007) *PPG* 31, 4); terracing technology is intimately related to irrigation and the availability of water. Ye and Yang (2002, *J. Chongqing U., Nat. Sci. Ed.* 25, 6, in Chinese, with English abstract) suggest that slopes of less than 8° are more suitable than steep slopes for changing into terraces. The terrace is a highly variable and only sometimes distinctive cultural artefact (Field (1966) *DTC* AD0679468).

terrain analysis (digital terrain analysis) The use of *remote sensing satellite data for mapping various aspects of terrain, such as land cover, land use, and soils. Software may then be utilized to derive terrain parameters, such as aspect, catchment area, and wetness index, which are then used to describe the morphology of the landscape and the influence of topography on environmental processes. See I. Florinsky (2011).

terrains of resistance The material and/or symbolic ground upon which action by social groups takes place.

'A terrain of resistance is simultaneously both metaphoric and literal' (Routledge (1996) *Pol. Geog.* 15, 6/7). See also Routledge (2000) *GeoJournal* 52, 1. R. Neumann (1998), for example, shows how Arusha National Park, Tanzania, 'embodies all the political-ecological dilemmas facing protected areas in Africa'.

terrain surface The landscape which may be simulated in a terrain surface display system. Smith and Mark (2003) *Env. & Plan. B* 30 identify six factors as topographic existences: elevation, terrain surface shape, topographic position, topographic context, spatial scale, and landform.

terrane In geology, a fragment or block of crust with an individual geological history that differs from the surrounding areas, and is usually bounded by faults. **Accreted terranes** are those that become attached to a continent by tectonic processes. If a relatively small crustal fragment is accreted to a larger continent, it is an **exotic terrane** (originating elsewhere). **Superterranes** are composite terranes, composed of individual terranes and other assemblages, which share a distinctive tectonic history; see Mezger et al. (2001) *J. Metam. Geol.* 19, 2.

((⊕)) SEE WEB LINKS
• Cordilleran accretionary terranes.

terra rossa A red intrazonal soil developed in Mediterranean regions on weathered limestone.

terrestrial magnetism *See* GEOMAGNETISM.

terrestrial radiation The heat radiated from the Earth. Short-wave *solar radiation reaching the Earth does not heat the atmosphere it passes through, but does heat the Earth's surface. In turn, and particularly on clear nights, much of this heat is radiated out from the Earth. The Earth also absorbs terrestrial radiation reflected from the overlying opaque atmospheric layer. It is by long-wave terrestrial radiation that the *atmosphere is heated. Almost one-third of the solar radiation intercepted by Earth is radiated back into space. In urban areas, a reduction in the *sky view factor decreases the loss of terrestrial radiation, thereby contributing to the creation of *urban heat islands (Grimmond (2007) *Geography* 30, 3).

terrier A written survey of an estate, recording the value and extent of the land. The earliest English example dates from AD 900, but they were extensively produced from the 16th century onwards. See Fletcher (1998) *TIBG* 23, 2.

territoriality The need by an individual or group to establish and hold an area of land is an urge, fuelled by aggression, to define a territory for mating and food supply. In human beings, territoriality is an organization of space in order to make sense of it. The individual needs security and identity, and this is shown most clearly in relation to the home, and the community requires a suburb or small town with which to associate. The importance of territory extends to larger units; the reorganization of the counties of Britain always causes distress, and the term is most often associated with nation-states which have the formal power to demarcate and control their borders. Hensel in J. Vasquez, ed. (2000) notes that, between 1816 and 1992, over half of all disputes began between neighbours, over a quarter involved territorial issues, and both contiguity and territory were involved in over half of all full-scale wars. See

D. Storey (2001) for theories of territoriality and the geographical outcomes of territorial control, Johnston (2001) *Polit. Geog.* 20, 6 on territoriality in political geography, and Oosten (2003) *Acta Politica* 38, 2 on territoriality and globalization. Heeg and Ossenbrügge (2002) *Geopolitics* 7, 3 provide a concise overview of the history of the EU territoriality; see also K. Cox (2002). See Johnson (2001) *Pol. Geog.* 20, 6 on theories of territoriality.

Agnew (2005) *AAAG* 9 defines **political territoriality** as the administration of infrastructural power, and identifies four 'sovereignty regimes' that result from distinctive combinations of central state authority and degree of political territoriality. Lewis et al. (2002) *PHG* 26, 4 describe a **territorial point of entry**, which 'emphasizes local social institutions, local government, the territoriality of resources, the articulation of local and global knowledge, and the spatial strategy of the enterprise'.

territorial justice The application of ideas of *social justice to an area of territory; that is, the identification by a government of areas of need, followed by a deliberate policy of redressing an imbalance. Ideas of social justice vary according to the *mode of production and the prevailing ideology, and so territorial justice varies globally. See Morgan (2006) *Publius* 36 on territorial justice and the North–South divide.

territorial seas (territorial waters) The coastal waters together with the sea bed beneath them and the air space above them, over which a state claims *sovereignty. Most countries claim twelve nautical miles. In 1983, the Law of the Sea Convention proposed a 200-nautical-mile exclusive economic zone, but most nations are less than 400 nautical miles apart, so a median line is drawn between the baselines of the states concerned.

territory 1. In ecology, the living space of an animal which it will defend from the forays of other territorial animals. Animals need space in which to reproduce and their territory can be some or all of the following: a source of food, a source of mates, and a breeding area. Pakeman (2001) *J. Biogeog.* 28, 6 investigates possible links between simple migration rates and herbivore territory size.

2. The area claimed or controlled by a state or other political actor. The contested basis for such claims in large part shapes the politics of conflict. Political geographers study *territoriality as the means by which political actors translate their control over a territory into control over people, resources, and relationships (Dahlman in C. Gallaher et al. 2009). A 'national territory'—the area of land seen to be inhabited by a nation—is based on claims to a particular space, usually backed up by references to history (actual or invented) and central to the nation's being. 'Territory comprises ancestors, knowledge, the use of plants, their evolution, the perception of the cosmos, customs and community and living history' (Ceceña (2004) *Antipode* 36, 3). Elden (2011) *PHG* 34, 6 outlines the issues at stake in grasping how 'territory' has been understood in different historical and geographical contexts. Competing territory claims are a major source of conflict. In a paper that well rewards careful study, Alatout (2006) *Pol. Geog.* 25, 6 shows that different perceptions of territory between Palestinians and Israelis lead to forms of resistance that are focused on property rights and questions of sovereignty in Palestine and on concerns over quality of life of the population in Israel.

terroir 'An area or terrain, usually rather small, whose soil and microclimate impart distinctive qualities to food products' (Barham (2003) *J. Rural Studs* 19), and which is tied in with the marketing and cultural branding of food through its association with place. Ilbery and Kneafsey (1998) *Eur. Urb. Reg. Studs* 5 find that 'terroir' represents a more embedded set of relations between producers and consumers. Gade (2004) *AAAG* 94 uses the French term *patrimonialisation* to describe 'this mesh of authenticity, heritage and food as manifested in regional cuisine, [and] the protection of rural landscapes'.

terrorism The use of unauthorized and spectacular violence and intimidation for political purposes.

(((●))) SEE WEB LINKS
• Website of the Handa Centre for the Study of Terrorism and Political Violence.

tertiary The first period of the *Cenozoic era, from about 65 to 1.8 million years ago, subdivided into the Palaeocene, Eocene, Oligocene, Miocene, and Pliocene *epochs.

tertiary industry Economic activity concerned with the sale and use of economic goods and services; service industry. Economically advanced countries have a high proportion of employment in services: USA 75.3%; Britain 72.6%, France 72.4%, Germany 69.2%, Japan 68.1%. See Daniels, O'Connor, and Hutton (1991) *Growth and Change* 22, 4

on the planning response to service-sector growth in urban areas.

China's employment statistics indicate that industrialization in developing countries may also be accompanied by a percentage rise in services.

The corresponding decline of primary and secondary industries in advanced economies has ushered in the *post-industrial era.

Tethys An ocean which developed during Palaeozoic and Mesozoic times, running from the coast of southern Spain to South-East Asia. Great thicknesses of sediment were formed since the sea floor kept subsiding at the same rate as deposits were laid down. These sediments were compacted, subjected to volcanism, and then uplifted and deformed by the earth movements; see Pivnik and Wells (1996) *GSA Bull.* 108, 10 on the transition from Tethys to the Himalayas, and Schreiber et al. (2011) *Geodin. Acta* 24, 1, 21 on the subduction of the Alpine Tethys.

text A group of practices for signalling meaning(s). This commonly means written texts, but has recently included economic, political, and social institutions, paintings, landscapes, and maps. Anthropologists view culture as a text. These texts have been exposed to the *semiotic analyses associated with structuralism, and the *deconstruction and *discourse analyses of *poststructuralism. The main foci of texts in geography have been the interpretation of maps as cultural texts (see Kokkonen (1998) *Cartographia* 35, 3–4 on

	1978	1989	1997	2002
% employed in primary industry	70.5	60.1	49.9	50.1
% employed in secondary industry	17.3	21.6	23.7	21.4
% employed in tertiary industry	12.3	18.3	26.4	28.6

Source: National Bureau of Statistics of China, *China Statistics Yearbook* (2003).

cartography in the Baltic Sea) and the nature and usefulness of texts as a metaphor: '[a] metaphorical understanding of geography is popular because it suggests an alternative to both relativism and to naturalist conceptions of mimetic truth. These alternatives often involve rather different understandings of geography' (Demeritt and Dyer (2002) *Area* 34, 3).

thalweg The line of the fastest flow along the course of a river. This usually crosses and recrosses the stream channel. Channel morphology can undergo rapid transitions along the thalweg, depending on changes in topography, bedrock structure, water, and sediment supplies; see Ramonell et al. (2002) *Zeitschrift* NF 192. For techniques to quantify thalweg profiles, see Bartley and Rutherfurd in F. Dyer et al. (2002).

thermal erosion *See* THERMO-EROSION.

thermal expansion (insolation weathering) The rupturing of rocks and minerals mainly as the result of large daily temperature changes. The exterior of the rock expands more than the interior. Whether thermal expansion is effective in an environment with no water is open to question; see Koch and Siegesmund (2002) *Spec. Publ. Geol. Soc. Lond.* 205. See also Smith et al. (2005) *Geomorph.* 67, 1–2 and Wigley and Raper (1987) *Nature* 357 on the thermal expansion of sea water associated with global warming.

thermal inertia The slowness with which the temperature of a body approaches that of its surroundings. 'The Earth's climate system has considerable thermal inertia. The effect of the inertia is to delay the atmospheric and oceanic response to various climate forcings. The existence of thermal inertia implies that still greater climate changes will be in

store for present increased atmospheric CO_2 levels, which may be difficult or impossible to avoid' (Lockwood in J. Holdem 2012).

thermal low An intense, low-pressure system caused by local heating of the Earth's surface, and leading to the rising of air by convection. Heavy rainfall will result if the air rises and cools enough for condensation to occur. The Iberian thermal low is thought to be the most prominent example of its type; see Portela and Castro (2006) *Qt. J. Royal Met. Soc.* 122, 529.

thermal pollution The contamination of cold water by warm water. Sources of heat include water used for cooling in electricity stations, the *urban heat island, and the construction of reservoirs. That warmer temperatures lower dissolved oxygen in the water, increase respiration rates of organisms, and increase fish and wildlife susceptibility to disease, parasites, and toxic chemicals is generally agreed; see B. Phillips, ed. (2001) on thermal pollution in the Murray-Darling Basin Waterways.

Interestingly, Beser's 2007 paper (*Digital Repository U. Maryland*), in a study of the effects of the thermal effluent in Chesapeake Bay, finds that thermal effluent does 'not cause diversity differences between submersed aquatic macrophyte communities in different areas of the thermal regime'.

thermal stress The strains set up as a result of temperature changes. Hall (1999) *Geomorph.* 31, 1–4 outlines the importance of thermal stress in cold regions; see also Hall and Andre (2001) *Geomorph.* 41, 1 on weathering by thermal stress in Antarctica. The term is also used to describe the impact of temperature changes on humans: Auliciems and Kalma (1979) *J. Applied Met.* 18, 5 propose a climatic classification of human thermal stress in Australia, and Yan (1997) *Sing. J. Trop. Geog.* 18, 2 analyses climatic thermal stress on the people of Hong Kong.

thermal wind Not a real wind, but an expression of *wind shear for a given layer of *atmosphere; the *vector expressing the difference between the *geostrophic winds at the bottom and top of the layer. It is proportional to the thickness of the layer, and is directed along the isotherms, with cold air to the left in the Northern Hemisphere, and to the right in the Southern. However, the term is used to denote a wind developing as follows: the *pressure gradients which produce surface winds may be due to the presence of cold and warm air masses. The fall in pressure with height is rapid in cold air, and much less rapid in warm air. Thus, at height, air pressure in the cold air will be less than that in the warm air. This creates a high-level pressure gradient and, therefore, a wind, often described as the 'thermal wind'. The strength of this wind is a function of its height and the temperature difference between air masses; the greater the difference, the stronger the wind.

Since there is a marked *meridional temperature gradient in the *troposphere, influenced at height by a powerful westerly factor, thermal winds are very strong at the point where the temperature gradient is greatest; at the polar *front. The result is the polar front jet. The force of a thermal wind may be strengthened by any pressure gradient at ground level. See Chu-shih (1963) *Acta Met.* 33 on thermal wind in a *baroclinic atmosphere, and Roth et al. (1999) *Boundary-Layer Met.* 92, 2.

thermocirque A large hollow on a hillside formed from the coalescence of *nivation hollows. See Kizyakov (2005) *Atlas Conferences Inc, Document #* capx-22 for the formation and retreat of thermocirques in the Arctic.

thermocline Also known as the metaliminion, the thermocline is the depth at which the temperature gradient of a large body of fluid water changes rapidly with depth. In an ocean, it is the transition layer between the mixed layer at the surface, which is of almost uniform temperature, and the deep water layer below. Within the lower atmosphere, a thermocline occurs at the boundary between the *troposphere and the *stratosphere.

thermo-erosion The combined thermal and mechanical activity of running

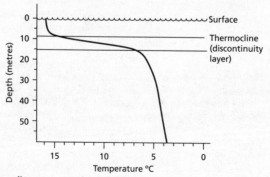

Oceanic thermocline

water in *periglacial conditions. See Etzelmüller (2000) *Zeitschrift Geomorph.* 4, on thermo-erosion in proglacial areas.

thermohaline circulation Also known as the **global thermohaline conveyor belt**, the Atlantic thermohaline circulation begins when the Gulf Stream (and its extension, the *North Atlantic Drift) bring warm, salty water to the north-east Atlantic, warming western Europe. The water cools, mixes with cold water coming from the Arctic Ocean, and becomes so dense that it sinks, both to the south and east of Greenland. This current is part of a larger system, connecting the Atlantic with the Indian and Pacific Oceans, and the Southern Ocean. Further sinking of dense water occurs near to Antarctica. Water from the two main sinking regions spreads out in the subsurface ocean affecting almost all the world's oceans at depths from 1 000 m and below. The cold, dense water gradually warms and returns to the surface, throughout the world's oceans. These surface and subsurface currents, the sinking regions, and the return of water to the surface form a closed loop.

Model simulations (for example, Srokozs (2003) *Royal Society*) suggest that a sudden shutdown of the thermohaline circulation will cause global climate changes. For an assessment of the risk of a collapse of the Atlantic thermohaline circulation, see Yin Yohe et al. (2005) *Change* 54.

thermokarst A landscape characterized by irregular hummocky topography and interconnected linear ridges, caused by the irregular heaving and melting of ground ice under periglacial conditions (Costard and Baker (2001) *Geomorph.* 37, 3–4). The exact form of the depressions depends on the original distribution of ice segregations, the subsurface movement of water during warmer periods, and the presence or absence of water in the hollows, since water-filled hollows tend to perpetuate themselves. (The term 'karst' is used to indicate the

Thermohaline circulation

numerous features formed by subsidence, and does not imply the presence or development of a limestone landscape.) See Kokelj and Lantz (2007) *Annual Conf. Geol. Ass. Canada* on thermokarst slumps and global warming, and Seppälä (1997) *Geomorph.* 20, 3–4 on piping in thermokarst.

thermosphere That part of the atmosphere, starting at about 85 km above the Earth (the top of the mesosphere), extending to the uttermost fringe of the atmosphere. Here, temperatures increase with height. See C. Lagos (1967) on the dynamics of the thermosphere.

thick description Describing a human action not merely to explain, but, additionally, to comment on and interpret it in the light of its context and structures; structures that are 'complexly layered one on top and into each other so that each fact might be subjected to intercrossing interpretations' (Anon., *the cultural studies reader* 2012). C. Geertz (1973).

Thiessen polygon A subdivision of a drainage basin, containing a rain gauge. Polygons are constructed by first siting the rain gauges, plotting them on a base map, and connecting the sites by straight lines. The lines are bisected with perpendiculars, which meet to form the polygons. The areas of the polygons are calculated and expressed as fractions of the total area. Each fraction is multiplied by the precipitation recorded by its rain gauge. The sum of these calculations represents total precipitation over the catchment area. For an account, see Yoo et al. (2007) *J. Hydrol.* 335, 3–4.

thin region As compared to a region with *institutional thickness, a less favoured region. A region may be 'thin' if it is located far from relevant knowledge organizations; for example, the Finnish sub-region of Seinäjoki (Kosonen, *ERSA Conference Papers* No. 3, 112). Functionally, it may have small industrial milieux, a lack of relevant organizations, or its firms may lack collective learning; see Kira and Frieling (2007) *J. of Wkplace Learn.* 19, 5. This functional thinness can be a consequence of a region's decision-making powers, financial resources, or policy orientation (Tödtling and Kaufmann (1999) *Eur. Plan. Studs* 7). See also Zimmerbauer (2013) *Eur. Urb. & Reg. Studs* DOI: 10.1177/0969776413512842.

Thiessen polygon

t

third food regime Also termed the *post-productivist transition, the third food regime is one in which 'strategies for profit capture are built around expediting internationally co-ordinated flows of production, commodities and capital' (Pritchard (1998) *Econ. Geog.* 74, 1). The third food regime is highlighted by the increasing globalization of production and consumption markets reflected, in part, by corporations which are 'geared to the global or regional markets that are replacing national markets' (McMichael (1992) *Sociolog. Perspectives* 35, 2), and characterized by flexible production and adaptation to niche markets, aided by research and development. Bursch and Lawrence (2009) *Agric. Hum. Values* 26, 267 stress the increasing influence of financial capital on the agri-food system, while LeHeron and Roche (1995) *Area* 27, 1 see the third food regime as centred around the expansion of the fresh fruit and vegetable complex. The development of organic food production (and its creation of social and economic space) is an expression of the third food regime. See McKenna et al. (2001) *Geoforum* 32, 2.

third space theory E. Soja (1996) proposes a different way of thinking about space and spatiality. First and second spaces are two different, and possibly conflicting, spatial groupings where people interact physically and socially: such as home (everyday knowledge) and school (academic knowledge). Third spaces are the in-between, or hybrid, spaces where the first and second spaces work together to generate a new third space. 'Soja is anxious to avoid the common dualities of the social and the individual, culture/nature, production/reproduction, the real versus the imagined, (which pervade geographical analysis), arguing "there is always another way"' (Wharf (2006) *PHG* 30, 819). Linehan and Gruffudd (2004) *TIBG* 29, 1

apply the term to the tented camps set up in inter-war Wales for retraining miners ('isolated from their wives and families, and separated from the rural society around them'); and, in a cheeky usage, Starbucks propose their premises as 'third space experience: the place between work and home'. See Moje et al. (2004) *Reading Res. Qt.* 39, 1 on third space in an educational context.

third world Originally, a synonym for those nations that aligned themselves with neither the West nor with the Soviet bloc during the Cold War. Today, the term is used to denote nations with the smallest UN Human Development Indices (HDI). There is no objective definition of Third World or 'Third World country', and these countries are also referred to as 'the South', developing countries, and least developed countries. H. Arendt (1970) argues that 'the Third World is not a reality but an ideology'. D. H. Aldcroft (2007) refers to 'Europe's third world' when considering the puzzling persistence of economic backwardness outside the north-western European core; but see Sidaway (2007) *PHG* 31, 3 on the decline of third worldism.

(((∰))) SEE WEB LINKS
• The Third World Network.

thread A river channel. *Anastomosing (braided) river channels are multi-threaded. Richardson and Thorne (2001) *Geomorph.* 38, 3–4 argue that the division of the water in a single-channel river into multiple velocity threads is a prerequisite for development of *braiding.

threshold In a system, the 'tipping point' which brings about lasting, non-reversible changes, after which the system must develop a new equilibrium condition, adjusted to the characteristics of the altered controlling factors. If, thereafter, the system reverts to the original, balanced condition, it has not crossed a threshold. **External thresholds** are crossed as a result of a change in one of the external variables. Schumm (1979) *TIBG* 4, 4 notes that the definition has been broadened to include abrupt landform change as a result of progressive change of external controls. Ritter et al. (1999) *Geomorph.* 29, 3 argue that 'if thresholds are accepted as boundaries between equilibrium and disequilibrium, they are required to be time-dependent phenomena. Thus, a steady-time cut and fill of a stream channel during the passage of a flood event is a different type of threshold crossing than the sustained, graded-time entrenchment of a channel due to distinct climate change.'

Most work on geomorphic thresholds has been on fluvial processes (see, for example, Ritter et al. (1999) *Geomorph.* 29, 3 and Tucker and Stokes (2004) *Amer. Geophys. Union*, Abstr H53E) and mass movement (see Caine (1980) *Geograf. Annal. A* 62, 1–2 and Friend (2000) *Prof. Geogr.* 52, 2). 'If the threshold conditions can be recognized, not only will different explanations for some landforms emerge, also the ability to identify incipiently unstable landforms and to predict their change will be of value to land managers and engineers' (Schumm, op. cit.).

threshold hillslope A hillslope steepened to its maximum stable, or threshold, angle. According to the **threshold hillslope model**, landscapes respond to river dissection into the bedrock by adjusting landslide erosion rates on threshold hillslopes. This model assumes that landslide erosion rates are related to river incision rates, and that landslide erosion rates increase non-linearly as hillslope gradients approach the threshold angle. See DiBiase et al. (2012) *ESPL* 37, 8, 855. In

*steady-state landscapes, the threshold hillslope model additionally predicts that landslide erosion rates are spatially coupled with rock uplift and exhumation rates.

threshold population The minimum population needed to justify the provision of a certain good or service. This may be expressed crudely, in population numbers, but purchasing power may be a better yardstick for commercial goods or services. Note that threshold populations may vary regionally, and certainly vary nationally; the threshold population for a baker's shop is far lower in France than in the UK, for example. The actual evaluation of the threshold population for most goods and services is difficult. However, Coon and Leistritz (2002) *North Dakota State U., Agribus. & Appl. Econ. Misc. Rept* 191 have, with many caveats, attempted to do so.

throughfall *See* INTERCEPTION.

throughflow Also known as subsurface flow, this is the movement diagonally downslope of water through the soil, as opposed to the vertical movement known as *percolation. It may follow natural *percolines in the soil. Throughflow involves micro-*pores (less than 0.1 mm in diameter) and macropores, natural *soil pipes, displacement flow, and groundwater discharge. Throughflow is a major factor in the hydrology of a drainage basin where the rocks underlying the soil are impermeable. See Kirkby and Chorley (1967) *Bull. Int. Assoc. Sci. Hydrol.* 12. See also Borga et al. (2002) *Hydrol. Procs* 16.

throw Of a *fault, the vertical displacement of strata along a fault line.

thrust A movement causing the formation of a reverse *fault of a very low angle. The **thrust plane** is the low-angle fault face over which movement occurs.

Rubin et al. (2001) *Eos* 82, 47 write on Taiwanese **thrust faults**. A **thrust belt** is a geological formation caused by compressional tectonics, which ultimately results in the formation of large mountain ranges. The Patagonian fold and thrust belt is an example (Suarez et al. (2000) *Geol. Mag.* 137, 4). Kwon and Mitra (2004) *Geology* 32, 7 present a model of a fold-thrust belt wedge. Thrust belts present enormous potential for tapping into oil and gas deposits; see Johnson et al. (2007) *The Leading Edge* 26, 2.

thunder When a stroke of *lightning passes through the atmosphere, the air becomes intensely hot, perhaps to 30 000 °C. The violent expansion thus caused makes a shock-wave heard as thunder.

thunderstorm A storm including strokes of *lightning, which cause the thunder, and draw off electrons earthward as part of the atmospheric electrical cycle.

tidal energy Energy based on the motions of the tide. See Howell and Drake (*FERN technical report# 2012.1*) on the socio-economic effects of tidal energy.

tidal prism The volume of water exchanged between a lagoon or estuary and the open sea in the course of a complete tidal cycle. McBride et al. (2007) *Geomorph.* 88, 3–4 argue that tidal prism and the quantity of littoral sediment transport are the most important factors controlling inlet stability. The tidal prism is also significant in the size and configuration of ebb-tidal deltas (Fitzgerald et al. (2002) *Geomorph.* 48, 1–3).

tidal range In coastal geomorphology, the difference in height between high and low tides. **Micro-tidal range** is less than 2 m, **meso-tidal range** is 2–4 m, and **macro-tidal range** is more than 4 m.

tidal wave *See* TSUNAMI.

tide Tides result from the gravitational pull of the moon and sun on the Earth (which also affects land masses, but the reaction of the water is greater and more apparent). The greatest effect of this pull is on each side of the Earth as it faces the moon. The moon 'pulls out' two bulges of water, which are fixed—the Earth moves through them, causing high water twice daily. The sun also attracts water. When the effects of both sun and moon coincide, twice monthly in the second and fourth quarters of the moon, high **spring tides** occur. When the sun and moon seem to be at right angles to each other from the Earth, the forces of moon and sun are opposed to each other, and lower, **neap tides** result. The **dynamic tide theory** suggests that the global tidal water motion is broken up into a large number of tidal systems, constrained by the coastal topography, and known as amphidromes. The tides travel around the centre of the amphidrome as a wave.

till The sediment deposited directly below a glacier, which exhibits a wide range of particle sizes, from fine clay to rock fragments and boulders. The rate of **till flow** beneath a glacier is thought to depend on basal shear stress and on the effective pressure, defined as the difference between overburden pressure and the interstitial pore water pressure within the sediments. See Evans et al. (2006) *Earth Sci. Revs* 78, 1–2. Where glaciers overlie soft sediments, deformation may rearrange and reorientate particles to form **deformation till**, and the examination of these tills can reveal structures that allow the strain history to be reconstructed. See Murray (2012) in J. Holden, ed., p. 490.

tilt-block *See* BLOCK.

time-space compression The impact of new systems of transport and communications as experienced by the individual, and especially the emphasis given by modernity to 'the shock of the new', immediacy, and simultaneity. The major periods of time-space compression are from about 1850 to 1914, and from the late 20th century onwards, when there occurred 'a radical restructuring in the nature and experience of both time and space . . . with a dissolution or collapse of traditional spatial coordinates' (J. May and N. Thrift 2001). Janelle (2003) UCSB colloquium, provides the following data for Los Angeles to Santa Barbara:

- 500 minutes apart in 1901
- 100 minutes apart in 2001.

time–space convergence Places are separated by absolute distance and by time. With improvements in communication systems and methods of transport, this time-distance diminishes. In essence, time–space convergence means that the *friction of distance—a concept fundamental to conventional *central place theory, *diffusion theory, and *location theory—is lessening. In 1992, R. O'Brien predicted 'the end of geography'. He was wrong. Simson (2005) *J. Transp. Geog.* 26, 2 writes that 'speed permeates the history of transport and modernity, but it does so in multiple ways. The time–space compression narrative is not adequate to encompass this history.' Furthermore 'there are different kinds of technologies, different levels of technological change, and even the same technology can be used differently by different people in different places' (N. Coe et al. 2007).

time–space geography Time–space geography provides a method of mapping spatial movements through time. Hägerstrand (in T. Carlstein 1978) expressed the time and space needed for events as a web model, based on four propositions:

- space and time are scarce resources which individuals draw on to achieve their aims
- achieving an aim is subject to *capability constraints, *coupling constraints, and authority constraints
- that these constraints interact to demarcate probability boundaries
- that choices are made within these boundaries.

Ahmed and Miller (2007, *J. Transp. Geog.* 15) acknowledge Hägerstrand's conceptual framework, but argue that 'rigorous, analytical definitions of basic time geography entities and relationships do not exist'. Accordingly, they propose 'analytical formulations for basic time geography entities and relations, specifically, the space–time path, prism, composite path-prisms, stations, bundling, and intersections'. Raubal et al. (2004) *Geografiska B* 86 also propose a spatio-temporal theory of location-based services, which they claim to be closer to the individual user and more plausible with respect to her or his daily life. See also Miller (2005) *Geog. Analysis* 37, 1. A **time–space prism** is a representation of the constraints limiting the time within which the individual can act.

TNC *See* TRANSNATIONAL CORPORATION.

Tobin tax James Tobin (1978) proposed a tax on all conversions of currency, which would be applied worldwide and at the same rate, in proportion to the size of the transaction. Tobin's aim was to discourage currency speculations, in addition to raising revenue for development. McCulloch and Pacillo (2010) *U. of Sussex, Economics Dept* 16-2010, conclude that, 'contrary to what is often assumed, a Tobin Tax is feasible and, if appropriately designed, could make a significant contribution to revenue without causing major distortions. However, it would be unlikely to reduce market volatility and could even increase it'.

Tobler's first law of geography This 'law' holds that everything is related to everything else, but near things are more related than distant things. See Klippel et al. (2011) *AAAG* 101, 5, 1011 for an enquiry into this 'law'.

tolerance The ability of an organism to survive environmental conditions. The prefixes *eury-* and *steno-* refer to wide and narrow ranges of tolerance respectively. An organism can be widely tolerant of one factor, such as temperature, but narrowly tolerant of another, such as salinity. The **climatic tolerance thesis** is that boundaries to the distribution of life forms and species often coincide with isometric lines of climatological variables (Grace (1987) *New Phytologist* 106, 1); see Ruggiero (2001) *J. Biogeog.* 28, 10.

tombolo A spit, resulting from *longshore drift, which joins an offshore island to the mainland. See S. Codrington (2005).

topography The arrangement of the natural and artificial physical features of an area: a **topographic(al) map** indicates, to scale, the natural features of the Earth's surface, as well as human features, with features at the correct relationship to each other.

topological map A map designed to show only a selected feature, such as the stations on the London Underground. Locations are shown as dots, with straight lines connecting them. Distance, scale, and relative orientation are not important. Try Remolina and Kuipers (2004) *Artific. Intell.* 152 on a general theory of topological maps.

topology This explicitly defines spatial relationships between connecting or adjacent features in geographic data; the arrangement that defines how point, line, and polygon features share coincident geometry. In geography, topology generally refers to the location of nodes and their connectivity to other nodes to form *networks. The London Tube map is a topological map, for the connections are important, and not the precise location of nodes and routes. In a geographic information system, **topology** is a set of rules that define the relationship between points, lines, and polygons. ESRI enables topology generation within its geodatabase feature classes.

toponymy The study of place names, and what they can reveal about the history or nature of a place. See Rose-Redwood et al. (2011) *PHG* 33, 4, 453.

topophilia The feeling of affection which individuals have for particular places; 'the term topophilia couples sentiment with place' (Y.-F. Tuan 1974); an 'affective bond between people and place' (Y.-F. Tuan 1990). Environmental designers have long exploited the basic ideas of topophilia to create presumably attractive surroundings that restore mental health based on the use of materials, sensory stimuli, and arrangements that remind people of the place and environmental settings that are comforting and/or associated with healing potential; see A. Carlson (2000). Try also Ogunseitan (2005) *U. Cal. Postprints* 623. 'Topophilia can be employed to map national territories in a more nuanced manner that reflects nations' own understandings of their territories' (White (2007) Fac. Env. Earth, & Resources, U. Manitoba). See González (2005) *Space & Culture* 8, 193 on topophilia and its reverse, topophobia.

topple The forward rotation, out of the slope, of a mass of soil or rock about a point or axis below the centre of gravity of the displaced mass (Cruden (2003) *Geomorph.* 55, 1-4).

topsoil The cultivated soil; the surface soil as opposed to the subsoil. Xiao et al. (2006) *Int. J. Rem. Sensing* 27, 12 propose a new index for detecting topsoil grain size composition, and Katra et al. (2007) *Geomorph.* 86, 3-4 outline dynamics of topsoil moisture. **Topsoil erosion** occurs when the topsoil layer is blown or washed away; it takes approximately 500 years for one inch of topsoil to be deposited, but there are 25 billion tons of topsoil lost each year. See Lovett et al. in E. Buhmann et al., eds (2006) on modelling and visualizing soil erosion, and Evans (2006) *CABReviews* 1, 30 for a review of sustainable practices to limit soil erosion.

tor An upstanding mass of rocks or boulders which rises above the gentler slopes which surround it. Tors were probably formed as a result of deep weathering and stripping in preglacial times, with additional sediment removal by mass movement under periglacial conditions in interglacial and interstadial times (Stroeven et al. (2002) *Geomorph.* 44, 1-2; see also Hätterstrand and Stroeven in the same issue). See Knight (1987) *U. Aberdeen, Dept. Geog., discuss. paper* 9, on structural controls on tor location.

tornado A destructive, rotating storm under a funnel-shaped cloud which advances over the land along a narrow path. The tangential speed of the whirling air may exceed 100 m s^{-1}, the core may perhaps be 200 m across, the duration of the storm about 20 minutes. Within a tornado, the central pressure is around 100 mb below that of the exterior; this may cause buildings to explode. This type of storm is very common in 'tornado alley', extending from northern Texas through Oklahoma, Kansas, and Missouri, with as many as 300 tornadoes a year. The exact mechanism of its formation is not fully understood, but tornadoes are associated with intense local heating coupled with the meeting of warm, moist air from the Gulf of Mexico and cold air from the *basin and range area of the western United States. See H. B. Bluestein (2006).

(⊕) SEE WEB LINKS

• The enhanced Fujita scale, which uses 28 tornado damage indicators.

total fertility rate *See* FERTILITY.

tourism Making a holiday involving an overnight stay away from the normal place of residence. Tourism has grown to become the world's second largest industry, directly accounting for 3.8% of global GDP (World Travel and Tourism Council, 2005). The mushrooming of **international tourism** may be explained by high levels of disposable income and longer holidays in *more economically developed countries; the development of package holidays, which reduce risk; cheap, mass air transport; and place myths, which persuade the tourist that the local culture they see represents the 'real thing'. Of critical importance has been the internationalization of finance: credit and debit cards, travellers' cheques, and hotel vouchers. **Creative tourism** means allowing visitors to participate in events specific to the city or area visited; see G. Richards and J. Wilson (2007).

Kweka (2004, *Conf. Growth, Poverty & Human Dev. Africa*) finds that tourism expansion has a substantial impact on the Tanzanian economy, while Mitchell and Ashley (2006, *ODI Briefing Paper*), point out that the most important pro-poor impact of tourism is local employment: 'tourism employment relatively favours the marginalized sector as a higher proportion of women workers relative to other similar industries are employed.' However, the tourism industry generally exhibits a high degree of leakage—whereby part of the foreign exchange earnings generated by tourism is either retained by tourist-generating countries or repatriated to them in the form of profits, income and royalty remittances, repayment of foreign loans, and imports of equipment, materials, capital, and consumer goods (Diaz (2001) *Symposium on Tourism Services Paper*).

Dixon et al. (2001) *World Bank, Env. Dept Paper* 80 list some of the environmental impacts of tourism: the generation of substantial amounts of solid and liquid waste, high demand on energy and freshwater resources, and the damage to environmentally sensitive areas. Peeters et al. (2007) *J. Transp. Geog.* 15, 2 find that climate change generates more than half of the externalities of tourist transport. Lee et al. (2006) *Annals Tourist Res.* 33, 2 study environmentally friendly practices in the Vietnamese hotel businesses; Singh (2007) *Annals Tourist Res.* 34, 4 describes environmental management strategies for damage control in India's mountainous areas; and Holden (2005) *Annals Tourist Res.* 32, 3 considers the environmental ethics of tourism. Socio-cultural impacts include increased crime rate, higher levels of prostitution, and sex tourism; see Shah and Gupta (2000) *Overseas Development Institute, UK*. Shepherd (2002) *Tourist Studs* 2, 2 reflects on commodification, culture, and tourism, and Clarke (2004) *TIBG* 23, 4 thinks tourism makes the world and its people 'modern'. For part 1 of a review of geographies of tourism, see Gibson (2008) *PHG* 32, 3.

tower karst Limestone towers, 30–200 m high, with nearly vertical walls and gently domed or serrated summits. The towers stand above large, flat *flood plains and swamps and show undercutting from rivers and swamps. Tower karsts are thought to represent the last remnant of a limestone outcrop. See Zhu and Chen (2006) *Cave & Karst Sci.* 32, 2–3 on the tower karst (*fenglin*) in southern China.

trace As defined by J. Anderson (2010), traces are marks, residues, or remnants left in place. They may be material such as buildings, signs, statues, or graffiti; or non-material such as activities, events, performances, or emotions.

traction load *See* BED LOAD.

trade The movements of goods from producers to consumers. The classic explanation for trade is expressed in terms of *comparative advantage. Greenaway and Torstensson (1997) *TU. Inst. Econ. Res. W. Paper* 144 support comparative advantage models, observing that an abundant human capital endowment, as well as a large domestic market, increases the quality of OECD countries' manufacturing exports, with human capital increasingly important over time. Forslid and Wooton (2003) *Rev. Int. Econ.* 11 find that lowering trade costs leads initially to increased concentration and then to dispersion of production. When a pattern of comparative advantage exists, integration may lead to international specialization of production. 'This may be good news for peripheral countries, which may be able to retain industry despite the attraction of the core.'

Baier and Bergstrand (2005, *Atlanta W. Paper* 2005-3) answer the question 'do free trade agreements increase members' international trade?' with a *yes!* Bernard et al. (2007) *CEP Discuss. Paper* 795 urge that the number of products firms trade as well as the number of countries with which they trade are central to understanding the role of distance in dampening aggregate trade flows. See Hummels and Hillberry (2005) *NBER W. Paper* 11339 on the impact of distance on trade.

trade bloc A large area of *free trade, formed through tariff, tax, or trade agreements, such as the EU. Buckley et al. (2007) *World Econ.* 30, 2 examine the link between globalization and the growth of trade blocs, and Glebe (2006) *Technische U. München, Env. Econ. & Ag. Policy Gp, Discuss. Paper* 022006 analyses the way the enlargement of a trade bloc will affect national welfare. 'Mercosur' is an economic and political agreement among Argentina, Brazil, Paraguay (which is currently suspended), and Uruguay to promote the free movement of goods, services, and people among member states. Mercosur's primary interest has been eliminating obstacles to regional trade, such as high tariffs and income inequalities (Klonsky et al. (2012) *Council on Foreign Relations*). *See also* BLOC.

trade wind inversion A temperature *inversion, found in the tropics at heights of between 3 000 and 2 000 m, caused by descending air of the *Hadley circulation. This inversion acts as a ceiling for pollution; *see* PHOTOCHEMICAL SMOG.

trade winds The tropical *easterlies, blowing towards the equator from the *subtropical anticyclones at a fairly constant speed. They are at their strongest around the equatorial flank of these *highs. Trades are not true *zonal winds; their zonal character is weaker in summer because of the south-west *monsoon. They are most regular around 15°, and are associated with fine weather resulting from the *anticyclonic subsidence of the Hadley cell. Over the equator and the western oceans, **trade wind weather** is rainier; see Snodgrass et al.

(2006) *12th Conf. Cloud Phys.* Liu (2002) *Science Daily* suggests that shifts in Pacific winds may support *El Niño formation.

traffic The movement of people and vehicles along a route. **Traffic capacity** is the maximum number of vehicles which can pass over a route in a given time, while **traffic density** is the existing number of vehicles. See Barthelemy et al. (2007) *Advances Complex Sys.* 10, 1 on the role of geography and traffic in complex networks.

traffic segregation The subdivision of towns and cities into certain units where road traffic is restricted and pedestrians predominate. Beckmann (2001) *Env. & Plan. D* 19 comments that 'the automobile turns into a structural prerequisite for the organization of everyday life, while at the same time the variety of forms of everyday action becomes the structural prerequisite for the expansion of the automobile.' See Bickerstaff and Walker (2005) *Urb. Studs* 42, 5 on local transport planning.

tragedy of the commons G. Hardin (1968) described an increase in the use of common land by a number of graziers, with each grazier continually adding to his stock of animals for as long as the marginal return from each animal is positive, even though the average return for each animal is falling, and even though the quality of the grazing deteriorates. Hardin used this metaphor to describe any situation where the interests of the individual do not coincide with the interests of the community, and where no organization has the power to regulate individual behaviour. Bloom (2002) *Agric. Hist.* 76, 3, 497 writes movingly of an *American Tragedy of the Commons: the Cherokee Nation*, 1870–1900.

transferability The capacity of a good to be transported. Transferability is largely determined by transport costs: as economic distance increases, transferability decreases. Since *economic distance and *intervening opportunities vary, transferability may change over time. See Chang (2004) *J. Geographical Sys.* 6, 1.

transfer costs Total transport costs involved in moving a cargo including extra costs such as tariffs and insurance. Transfer costs are highest for people, because of the very steep cost of insurance. If there is just one market and just one material source, transfer costs can be minimized by locating the processing unit at one of those two points and not at any intermediate point; see McCann (2005) *J. Econ. Geog.* 5, 3.

transform fault *Faults which are parallel to the arc of sea-floor spreading. See Greg et al. (2006) *Geology* 34, 4.

transfrontier zone A space which straddles a land or sea frontier—whether political, environmental, or socio-economic. See Bunnell et al. (2006) *Int. J. Urb. Reg. Res.* 30, 1 on transfrontier zones and resistances in Bintan, Indonesia, and Duffy (2006) *Polit. Geog.* 25, 1 on the politics of **transfrontier conservation** areas.

transmission loss The loss of flow from a river channel, due to downward *percolation and high rates of *evaporation.

transnational capitalist class (TCC) 'That segment of the world bourgeoisie that represents transnational capital; the owners of the leading worldwide means of production as embodied in the transnational corporations' (W. Robinson 2003); an '"ungrounded and deterritorialized" class' (Ley (2004) *TIBG* 29, 2).

Carroll and Meindert-Fennema (2002, *Int. Sociol.* 17) claim that 'transnational corporate interlocking is less about intercorporate control than it is about the construction of an international business community', and Robinson and Harris (2000, *Sci. & Soc.* 64, 1) argue that 'the TCC became politicized from the 1970s into the 1990s and has pursued a class project of capitalist globalization institutionalized in an emergent transnational state apparatus and in a "Third Way" political program'. 'Transnational elites have long been associated with the global city: embedded at work, disembedded at home' (Beaverstock and Bostock (2000) *GaWC Res. Bull.* 27). However, Burris and Staples (2012) *Int. J. Comp. Soc.* 53, 4, 323 conclude that a transnational capitalist class is very far from being realized on a global scale.

transnational corporation (TNC) A firm which owns or controls production facilities in more than one country, through direct foreign investment. Transnationals are made possible by improved international communications which provide rapid containerized trans-shipment and foreign travel; easy communication of information; and international mobility of capital. TNCs have the potential to 'take advantage of geographical differences in the availability and cost of resources, and in state policies, and to switch and re-switch operations between locations' (Dicken (2004) *TIBG* 29, 1), and when one market is saturated, the transnational can rapidly develop others, since foreign investment cuts transport costs, and makes possible a rapid response to local markets. TNCs are probably the major force affecting worldwide shifts in economic activity, since the largest have a turnover greater than the *GNP of many *less economically developed nations.

Theoretical approaches to TNCs include Pitelis (2001) *Glob. Business &*

Econ. Rev. 3, 1 on the resource-based theory of the TNC, and Wrigley et al. (2005) *PHG* 29, 4 on the retail TNC. Hudson (in J. Peck and H. Yeung, eds 2003) argues that 'rather than seeing a dichotomy between "global outposts" and "embedded branch plants", we need to conceptualise these as opposite ends of a continuum'; A. Isaksen (1996) agrees: 'TNCs may act as interfaces between global production networks and regional economies.'

Discussions of the advantages and disadvantages of TNCs abound. UNCTAD's *World Investment* states that, since host developing countries tend to use more labour-intensive industries and technologies than do developed countries, the location of a TNC in the former is likely to have greater employment-generating potential than in the latter. Ivarsson (2002, *J. Econ. Geog.* 2, 2) finds that foreign-located affiliates of TNCs generate technological competencies, 'both internally as well as through organized cooperation with external business partners in the host country'.

However, the ability of transnational corporations to move capital across frontiers, and to manipulate the transfer prices at which their component firms exchange goods and services, has been held against them. Transnationals have also been criticized for undermining national cultures through intensive use of advertising to substitute their synthetic and standardized goods for natural and distinctively local alternatives; the powdered milk controversy comes to mind. Smith (1998, *Eur. Urb. Reg. Studs* 5) in a study of east and central European economies argues that reliance upon inward investment may not be the 'godsend' that many suggest it will be, and Herrmann (1995, *Hum. Ecol.* 23, 2) examines the 'culpability' of TNCs for human environmental crises.

Some claim that TNCs undermine states, or, indeed, the state system, but Dicken (2004, *TIBG* 29, 1) argues that 'states still have power over what they permit'.

transnationality There seems to be no general agreement on the definition of transnationality; its aspects include international cooperation on a level other than exclusively between nation states; cooperation not only with neighbouring countries, but rather with regions that suffer from similar difficulties and problems; and relating to a transnational region, such as the North Sea Region (see the Northern Periphery Programme).

For an individual, Mau et al. (2008) *Glob. Netwks* 8, 1 use a **transnationality index** composed of the number of private transnational relations, the number of times an individual had been abroad in the twelve months before the time of the interview, and the total time the individual had lived abroad for a longer period. For a company, the UNCTAD transnationality index (an indicator of the share of a company's activities taking place in a country other than its own) is the unweighted average of the following ratios: foreign assets/total assets, foreign sales/total sales, and foreign employment/total employment. In 2003, the most transnationalized economy was Hong Kong China, followed by Ireland.

(())) SEE WEB LINKS
• The Northern Periphery Programme.

transnational municipal networks (TMN) Networks of municipalities that operate nationally and transnationally, directly representing and involving cities in policy issues at the international and European levels, and across national borders: the transnational activities of municipal urban governments. TMNs became a pivotal feature of mainstream policy in the European Union from the mid-1980s;

P.-Y. Saunier and S. Ewen (2008) cover the history of TMNs. **Transnational urbanism** represents the ways in which cities are ever more defined by all sorts of connections to faraway places (A. Latham et al. 2009).

transport costs Krugman (1991) *J. Pol. Econ.* 99, 3 suggests that so long as transport costs are low, firms can benefit from concentrating in a single location and delivering products to farmers in another location. Alonso-Villar (2005) *J. Econ. Geog.* 5, 5 finds that the effects of cost reductions in transporting final goods are different from those in inter-mediate goods. For **feeder costs** (the costs of transferring freight from large ships/vehicles/aircraft to smaller ones) and **handling costs** (the costs of physically shifting the freight), and marginal cost/revenue analysis, see Konings (2007) *J. Transp. Geog.* 15, 6. Baier and Bergstrand (2001) *J. Transp. Geog.* 53, 1 find that, for the growth in world trade, 1992–2000, trade liberalization was three times more important than transport costs.

Knowles (2006) *J. Transp. Geog.* 14, 6 adds the concept of **cost/space convergence** to *time–space convergence: 'time/space and cost/space have both collapsed differentially and especially in the last 50 years. Small but incremental changes can have significant consequences for time/cost–space convergence or divergence whilst response lags and inefficient decision making retard change. Collapses in time/space and cost/space are due to innovations in air, rail, road and sea transport technology, communications and handling technology. The differential spatial effects reflect a more unequal world and are largely due to unequal investment in modal capacity, routes and terminals at local, national and international scales. This tends to enhance the importance of the largest demand centres in developed countries and of nodal centrality and intermediacy.' Knowles summarizes the result as a 'shrunken, misshapen world'. Combes and Lafourcard (2005) *J. Econ. Geog.* 5, 3 show that transport technology and market structure are responsible for most of the transport cost decrease. 'Infrastructure improvements only condition the spatial distribution of the gains.' See Link (2005) *J. Transp. Geog.* 13, 1 on the **social costs of transport**: pollution, noise, and accidents among them.

transport geography A branch of human geography concerned with: transport policy practice and analysis, especially the impacts of deregulation, privatization, and subsidy control; infrastructure impact on trip-making, the spatial economy, and regional development; technological innovation in transport and telecommunications and global and regional economic integration; the growing mobility gap between rich and poor and differential accessibility to jobs and services; transport, environment, and energy; travel, recreation, and tourism; and spatial and behavioural aspects of modelling transport demand. See Woudsma and Andrey (2004) in the special issue on transport geography, *Canad. Geogr./Géog. canad.* 48, 4. See also Keeling (2007) *PPG* 31, 2 and (2008) *PPG* 32, 2. Try B. Hoyle and R. Knowles, eds (1998); no longer new, but very well reviewed.

travel-to-work area (TTWA) In theory, a self-contained labour market area is one in which all commuting occurs within the boundary of the area. In practice, there are entirely separate labour market areas. For those involved in labour market analysis and planning, it is useful to be able to define zones in which the bulk of the resident population also work, so, by applying a multi-stage allocation process, the UK Office for National Statistics has defined 'Travel to Work Areas' as approximations of self-contained labour markets. The basic criteria used for defining a TTWA are that, of the resident economically active population,

- at least 75% actually work in the area,
- of everyone working in the area, at least 75% actually live in the area.

The definitive minimum working population in a TTWA is 3 500, but many are much larger—indeed, the whole of London and surrounding area forms one TTWA.

For the methodology, see appendix 1, Office for National Statistics 1991 Travel-to-Work Areas, and for an update, see M. Coombes, S. Raybould, and C. Wymer (2005).

(⊕) SEE WEB LINKS

- Office for National Statistics 1991-based Travel to Work Areas.
- National Statistics, Travel to Work Areas.

trellised drainage *See* DRAINAGE PATTERNS.

triangular irregular network (TIN) A representation of a surface in a *GIS created through digital data. It is made up of irregularly distributed nodes and lines with three-dimensional coordinates (x, y, and z). See Lee (1991) *Int. J. Info. Systems* 5, 3, 267 for methods of building TIN models from grid *digital elevation models.

triangulation In the social sciences, a form of cross-checking; the equivalent of 'validity and reliability' in scientific paradigm research (that is, using multiple people's perspectives, or research gathering methods, to check that your interpretation is robust and not wholly a subjective construct) (Ernest (2000) *U. Exeter*). Forms include data, investigator, method, and theoretical triangulation; see K. Hoggart et al. (2002) on researching human geography, and Farmer et al. (2006) *Qual. Health Res.* 16 on developing and implementing a triangulation protocol. Hobson (2006) *Area* 38, 3 uses the term for interdisciplinary work: 'triangulation from the human sciences literature with human geographers' arguments could strengthen both sub-disciplines' ability to have

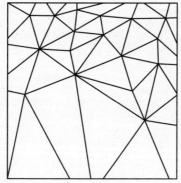

Triangular irregular network

the viable policy inputs that both have missed out on in recent times, as it may, in the eyes of policymakers at least, add legitimacy through inter-disciplinarity.'

triangulation of space A term used by Kim (1999), *Politics & Soc.* 27, 1 to move beyond the traditional divisions of American society as black or white, by considering an ethnic group as being located in the field of racial positions with reference to those traditional points. See Lai (2012) *AAAG* 102, 1 on the racial triangulation of space in San Francisco.

Triassic The oldest period of *Mesozoic time stretching approximately from 225 to 190 million years BP.

(((SEE WEB LINKS
• An overview of the Triassic period.

trigger In geomorphology, an event which puts an end to the balance between driving and resisting forces and thus sets the mass moving (T. Erismann and G. Abele 2001). P. Fookes et al. (2005) consider the main predisposing and triggering factors of geomorphological surface processes, including climate and weathering, sedimentology, tectonics, and stratigraphy. Jones and Omoto (2000) *Sedimentol.* 47, 6 outline the criteria that can be used to identify a **seismic triggering** agent. See also Moro et al. (2007) *Geomorph.* 89, 3–4. For an

example of atmospheric triggering, see Miniscloux et al. (2001) *J. Appl. Climat.* 40, 11. *See also* THRESHOLD.

trope A figurative or metaphorical turn of phrase, used in social constructions of gender, race, imperialism, and so on.

trophic level An individual layer on the *pyramid of numbers which represents types of organisms living at parallel levels on *food chains. All herbivores live at one level, all primary carnivores on the next level, all secondary carnivores on the next level, and so on. The animals on each level are remarkably distinct in size from those on other levels; there is a clear jump in size between an insect and a bird, for example.

The trophic biology of a taxon (species/family/class) fundamentally constrains its ability to convert productivity into individuals. Predator taxa require at least an order of magnitude more *net primary productivity (NPP) to support an individual than the trophic level upon which it feeds; the more trophic links between NPP and a consumer taxocene (taxonomically related set of species within a community), the more energy is necessary to support viable populations of its individuals. If taxocene abundance is energy limited, consumer taxocenes that occupy broad NPP gradients should show a predominance of lower trophic levels at low NPP. Higher trophic levels should accrue as NPP increases, and

Trophic level

taxocenes comprised largely of predators should be under-represented at lower NPP (Kaspari (2001) *Glob. Ecol. & Biogeog.* 10, 3).

tropical cyclone Also known as a hurricane, or tropical storm, this is a disturbance about 650 km across, spinning about a central area of very low pressure, with winds over 140 km per hour. The violent winds are accompanied by towering clouds, some 4000 m high, and by torrential rain; 150 mm (6 inches) frequently fall within the space of a few hours. There is, as yet, no complete understanding of how these storms develop; they can begin when air spreads out at high level above a newly formed disturbance at low levels. The upper-level outflow acts rather like a suction pump, drawing away the rising air at height and causing low-level air to be pulled in. The winds spiral into the centre because they are affected by the Earth's rotation. The intense energy of these storms comes from the warmth of the tropical seas over which they develop. Thus, an extensive ocean area with surface temperatures of over 27 °C is necessary for hurricane formation.

The source regions must be far enough away from the equator—5° at least—for the *Coriolis force to have an effect. The removal of air at height may be along the eastern limb of an upper air trough. Moisture-laden air spirals into the centre and rises, condensing to form a ring-like tower of *cumulo-nimbus clouds. With this condensation, *latent heat is released which causes the air to rise further and faster. The condensation also causes torrential rain. In the upper *troposphere water droplets freeze and form cirrus clouds, which are thrown outwards by the spin of the storm.

The structure of a tropical hurricane

At ground level, the temperature at the centre, or eye, of the storm is only slightly warmer than that at the margins, but, at heights of around 5000 m, the centre can be 18 °C warmer than the margins. This warm core maintains the low pressure which drags in the winds. Senkbeil et al. (2011) *Geog. Compass* DOI: 10.1111/j.1749-8198.2011.00439.x review topics that summarize the current state of hurricane hazards in the USA, discussing wind and water hazards associated with land-falling tropical cyclones; and review societal impact such as mortality and morbidity, hazard perception, communication, evacuation, and risk. Ideas are presented that introduce suggestions on how possibly to change processes. *See* TROPICAL STORM.

tropical dry forest A tropical forest type that goes through an annual dry season. Drier soils sometimes produce pockets of tropical dry forest within a tropical rain forest. R. Robichaux and D. Yetman, eds (2002) provide the ur-text; see also Miles et al. (2006) *J. Biogeog.* 33, 3 for an overview of the conservation of tropical dry forests.

tropical forest Forested areas which often extend beyond the tropics, and consist of *tropical rain forest and *mangrove forest.

SEE WEB LINKS
• The European Tropical Forest Network.

tropical montane forest Forest found in mountainous regions of the tropics, where rainfall can be heavy and persistent. The trees are typically short and crooked, and sites for mosses, climbing ferns, lichens, and *epiphytes. See Bruijnzeel and Veneklaas (1998) *Ecology* 79, 1 on climatic conditions and tropical montane forest productivity. Martin et al. (2007) *J. Biogeog.* 34, 10 examine relationships between climate-disturbance gradients, vegetation

patterns, and *ecotones on subtropical mountains.

tropical rain forest A tropical forest of trees characterized by buttress roots, long, straight, lower trunks, and leathery leaves (see I. Turner 2001). The vegetation shows distinct layering: the canopy, or upper layer, at around 30 m; the intermediate layer at 20–25 m; and the lower layer at around 10–15 m. Undergrowth is poorly developed but *epiphytes and lianas are common. Davidar et al. (2005) *J. Biogeog.* 32, 3 indicate that seasonality, and not annual rainfall, is the variable that drives tropical rain forest tree diversity. The range of plant and animal species is immense. Hill and Hill (2001) *PPG* 25, 3 evaluate the following theories explaining the species richness of tropical rain forests: cumulative evolution, productivity, genetic drift and isolation, pleistocene refugia, intermediate disturbance, *niche packing, *habitat heterogeneity, seedling and foliar herbivore, recruitment limitation, and temporal resource partitioning.

The felling of the rain forest causes soil erosion and the destruction of potentially useful species (Voeks in D. S. Edwards et al. 1996), and the reduction of oxygen from photosynthesis, while smoke from burning logs increases the quantity of *aerosols and carbon dioxide in the atmosphere (Stein (1982) *GeoJ.* 6, 2). Clark et al. (2003) *PNAS* 100, 10 report strong reductions in tree growth and large inferred tropical releases of CO_2 to the atmosphere during the record-hot 1997–8 *El Niño; 'these and other recent findings are consistent with decreased net primary production in tropical forests in the warmer years of the last two decades.' Lettau et al. (1979) *Monthly Weather Rev.* 107, 3 find that modest deforestation in the Amazon basin will bring slightly warmer, moister weather with drier soils, but that more radical deforestation will result in a drier, cooler climate. See E. Guhardja, ed. (2000) on human impacts on tropical rain forest in Kalimantan.

tropical storm An intense low-pressure cyclonic system that does not contain *fronts between warm and cold air masses, developing when the upper 50 m of tropical ocean warms to over 26.5 °C in regions over 500 km from the equator. See DiMego and Boshart (1982) *Monthly Weather Rev.* 110, 5 on the transformation of a tropical storm into a *tropical cyclone.

tropics The *Tropic of Cancer* lies approximately along latitude 23° 30′ N. Around 21–2 June, the sun's rays are perpendicular to the ground along this line and the sun exerts its maximum strength in the Northern Hemisphere. Conversely, the sun is overhead at the approximate latitude of 23° 30′ S, the *Tropic of Capricorn*, on 22–3 December when the sun's heat is at its maximum in the Southern Hemisphere. Between these two lines of latitude lie the tropics. In America this zone is known as the **neotropics** and in Africa and South-East Asia as the **palaeotropics.**

tropopause The upper limit of the *troposphere, above which very few clouds, except for the nacreous and noctilucent, form.

troposphere The lower layer of the *atmosphere, extending to 16 km above ground level at the equator, 11 km at 50 °N and S, and 9 km at the poles. Most clouds and *precipitation, and, indeed, weather events, occur within this layer. Increasingly, it is understood that air movements in the upper troposphere greatly influence weather systems in the lower troposphere. *See also* JET STREAM; ROSSBY WAVES.

Within the troposphere, temperatures of rising air fall, at varying *lapse rates, because the air expands and, therefore,

cools. Sinking parcels of air experience a corresponding heating. Most of the water vapour in the troposphere is concentrated in the lower, warmer zone; there is little where the temperature falls below about –40 °C because moist air, rising through *convection or *turbulence, condenses out as ice crystals form.

truncated spur A steep bluff on the side of a *glacial trough, protruding between tributary, possibly *hanging, valleys. This landform is the result of mainly vertical *glacial erosion.

tsunami A huge sea wave, or series of waves. Most are formed from earthquakes of 5.5 or more on the *Richter scale. Other causes include the eruption of submarine volcanoes, very large landslides off coastal cliffs, or the calving of very large icebergs from glaciers in fiords. Fault-generated tsunamis may spread thousands of kilometres across ocean basins with velocities up to 800 km h^{-1}. In mid-ocean the height of a tsunami may be no more than 30–50 cm, but this can increase up to 500 m when the tsunami hits the shore. The most active source region of tsunamis between 1900 and 1983 was along the Japan–Taiwan island arc, where over a quarter of all tsunamis were generated. Wikipedia is very good on the Japanese 2011 Tohoku earthquake, which triggered tsunami waves up to 40.5 m high, and travelled up to 10 km inland. The earthquake which caused the tsunami on 26 December 2004 was caused when a portion of the Indian tectonic plate slid under the Burma plate. The US Geological Survey estimates that the ensuing rupture was more than 1000 km long, displacing the sea-floor above by perhaps 10 m horizontally. The great volume of the ocean displaced thereby created the tsunami, which hit the coastline of eleven Indian Ocean countries, with wave heights up to 15 m.

Estimates of the death toll range between 188000 and 320000. See Krom in J. Holden (2012), pp. 42–3 for an excellent summary of this tsunami.

The largest recorded earthquake in the world occurred in May 1960, with a magnitude of 9.5. The damage caused by the resulting tsunami caused:

- approximately 1655 deaths and $550 million damage in southern Chile;
- 61 deaths, $75 million damage in Hawaii;
- 138 deaths and $50 million damage in Japan;
- 32 dead and missing in the Philippines;
- $500000 damage to the west coast of the United States.

These statistics reflect the population densities in the affected zones, the economic value of the properties affected, and the importance of warning systems and evacuation planning (Keys et al. (2004) *Antipode* 38, 2).

tsunami deposits Many former tsunamis have left sedimentary fingerprints; see Dawson and Stewart (2007) *PPG* 31, 6. See also Scheffers and Kelletat (2003) *Earth-Sci. Rev.* 63, 1–2 and Scheffers and Scheffers (2007) *Earth Sci. & Planet. Letts* 259.

tundra The barren plains of northern Canada, Alaska, and Eurasia. Temperatures and rainfall are low, so tundra vegetation is restricted by the intense winter cold, insufficient summer heat, and waterlogged soil, and characteristically comprises herbaceous perennials with scattered trees, mosses, lichens, and sedges. Arctic tundra ecosystems are warmth-limited and can respond to temperature dynamics on a relatively short time scale. They therefore are highly sensitive to both environmental fluctuation and directional climate change, and play a significant role in biospheric feedbacks to global climate (Gensuo et al. (2006) *Glob. Change Biol.* 12, 1). Alpine tundra does not contain trees because it is at high altitude, and differs from arctic tundra, as its soils are generally better drained than arctic soils. See Mallinson (2011) *Phys. Geog.* 32, 2.

tundra soil A dark soil with a thick, peat layer of poorly decomposed vegetation, which is usually underlain by a frozen layer of soil. Translocation is limited and there is, therefore, little development of *horizons. Tundra soils range from *brown earths in the more humid areas to polar desert soils in arid areas. Climate change may cause tundra soil to dry—accelerating oxygen uptake and hence heterotrophic respiration, which would result in increased CO_2 and methane emissions (Grant et al. (2003) *Glob. Change Biol.* 9, 1).

tunnel valley A large, elongate, *overdeepened depression cut into bedrock or glacigenic sediment, up to 4 km wide, over 100 m deep, and 30–100 km long. In long profile they contain enclosed hollows or isolated and elongated basins, sometimes displaying convex-up or undulating longitudinal channel profiles and anabranching channel networks (Beaney (2002) *Qt. Int.* 90). Ó'Cofaigh (1996, *PPG* 20) attributes their formation to the creep of deformable subglacial sediment into a subglacial conduit, followed by the removal of this material by meltwater flow.

turbosphere That part of the *troposphere where the mixing of the atmosphere is achieved more by *convection than by *diffusion. Its upper limit is the **turbopause**, an ill-defined layer about 100 km above the Earth's surface.

turbulence (turbulent flow) A gustiness in the three-dimensional flow of a fluid, irregular in both space and time, and characterized by local, short-lived rotation currents known as vortices.

Turbulence is hierarchical; large *eddies produce smaller ones, and so on, down a series of smaller and smaller scales. See Marquis and Roy (2011) *ESPL* 36, 4, 563, 2. In meteorology, it develops because of disturbances in air flows, the most important of which is *wind shear; air parcels caught in a wind shear tend to roll over and over. The *Reynolds number for turbulent flow is 2000 to 2500. In geomorphology, turbulent flow is classified according to the *Froude number of a stream.

turbulent boundary layer *See* BOUNDARY LAYER.

ubac That side of a valley which receives less *insolation; the shaded side, termed **back ridge** in Jamaica (Beckford and Barker (2007) *Geog. J.* 173, 2).

ultisol A soil of the *US soil classification. *See* FERRUGINOUS.

umland A *sphere of influence, *catchment area, tributary area, or urban field.

unauthorized colony (UC) In India, a residential area that has evolved mostly on private land. Roughly a quarter to one-third of Delhi's population live on UCs. These are technically illegal locations for residential development, and consequently are excluded from Delhi's network of basic urban services (water, roads, and electricity), and face potential demolition. See Lemanski et al. (2013) *TIBG* 38, 1, 91.

uncertainty A term which has subtle differences in meaning according to its academic context; F. Knight (1921) distinguishes between *risk* (randomness with knowable probabilities) and *uncertainty* (randomness with unknowable probabilities). The measurement of uncertainty is imperative for the production of knowledge; as Goodchild in S. Shekhar and H. Xiong (2008) notes: 'all geographic information leaves some degree of uncertainty about conditions in the real world'. The two broad questions of uncertainty in geography are the estimation of experimental error, and the question of 'how we come to know'.

The first theme is less problematic: D. W. Hubbard (2007) provides useful guidelines for assessing the value of information, and Schroeder (2007) *Geogr. Analysis* 39, 3, for example, outlines a model used to assess uncertainty in analyses of US census tracts, using aggregated data from incompatible zonal systems.

The second theme relates to the production and status of 'knowledge'. 'In assessing uncertainty, there is a need to question both "what is known" in geography, where knowledge and reflections about the "status" of knowledge differ between individuals and groups of scientists, and "how we came to know", where knowledge and uncertainty are closely related to research tradition and practice' (Brown (2004) *TIBG* 29, 3). Reed and Peters (2004, *ACME* 31, 1) draw on contemporary systems ecology to design research practices which 'embrace the uncertainty and partiality of knowledge creation'—one of their wise practices is 'preparing for surprise'. See also Ekinsmyth et al. (1995, *Area* 27, 4) on the uncertainty of *economic geography, G. Foody and P. Atkinson's e-book on *GIS and uncertainty, Murphy et al. (2006) *Area* 38, 1 on uncertainty and climate change, and Hunsaker et al. eds (2001) on spatial uncertainty in *ecology.

underclass The poor, the unemployed, the lowest social class. Wilson (1991–92) *Pol. Sci. Quarterly* 106, 4, 639 writes of the underclass 'whose primary predicament is joblessness reinforced by growing social isolation. Outmigration has decreased the contact between groups of different class and racial backgrounds and thereby concentrated the adverse effects of living in impoverished neighbourhoods. These concentration effects, reflected, for example, in the residents' self-limiting social dispositions, are created by inadequate access to jobs and job networks, the lack of involvement in quality schools, the unavailability of suitable marriage partners, and the lack of exposure to informal mainstream social networks and conventional role models'.

underdevelopment The original meaning of the term indicated that existing resources had not been exploited. The word is now close in meaning to 'poverty', although some oil-rich underdeveloped countries have high incomes that are enjoyed by the few. Indicators of underdevelopment include: high birth rates, high infant mortality, undernourishment, a large agricultural sector and small industrial sector, low per capita *GDP, high levels of illiteracy, and low *life expectancy.

Mawdesly and Rigg (2003) *Prog. Dev. Studs* 3, 4 identify, between 1978 and 2001, a gradual weakening in the view that development is a global responsibility, with increasing emphasis placed on the choices (and shortcomings) of national governments; 'the international donor community has conveniently emphasised national governance as the principal . . . explanation for underdevelopment, rather than factors such as unfair trade and agricultural dumping of subsidized products (global governance), in which they are implicated' (Carmody (2008) *Geog. Compass* 2, 1). Power and Sidaway (2004) *AAAG* 94, 3 provide a review of explanations for underdevelopment; easy to read with useful references and footnotes. Desmet

and Ortán (2007) *Scand. J. Econ.* 109, 1 describe what they call **rational underdevelopment**: 'the advanced region gives transfers to protect itself against low-wage competition, thus preventing the backward region from taking off, and ensuring its dominant position; the backward region accepts these transfers—and rationally chooses to remain backward—because even without transfers it is not sure to benefit from the new technology.' Bhattacharyya (2007) Padjadjaran University WoPEDS 200704 looks for the root cause of Africa's current state of underdevelopment.

underfit stream (misfit stream)
A stream that flows with narrower meander belts and shorter meander wavelengths than are appropriate to the valley. Davis et al. (2006) *Geomorph.* 75, 3–4 suggest that the valley was initially formed by a river of much greater *discharge, so that an underfit stream is evidence of climate change. However, while Dury (in K. J. Gregory, ed. 1977) developed mathematical relationships between the valley meander wavelength and the stream meander wavelength on the Ozark Plateaus to define the underfit condition, Shepherd et al. (2011 *Phys Geog* 32, 2) were unable to replicate his findings in the same locations.

underpopulation A situation where there are too few people to develop fully the economic potential of an area or nation; a larger population could be supported on the same resource base (see M. Schouten (1998) on Minahasa). Christophers (2007) *Geografiska B* 89, 1 finds a tendency for underpopulation being espoused, or more often predicted, mostly by individuals with a political, ideological, or economic agenda: 'it can never be "proven"; but that does not diminish its potential power.' Underpopulation is a largely hypothetical concept; 'the concept of underpopulation ordinarily relates numbers of people not to physical resources but to economic systems. In short, the argument goes, as fertility falls and populations age, the primary "inputs" to national economic systems (the generation of wealth and of taxation receipts by those of working age) decline, making it increasingly difficult to fund the public expenditure "outputs" demanded of those systems, *particularly* those that inflate in line with an ageing population'.

unemployment Being involuntarily out of work. **Structural unemployment** occurs when the labour market no longer requires a particular skill; in the case of **fractional unemployment**, jobs are available but not taken up because of immobility or the lack of information. Suedekum (2005) *Papers Reg. Sci.* 84, 2 shows that a *core–periphery in real wages is associated with and magnified by regional unemployment disparities. For Italy, M. F. Cracolici et al. (2007) explain spatial variations in unemployment differentials in terms of spatial equilibrium and disequilibrium factors, and a significant degree of spatial dependence among labour markets at the provincial level in Italy. Additionally, provinces marked by high, or low, unemployment tend to be spatially clustered. Writing on Spain, López-Bazo et al. (2002) *Papers Reg. Sci.* 81, 3 also find evidence of clustering. Hämäläinen (2002) *Papers Reg. Sci.* 81, 4 suggests that regional measures to reduce unemployment may cut out-migration, but only during an era of low unemployment. Christopoulos (2004) *Papers Reg. Sci.* 83, 3 confirms Okun's law: that unemployment falls with increases in the ratio of real output (GDP) growth to *potential* real output growth. The **unemployment rate** is the number unemployed as a percentage of the total population of working age.

uneven development On a multitude of scales, the condition of an economy that has not benefited equally from development either in a spatial sense and/or within classes in society. Henderson et al. (2000) *World Bank Working Paper* WPS 2456: 'The most striking fact about the economic geography of the world is the uneven spatial distribution of economic activity, including the co-existence of economic development and underdevelopment. High-income regions are almost entirely concentrated in a few temperate zones, half of the world's GDP is produced by 15% of the world's population, and 54% of the world's GDP is produced by countries occupying just 10% of the world's land area. The poorest half of the world's population produces only 14% of the world's GDP, and 17 of the poorest 20 nations are in tropical Africa.'

Global uneven development may be seen to be a result of capitalism, based as it is on competition and *accumulation, but inequality is not unique to capitalism. For a national case study, see Hsu and Cheng (2002) *Reg. Studs* 36, 8 on uneven development in post-war Taiwan. Gallup et al. (1998) *NBER Working Paper* 6849 find that location and climate have large effects on income levels and income growth, through their effects on transport costs, disease burdens, and agricultural productivity, among other channels, and geography seems to be a factor in the choice of economic policy itself. 'Many geographical regions that are not conducive to modern economic growth have a combination of dense population and rapid population increase. This is especially true of populations that are located far from the coast, and thus that face large transport costs for international trade, as well as populations in tropical regions of high disease burden. Furthermore, much of the population increase in the next thirty years is likely to take place in these geographically

Uneven Development World per capita GDP
Source: World Bank

GDP per capita
Not available
Less than $1,000
$1,000 to $5,000
$5,000 to $10,000
$10,000 to $20,000
More than $20,000

Uneven Development in China

disadvantaged regions' (Gallup et al. (1999) *Harvard U. CID W. Papers*, 1).

Business development is still very uneven within the UK, because of local transaction costs, variations in *institutional thickness, *agglomeration, *external and urbanization economies, and the continued importance of face-to-face contacts. Transport costs, despite dissenting views, are still important (Bennett, Graham, and Bratton (1999) *TIBG* 24, 4). Peripheral location appears to affect economic development strongly in China. Wei has shown that the per capita gross value of industrial output of the coastal provinces, such as Guangdong,

was double that of provinces deep inland, such as Xizang (Tibet) (2006, *J. Planning Lit.* 20). J. Y. Lin (2005, East Asian Bureau of Economic Research *Development Economics Working Papers* 656) proposes that flawed development strategy is responsible for the increasing disparities in economic development among Chinese provinces.

uniformitarianism The view that the interpretations of Earth history can be based on the present-day evidence of natural processes. From this comes the maxim 'the present is the key to the past'. See Ager (1989) *J. Geol. Soc.* 146, 4.

universal soil loss equation An empirical model of the relationships between the various components of soil loss and rates of erosion. It has the form:

$$A = R \times K \times L \times S \times C \times P$$

where A is the mean annual soil loss, R is a rainfall erosivity factor, L a slope length factor, S a slope steepness factor, C a crop management factor, P an erosion control practice factor, and K the soil erodibility index.

Universal Transverse Mercator (UTM) A two-dimensional coordinate system that divides the Earth into sixty zones, each a six-degree band of longitude. The system is then used to pinpoint a location. It is not a single map projection, and *not* the same as the traditional method of latitude and longitude.

unloading The removal, by erosion, of rock or, by ablation, of ice. See Bonow et al. (2006) *Geomorph.* 80, 3–4; and Cossart et al. (2008) *Geomorph.* 95, 1–2 on paraglacial stress release in the immediate aftermath of glacial unloading.

Upton Warren An *interstadial, between 43000 and 40000 years BP. Faunal and floral evidence suggests that the landscape was devoid of trees. *See also* PLEISTOCENE.

upwelling The rise of sea water from depths to the surface, bringing nutrients for plankton. Many of the world's best fishing grounds are located at such points. See Ramp and Bahr (2008) *J. Phys. Oceanog.* 38, 1 on upwelling-favourable winds. *See also* THERMOHALINE CIRCULATION.

urban In medieval Europe, for example, there was a clear-cut boundary between urban and rural, usually in the form of a wall. Subsequently, towns have progressively overflowed their municipal boundaries. In the 1960s, UNO introduced the concept of the **urban agglomeration** (UK—urban area; France—*unité*, Germany—*Verdichtungsraum*) in order to harmonize the statistical definitions of urban settlements from one country to the next. The extensive suburban sprawling based on mass car ownership led to the concepts of *daily urban systems, or *SMSA (Standard Metropolitan Statistical Areas). Later, the French statistical system defined the *aire urbaine*: a contiguous urban core and its commuter belt. About 360 *aires urbaines* were defined (Le Jeannic (1996) *Cybergeo écon. & stat.* 294/295).

urban assets The urban talent in the knowledge economy, believed to 'energize a new capitalist epoch as well as urban economic renewal' (Amin and Thrift (2007) *PHG* 31, 2).

urban climate Urban centres are subject to climatic conditions that represent a significant modification of the pre-urban climatic state, arising as a result of the modification of radiation, energy, and momentum exchanges resulting from the built form of the city, together with the emission of heat, moisture, and pollutants from human activities. 'For the increasingly urbanized population of the world these effects are already of similar to, or of greater magnitude than, the climatic changes predicted to occur at larger scales as a result of the enhanced *greenhouse effect' (McKenry (2002) *PPG* 37, 4).

Increasingly, the focus of urban climate research is on understanding the fundamental processes that generate urban climates, not just the resultant effects. 'While the stated rationale for much urban climate research is human health and well-being or energy and water consumption, urban climatologists often note the lack of communication of new knowledge and its implications to end-users, such as planners, architects and engineers' (Souch and Grimmond (2006) *PPG* 30, 2). See Assimakopoulos et al. (2003) *Atmos. Env.* 37, 29 on pollution in 'street canyons'. Mills (2006) *Theoretical and Applied Climatology* 84, 69–76 provides a useful summary of ways to modify urban climates, including outdoor landscaping, street orientation, zoning, and transport policy. See Chow and Hung (2014), *SEAGA* on the geography of urban climate change in Southeast Asia. *See* URBAN HEAT ISLAND.

urban containment Limiting *urban sprawl. New Zealand's Regional Growth Forum has developed a strategy for regional urban containment, matched by urban intensification (Memn et al. (2007) *NZ Geogr.* 63, 1). Urban containment has been promoted by US state governments since the 1980s; Wassmer (2006) *J. Reg. Sci.* 46, 1 finds that only statewide growth management programmes, with the stipulation that local plans must coincide with a state plan or with geographically contiguous local plans, have been effective at reducing the square mile size of US urban areas. Nelson et al. (2004) *J. Amer. Planning Assoc.* 70, 4, 411 surmise that containment shifts development from exurban and rural areas to suburban and urban areas because of containment boundaries.

urban design A. Cuthbert (2006) sees urban design as primarily concerned with social interaction and communication in the public realm, using architectural elements and ambient space as its basic vocabulary. Yow (2007) *Geog. Compass* 1, 6 believes that smart urban design makes it possible to create sustainable cities. See Charney (2007) *Area* 39, 2 on very tall buildings in London.

urban diseconomies The financial and social burdens arising from an urban location. These include constricted sites, high land prices/local taxation/commuting costs, traffic congestion, and pollution. Firms concentrate geographically in order to exploit benefits of *agglomeration which are traded off against urban diseconomies in determining city sizes (Black and Henderson (2003) *J. Econ. Geog.* 3, 4).

urban ecology The application of the principles of *ecology to a study of urban environments. *See also* HUMAN ECOLOGY. Urban ecologists look at individual areas of the city in the context of the whole city, and focus on the way a population organizes itself, and the way it adapts to change. Just as ecologists study the way in which an ecosystem seeks to re-establish equilibrium after a sudden alteration, so urban ecologists assume that people will try to re-establish equilibrium after sudden change. Urban ecology has been criticized by many; among them E. Fowler (1996), but Longley (2005) *PHG* 29, 1 detects some rehabilitation in urban ecology and social area analysis.

urban geography The spatial study of the city, including urban policy; race, poverty, and ethnicity in the city; international differences in urban form and function; historical preservation; the urban housing market; and provision of services and urban economic activity. Pacione (2003) *Urb. Geog.* 24, 4 writes that 'it is now difficult, if not impossible, to place analytical boundaries around the city. Even those . . . who maintain the value of a distinctly urban focus for their studies must integrate their inquiries into broader social processes . . . The strengths of urban geography lie in its unique spatial perspective and its integrative approach to urban analysis.'

R. Johnston (2000) observes that many of the concerns formerly encapsulated within urban geography are now studied under different banners, but Lees (2002) *PHG* 26, 1 argues that urban geography has taken on board the interpretative turn, engaging with language, discourse, culture, the immaterial, and so on. Wolch (2007) *AAAG* 97, 2 argues for 'an urban geography that is not solely a human urban geography but an urban biogeography, an urban geomorphology, an urban hydrology, and an urban GIScience'.

urban governance The exercise of political, economic, and administrative authority in the management of a city. Proudfoot and McCann (2008) *Urb. Geog.* 29, 4 discuss how street-level bureaucrats negotiate the constraints and pressures inherent to their practice while also exercising a degree of discretion, arguing that these micro-level concerns are important to understanding how cities are produced, but must also be linked with analyses of wider processes that shape contemporary urban development.

urban heat island (UHI) The temperature differentials between urban and rural areas, occurring on the ground, at the surface, and at various heights in the air. The UHI may actually be described in multiple ways with various methodological approaches to investigate each type, and is divided into: the **subsurface heat island** (Allen et al. (2003) *Geothermics* 32 state that shallow aquifers existing below many northern European cities have groundwater temperatures 3-4 °C higher than in surrounding rural areas) and urban heat islands in the air. Each city's unique land cover patterns will influence the spatial character of its UHI; see Roy et al. (2011) *Phys. Geog.* 32, 2, 114. Spatial and temporal patterns of CO_2 concentrations are useful in understanding urban climate; see Wentz et al. (2002) *AAAG* 92, 1.

urban hydrology The study of the consistent transformation of hydrological regimes that results from urban development. These changes have been attributed to the construction of impervious surfaces and the compaction of pervious surfaces, which promote rapid surface run-off from rainfall, and the efficient drainage of surface run-off through storm sewers to the river network. The percentage urban land cover/ percentage impervious cover/percentage connected impervious cover within a catchment provide simple but very effective predictors of changes in streamflow characteristics resulting from urban development (A. O. Akan and R. Houghtalen 2003). See Gurnell et al. (2007) *Geog. Compass* 1, 5 for an excellent review of the topic.

urbanism Wirth (1938, *Am. J. Soc.*) used this term to discuss a way of life associated with urban dwelling (suggesting that urban dwellers follow a distinctly different way of life from rural dwellers). Sheppard (2013) *Urb. Geog.* 34, 7, 893 notes that 'urbanism' is seen as 'a hallmark of modernism, progress, development and the metropole—the opposite of provincialism'.

Kern (2007) *Urb. Geog.* 28, 7 defines **neoliberal urbanism** as the production of self-governing, consumption-oriented, autonomous urban citizens. Wilson (2008) *Urb. Geog.* 29, 3 finds that a key underpinning of neoliberal urbanism is 'the widespread acceptance of distinctive, simple sensibilities about people, places, and processes that underpin extensions of neoliberalism into urban redevelopment'. These include the beliefs that hierarchical power arrangements in cities and beyond are natural; that power arrangements reflect

u

differences in people's levels of aspirations, abilities, and desires to determine their own destinies; and that human nature is prone to conflict spurred by the 'naturalness' of racial, ethnic, and class differences. This is an illuminating paper. See Ward (2003) *Area* 35, 2 on **entrepreneurial urbanism**.

However, it is clear from the literature that urbanism can mean *urbanization in North America: E. Talen (2005) identifies four traditions in the United States: incrementalism, the urban plan-making culture, planned communities, and regionalism (the last two outside the existing infrastructure of the city). Talen suggests that new urbanists, 'unlike practitioners of past planning movements', are attempting to decipher and implement the 'successes' of these four.

urbanization The increase in the proportion of the population residing in towns, brought about by migration of rural populations into towns and cities, and/or the higher urban levels of natural increase resulting from the greater proportion of people of childbearing age in cities. R. Bilsborrow (1997) discusses the problems in the definition of urbanization, and the difficulties of making international comparisons. Start with Solecki and Leichenko (2006) *Environment* 48, who give an excellent overview, while also discussing urbanization and sustainability. See also T. Hall (2006).

Urbanization indicates a change of employment structure from agriculture and *cottage industries to mass production and *service industries. This backs up the view that urbanization results from, rather than causes, social change. This is most notable in the development of *capitalism and its attendant *industrialization (see Fields (1999) *Berkeley Plan. J.* 13; try E. Soja 1989). Others argue that urbanization is the inevitable result of economic growth, with the rise of specialized craftsmen, merchants, and

administrators (Fafchamps and Shilpi (2001) *U. Oxford Econ. Series W. Papers* 139). A further view stresses the importance of *agglomeration economies; cities offer markets, labour, and capital with a well-developed infrastructure, all of which increase their *comparative advantage. J. Jacobs (1969) argues that urbanization economies have been the primary driving force for the geohistorical development of human societies over the past 12000 years.

N. Brenner and R. Keil, eds (2006) state that globalizing cities in the 'south', as with those in the 'north', although being distinctly different have a number of common traits, in particular: a colonial legacy; imposed neoliberalism and an associated socio-spatial polarization often driven by international organizations such as the World Bank and United Nations, and mass urban migration, creating extreme pressure on infrastructure and service provision.

T. Champion and G. Hugo (2003) argue that there is no longer any difference between the urban and the rural; instead, there are new forms of urbanization. Gandy (2005) *Int. J. Urb. & Reg. Res.* 29, 1 argues that an emphasis on **cyborg urbanization** (a 'dialectically conceived version of urban metabolism relating technical developments to a broader political and cultural terrain') 'extends our analysis of flows, structures and relations beyond so-called "global cities" to a diversity of ordinary or neglected urban spaces'.

urbanization curve A model of the progress of urbanization, based on empirical evidence from Europe. In a traditional society, urbanization is below 20%, and the rate of urbanization is slow, so the curve starts gently. With industrialization, and a rise in the importance of manufacturing and services, the pace of urbanization quickens, but the curve slackens after about 75%. While most

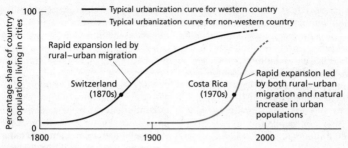

Urbanization curve

developed countries have reached this third stage, the countries of the developing world are still on a rising curve of urbanization, often with a steeper gradient than is characteristic of advanced economies.

Some geographers believe that *counter-urbanization constitutes a fourth stage, when urban populations percentages fall; others that counter-urbanization is temporary.

urbanization economies The benefits of locating in an urban environment; the '[positive] effects on productivity and utility created endogenously by larger cities' (Krupka (2009) *IZA DP* 4573). These economies include proximity to a market, labour supply, good communications, and financial and commercial services such as auditing, stockbroking, advertising, investment, industrial cleaning, and maintenance. Desmet and Fafchamps (2005) *J. Econ. Geog.* 5, 3 distinguish between *localization economies*, which derive from being located close to other firms in the same industry, and urbanization economies, associated with closeness to overall economic activity. Larger cities have a greater *comparative advantage than small, and the relationship between city size and urbanization economies is non-linear. Demand-side explanations for this argue that the

bigger the city, the more likely a match between specific demands and suppliers (an explanation with echoes of *threshold population), while supply-side explanations cite the provision of traded, and non-traded, inputs which increase innovation and variety, and lower costs.

urban morphology The form, function, and layout of the city. Geographic attributes—including shape, density, and pattern of land use categories—all reflect the outcome of the *urbanization process (Longley and Mesev (2002) *Papers Reg. Sci.* 81, 1). C. Boone and A. Modarres (2006) provide an account and explanation of urban morphology through time.

urban planning An attempt to manage the city, often in order to avoid, or alleviate, common urban problems such as inner-city decay, overcrowding, traffic and other forms of congestion; 'the management of spatial interdependence' (Castells in Dunford and Pickvance (2007) *PHG* 31, 4). McCann (2001) *Prof. Geogr.* 53, 2 argues that, in order to understand the role of visioning in contemporary urban politics and in policy-making outcomes, 'we must recognize the sociospatial context in which it is deployed'. Soja, quoted in Warf (2006) *PHG* 30, saw urban planning as the attempt by the state to contain the contradictions of *capitalism.

urban political ecology If one component of *political ecology is the study of political struggles for control over natural resources, urban political ecology is easy to define, and Swyngedouw and Heynen are its begetters: 'the political programme of urban political ecology is to enhance the democratic content of socio-environmental construction' (Swyngedouw and Heynen (2003) *Antipode* 35, 5). N. Heynen (2006) calls for 'more equitable distribution of social power and a more inclusive mode of the production of nature'; see Myers (2008) *Urb. Geog.* 29, 3.

urban regeneration (urban renewal, urban revival) The attempt to reinvigorate a run-down urban area, through strategies including the redevelopment of *brownfield sites, rehabilitation of the existing building stock, and enhancement of public spaces and infrastructure (see Temelová (2007) *Geografiska B* 89, 2). In the *slums of the developed world, substandard housing is broken up by urban regeneration projects, and low-income households, 'unable to gain access to the shrinking stock of public rental housing find themselves increasingly restricted to pockets of low quality housing in the private rental sector' (Randolph and Holloway (2005) *Urb. Policy & Res.* 23). See Forster (2006) *Geog. Res.* 44, 2. Hall (2007) *Geog. Compass* 1, 6 notes that public art 'has become wedded' to urban regeneration in developed world cities, and Bell (2007) *PHG* 31, 1 finds that hospitality has become central to regeneration 'as cities increasingly rebrand themselves as pleasure'. Podagrosi and Vojnovic (2008) *Urb. Geog.* 29, 4 report that, in Houston's urban revival, physical upgrading has been accompanied by the displacement of the community's traditional population and the destruction of an historic neighbourhood.

urban social geography The study of social and spatial dimensions of city life; the study of the social patterns and processes of cities; the social interpretation of the spatial structure and functioning of the urban environments. See P. Knox and S. Pinch (2000).

urban sprawl The extension of the city into the countryside, particularly associated with improvements in mass transport. Torrens (2006) *AAAG* 96, 2 argues that population growth is one of the most important engines of urban sprawl, and Irwin and Bockstaehl (2002) *J. Econ. Geog.* 2 think that sprawl may be caused by negative *externalities between developments. The implication is that the undeveloped land that is adjacent to sprawl is less valuable in residential use and less likely, *ceteris paribus*, to be developed residentially in the future. See J. Wolch et al. (2004) on urban sprawl in California, and Glaster and Cutsinger (2007) *Urb. Geog.* 28, 6 on sprawl and racial segregation. Bartl (2011) *Urbaniz. & Water* has some interesting thoughts on urban sprawl in Indian cities. *See* URBAN CONTAINMENT.

urban system Any network of towns and cities, and their hinterlands, which can be seen as a system, since it depends on the movements of labour, goods and services, ideas, and *capital through the network. Freestone et al. (2003) *Tijdschrift* 94, 2 illustrate the way Australian urban systems reflect broader economic, demographic, and social trends. Abdel-Rahman (2002) *J. Reg. Sci.* 42 links the structure of an urban system with income disparities: economies that are dominated by a large metropolis, with a number of low-technology cities, will have large income disparities.

urban theory Theorizations of *urbanization which stress the logic of capital in *uneven urban development, appraise the tensions between modernity and postmodernity as conceptual frameworks for understanding 'the urban', explore the *globalization and the concept of world cities, and analyse the complexities of multiple and hybrid urban identities. P. Hubbard (2006) concludes that urban theory can only progress 'by taking the distinctive and excessive materiality of cities more seriously'. A. Amin and N. Thrift (2002) stress instead the 'significant banality' of everyday life. See Scott and Storper (2013) *Int. J. Urban & Reg. Res.* on the range and substance of urban theory, and Beauregard (2003) *Antipode* 35, 5, although he is somewhat diffuse.

urban village A residential area of the city containing a cluster of individuals of similar culture and/or interests. Ghazali (2003) *Area* 35, 2 notes that many rural villages in Malaysia have been absorbed by urban expansion and become urban villages. Hao et al. (2013) *Int. J. Urban & Reg. Res.* 37, 6, 2177 report on the urban village development process in Shenzhen, China, while Franklin and Tait (2002) *Planning Theory* 1, 3, 250 report that the concept of the urban village in the UK was first promoted by the Urban Villages Group in the late 1980s as a means to achieve more human-scale, mixed-use, and well-designed places: 'The term urban village has since entered the planning discourse'.

USGS DEM A *geospatial file format developed by the United States Geological Survey for storing a raster-based digital elevation model. It is an open standard and is used throughout the world (K.-T. Chang (2014)).

U-shaped valley Most U-shaped valleys—valleys with a parabolic cross-section—are *glacial troughs. Brook et al. (2008) *Geomorph.* 97, 1–2 find that valley cross-profiles become U-shaped with *c.* 400,000–600,000 years of glacial occupancy. When no longer supported by ice, the steep sides of U-shaped valleys are prone to rock fall (Blair (1994) *Mountain Res. & Dev.* 14). Valleys with this form are also encountered in non-glaciated chalk topography.

US soil classification The US Department of Agriculture recognized ten major soil groups in the Seventh Approximation. Alfisols are relatively young and acid soils with a clay B horizon. Aridisols are semi-desert and *desert soils. Entisols are immature, mainly *azonal soils. Histosols are primarily organic in content, developing in marshes or peat bogs. Inceptisols are young soils with weakly developed horizons. Mollisols are characteristic of grassland, high in bases, and with a thick, organically rich A horizon. Oxisols occur in tropical and subtropical areas. They are well weathered and often have a layer of *plinthite near the surface. Spodosols have been *podzolized. Ultisols develop where summers are wet and winters are dry. They are quite deeply weathered and are often reddish-yellow in colour. Vertisols are clay soils characterized by deep, wide cracks in the dry season.

utility The satisfaction given to an individual by the goods and services used.

vadose Describing the zone immediately below the ground surface and above the *water-table in which the water content varies greatly in amount and position.

validation In academic life, the methods used to check the validity of a finished piece of research. 'Validation in quantitative methods involves the calculation of statistical errors, whereas in qualitative methods validation methods include cross-checking or triangulation with other methods; theoretical saturation, or the point at which no new information is gained; and adherences to a code of conduct believed to ensure accurate results' (Nightingale, in R. Kitchin and N. Thrift 2009).

valley train *See* SANDUR.

valley wind An *anabatic air flow generated as a valley floor is heated by the sun, and the warm air moves up-slope. Valley winds are at their strongest in valleys of a north–south orientation. (See Weigel et al. (2007) *Boundary-Layer Met.* 123, 1.)

value The importance, worth, or usefulness of something; its material or monetary worth. 'The crucial distinction between poor and rich countries is in the relative value of the commodities produced in each area' (Gereffi et al. in G. Gereffi and M. Korzeniewicz, eds 1994). Gidwani and Chari (2004) *Env. & Plan. D* 22, 4 describe the **governmentalization of value**: the proliferation of tactics (programmes, laws, codes, standards, and institutionalized knowledges) that '*interpellate subjects of value who are in subjection to the Darwinian logic of value, yet are animated by its almost spiritual promise of self-betterment'. Smith et al. (2002) *PHG* 26, 1 argue that a focus on *value chains and networks* is more useful than a focus on *commodities* ('which implicitly embody that value') per se, in understanding territorial economies in increasingly integrated macro-regions.

vapour pressure In meteorology, the pressure exerted by the water vapour in the air. This is a partial pressure since pressure is also exerted by the air itself. Moisture content is commonly measured by *relative humidity. The absolute value of the water content is not usually the most important issue but its value compared with saturation vapour pressure; thus relative humidity is more relevant than specific humidity.

variable source areas *See* *DYNAMIC CONTRIBUTING AREAS.

variegated capitalism (varieties of capitalism) *Capitalism varies according to the way it is embedded in different sorts of states. The major varieties of capitalism are: *welfare states, *neoliberal states, *developmental states, and *socialist states; see P. A. Hall and D. Kosice (2001) for the seminal work,

and R. Hancke, M. Rhodes, and
M. Thatcher (2008).

varve A pair of coarse and fine deposits
that reflect seasonal deposition in *proglacial lakes. In summer, *meltwater
streams deposit silt- to sand-sized coarse
sediments; in winter the lake surface is
frozen, the water is calm, and the fine
deposits settle out in a thinner layer than
the summer sediments. A **varve couplet**
represents the total *fluvio-glacial deposition on a lake floor for one year. Chu et
al. (2012) *Holocene* 22, 785 reconstruct
the growing season over the past 1600
years in northeastern China from varved
sediments. At Lake Sihailongwan,
Snowball et al. (2002) *Holocene* 12, 1,
analyse a series of varves to reconstruct
climatic changes during glaciations.

vector In *geographic information systems (GIS), **vector data** represent features as discrete points, lines, and
polygons. In computer graphics, **vectorization** refers to the process of converting raster graphics into vector
graphics. Vectorization is not well defined, meaning there is not a single correct method. Many different algorithms
exist, and each gives different results, as
vector representations are more abstract
than pixels (K.-T. Chang 2014).

veering Of winds in the Northern
Hemisphere, changing direction in a
clockwise motion, e.g. from westerly to
northerly. The *Coriolis effect causes
freely moving objects to veer towards the
right in the Northern Hemisphere and to
the left in the Southern Hemisphere.

**vegetation index (spectral vegeta-
tion index)** Satellite maps of vegetation
show the density of plant growth over
the entire globe. To determine the density of green on a patch of land, researchers must observe the distinct
colours (wavelengths) of visible and
near-infrared sunlight reflected by the
plants. Using the United States' National
Oceanic & Atmospheric Administration
AHVRR (Advanced Very High Resolution
Radiometer), researchers can measure
the intensity of light coming off the Earth
in visible and near-infrared wavelengths
and quantify the photosynthetic capacity
of the vegetation in a given pixel of land
surface. In general, if there is much more
reflected radiation in near-infrared
wavelengths than in visible wavelengths,
then the vegetation in that pixel is likely
to be dense and may contain some type
of forest. If there is very little difference
in the intensity of visible and near-infrared wavelengths reflected, then the
vegetation is probably sparse and may
consist of grassland, tundra, or desert.
This is the basis of the Normalized Difference Vegetation Index (NDVI). Very
low values of NDVI (0.1 and below) correspond to barren areas of rock, sand, or
snow. Moderate values represent shrub
and grassland (0.2 to 0.3), while high
values indicate temperate and tropical
rainforests (0.6 to 0.8).

velocity reversal hypothesis An
explanation of the occurrence of coarse
sediments in riffles, and fine sediments
in pools, by variations in *discharge. At
high discharge, high velocities would
move coarse sediment rapidly through
the pools, while relatively low velocities
in riffles would cause its deposition
there. The fine particles are winnowed
from riffles and deposited in the pools.
Booker et al. (2001) *ESPL* 26 think that
velocity reversal is an inadequate explanation of pool-riffle sequences, and
Thompson (2011) *PPG* 35, 1, 123 agrees:
'the velocity reversal model by itself does
not explain pool–riffle maintenance'.

vent In geomorphology, an opening in
the *crust through which volcanic material flows. **Vent conglomerate** consists
of rounded blocks of juvenile material,
derived from the collapse of the cone

v

walls during a volcanic eruption, filling a volcanic vent. See Oppenheimer (2003) *PPG* 27, 2.

venture capital Private equity (finance) offered to outside investors in new/growing businesses. Venture capital investments are usually high risk, but offer the possibility of above-average returns. Mason and Harrison (2002) *TIBG* 27, 4 find that, in the UK, venture capital investments were concentrated in Greater London and the South-East in the 1980s, but that this regional concentration has been greatly reduced since the 1980s. Martin et al. (2002) *J. Econ. Geog.* 2, 2 find that, although it has recently experienced a marked increase in activity, the venture capital industry is much less developed in Europe than it is in the USA.

vertex In *network analysis, this is the place joined by two or more *links (also called *node*).

vertisols Soils of the *US soil classification found in regions of high temperature where bacteria destroy organic residues; hence the humus content is low. Alternate wet and dry seasons lead to the alternate swelling and shrinking of these soils. By these processes, horizons become mixed or inverted.

vicariance (vicariance biogeography) The geographical separation and isolation of a subpopulation, resulting in the original population's differentiation as a new variety or species. When a continuous population (or species) is divided by a **vicariance event**, such as the formation of a new river or a new mountain range, two populations or species are created. See Wiley (1988) *Ann. Rev. Ecol. Syst.* 19, 513, but Wallis and Trewick (1998), now published in *Systematic Biol.* 50, 4, 602, find fault with vicariance.

violence, geographies of Violence can be seen as social practice where human needs and human securities are contested and fought over. This would see the geographies of violence as places where human freedoms and rights are struggled for, negotiated, lost, and won.

viscosity The resistance to flow exhibited by a material; the parameter describing the flow characteristics of a fluid. Easy flows result from low viscosity.

vital rates Rates of those components, such as birth, marriage, fertility, and death, which indicate the nature of and possible changes in a population.

volcanic ash Finely pulverized fragments of rock and lava which have been thrown out during a volcanic eruption. Volcanic ash forms a thick coating on the landscape adjacent to a volcano, and can travel great distances. Adey et al. (2011) *TIBG* 36, 3, 338 examine the way that the eruption of Icelandic volcano Eyjafjallajökull in 2010, revealed 'the fragility of a tightly coupled, complex, and quite fragile network of airline movements, logistics chains, insurance products, and the complex supra-national organization of European airspace'. See also Ort et al. (2008) *GSA Bull.* 120, 3. Atwater et al. (2004) *Holocene* 14, 4 use volcanic ash as a dating method. A **volcanic bomb** is a block of lava ejected into the air from a volcano. As it is thrown out, it cools and spins; see Francis (1973) *GSA Bull.* 84, 8 on cannonball bombs.

volcanism (vulcanicity, vulcanism) All the processes associated with the transfer of magma and volatiles from the interior of the Earth to its surface. *Magma beneath the crust is under very great pressure. Deep in the crust, faults and joints develop downward, reach the magma, and allow it to rise up and intrude the crust. The magma then rises in conduits, depressurizes, forms bubbles, and can give rise to explosive volcanism; see Gaonac'h et al. (2003) *Geophys. Res. Letters* 30, 11, but also see Gonnermann and Manga (2003) *Nature* 426. **Intrusive volcanicity** describes magma injected into the crust which fails to reach the surface, and cools slowly underground forming *batholiths, *laccoliths, *dykes, *pipes, and *sills. **Extrusive volcanicity** describes magma which forces its way to the Earth's surface and erupts as *lava from volcanoes or fissures, forming cones or plateaux.

Current volcanicity is confined to regions of the Earth where lithospheric *plates converge, diverge, or pass over possible mantle *hot spots. In a *subduction zone, water is trapped in the descending plate. As the plate is heated, water is released, rising into the wedge of mantle above the plume, causing it to melt. This process forms chains of volcanoes; the cordillera of North America is an area of volcanism related to the subducting eastern margin of the Pacific Ocean. Where one oceanic plate is subducted beneath another, a chain of volcanic islands forms; Japan is an example (Bunbury in P. L. Hancock and B. J. Skinner, eds 2000). When a mantle plume rises through the lithosphere, and eventually erupts at the surface, it is often called *hot-spot volcanism*. About 10% of volcanoes arise through hot-spot activity. Mass movements on volcanic islands may generate large-magnitude tsunamis; see Whelan and Kelletat (2003) *PPG* 27, 2. Hillier (2007) *Geophys. J. Int.* 168, in a fairly technical paper, notes that *seamounts constitute some of the most direct evidence about intra-plate volcanism, reflecting the ways that the properties of the lithosphere interact with magma generation in the fluid mantle beneath.

In recent years the focus of volcanology has shifted from the physical aspects of why, when, and how volcanoes erupt to incorporate studies on the short- and long-term effects volcanic eruptions have on society and the atmospheric, aquatic, and terrestrial environments; see J. Marti and G. Ernst, eds (2005).

volcano An opening in the crust out of which *magma, ash, and gases erupt. The shape of the volcano depends very much on the type of lava. *Cone volcanoes are associated with thick lava and much ash. *Shield volcanoes are formed when less thick lava wells up and spreads over a large area, creating a wide, gently sloping landform. Most volcanoes are located at *destructive or *constructive plate margins.

volume On a *GIS triangulated irregular network, three-dimensional space measured by length, width, and depth, or height.

von Thünen models Johann von Thünen (1783–1850) had two basic models: the first postulates that the intensity of production of a particular crop declines with distance from the market; the second is concerned with land use patterns as they relate to transport costs. See Barnes (2000) *PHG* 27, 1 for a summary.

vortex *(pl. vortices)* A whirling mass of fluid or air. See Walker and Nickling (2002) *PPG* 26, 1 on helical vortices in dunes; Wiggs (2001) *PPG* on the roll-vortex theory of dune formation; Haines (1982) *J. Applied Met.* 21 on horizontal roll vortices and crown fires; and Davies and Thorne (2005) *J. Geophys. Res.* 110, C05017 on waves in the vortex ripple regime. Huang et al. (2000) *Atmos. Env.* 34 simulate a stable canyon vortex, and MacDonald (2000) *Boundary-Layer Met.* 97 has a numerical model to describe the spatially averaged wind speed profile within the urban canopy layer.

Power (2003) *PPG* 27 explains that the strong **polar vortex** in the Antarctic atmosphere isolates the polar air mass for

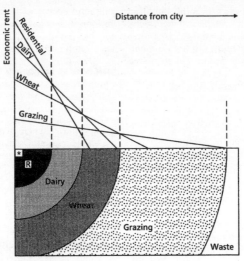

von Thünen model

much of the year. Over the Arctic region the polar vortex is weaker—resulting, in part, from the uneven distribution of land and water. Perlwitz and Graf (1995, *J. Climate* 8) show that the *North Atlantic Oscillation tends to be positive during winters with an anomalously strong **stratospheric polar vortex**.

vorticity In meteorology, a measure of the local spin of a part of the *atmosphere. The local spin of the atmosphere relative to the Earth has opposite signs in *cyclones and *anticyclones; conventionally the cyclonic direction is taken as positive. A major principle governing the vorticity change of flowing air is the *conservation of angular momentum: as air spreads out horizontally, the rate of

spin falls; as it contracts horizontally, its rate of spin rises. Bornstein et al. (1996, *J. App. Met.* 35) and Schayes et al. (1996, *J. App. Met.* 35) describe and evaluate a general-purpose vorticity-mode meso-scale model. **Absolute vorticity** is the sum of the **planetary vorticity**, which is associated with the rotation of Earth, and **relative vorticity**, which is associated with the fluid motion relative to the surface of Earth.

vulcanian eruption Also known as a **Vesuvian eruption**, this is marked by periodic lulls during which gas pressure builds up behind the lavas that clog the vent. This blockage is removed by an explosion which throws off masses of *pyroclasts.

wadi In a hot desert, a steep-sided, flat-floored valley, very occasionally occupied by an intermittent stream. Wadis were probably cut during the rainier *pluvials; see Morin et al. (2007) *Adv. Geosci.* 12 on wadis. See also F. El-Baz, ed. (2003).

Walker circulation 'One of the world's most prominent and important atmospheric systems, closely linked to the mean-state of the equatorial Pacific Ocean. It extends across the entire tropical Pacific Ocean, encompassing: the trade winds blowing from east to west; air forced to rise over the western Pacific, south-east Asia and northern Australia through enhanced convection; winds blowing counter to the trades aloft, and air descending over the eastern Pacific Ocean' (Power and Smith (2007) *Geophys. Res. Abstr.* 34, L18702). See Lockwood in J. Holden (2012), p. 104. Changes in the Walker Circulation are linked to the *El Niño–Southern Oscillation; see Power et al. (1999) *Climate Dyn.* 15.

wandering river A fluvial channel with alternate stable single-channel reaches, and unstable multi-channel sedimentation zones. This term has also been applied to channels which are midway between meandering and braiding. Wang et al. (2007) *Geomorph.* 91, 1–2 describe changes from braided to wandering in the Weihe River, China.

waning slope *See* PEDIMENT.

war on terrorism (war on terror) An operation initiated by the United States government under George W. Bush, using legal, military, personal, and political actions to limit the spread of terrorism after 9/11. Graham (2006, *Int. J. Urb. & Reg. Res.* 30, 2) describes the reworking of US cities in order to

Walker circulation

construct them as 'homeland' spaces, re-engineered in the interests of 'national security', while Dalby (2007, *Geopolitics* 12, 4) outlines the 'dichotomous mapping of the world into civilized core and dangerous periphery'. Thobani (2007) *Feminist Theory* 8, 2 examines the practices that constitute whiteness, triggered by the war on terror, and Macrae and Harmer (2003) *Overseas Dev. Inst.* provide a clear review of the issues. See Amore (2009) *Antipode* 41, 1, 49.

wash board moraine Synonymous with De Geer *moraine.

wash load An alternative term for *suspended *load.

water balance The balance at any location between the input (*precipitation, *P*), and the outputs: *evapotranspiration (*E*) and *run-off (*R*):

$$P = E + R$$

If water balance is computed for a number of years, *groundwater storage (*S*) is held to be constant, but for a single year, groundwater fluctuations are taken into account:

$$P = E + R \pm S$$

Storage and run-off tend to be higher in winter, and evapotranspiration is higher in summer. See Arnell (1999) *J. Hydrol.* 217 on a simple water balance.

waterfall A site on the *long profile of a river where water falls vertically. Waterfalls may be found at a band of more resistant rock (Alexandrowicz (1994) *Geomorph.* 9, 2), at a *knick point (Crosby and Whipple (2006) *Geomorph.* 82, 1–2), or where deposition has occurred. Hyakawa and Matsukora (2003) *ESPL* 28, 6 note that the principal factor in determining the rate of waterfall recession is the ratio of the erosive force of stream to the bedrock resistance, and derive an empirical equation connecting the force/resistance ratios and the rates of waterfall recession.

water-level weathering The development and enlargement of tidal pools by *weathering, and by the action of rock-grinding animals. Water-level weathering is seasonal, with higher rates in the summer months. See Inkpen (2007) *Area* 39, 1.

water provision The provision of water. Agnew (2011) *AAAG* 101, 3, 463 shows how politics is at work around the world in managing water problems and identifying the challenges that water provision poses for politics.

water quality This can be registered as a measure of the ecological health of the water, determined by the level of indicator species in relation to a predetermined critical threshold, derived from toxological data; the biochemical oxygen demand can also be used. Land cover change and the expansion of modern agricultural practices has significantly increased leaching of chemicals to the surface waters, leading to the degradation of aquatic ecosystems worldwide (Donner (2003) *Glob. Ecol. & Biol.* 12, 4). See Beasley and Kneale (2002) *PPG* 26, 2 on pollution from urban surface run-off.

watershed In the UK, the boundary between two river systems. Horton (1945) *GSA Bull.* 56, 3 uses the term *watershed line*, and refers to drainage basins as separate entities, but watershed is now commonly used in North America to mean drainage basin; 'the Mississippi River has the largest of all North American watersheds' (Galler and Anderson (2008) *GSA Bull.* 120, 3).

water-table The level below which the ground is saturated; the upper surface of the *groundwater. See Quade et al. (1995) *GSA Bull.* 107, 2 and Romanelli et al. in Massimo Greco et al., eds (2004).

water transfer The movement of water from a *drainage basin or catchment where there is more than enough water for human needs, to another catchment where there is a deficit.

wave A ridge of water between two depressions. **Wave height** is the difference in elevation between the wave crest and the wave trough. The height of a wave is generally proportional to the square of wind velocity.

The **wavelength** is the distance between successive crests (or troughs), and the **wave period** is the time it takes for the wave to travel a distance equal to its wavelength. **Wave steepness** is given by the ratio of wave height to wavelength H/L. **Significant wave height** H_s is the average of the highest one-third of the waves, and can be obtained by recording a time series of water depth and multiplying the associated standard deviation by four. *See* LINEAR WAVE THEORY.

Wave shoaling is the process whereby waves change in height as they travel into shallower water and make contact with the seabed. Short and Brander (1999) *J. Coastal Res.* 15, 3 suggest that distinct morphodynamic scaling exists between various wave energy environments.

Waves are the moving force in *longshore drift. See Watt et al. (2005) *Procs Coast. Dynam.* on surface sediment distribution patterns in response to wave conditions.

wave convergence The focusing of wave rays so that they come together, increasing in energy and height. This occurs when waves travel over a localized area of relatively shallow water such as a shoal on the sea floor, causing an increase in *wave energy and wave height (Masselink in J. Holden, ed. 2012).

wave-cut platform *See* SHORE PLATFORM.

wave energy Energy generated by the force of ocean waves. Wave energy is expressed as the amount of energy per unit area (N m^{-2}) and is more appropriately referred to as **wave energy density**. Wave energy depends on the square of the wave height. A doubling of the wave height therefore results in a four-fold increase in wave energy. Total wave energy E is given by:

$$E = \frac{1}{8} \rho g H^2$$

where ρ is the density of water, g is gravity, and H is the wave height. See Masselink in J. Holden, ed. (2012). **Wave energy flux** is the rate at which wave energy is carried along by moving waves and is the product of the amount of energy associated with the waves and the speed at which this energy travels.

The University of Hull has a group researching the flow dynamics of wave and tidal power devices, and analysing marine energy resources; see Hardisty (2006) *J. Mar. Sci. and Env.* C4.

wavelets Mathematical functions that cut up data into different frequency components, and then study each component with a resolution matched to its scale. **Wavelet analysis** is a statistical tool for analysing such localized variations of power within a time series. By decomposing a time series into time-frequency space, it is possible to determine the dominant modes of variability and how those modes vary in time. Torrence and Compo (1998) *Bull. Amer. Meteor. Soc.*, 79, 61 provide a practical step-by-step guide to wavelet analysis, and Kumar (1997) *Rev. Geophys.* 35, 4 outlines the possibilities of wavelet

w

analysis in physical geography (see Barrett and Houser (2012) *Phys. Geog.* 33, 1, 32).

wave refraction The change in the approach angle of a wave as it moves towards the shore. As water becomes shallow, waves slow down. This change in speed causes the *orthogonals of a wave to 'bend' so that the line of the wave mirrors the submarine contours. Refraction causes waves to converge on headlands and diverge in bays. This means that the energy of the waves is concentrated on the headlands rather than on the beaches. See Bristow and Pucillo (2006) *Sedimentol.* 53, 4 and Gamito and Musgrave (2002) *Computers & Graphics* 26 on calculating shallow-water wave refraction.

wave set-up When waves break, water piles up against the shore. This results in a slope of water, with higher water pushed nearer the shore and this 'set-up' is sufficient to oppose the shoreward wave stresses (Masselink in J. Holden, ed. 2012).

waxing slope The *convex element at the foot of a hillslope.

waypoint A reference point in physical space used for purposes of navigation, including GPS and *GIS.

weathering The breakdown, but not the removal, of rocks. **Chemical weathering** involves chemical change, and includes *carbonation, *hydration, *hydrolysis, *oxidation, and some *organic weathering. See A. Turkington et al. (2005); see also the special issue of *Geomorphology* 67, 1–2. **Mechanical weathering** is the physical

Headland

Wave

Orthogonal

Wave refraction

W

disintegration of the rock, as in *pressure release, crystal growth, *salt weathering, *thermal expansion, and some organic weathering such as *chelation, and bacterial reduction as in *gley soils. See Trenhaile in D. E. Smith and A. Dawson, eds (1993) on freeze–thaw and shore-lines. Hall et al. (2002) *PPG* 26, 4 find that in summer, and often in winter, rock temperatures support mechanical *and* chemical weathering, if water is present. Thorn (2004) *Polar Geog.* 28, 1 summa-rizes periglacial weathering research. **Biological weathering** is the disinte-gration of rocks through the chemical and/or physical agency of an organism. Etienne (2002) *Geomorph.* 47, 1 considers biogeochemical processes in the morphogenic system of periglacial environments.

Weathering rates vary with the physi-cal structure of the rock (particularly on crystalline rocks—see Borrelli et al. (2006) *Geomorph.* 87, 3); bond strength (Velbel (1999) *Am. J. Sci.* 299), and chemical composition (Yokoyama and Matsukura (2006) *Geology* 34).

The **weathering front** is the zone of contact of the *regolith with the under-lying rock. In humid environments, in-creasing soil thickness increases weathering rates, primarily due to mois-ture storage and biological effects as soil develops. Eventually the rock weather-ing rate slows down with increasing dis-tance from surface moisture (Gracheva et al. (2001) *Quat. Int.* 78). **Weathering pits** are depressions on a flat surface, usually on very soluble rocks, and vary-ing in shape and ranging in size from a few centimetres to several metres in width (Domínguez-Villar (2006) *Geomorph.* 76, 1–2). **Honeycomb weathering** is a grouping of many, closely spaced pits (Pye and Motters-head (1995) *Int. J. Rock Mechan. Mining Scis & Geomech.* 33, 8). *Tafoni are weathering pits which are cut into near

vertical rock faces. **Weathering rind** is the chemically altered 'skin' of rock around an unaltered core (Etienne (2002) *Geomorph.* 47, 1).

welding The bonding of grains; see, for example, Anthony et al. (2006) *Geo-morph.* 81, 3–4 on dune accretion.

welfare While welfare might be equated with well-being, within human geography it refers mostly to factors within the control of societies: environ-mental quality, security, and access to commodities and services. It therefore incorporates income, standard of living, housing, employment, and access to educational, health, and social services; see Sen's **welfare index** (1976) *Rev. Econ. Studs* 43. Ezcurra et al. (2006) *Tijdschrift* 98, 4 find that, within the EU, regions with high welfare tend to have advanced services, high productivity, and good human capital. Low-welfare regions tend to higher unemployment, and a larger agricultural sector.

DeVerteuil (2007) *AAAG* 97, 1 thinks that 'the current decline of the welfare state within a context of *neoliberalism in most developed nations is a key factor in exacerbating the situation of those worst-off, thereby widening the socio-economic gaps even further'.

Welfare capitalism is the provision of welfare by corporations aiming to fos-ter loyalty among the workforce. S. Jacoby (1998) provides a comprehensive history of 20th-century welfare capitalism.

Welfare geography studies *in-equalities in *social well-being and *social justice, looking at the areal differentiation and spatial organization of human activity from the point of view of the welfare of the people involved. It focuses on those factors which affect the quality of human life: crime/lack of crime, poverty/wealth, housing/homelessness, and the provision/lack of

First quartile (lowest)
Second quartile
Third quartile
Fourth quartile (highest)

Welfare variations in the EU

educational, health, leisure, and social services. The first line of enquiry examines inequality in the distribution of welfare indicators, which provides a base from which to evaluate the impact of past or proposed changes; see Falah (1999) *Urb. Geog.* 20, 5 on Israel's Arab citizens. From this, the second aim is an attempt to explain these inequalities. Current explanations link inequality with *uneven development as an inevitable consequence of *capitalism; or with an uneven spatial pattern of service provision; see DeVerteuil (2001) *PHG* 24, 1

and I. M. Young (1990). The third aim is the formulation of measures to bring about a fairer distribution of resources and opportunities. Haylet (2001) *TIBG* 26, 1 argues for a welfare geography attuned to 'the languages and practices through which dominant systems of social and economic distribution are constituted'.

westerlies (westerly winds) Winds blowing from the west, most often in mid-*latitudes.

Westphalia The Peace of Westphalia treaty, 1648, is considered a watershed event in traditional *geopolitics. It led to the creation of the modern nation-state system by allowing *hegemons to connect their power to discrete, definable territorial units for the first time in history. As the system was concretized and formalized, it also paved the way for inter-state struggles for power. See Gallaher, in C. Gallaher et al. (2009).

wetland Any land which is intermittently or periodically waterlogged, including *salt marshes, tidal *estuaries, marshes, and bogs. Wetlands are rapidly disappearing habitats; see M. Williams (1993) and Liu et al. (2004) *Ambio* 33, 6 on the loss of wetland and *biodiversity.

In the United States a **constructed wetland** is intentionally created from non-wetland sites for wastewater or storm water treatment; the Constructed Wetland Association (CWA) website has a reading list. The concept of a **designer wetland** emphasizes the life history strategy of species as the major factor in restoring a wetland type, with no fixed endpoint; see M. Reuss (1998). For **wetland restoration**, see Tooth and McCarthy (2007) *PPG* 31, 1 and Holden et al. (2004) *PPG* 28, 1; see also Van Lonkhuyzen et al. (2004) *Env. Management* 33, 3. A **wetland mitigation bank** site is a property purchased and developed by a public agency or utility to earn credits to compensate for adverse impacts to wetlands due to development activities of other agencies, utilities, or in specific instances, private-sector developers; see Gilman (1997) *Ocean & Coastal Manage.* 34, 2.

(⊕) SEE WEB LINKS

• The Constructed Wetland Association reading list.

wetness index (*wi*) Also called the *compound topographic index*, the wetness index describes the effect of topography (slope and aspect) on the location and size of areas of water accumulation in soils. In its simplest form, the wetness index is calculated as: $wi = \ln(\alpha/\tan\beta)$, where α is upslope contributing area and β the local surface gradient. See Sulebak et al. (2000) *Geografiska A* 82, 1.

wetted perimeter In a cross-section of a river channel, the line of contact between water and bed. *See also* HYDRAULIC RADIUS.

whaleback A streamlined rock knob with symmetrical longitudinal profiles caused by abrasion of both stoss and lee sides. Small whalebacks can form under only a few hundred metres of ice; larger ones under deep ice streams. In theory, whalebacks tend to form in areas of deep ice with rapid flow velocities (James (2003) *Geomorph.* 55, 1–4). See also Glasser and Harrison (2005) *Geograf. Annal. A* 87, 3.

whiteness The normalization of white people's social norms and cultural practices as the natural and proper way of society, where 'white' is used as a synonym for people of European origin who have a light-coloured skin. Simply put, a higher cultural value has been attached to people who are classed as white, and S. Garner (2007) argues that the meanings attached to 'race' are always time- and place-specific (although there are certain overlapping truths). Berg (2012) *PHG* 36, 4, 508 argues that 'white supremacy is invisible to most geographers, and especially so to critical geographers, because of the way that (neo) liberalism provides us white (critical) geographers (and others) with the means to disaffiliate ourselves from "racism"—to claim innocence—all the while benefiting in very real material ways from ongoing white supremacy'.

For anti-racists, whiteness is a problem because it works to discriminate against, either implicitly or explicitly, non-whites and to suppress other cultural forms. Garner stresses that the analysis of whiteness must contribute to anti-racist scholarship by emphasizing the social relationships referred to as racism, rather than slipping into an identity-based paradigm.

Geographies of whiteness consider that spaces, places, landscapes, natures, mobilities, and bodies are assumed to be white or are in some way structured, though often implicitly, by some notion of whiteness.

wilderness An area which has generally been affected more by natural forces than by human agency; a region little affected by people. Some 77000 km^2 of the USA have been designated as **wilderness areas**, under the Wilderness Act of 1964. These are, ideally, areas which have never been subject to human manipulation of the ecology, whether deliberate or unconscious, and which are set aside as nature reserves to which human access is very severely restricted. Roads, motor vehicles, aircraft (except in an emergency), and any economic use are all forbidden.

The idea of wilderness is socially constructed: Baker (2002) *Space & Cult.* 5 argues that 'the spatial production of wilderness is enmeshed in late capitalist economics and politics, whereas its consumption is structured to reproduce class distinction'. Hintz (2007) *Eth. Place. Env.* 10, 2 writes that 'wilderness, we are told, can no longer be seen as a scenic playground for weary humans—it is, rather, an ecological necessity for the conservation of biodiversity'.

Some constructions of wilderness omit the works of people. P. Gobster and B. Hull, eds (2000) write of 're-wilding' the landscape, an ideal shared by Noss (2000) USDA, *Rocky Mountain Research Station Proceedings* 15, 1: 'the goal is not just to save existing wilderness, but to 're-wild' much of what has been lost.' This view has led to the expulsion of indigenous people from designated wilderness areas (W. Cronon 1995); Warren (2007) *PHG* 31, 4 observes that 'native species and wilderness areas are accorded high value precisely because of their naturalness, and "natural" typically carries the force of a moral imperative: this is the way things *ought* to be. But the belief that nature, to be natural, must be human-free, is rooted in a binary separation of nature and culture which is now widely regarded as false.' Whatmore (1998) *TIBG* 23, 4 observes that animals have not been helped by the term 'wild': 'Even as they are caught up in the assemblage of global regulatory networks designed to "protect" them, they find themselves objectified again in the urgent business of "wildlife management".'

Wilson cycle A description of the aggregation and break-up of supercontinents. Initially, a *hot spot rises up under a *craton, heating it, causing it to swell upward, stretch, and thin, not only splitting a continent in two, but also creating a new divergent plate boundary. A new ocean basin is generated between the two new continents, and as this widens (possibly by thousands of miles), wedges of divergent continental margins' sediments accumulate on both new continental edges. All of the above is the opening phase of the Wilson cycle. The closing phase of this cycle begins when a new *subduction zone forms at the margin of one of the continents.

Once this zone is active, the ocean basin begins to be subducted, causing the two continents to draw nearer.

Magma is generated deep in the subduction zone, rising to the surface to form a cordillera of volcanoes, accompanied by *metamorphism, folding, and faulting. When the two continents finally collide, the closing phase of the Wilson cycle is technically over. Because the subduction zone acts as a ramp, the continent with the subduction zone slides up over the edge of the other one. After this collision, the cordillera will be eroded to a *peneplain. Most of the upper continent will be eroded away, and the lower continent will eventually return to the Earth's surface. See Russo and Silver (1996) *Geology* 24 for further explanation on cordillera formation and the Wilson cycle.

(⊕) SEE WEB LINKS
• Information about the Wilson cycle.

wind energy Power generated by harnessing the wind, including electricity generation. Drawbacks include the inconstant nature of the wind, construction difficulties, and finding a suitable site; see Wolsink (2006) *TIBG* 31, 1. 'Installed wind energy capacity has been increasing rapidly worldwide over the last 5 years and now exceeds 70 GW. Wind energy contributes up to 18.5% of electricity consumption in Denmark and about 3% in Europe. Wind energy is economically competitive at good wind sites, can be rapidly deployed compared to conventional sources, and contributes to the goal of meeting increasing electricity demand while reducing emissions of greenhouse gases' (Barthelmie (2007) *Geog. Compass* 2). Globally, wind power had the fastest rate of growth of any source for electricity (Heiman in S. Majumdar et al., eds 2000). See also Heiman and Solomon (2004) *AAAG* 94, 1. See Bassi et al. (2012) on the case for and against onshore wind energy in the UK.

(⊕) SEE WEB LINKS
• The London School of Economics/Grantham Institute pages on Climate Change and the Environment where Bassi et al. can be accessed.

Windermere interstadial An interstadial of about 13000 to 11000 years BP, including the *Bølling, Older *Dryas, and *Allerød chronozones of Scandinavia. See Holloway et al. (2002) *Scot. J. Geol.* 38, 4.

wind shear The local variation of the wind vector, or any of its components, in a given direction. A change in wind speed and/or direction with height is the **vertical shear**, given by the *thermal wind equation, if the wind is *geostrophic. See Shepherd and Knutson, (2007) *Geo. Compass* 1, 1, and Vecchi and Soden (2007) *Geophys. Res. Letts* 34 L08702 on vertical wind shear and hurricanes.

window In *meteorology, the ability of radiation of wavelength around 10 μm to escape absorption by the Earth's *atmosphere. This wavelength nearly coincides with Earth's peak radiation, and allows some of the outflow of terrestrial radiation to be lost to space, thus, in part, upholding the thermal equilibrium of the atmosphere, which is also achieved by *convection. See, for example, Gube et al. (1996) *J. Climatol.* 3, 4.

work, geographies of The geography of work, whether at global, national, regional, or local scale, varies over time and space with the nature of the economy, the level of technology, and spatial variations in *class structure. 'To stake an agenda for "geographies of work" rather than for "labor geographies" means, in the most general sense, mounting a critique of the putative unitariness of value on three fronts: its basis—ostensibly, the production

w

process; the homogeneity of its extraction—ostensibly, the wage/labor relation; and its frequently posited autonomy from use-values—ostensibly, because the qualities of things, including labor, are effaced when values truck as commodities or exchange-values' (Gidwani and Chari (2004) *Env. & Plan. D* 22, 4). 'We are encouraged to extend geographies of work beyond the work-place, to contest definitions of work, to repeatedly and insistently connect work to lives beyond work and to see workers as parents, partners, consumers, activists and much more besides' (Stenning (2008) *Antipode* 40, 1). See James (2008) *Geog. Compass* 2, 1 on gendered geographies of labour, and Jennings et al. (2006) *Area* 38, 3 on children's geographies of work. See Ward et al. (2007) *Geoforum* 38 on living and working in working-class communities; see also A. Herod (2001).

World Bank The International Fund for Reconstruction and Development, popularly known as the World Bank, and now generally concerned with aid to the developing world. The Bank acquires its funds partly through government sub-scription, but mainly through borrowing, and voting power is proportional to capital subscription, so the bank is ef-fectively controlled by the rich countries. It has made loans of over $US 200 billion. Try Bunnell (2007) *Pol. Geog.* 26, 5 on the International Monetary Fund and World Bank in Singapore, Schec (in D. Moore, ed. 2007), on the World Bank and gender justice, Coleman's (2002, *Pol. Geo.* 21, 4) charge that, with reference to the debt crisis, and structural adjustment, the Bank 'struggles to explain contemporary developments in the geography of money', and Shriar (2005) *J. Lat. American Geog.* 42 on the impact on Jamaica of structural adjustment, trade liberaliza-tion, and economic globalization.

⊕ SEE WEB LINKS
• The *World Bank Economic Review* and the *PovertyNet Newsletters*.

world city A city that has a substantial, straightforward effect on world affairs: economically, socially, and politically. Key descriptors include: centrality—a concentration of international activities; externality—a relatively high global role, compared to its national role; and con-nectivity—as measured by the numbers and intensity of transactions. 'World cit-ies are not merely outcomes of a global economic machine; rather, they are key structures of the world economy itself' (Taylor et al. (2002) *Urb. Studs* 39, 13; see also P. Taylor 2003). Gugler (2003) *Int. J. Urb. & Reg. Res.* 27, 3 notes that world cities in the developing world are more subject to the economic, political, and cultural impact of powerful foreign ac-tors: governments, corporations, inter-national organizations, universities, the media. The resources they can draw on are much more limited, even as they have to make heavier investments in in-frastructure to meet the standards re-quired by foreign investors. Beaverstock et al. (1999) *Cities* 16, 6 propose a hier-archy of world cities: 'alpha world cities', such as London, New York, and Frank-furt; 'beta world cities', such as Zurich, São Paulo, and Mexico City; and 'gamma world cities' such as Buenos Aires, Montreal, and Santiago. Ewers (2007, *Geog. J.* 68, 2–3) also examines mea-surement issues and the methodological problems in creating indices of world cities. See Brown et al. (2007) *GaWC Research Bulletin* 236 on *world city net-works, and Ewers (loc. cit.) on the role played by world cities in global migra-tions. Smith (2003, *PHG* 27, 1) holds that further progress in the study of world cities and their networks can be made through actor-network theory and *non-representational theory. See Taylor (2007) *GaWC Res. Bull.* 238 for an admirably clear summary of GaWC's work on world cities.

(⊕) SEE WEB LINKS
• GaWC Inventory of World Cities.

world city network An interlocking network with cities at a nodal level, the network at inter-nodal level, and business service firms as a sub-nodal level. These last 'interlock' cities through their myriad networks of offices. In this new city system, it is the functions of a city that largely determine its role and importance nationally, regionally, and globally (Taylor et al. (2002) *Urb. Studs* 39, 13). See Beaverstock et al. (2000) *Applied Geog.* 20 on the conceptualization and measurement of the network connectivity of cities.

world-systems theory A theory arguing that macro-level patterns govern social and economic change; that history is not about singular events but about materially structured ways of life. See I. Wallerstein (1974, 1983) and S. Babones and C. Chase-Dunn (2012).

The *core* consists of the most economically developed countries, the *periphery* the less economically developed countries, and the nations of the *semi-periphery,* such as the newly industrialized countries, have aspects of both core and periphery—actively part of the capitalist world system but with limited authority therein. J. M. Blaut (2000) has used the theory, for example, to criticize historians who argue that the Industrial Revolution happened in Europe rather than Africa, Latin America, or Asia because it possessed special features these other regions did not. Rather, the extraction of peripheral countries' surplus allowed core countries to develop, at the expense of those places whose surplus they took.

These three economic classes are continually changing, and nations move up and down in the system. See A. Ash (2003) and Sheppard (2005) *Antipode* 37, 5. Furthermore, core–periphery relationships operate not only at the global scale, but also at national and even local levels. This theory has been criticized for representing the global economic system as a predominantly capitalist phenomenon, for neglecting the historical and cultural differences of developing countries, and for overlooking the core–periphery inequalities existing within nations. 'World systems approaches... tend to theorize hegemony chiefly on a global scale' (Sparke (2004) *PHG* 28, 779).

World Trade Organization (WTO)
An international organization established in 1995 to replace *GATT. Its stated aims are: expanding free-trade concessions equally to all members; establishing freer global trade with fewer barriers; making trade more predictable through established rules; and making trade more competitive by removing subsidies (see Mullins (2004) *J. Econ. Geog.* 4 on the WTO's dismantling of the protocols that gave Caribbean, African, and Pacific ex-colonies preferential access to EU markets). Since 2005, WTO members have been unable to place quotas on imports of textiles and garments from other members (Thomsen (2007) *J. Econ. Geog.* 7).

Potter and Tilzey (2005) *PHG* 29, 5 argue that 'the entrenched position of neoliberal ideas and framings within the WTO means that the ideology of the free market is coming to define the terms of international trade'. Wade (2003) *Rev. Int. Polit. Econ.* 10, 4 claims that the 'development space' and 'self-determination space' for diversification and upgrading policies in

w

developing countries is being shrunk behind the rhetorical commitment to universitization and privatization. Woods (2007) *PHG* 31, 4 argues that conclaves of the WTO have created new sites of political authority for the global countryside. Uitermark (2002) *PHG* 26, 6 contends that global institutions, such as the WTO, have facilitated competition between different nations such that the nation-state has been 'localized'. As against these tirades, Zhu and Sarkis (2006) *J. Cleaner Prod.* 14, 5 confirm that *globalization and China's entry into the WTO have helped promote green supply chain management practices.

xeroll A soil of the *US soil classification. *See also* CHESTNUT SOIL.

xerophyte A plant which is able to grow in very arid conditions because it has adapted to restrict any water loss. Such adaptations include dense hairs or waxy leaves and shedding leaves at the start of the arid season. **Succulent xerophytes** incorporate water into their structure.

xerosere A plant succession developed under dry conditions such as bare rock, or sand.

yardang In a *desert landscape, a long ridge which has been isolated by the removal of rocks on either side. Yardangs can be 100 m or more in height and can stretch for many kilometres (Goudie (2007) *Geog. Compass* 1, 1, 65).

yazoo A tributary stream which does not join the main stream directly but runs parallel to it for some distance, usually because it cannot breach the *levées which flank the main stream.

Younger Dryas A phase of reduced temperatures and glacial advance in north-west Europe between 11 000 and 10 000 years BP. There is evidence that other regions, such as the Mediterranean, Russia, and North America, also underwent climatic deterioration. During this phase, these non-glaciated areas experienced periglacial conditions. This climatic cooling has been attributed to increased *precipitation, and therefore cloudiness, resulting from changes in the locations of the polar and oceanic Atlantic fronts.

young fold mountain *See* FOLD MOUNTAIN.

Zelinsky *See* MOBILITY.

zero population growth The ending of population growth when birth and death rates are equal. This would require an average number of 2.3 children per family. In 1993, 58 members of the World's Scientific Academies stated that 'there is no doubt that the threat to the [global] ecosystem is linked to population size and resource use... Family planning could bring more benefits to more people at less cost than any other single technology... Success in dealing with global social, economic and environmental problems cannot be achieved without a stable world population... We must achieve zero population growth within the lifetime of our children' (1993, Royal Society, London). This is a contentious subject.

zero-sum game A *formal game whereby, on choosing a particular strategy, one competitor's gain is his opponent's loss. Liard-Muriente (2007) *Area* 39, 2 writes that 'regional development policies could be described as a zero-sum game, with local job reshuffling as the outcome. After all, if one area accomplishes growth, it will be at the expense of another area. Thus, overall welfare is unchanged.' Polèse and Shearmur (2006) *Papers Reg. Sci.* 85, 1 agree: 'stable national populations means a zero-sum demographic game for interregional systems... Assuming (as most migration models do), that population movements largely follow the spatial distribution of economic

opportunities, the key factor becomes the spatial distribution of employment. In such an environment, if shifts in employment continue, on balance, to favour the centre over the periphery, the latter must decline.'

zeuge (pl. *zeugen*), An upstanding rock in a *desert landscape capped with a harder stratum and undercut by wind at the base, indicative of *differential erosion.

Zipper-rift model A model of rifting across Rodinia, proposed by Eyles and Januszczak (2004) *Earth Science Revs* 65, 1, who hypothesize that Rodinia became progressively 'unzipped' after c.750 Ma by two large 'megarift' systems that developed along the western and eastern margins of Laurentia, and that the duration of rifting varied from one part of the rift system to another.

zonal Referring to phenomena occurring in bands roughly parallel with lines of *latitude.

zonal index Meridional displacements of the zonally averaged zonal jet (Feldstein and Lee (1998) *J. Atmos. Scis* 55, 19). Robinson (2005) *J. Atmos. Scis* 57, 3 proposes a *baroclinic mechanism for the positive eddy feedback on the zonal index. *See also* ROSSBY WAVES.

zonal model *See* CONCENTRIC ZONE MODEL.

zonal soil A soil where differences in local rock formation and *lithology are largely masked by the overriding effects of climate. The major zonal soils are *tundra soils, *podzols, Mediterranean soils, *chernozems, *chestnut soils, and ferallitic soils (*see* FERRALLITIZATION).

zonal winds Winds, such as the *trade winds or the *westerlies, which are associated with particular *latitudinal zones.

zone of assimilation The area which increasingly develops the functions of the *CBD; the CBD of the future, characterized by whole scale redevelopment of shops, offices, and hotels. The **zone of discard** is that area that was once a part of the CBD but is now in decline and characterized by low-status shops and warehouses, and vacant property.

zone of overlap 1. An area served by more than one urban centre, i.e. within two or more different urban fields.

2. A transition zone between two floral/faunal communities. See, for example, Wolf and Flamenco-S (2003) *J. Biogeog.* 30, 11.

zoogeomorphology The study of animals as geomorphic agents—as in the creation and collapse of gopher tunnels (Knight (2009) *ESPL* 34, 4, 590). The classic text is by D. R. Butler (1995).

z

More History titles from OUP

The Oxford Companion to Black British History
David Dabydeen, John Gilmore, and Cecily Jones

The first reference book to explore the full history of black people in the British Isles from Roman times to the present day.

'From Haiti to Kingston, to Harlem, to Tottenham, the story of the African Diaspora is seldom told. This Companion will ensure that the history of Black Britain begins to take its rightful place in mainstream British consciousness.'

David Lammy, MP, former Minister for Culture

A Dictionary of World History

Contains a wealth of information on all aspects of history, from prehistory right up to the present day. Over 4,000 clear, concise entries include biographies of key figures in world history, separate entries for every country in the world, and subject entries on religious and political movements, international organizations, and key battles and places.

The Concise Oxford Dictionary of Archaeology
Timothy Darvill

The most wide-ranging, up-to-date, and authoritative dictionary of its kind.

'Comprehensive, proportionate, and limpid'

Antiquity

Oxford Quick Reference

A Dictionary of Marketing
Charles Doyle

Covers traditional marketing techniques and theories alongside the latest concepts in over 2,000 clear and authoritative entries.

'Flick to any page [for] a lecture's worth of well thought through information'

Dan Germain, Head of Creative, innocent ltd

A Dictionary of Media and Communication
Daniel Chandler and Rod Munday

Provides over 2,200 authoritative entries on terms used in media and communication, from concepts and theories to technical terms, across subject areas that include advertising, digital culture, journalism, new media, radio studies, and telecommunications.

'a wonderful volume that is much more than a simple dictionary'

Professor Joshua Meyrowitz, University of New Hampshire

A Dictionary of Film Studies
Annette Kuhn and Guy Westwell

Features terms covering all aspects of film studies in 500 detailed entries, from theory and history to technical terms and practices.

A Dictionary of Journalism
Tony Harcup

Covers terminology relating to the practice, business, and technology of journalism, as well as its concepts and theories, organizations and institutions, publications, and key events.

Oxford Quick Reference

Concise Medical Dictionary

Over 12,000 clear entries covering all the major medical and surgical specialities make this one of our best-selling dictionaries.

'"No home should be without one" certainly applies to this splendid medical dictionary'

Journal of the Institute of Health Education

'An extraordinary bargain'   *New Scientist*

A Dictionary of Nursing

Comprehensive coverage of the ever-expanding vocabulary of the nursing professions. Features over 10,000 entries written by medical and nursing specialists.

A Dictionary of Dentistry
Robert Ireland

Over 4,000 succinct and authoritative entries define all the important terms used in dentistry today. This is the ideal reference for all members of the dental team.

A Dictionary of Forensic Science
Suzanne Bell

In over 1,300 entries, this new dictionary covers the key concepts within Forensic Science and is a must-have for students and practitioners of forensic science.

OXFORD

Oxford Quick Reference

The Oxford Dictionary of Dance
Debra Craine and Judith Mackrell

Over 2,600 entries on everything from hip-hop to classical ballet, covering dancers, dance styles, choreographers and composers, techniques, companies, and productions.

'A must-have volume ... impressively thorough'
Margaret Reynolds, *The Times*

The Oxford Guide to Plays
Michael Patterson

Covers 1,000 of the most important, best-known, and most popular plays of world theatre.

'Superb synopses ... Superbly formatted ... Fascinating and accessible style'
THES

The Oxford Dictionary of Music
Michael & Joyce Kennedy & Tim Rutherford-Johnson

The most comprehensive, authoritative, and up-to-date dictionary of music available in paperback.

'clearly the best around ... the dictionary that everyone should have'
Literary Review

Oxford Quick Reference

A Dictionary of Chemistry

Over 4,700 entries covering all aspects of chemistry, including physical chemistry and biochemistry.

'It should be in every classroom and library ... the reader is drawn inevitably from one entry to the next merely to satisfy curiosity.'

School Science Review

A Dictionary of Physics

Ranging from crystal defects to the solar system, 4,000 clear and concise entries cover all commonly encountered terms and concepts of physics.

A Dictionary of Biology

The perfect guide for those studying biology — with over 5,500 entries on key terms from biology, biochemistry, medicine, and palaeontology.

'lives up to its expectations; the entries are concise, but explanatory'

Biologist

'ideally suited to students of biology, at either secondary or university level, or as a general reference source for anyone with an interest in the life sciences'

Journal of Anatomy

Oxford Quick Reference

A Dictionary of Psychology
Andrew M. Colman

Over 9,000 authoritative entries make up the most wide-ranging
dictionary of psychology available.

'impressive ... certainly to be recommended'
Times Higher Education Supplement

'probably the best single-volume dictionary of its kind.'
Library Journal

A Dictionary of Economics
John Black, Nigar Hashimzade, and Gareth Myles

Fully up-to-date and jargon-free coverage of economics. Over 3,400
terms on all aspects of economic theory and practice.

'strongly recommended as a handy work of reference.'
Times Higher Education Supplement

A Dictionary of Law

An ideal source of legal terminology for systems based on English law.
Over 4,200 clear and concise entries.

'The entries are clearly drafted and succinctly written ... Precision for the
professional is combined with a layman's enlightenment.'
Times Literary Supplement

A Dictionary of Education
Susan Wallace

In over 1,250 clear and concise entries, this authoritative dictionary
covers all aspects of education, including organizations, qualifications,
key figures, major legislation, theory, and curriculum and assessment
terminology.

OXFORD

Oxford Quick Reference

A Dictionary of Sociology
John Scott

The most wide-ranging and authoritative dictionary of its kind.

'Readers and especially beginning readers of sociology can scarcely do better ... there is no better single volume compilation for an up-to-date, readable, and authoritative source of definitions, summaries and references in contemporary Sociology.'

A. H. Halsey, *Emeritus Professor, Nuffield College, University of Oxford*

The Concise Oxford Dictionary of Politics
Iain McLean and Alistair McMillan

The bestselling A–Z of politics with over 1,700 detailed entries.

'A first class work of reference ... probably the most complete as well as the best work of its type available ... Every politics student should have one'

Political Studies Association

A Dictionary of Environment and Conservation
Chris Park and Michael Allaby

An essential guide to all aspects of the environment and conservation containing over 8,500 entries.

'from *aa* to *zygote*, choices are sound and definitions are unspun'

New Scientist

Oxford Quick Reference

The Kings and Queens of Britain
John Cannon and Anne Hargreaves

A detailed, fully-illustrated history ranging from mythical and pre-conquest rulers to the present House of Windsor, featuring regional maps and genealogies.

A Dictionary of World History

Over 4,000 entries on everything from prehistory to recent changes in world affairs. An excellent overview of world history.

A Dictionary of British History
Edited by John Cannon

An invaluable source of information covering the history of Britain over the past two millennia. Over 3,000 entries written by more than 100 specialist contributors.

Review of the parent volume
'the range is impressive ... truly (almost) all of human life is here'
Kenneth Morgan, *Observer*

The Oxford Companion to Irish History
Edited by S. J. Connolly

A wide-ranging and authoritative guide to all aspects of Ireland's past from prehistoric times to the present day.

'packed with small nuggets of knowledge'   *Daily Telegraph*

The Oxford Companion to Scottish History
Edited by Michael Lynch

The definitive guide to twenty centuries of life in Scotland.
'exemplary and wonderfully readable'

Financial Times